Lecture Notes in Computer Science 2606

Edited by G. Goos, J. Hartmanis, and J. van Leeuwen

Springer
Berlin
Heidelberg
New York
Barcelona
Hong Kong
London
Milan
Paris
Tokyo

Andy M. Tyrrell   Pauline C. Haddow
Jim Torresen (Eds.)

Evolvable Systems: From Biology to Hardware

5th International Conference, ICES 2003
Trondheim, Norway, March 17-20, 2003
Proceedings

Springer

Series Editors

Gerhard Goos, Karlsruhe University, Germany
Juris Hartmanis, Cornell University, NY, USA
Jan van Leeuwen, Utrecht University, The Netherlands

Volume Editors

Andy M. Tyrrell
The University of York, The Department of Electronics
Heslington, York YO10 5DD, United Kingdom
E-mail: amt@ohm.york.ac.uk

Pauline C. Haddow
The Norwegian University of Science and Technology
Department of Computer and Information Science
Sem Sælands Vei, 7491 Trondheim, Norway
E-mail: pauline@idi.ntnu.no

Jim Torresen
University of Oslo, Department of Informatics
P.O. Box 1080 Blindern, 0316 Oslo, Norway
E-mail: jimtoer@ifi.uio.no

Cataloging-in-Publication Data applied for

A catalog record for this book is available from the Library of Congress

Bibliographic information published by Die Deutsche Bibliothek
Die Deutsche Bibliothek lists this publication in the Deutsche Nationalbibliographie;
detailed bibliographic data is available in the Internet at <http://dnb.ddb.de>.

CR Subject Classification (1998): B.6, B.7, F.1, I.6, I.2, J.2, J.3

ISSN 0302-9743
ISBN 3-540-00730-X Springer-Verlag Berlin Heidelberg New York

Springer-Verlag Berlin Heidelberg New York
a member of BertelsmannSpringer Science+Business Media GmbH

http://www.springer.de

© Springer-Verlag Berlin Heidelberg 2003
Printed in Germany

Typesetting: Camera-ready by author, data conversion by Olgun Computergraphik
Printed on acid-free paper   SPIN: 10872784   06/3142   5 4 3 2 1 0

Preface

The idea of evolving machines, whose origins can be traced to the cybernetics movement of the 1940s and 1950s, has recently resurged in the form of the nascent field of bio-inspired systems and evolvable hardware. The inaugural workshop, Towards Evolvable Hardware, took place in Lausanne in October 1995, followed by the First International Conference on Evolvable Systems: From Biology to Hardware (ICES), held in Tsukuba, Japan in October 1996. The second ICES conference was held in Lausanne in September 1998, with the third and fourth being held in Edinburgh, April 2000 and Tokyo, October 2001 respectively. This has become the leading conference in the field of evolvable systems and the 2003 conference promised to be at least as good as, if not better than, the four that preceeded it.

The fifth international conference was built on the success of its predecessors, aiming at presenting the latest developments in the field. In addition, it brought together researchers who use biologically inspired concepts to implement real systems in artificial intelligence, artificial life, robotics, VLSI design and related domains. We would say that this fifth conference followed on from the previous four in that it consisted of a number of high-quality interesting thought-provoking papers.

We received 58 papers in total. All of these papers were reviewed by three independent reviewers. As such, we feel that we compiled an excellent package for ICES 2003. The conference included 3 keynote talks titled: "Nano- and biotechnology," "From wheels to wings with evolutionary spiking neurons," and "Machine design of quantum computers: A new frontier." We had 41 technical presentations, a panel debate, and 3 tutorials in the areas of: evolutionary algorithms, evolvable hardware and reconfigurable devices, and nanotechnology. In addition to the technical program, there was a strong and varied social program both during and after the conference.

We would like to thank the reviewers for their time and effort in reviewing all of the submitted papers. We would also like to thank the other members of the organizing committee, including the local chair Keith Downing and publicity chair Gunnar Tufte. We are most grateful to Frode Eskelund, Diego Federici, and Karstein A. Kristiansen for their great help in developing the web-based paper submission and registration tool. We wish to thank the following for their contribution to the success of this conference: The Research Council of Norway; European Community IST programme; Norwegian University of Science and Technology; University of Oslo; European Office of Aerospace Research and Development, Air Force Office of Scientific Research, United States Air Force Research Laboratory; Telenor ASA; Atmel; and Siemens. Finally, we would like to thank all of those authors who put so much effort into their research and decided to publish their work at our conference.

What topics might we consider important for the next few years? It is clear that the field is still developing, evolving, and that many of the "hot" topics at this conference will be seen again next time. These will include: evolutionary hardware design; co-evolution of hybrid systems; evolving hardware systems; intrinsic, and on-line evolution; hardware/software co-evolution; self-repairing hardware; self-reconfiguring hardware; embryonic hardware; morphogenesis; novel devices; adaptive computing; and of course we are always looking for, and interested in, real-world applications of evolvable hardware. Will we see breakthroughs relating to nanotechnology, new reconfigurable FPGAs and new models of reliability for long space missions? Only time will tell – as with all evolution.

We hope you enjoy reading these proceedings as much as we enjoyed putting them together.

January 2003

Andy M. Tyrrell
Pauline C. Haddow
Jim Torresen

Organization

Organizing Committee

General Chair:	Andy M. Tyrrell (University of York, UK)
Program Co-chair:	Pauline C. Haddow (Norwegian University of Science and Technology, Norway)
Program Co-chair:	Jim Torresen (University of Oslo, Norway)
Local Chair:	Keith Downing (Norwegian University of Science and Technology, Norway)
Publicity Chair:	Gunnar Tufte (Norwegian University of Science and Technology, Norway)

International Steering Committee

Tetsuya Higuchi, National Institute of Advanced Industrial Science
and Technology (AIST), Japan
Daniel Mange, Swiss Federal Institute of Technology, Switzerland
Julian Miller, University of Birmingham, UK
Moshe Sipper, Ben-Gurion University, Israel
Adrian Thompson, University of Sussex, UK

Program Committee

Juan Manuel Moreno Arostegui, Universitat Politecnica de Catalunya, Spain
Wolfgang Banzhaf, University of Dortmund, Germany
Peter J. Bentley, University College London, UK
Gordon Brebner, Xilinx, USA
Richard Canham, University of York, UK
Stefano Cagnoni, Universita' di Parma, Italy
Prabhas Chongstitvatana, Chulalongkorn University, Thailand
Carlos A. Coello Coello, LANIA, Mexico
Peter Dittrich, University of Dortmund, Germany
Marco Dorigo, Université Libre de Bruxelles, Belgium
Keith Downing, Norwegian University of Science and Technology, Norway
Rolf Drechsler, University of Bremen, Germany
Marc Ebner, Universität Würzburg, Germany
Stuart J. Flockton, University of London, UK
Dario Floreano, Swiss Federal Institute of Technology, Switzerland
Terence C. Fogarty, South Bank University, UK
David B. Fogel, Natural Selection, Inc., USA
Andrew Greenstead, University of York, UK

Hugo deGaris, Utah State University, USA
Tim Gordon, University College London, UK
Darko Grundler, Univesity of Zagreb, Croatia
Pauline C. Haddow, Norwegian University of Science and Technology, Norway
David M. Halliday, University of York, UK
Alister Hamilton, University of Edinburgh, UK
Arturo Hernandez Aguirre, Tulane University, USA
Francisco Herrera, University of Granada, Spain
Jean-Claude Heudin, Pule Universitaire Léonard de Vinci, France
Tetsuya Higuchi, National Institute of Advanced Industrial Science
    and Technology (AIST), Japan
Hitoshi Iba, University of Tokyo, Japan
Masaya Iwata, National Institute of Advanced Industrial Science
    and Technology (AIST), Japan
Alex Jackson, University of York, UK
Tatiana Kalganova, Brunel University, UK
Didier Keymeulen, Jet Propulsion Laboratory, USA
Michael Korkin, Genobyte, Inc., USA
Sanjeev Kumar, University College London, UK
William B. Langdon, University College London, UK
Yong Liu, University of Aizu, Japan
Jason Lohn, NASA Ames Research Center, USA
Michael Lones, University of York, UK
Evelyne Lutton, INRIA, France
Daniel Mange, Swiss Federal Institute of Technology, Switzerland
Karlheinz Meier, Heidelberg University, Germany
Julian Miller, University of Birmingham, UK
David Montana, BBN Technologies, USA
Masahiro Murakawa, National Institute of Advanced Industrial Science
    and Technology (AIST), Japan
Marek A. Perkowski, Portland State University, USA
Matteo Sonza Reorda, Politecnico di Torino, Italy
Mehrdad Salami, School of Biophysical Sciences and Electrical Engineering,
    Australia
Eduardo Sanchez, Swiss Federal Institute of Technology, Switzerland
Lukas Sekanina, Brno University of Technology, Czech Republic
Moshe Sipper, Ben-Gurion University, Israel
Giovanni Squillero, Politecnico di Torino, Italy
Andre Stauffer, Swiss Federal Institute of Technology, Switzerland
Adrian Stoica, Jet Propulsion Laboratory, USA
Kiyoshi Tanaka, Shinshu University, Japan
Gianluca Tempesti, Swiss Federal Institute of Technology, Switzerland
Christof Teuscher, Swiss Federal Institute of Technology, Switzerland
Jonathan Timmis, University of Kent, UK
Adrian Thompson, University of Sussex, UK
Peter Thomson, Napier University, UK

Sponsoring Institutions

We wish to thank the following for their contribution to the success of this conference:

The Research Council of Norway
European Community IST programme (Fifth Framework Programme)
Department of Computer and Information Science, Norwegian University of
    Science and Technology
Department of Informatics, University of Oslo
European Office of Aerospace Research and Development, Air Force Office of
    Scientific Research, United States Air Force Research Laboratory
Telenor ASA
Atmel
Siemens

Table of Contents

Evolution

Fault Tolerance and Fault Recovery

Development

POEtic

Applications 1

Evolution of Digital Circuits

Hardware Challenges

Applications 2

Evolutionary Hardware

Neural Systems

Logic Design

Evolutionary Strategies

Author Index

On Fireflies, Cellular Systems, and Evolware

Christof Teuscher[1] and Mathieu S. Capcarrere[2]

[1] Logic Systems Laboratory, Swiss Federal Institute of Technology,
CH-1015 Lausanne, Switzerland
`christof@teuscher.ch`, `http://lslwww.epfl.ch`
[2] Computer Science Institute, Collège Propédeutique, University of Lausanne
CH-1015 Lausanne, Switzerland
`mathieu.capcarrere@philo.unil.ch`, `http://www-iis.unil.ch/~mcapcarr`

Abstract. Many observers have marveled at the beauty of the synchronous flashing of fireflies that has an almost hypnotic effect. In this paper we consider the issue of evolving two-dimensional cellular automata as well as random boolean networks to solve the firefly synchronization task. The task was successfully solved by means of cellular programming based co-evolution performing computations in a completely local manner, each cell having access only to its immediate neighbor's states. An FPGA-based *Evolware* implementation on the BioWall's cellular tissue and different other simulations show that the approach is very efficient and easily implementable in hardware.

1 Introduction

The mutual synchronization of oscillators – a both surprising and interesting phenomenon – in living things and nature is ubiquitous (see for example [12]): dancing (natural response to music), pacemaker cells in the heart, the circadian clock of multicellular organisms (consists of multiple autonomous single-cell oscillators) [31], the nervous system that controls rhythmic behavior such as breathing, running and chewing, synchronous flashing of fireflies, choruses of cicadas or crickets [32], etc.

An oscillator can be defined in a loosely manner as a system that executes a periodic behavior. The pendulum is probably the best example of a periodic behavior in space and time. Electronic circuits offer many different types of oscillators that produce an output signal of a specific frequency. Often, a very stable mechanical oscillator, such as a specially prepared quartz crystal, may be coupled to an electronic oscillator to enhance its frequency stability.

Once the behavior of a single oscillator is understood, the more complex behavior of coupled oscillators can be investigated, although the equations governing their behavior usually become quickly intractable. Strogarz and Stewart [29] write that the most familiar mode of organization for coupled oscillators is synchrony. "One of the most spectacular examples of this kind of coupling can be seen along the tidal rivers of Malaysia, Thailand and New Guinea, where thousands of male fireflies gather together in trees at night and flash on and off

A.M. Tyrrell, P.C. Haddow, and J. Torresen (Eds.): ICES 2003, LNCS 2606, pp. 1–12, 2003.

in unison in an attempt to attract the females that cruise overhead. When the males arrive at dusk, their flickerings are uncoordinated. As the night deepens, pockets of synchrony begin to emerge and grow. Eventually whole trees pulsate in a silent, hypnotic concert that continues for hours" [29]. Synchronization in living things usually emerges spontaneously and through cooperative behavior: if a few individuals happen to synchronize, they tend to exert stronger influence on their common neighbors. See Buck and Buck [2] for a review of various theories and information about synchronous fireflies.

Another well known example is the synchronization of menstrual cycles among women friends or roommates [20]. There exist various ideas about the mechanism of synchronization, but an experiment in 1980 has shown that it might have something to do with sweat [24].

In this paper, we concentrate on the two-dimensional firefly synchronization task for non-uniform CAs (each automaton may have a different rule) and random boolean networks. The two-dimensional version of the task is in principle similar to the one-dimensional version, although the speed of synchronization is in general much faster.

The organization of the paper is as follows: Section 2 presents the synchronization task for synchronous cellular automata. Asynchronous automata are presented in Section 3. Section 4.1 briefly presents some measures that allow a better comparison of the experiments. A hardware-based implementation of the firefly synchronization task on the BioWall's cellular tissue is presented in Section 4.2. The CA implemented on the BioWall uses co-evolution in a completely local manner, each cell having access only to its immediate neighbors' states, to find a solution to the task. Section 4.3 presents an alternative to the synchronous application of the genetic algorithms to each cell: the algorithm works asynchronously. Finally, Section 4.4 successfully applies the co-evolutionary approach to random boolean networks (RBN). Section 5 concludes the paper.

2   The Synchronization Task for Synchronous Cellular Automata

The one-dimensional synchronization task for synchronous CA was introduced by Das et al. [9] and studied among others by Hordijk [14] and Sipper [27]. In this task the two-state one-dimensional CA, given any initial configuration, must reach a final configuration, within M time steps, that oscillates between all 0s and all 1s on successive time steps. The whole automaton is then globally synchronized. Synchronization comprises a non-trivial computation for a small-radius CA: all cells must coordinate with all the other cells while having only a very local view of its neighbors. The existence of the global clock, though not without consequences as we will see later, should not lure us into thinking that the task is straightforward. Obviously, the fact that the whole computation should only occur within a two-state machine prevents any counting.

In this paper, we concentrate on the the two-dimensional version of the task for non-uniform CAs and random boolean networks. The two-dimensional ver-

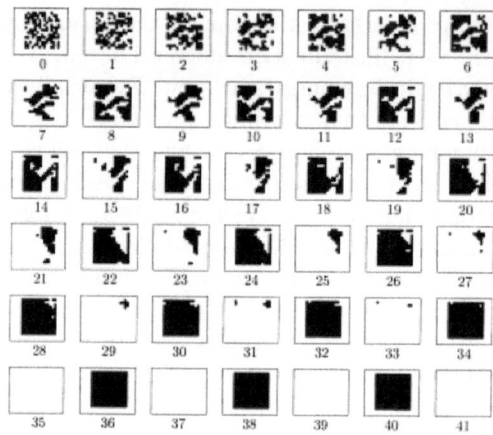

Fig. 1. The two-dimensional synchronization task for synchronous CA. Source: [27].

sion of the task, see Figure 1, is identical to the one-dimensional version except that the necessary number of time steps granted to synchronize is not any more in the order of N, the number of cells in the automaton, but in the order of $n + m$ where n, m are the size of each side of the CA. Thereby, the speed of synchronization is much faster. Non-uniform CAs are CAs where each automaton may have a different rule.

Obviously, in the non-uniform case there is an immediate solution consisting of a unique 'master' rule, alternating between '0' and '1', whatever the neighborhood, and all other cells being its 'slave' and alternating according to its right or up neighbor state only. However, Sipper [27] used non-uniform CAs to find perfect synchronizing CAs only by means of evolution. It appeared that this "basic" solution was never found by evolution and, in fact, the "master" or "blind" rule 10101010 (rule 170) in one dimension, was never part of the evolved solutions. This is simply due to the fact that this rule has to be unique for the solution to be perfect, which is contradictory to the natural tendency of the evolutionary algorithm used, as we demonstrated in [3]. In this paper, we only use evolved solutions to this task.

3   Asynchronous Cellular Automata

In the preceding section the CA model is synchronous as is traditionally the case. But this feature is far from being realistic from a biological point of view as well as from a computational point of view. As for the former, it is quite evident that there is no accurate global synchronization in nature. As for the latter, it turns out to be impossible to maintain a large network of automata globally synchronous in practice. Besides, it is interesting to try to delineate what part of the computation relies on the synchrony constraint.

Ingerson and Buwel [15] compared the different dynamical behavior of synchronous and asynchronous CAs; they argued that some of the apparent self-organization of CAs is an artifact of the synchronization of the clocks. Wolfram [33] noted that asynchronous updating makes it more difficult for information to propagate through the CA and that, furthermore, such CAs may be harder to analyze. Asynchronous CAs have also been discussed from a problem-solving and/or Artificial Life standpoint in references [22, 1, 25, 16, 13, 28]. All these works devoted to asynchronous cellular automata only concentrated on the study of the effects but not on correcting asynchrony or dealing with it. From a theoretical computer science perspective, Zielonka [34, 35] introduced the concept of asynchronous cellular automata. Though the question attracted quite some interest [18, 23, 8], the essential idea behind them was to prove that they were "equivalent" to synchronous CA in the sense that, for instance, they recognize the same trace languages or could decide emptiness. From these two fields, we thus knew that asynchronous CA are potentially as powerful as synchronous CA, computationally speaking[1], and, nevertheless, that most of the effects observed in the synchronous case are artifacts of the global clock.

In [5, 4] we proposed asynchronous CAs exhibiting the same behavior as synchronous CAs through both design and evolution. The main idea behind these models was that time is part of the information contained in a CA configuration. Hence suppressing the global clock constraint is equivalent to suppressing information. Thereby, and quite naturally, if we are to maintain the capability of the CA, then we must compensate for that loss of information by adding extra states. As shown then, it is possible to design a CA with $3 * q^2$ states working asynchronously which simulates perfectly a given q-state CA. Tolerating some loss of information, we have shown that it was possible to evolve self-correcting 4-state CA to do the synchronization task under low asynchrony.

As Gacs [10, 11] reminds us, asynchrony may be considered as a special case of fault-tolerance. However, even though this consideration is nice in its generalization (i.e., a fault-tolerant CA is also asynchronous), it eschews a lot of potential optimization.

4   Implementations and Experiments

4.1   Genotypic Measures for Non-uniform CAs

To better compare the different simulations and implementations, we define a number of metrics (some of them adapted from [3]).

Let's first define N to be the total number of cells. We have $N = n * m$ where m, n are the length of the sides of the two-dimensional CA considered. We can now define the frequency of transitions, ν, as the number of borders between homogeneous blocks of cells having the same genotype, divided by the number of distinct couples of adjacent cells. Adjency is here taken along the neighborhood

---

[1] In the traditional sense, not in the sense of visual computation.

used, that is the von Neumann neighborhood. Thus ν is the probability that two adjacent cells have a different rule (Equation 1).

$$\nu = \frac{1}{2(n*m)} \sum_{i=1}^{n} \sum_{j=1}^{n} \left[R_{i,j} \neq R_{(i+1 \bmod n, j)} + R_{i,j} \neq R_{(i,j+1 \bmod m)} \right] \tag{1}$$

can also define a more global measure of the diversity of the different genotypes encountered, an entropy, H (Equation 2).

$$H = \sum_{r \in \Gamma} q(r) * log \left(\frac{1}{q(r)} \right) \tag{2}$$

where Γ is the set of all possible genotypes (rules), and $q(r)$ is the global proportion of the genotype (rule) r in the cellular automata. Obviously, H takes on values in the interval $[0, \log n]$ and attains its maximum, $H(x) = \log n$, when x comprises n different genotypes. H is usually normalized to take values in the interval $[0, 1]$.

4.2 Firefly Hardware

Fireflies synchronize according to the "pulse coupled" system [21], i.e., they interact only when one sees the sudden flash of another. The firefly then shifts its rhythm accordingly. "Pulse coupling is difficult to handle mathematically because it introduces discontinuous behavior into the otherwise continuous model and so stymies most of the standard mathematical techniques" [29]. Mathematicians have turned to the theory of symmetry breaking to tackle the complex problems that arise when identical oscillators are coupled. Computer scientists, on the other hand, often use cellular automata, irregular networks and grid arrays or more complex agent-based models to analyze the behavior of interacting elements.

In 1997, the Logic Systems Laboratory presented an evolving hardware system called Firefly [26] that successfully solved the synchronization task in one dimension using cellular programming based co-evolution. The novelty of Firefly was that it operates with no reference to an external device (such as a computer that carries out genetic operators) thereby exhibiting online autonomous evolution.

Recently, the synchronization task in two dimensions has successfully been evolved on the BioWall's 3200 FPGAs. The BioWall [30] is a giant reconfigurable computing tissue developed to implement machines according to the principles of the Embryonics (embryonic electronics) project [19]. The BioWall's size and features are designed for public exhibition, but at the same time it represents an invaluable research tool, particularly since its complete programmability and cellular structure are extremely well adapted to the implementation of many different kinds of bio-inspired systems. The implementation on the BioWall consists of a two-state, non-uniform CA, in which each cell (i.e., each FPGA of the BioWall) may contain a different rule. The cells rule tables are encoded as a

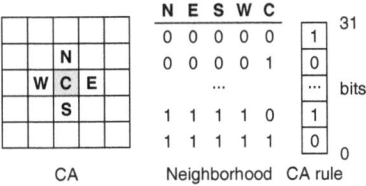

Fig. 2. Von Neumann neighborhood of our two-dimensional CA and cell rule encoding.

bit-string, known as the genome, that has a length of $2^5 = 32$ bits for our 2D CA since the binary CA has a neighborhood of 5 (see also Figure 2).

Rather than to employ a population of evolving CAs, our algorithm (see Algorithm 1) evolves a single, non-uniform CA of the size of the entire BioWall (one cell of the CA in each unit of the BioWall, that is, 3200 cells), whose rules are initialized at random. Initial configurations are then randomly generated and for each configuration the CA is run for M time steps (M has been set to $2(n + m) = 2(160 + 20) = 360$, where n, m are the size of each side of the CA). Each cell's fitness is accumulated over C (C has been set to 300) initial configurations. The (local) fitness score for the synchronization task is assigned to each cell by considering the last four time steps ($M + 1$ to $M + 4$): if the sequence of states over these steps is precisely $0 - 1 - 0 - 1$, the cell's fitness score is 1, otherwise this score is 0.

After every C configurations the rules are evolved through crossover and mutation. This evolutionary process is performed in a completely local manner, that is, genetic operators are applied only between directly connected cells. Unlike standard genetic algorithms, where a population of independent problem solutions globally evolves, our approach involves a grid of rules that co-evolves locally. The CA implemented on the BioWall performs computations in a completely local manner, each cell having access to its immediate neighbors' states only. In addition, the evolutionary process is also completely local, since the application of the genetic operators as well as the fitness assignment occurs locally. Using the above-described cellular programming approach (see also Algorithm 1) on the BioWall, we have shown that a non-uniform CA of radius 1 can be evolved to successfully solve the synchronization task. In addition, once a set of successful rules has been found, our machine allows the state of each CA cell to be changed by pressing on its membrane. The user can then observe how the machine synchronizes the 3200 cells.

In contrast to the original Firefly machine [26], which used a slightly simplified algorithm to facilitate the implementation, the BioWall implements exactly the algorithm as described above. The BioWall implementation, however, contains no global synchronization detector, therefore, it is not 100% guaranteed that the CA always synchronizes perfectly (i.e., for any initial configuration). Each cell contains a pseudo-random generator, realized by means of a linear feedback shift register, to initialize randomly the cell's state and its rule table. The register receives an initial random seed from an external PC that also con-

---

Algorithm 1 Cellular programming pseudo-code for the synchronization of a two-dimensional cellular automaton

---

for each cell i in CA [in parallel] **do**
   Initialize rule table of cell i randomly
   $f_i = 0$ [fitness value]
end for
$c = 0$ [initial configurations counter]
while not done **do**
   Generate a random initial configuration
   Run CA on initial configuration for $M = 2(m + n)$ time steps
   for each cell i in CA [in parallel] **do**
      if cell i is in the correct final state **then**
         $f_i = f_i + 1$
      end if
   end for
   $c = c + 1$
   if $c \bmod C = 0$ [evolve every C configurations] **then**
      for each cell i [in parallel] **do**
         Compute $nf_i(c)$ [number of fitter neighbors]
         if $nf_i(c) = 0$ **then**
            Rule i is left unchanged
         else if $nf_i = 1$ **then**
            Replace rule i with the fitter neighboring rule, followed by mutation
         else if $nf_i > 1$ **then**
            Replace rule i with the crossover of the two fittest neighboring rules (randomly chosen, if equal fitness), followed by mutation
         end if
         $f_i = 0$
      end for
   end if
end while

---

figures the FPGAs. The actual implementation easily fits into a Xilinx Spartan XCS10XL FPGA.

As an example, Figure 3 shows the evolution of the fitness as well as the number of rules in function of the number of evolutionary phases for a small 4×4 cell CA (toroidal). A solution that perfectly synchronizes the 16 cells from any initial configuration (there are $2^{16} = 65536$) has been found with the following five rules:

99ea2e68	81f2a5fa	fe7aeb98	bf82fa06
99ea2e68	81f2a5fa	fe7aeb98	bf82fa06
99ea2e68	81f2a5fa	fe7aeb98	bf82fa06
99e82e68	81f2a5fa	fe7aeb98	bf82fa06

As stated in Section 2, there is a trivial solution with two rules (one cell being the 'master' and all the other cells being 'slaves'), but it is very unlikely that this configuration is found by evolution since this rule has to be unique for

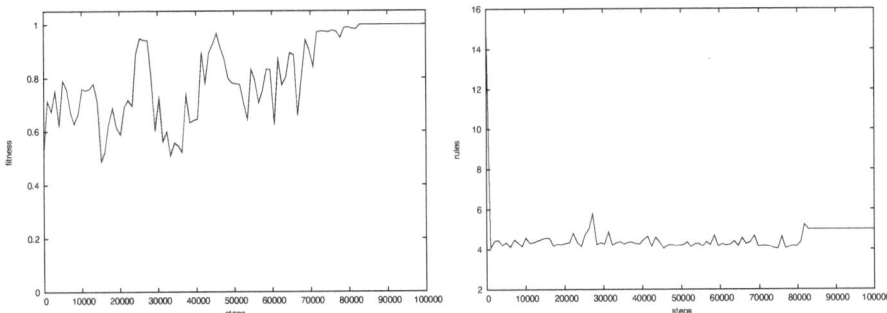

Fig. 3. The CA's fitness (on the left) and the number of different rules (on the right) of a small 4×4 cell CA (means over several runs). The CA successfully synchronizes for any initial configuration using five different rules.

Algorithm 2 Cellular programming pseudo-code for the synchronization of a two-dimensional cellular automaton using asynchronous co-evolution

---

   for each cell i in CA [in parallel] **do**
     Initialize rule table of cell i randomly
   end for
   while not done **do**
     Generate a random initial configuration
     for D steps **do**
       Run CA on initial configuration for T time steps
       Choose a cell i at random
       Compute fitness f_i and fitness of adjacent cells
       Apply GA to cell i [crossover and mutation]
     end for
   end while

---

the solution to be perfect. The above example has a ν of 0.563, the entropy H is 0.562.

4.3   Asynchronous Co-evolution

In the previous section we described the co-evolution of a synchronous CA where the genetic algorithm in each cell is synchronized by means of a global signal: all cells evaluated the fitness at the same time, replaced the rules at the same time, etc. It would obviously be interesting to use some kind of 'asynchronous' co-evolution where cells still operate synchronously, but where the genetic algorithm operates asynchronously and randomly in time. This would allow to remove the GA's global synchronization signal.

We used the method as shown in Algorithm 2.

Our simulations (with $T = 10N$, $D = N$) have revealed that the CA is still able to synchronize perfectly, although more different rules are required and synchronization seems slower in general. Evolution found eight different rules for

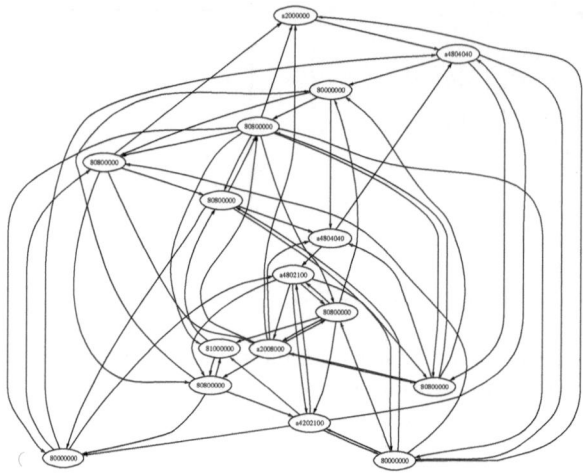

Fig. 4. A random graph with 4 incoming links per node/cell. The eight rules that allow to perfectly synchronize any initial configuration are written on the nodes.

the example of the previous section. Naturally, the probability that two adjacent cells have a different rule was higher: $\nu = 0.969$. The entropy H was 0.938.

4.4   The Synchronization Task for Random Boolean Networks

Solving the synchronization task on a completely regular grid abstracts from the original firefly synchronization task in nature and it would thus be more interesting and biologically plausible to use a random graph. To avoid problems during crossover between rules having different lengths, the number of incoming links per node/cell was held constant. Figure 4 shows a random graph with four incoming links per node. Using Stuart Kauffman's terminology (see for example [17]), the graph is a $N = 16$, $K = 4$ random boolean network. In this context, K specifies the exact number of inputs per node.

Using Algorithm 1, our co-evolutionary approach was able to find a perfect solution using eight different rules ($\nu = 0.859$, $H = 0.652$) for the $K = 4$ random boolean network (RBN) of Figure 4. Figure 5 shows the evolution of the number of rules for a $K = 4$ (on the left) and for a $K = 2$ RBN (on the right). It is astonishing that one rule is in general sufficient for the $K = 2$ network ($\nu = 0$, $H = 0$). Further investigations would be needed, but it almost certainly has something to do with Kauffman's findings that the most highly organized behavior appeared to occur in networks where each node receives inputs from two other nodes ($K = 2$). It turned out that Kauffman's networks exhibit three major regimes of behavior: *ordered* (solid), *complex* (liquid), and *chaotic* (gas). The most complex and interesting dynamics correspond to the liquid interface, the boundary between order and chaos. In the ordered regime, little computation can occur. In the chaotic phase, dynamics are too disordered to be useful.

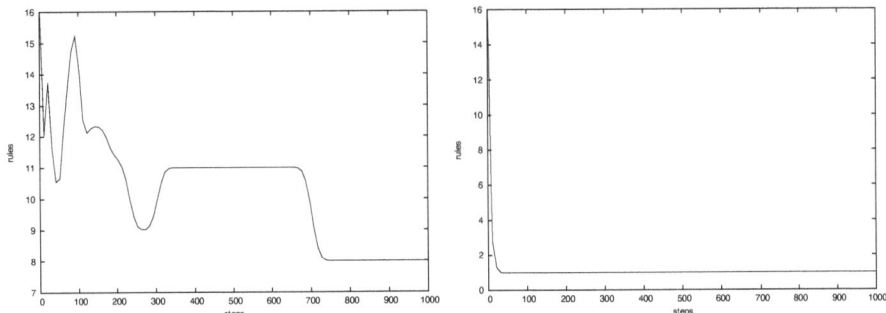

Fig. 5. The number of different rules (on the left) of a $K = 4$ random boolean network (RBN) and of a $K = 2$ RBN (on the right). The $K = 2$ network synchronizes with one rule only, the $k = 4$ network with eight rules.

5 Conclusion and Future Work

In this paper we considered the issue of evolving two-dimensional cellular automata as well as random boolean networks to solve the firefly synchronization task. The task has been solved using a co-evolutionary genetic algorithm that performs computations in a completely local manner, each cell having access to its immediate neighbors' states only.

The FPGA-based implementation on the BioWall's cellular tissue exhibits complete online co-evolution using local interactions only between neighboring cells. Besides the global clock signals, no other global signal is needed.

Our experiments have shown that the cellular programming based approach is very efficient, easily finds a solution for classical two-dimensional cellular automata as well as for random boolean networks, and is easily implementable on cellular hardware (e.g. FPGAs or arrays of FPGAs) since there are – except the global clock signal – no other global signals needed.

Future work will be focused on the firefly synchronization of completely asynchronous cellular automata and of random boolean networks having a different number of inputs for each cell. This would allow to remove all global signals and to use local clocks only, which would be an important additional advantage for Evolware implementations. Another focus point will be the synchronization of imperfect (lossy) 2D CAs and networks similar to the recent work of Challet and Johnson [6] (see also [7]) who have shown how imprecise clocks can keep good time by working together.

Acknowledgments

We are grateful to Philippe Raoult, Hans Jäckle, and Pierre-André Mudry for their help.

References

1. H. Bersini and V. Detours. Asynchrony induces stability in cellular automata based models. In R.A. Brooks and P. Maes, editors, *Proceedings of the Artificial Life IV conference*, pages 382–387, Cambridge, MA, 1994. MIT Press.
2. J. Buck and E. Buck. Synchronous fireflies. *Scientific American*, 234:74–85, May 1976.
3. M. S. Capcarrere, A. Tettamanzi, M. Tomassini, and M. Sipper. Statistical study of a class of cellular evolutionary algorithms. *Evolutionary Computation*, 7(3):255–274, 1998.
4. M. S. Capcarrere. *Cellular Automata and Other Cellular Systems: Design & Evolution*. PhD Thesis No 2541, Swiss Federal Institute of Technology, Lausanne, 2002.
5. M. S. Capcarrere. Evolution of asynchronous cellular automata. In J. J. Merelo, A. Panagiotis, and H.-G. Beyer, editors, *Parallel Problem Solving from Nature - PPSN VII 7th International Conference, Granada, Spain, September 7-11, 2002, Proceedings*. Springer-Verlag, 2002.
6. D. Challet and N. F. Johnson. Optimal combinations of imperfect components. *Physical Review Letters*, 89(2):028701-1–028701-4, July 8 2002.
7. A. Cho. Collective effort makes the good times roll. *Science*, 297:33, July 5 2002.
8. R. Cori, Y. Métivier, and W. Zielonka. Asynchronous mappings and asynchronous cellular automata. *Information and Computation*, 106:159–202, 1993.
9. R. Das, J. P. Crutchfield, M. Mitchell, and J. E. Hanson. Evolving globally synchronized cellular automata. In L. J. Eshelman, editor, *Proceedings of the Sixth International Conference on Genetic Algorithms*, pages 336–343, San Francisco, CA, 1995. Morgan Kaufmann.
10. P. Gács. Self-correcting two-dimensionnal arrays. In Silvio Micali, editor, *Randomness in computation*, volume 5 of *Advances in Computing Research*, pages 223–326, Greenwich, Conn, 1989. JAI Press.
11. P. Gács. Reliable cellular automata with self-organization. In *Proceedings of the 38th IEEE Symposium on the Foundation of Computer Science*, pages 90–99, 1997.
12. L. Glass and M. C. Mackey. *From Clocks to Chaos: The Rhythms of Life*. Princeton University Press, 1988.
13. H. Hartman and Gérard Y. Vichniac. Inhomogeneous cellular automata (inca). In E. Bienenstock et al., editor, *Disordered Systems and Biological Organization*, volume F 20, pages 53–57. Springer-Verlag, Berlin, 1986.
14. W. Hordijk. The structure of the synchonizing-CA landscape. Technical Report 96-10-078, Santa Fe Institute, Santa Fe, NM (USA), 1996.
15. T. E. Ingerson and R. L. Buvel. Structures in asynchronous cellular automata. *Physica D*, 10:59–68, 1984.
16. Y. Kanada. Asynchronous 1D cellular automata and the effects of fluctuation and randomness. In R. A. Brooks and P. Maes, editors, *ALife IV: Proceedings of the Fourth Conference on Artificial Life*, page Poster, Cambridge, MA, 1994. MIT Press.
17. S. A. Kauffman. *The Origins of Order: Self–Organization and Selection in Evolution*. Oxford University Press, New York; Oxford, 1993.
18. D. Kuske. Emptiness is decidable for asynchronous cellular machines. In C. Palamidessi, editor, *CONCUR 2000*, Lecture Notes in Computer Science, LNCS 1877, pages 536–551, Berlin, 2000.

19. D. Mange, M. Sipper, A. Stauffer, and G. Tempesti. Toward robust integrated circuits: The embryonics approach. *Proceedings of the IEEE*, 88(4):516–540, April 2000.

20. M. K. McClintock. Menstrual synchrony and suppression. *Nature*, 229:244–245, 1971.

21. R. E. Mirollo and S. H. Strogatz. Synchronization of pulse-coupled biological oscillators. *SIAM Journal on Applied Mathematics*, 50(6):1645–1662, December 1990.

22. M. A. Nowak, S. Bonhoeffer, and R. M. May. Spatial games and the maintenance of cooperation. *Proceedings of the National Academic of Sciences USA*, 91:4877–4881, May 1994.

23. G. Pighizzini. Asynchronous automata versus asynchronous cellular automata. *Theoretical Computer Science*, 132:179–207, 1994.

24. M. J. Russel, M. G. Switz, and K. Thompson. Olfactory influences on the human menstrual cycle. *Pharmacology Biochemistry and Behavior*, 13:737–738, 1980.

25. B. Schönfisch and A. de Roos. Synchronous and asynchronous updating in spatially distributed systems. *BioSystems*, 51:123–143, 1999.

26. M. Sipper, M. Goeke, D. Mange, A. Stauffer, E. Sanchez, and M. Tomassini. The firefly machine: Online evolware. In *Proceedings of the 1997 IEEE International Conference on Evolutionary Computation (ICEC'97)*, pages 181–186. IEEE, 1997.

27. M. Sipper. *Evolution of Parallel Cellular Machines: The Cellular Programming Approach*. Springer-Verlag, Heidelberg, 1997.

28. R. W. Stark. Dynamics for fundamental problem of biological information processing. *International Journal of Artificial Intelligence Tools*, 4(4):471–488, 1995.

29. S. H. Strogatz and I. Stewart. Coupled oscillators and biological synchronization. *Scientific American*, 269:68–75, December 1983.

30. G. Tempesti, D. Mange, A. Stauffer, and C. Teuscher. The Biowall: An electronic tissue for prototyping bio-inspired systems. In A. Stoica, J. Lohn, R. Katz, D. Keymeulen, and R. S. Zebulum, editors, *Proceedings of the 2002 NASA/DoD Conference on Evolvable Hardware*, pages 221–230. IEEE Computer Society, Los Alamitos, CA, 2002.

31. H. R. Ueda, K. Hirose, and M. Iino. Intercellular coupling mechanism for synchronized and noise-resistant circadian oscillators. *Journal of Theoretical Biology*, 216:501–512, 2002.

32. T. J. Walker. Acoustic synchrony: Two mechanisms in the snowy tree cricket. *Science*, 166:891–894, 1969.

33. S. Wolfram. Approaches to complexity engineering. *Physica D*, 22:385–399, 1986.

34. W. Zielonka. Notes on finite asynchronous automata. *Informatique théorique et Applications/Theoretical Infomatics and Applications*, 21(2):99–135, 1987.

35. W. Zielonka. Safe executions of recognizable trace languages by asynchronous automata. In Albert R. Meyer and Michael A. Taitslin, editors, *Logic at Botik'89*, Lecture Notes in Computer Science, LNCS 363, pages 278–289, Berlin, 1989. Springer-Verlag.

A Comparison of Different Circuit Representations for Evolutionary Analog Circuit Design

Lyudmilla Zinchenko[2], Heinz Mühlenbein[1], Victor Kureichik[2], and Thilo Mahnig[1]

[1] FhG-AiS Schloss Birlinghoven 53754 Sankt – Augustin Germany
{muehlenbein,mahnig}@gmd.de
[2] Taganrog State University of Radio Engineering, l. Nekrasovsky, 44, GSP-17A, Taganrog, 347928, Russia
{kur,toe}@tsure.ru
Ph. +7-8634-371-694

Abstract. Evolvable hardware represents an emerging field in which evolutionary design has recently produced promising results. However, the choice of effective circuit representation is inexplicit. In this paper, we compare different circuit representations for evolutionary analog circuit design. The results indicate that the design quality is better for the element-list circuit representation.

1 Introduction

Circuit design is a difficult problem because of huge design space dimensions [1]. Different circuits are generated by a variation of parameters and topologies. Mainly they are useless, while the small part of them corresponds to a design goal. An effective search in such large complex spaces is impossible without smart tools [5]. One of the approaches to circuit design is a restriction of design space by means of expert knowledge. However, an expert circuit designer uses the part of possible circuit solutions that have been found by chance or by a search within tiny regions.

Recently, automated analog circuit design techniques have been proposed. Their advantage is a search within the whole space, but it requires large computational efforts. There are several approaches to a reduction of design space by means of macromodelling, symbolic analysis etc. [3]. However, the application of evolutionary algorithms for circuit design seems the most attractive approach, which allows us to obtain novel circuit solutions. Many positive results have been published [1-15]. However, the choice of the best structure of evolutionary design is unclear. Circuit representation, fitness evaluation technique, and search method have to be chosen for effective evolutionary circuit design carefully. However, their choice is inexplicit in practice.

In this paper we aim to compare different circuit representations for static fitness schedule and distribution algorithms. Univariate and bivariate distribution algorithms use estimations of the probability. Hence, a reasonable choice of circuit representation has a profound result for effective circuit design. We compare the effectiveness of branch-list circuit representation and element-list circuit representation for univariate algorithms. We use lowpass filter benchmark.

The remainder of the paper is organized as follows. Section 2 summaries briefly recent results obtained for evolutionary circuit design. Section 3 elaborates on the

A.M. Tyrrell, P.C. Haddow, and J. Torresen (Eds.): ICES 2003, LNCS 2606, pp. 13–23, 2003.

circuit representations. The experimental setup and results are presented in Sections 4 and 5. We discuss the results in Section 6.

2 Approaches to Evolutionary Circuit Design

Evolutionary algorithms are used both for digital circuit design and analog circuit design [1]. Circuit representation, fitness evaluation technique, and search algorithms are crucial ingredients of evolutionary circuit design. In this paper, we restrict our research to analog circuit design, although there are the same problems for digital circuit design as well.

A correct circuit representation is the base for effective design. There are many approaches to circuit description. Tree representation is used by the growing technique [8], which allows us to obtain a sequence of circuits by an addition or a deletion of circuit elements. The deficiencies of the approach are circuits of illegal topologies that are generated at each step. Analysis of useless solutions requires computational efforts, while solutions have the right topology only. A generation of certain topologies using cc-bot instructions restricts the design space [4]. This limitation has both advantages and deficiencies. The computation efforts are reduced, whilst a circuit variety is restricted as well. The matrix representation [11] requires preliminary knowledge. It should be noted that it is used for fixed circuit sizes only. The integer representation, which is based on the linear string, has been used effectively [7, 11].

The complexity of evolutionary design is defined by the ruggedness of fitness landscape. Some investigations of fitness landscape changes have been performed. Dynamic fitness schedules change the fitness landscape [6], but their deficiency is a varying gradient of the fitness function landscape. For the following generations the information obtained is false. The fitness evaluation technique [13] varies for small and large circuits by means of penalty coefficients. Therefore, more efficient solutions are selected at each step.

Note that the right choice of search algorithm represents the most important decision for effective design. There are many publications that represent the applications of genetic programming, genetic algorithms, and evolutionary strategies for circuit design [8 - 10]. Specific mutation operators for analog circuit design have been used in [12]. Some researchers combine genetic search and annealing optimisation [3], genetic algorithms and Davidon-Fletcher-Powell optimisation method [14]. The positive results have been obtained for the hardware immune system [2]. Recently, ant colony algorithm has been used for analog circuit design successfully [15]. Our focus is on the application of probabilistic evolutionary models. The Estimation of Distribution Algorithms EDA has been proposed by Mühlenbein and Paaß [16] as an extension of genetic algorithms. Instead of performing recombination of strings, EDA generates new points according to the probability distribution defined by the selected points. In [17] Mühlenbein showed that genetic algorithms can be approximated by an algorithm using univariate marginal distributions only (UMDA). UMDA is an evolutionary algorithm, which combines mutation and recombination by means of distribution. The distribution is estimated from a set of selected points. It is used then to generate new points for the next generation. It should be noted that we use truncation selection because of its simplicity. Some experimental results of the distribution algorithms application for several test functions have been published in [18]. Recently, we have proposed the dynamic coding, the hierarchy circuit representation, and the application of UMDA for analog circuit design that allow us to use topological and parametrical search simultaneously [19]. Observe that the hierarchical genotype structure reduces the search space without the restriction of circuit variety. In order to improve design abilities, mutation has been introduced into UMDA by a concept called Bayesian prior [18]. The choice of hyperparameter r has been considered for test functions [18]. We have improved the effectiveness of circuit design by means of mutation too [20]. UMDA with Bayesian prior is able to overcome local minima. Furthermore, we used the experimental research in order to make a reasonable choice of Bayesian prior for analog circuit design [21]. Moreover, we synthesized the circuits that are better than those produced by an expert circuit designer [20, 21].

The probabilistic approach evaluates the direct relation between the genotype and the fitness. In this paper we compare different circuit representations for the application of UMDA for analog circuit design. It should be noted that we restrict our research to the static technique of fitness function evaluation only.

3   Circuit Representations

There are many circuit representation techniques for evolutionary circuit design. The hierarchical circuit representation [19 - 21] was chosen as the basis for the investigation for the following reasons. First, as a large number of circuits are needed to evaluate, the search within the whole space will be very slow. Second, the relations between the circuit elements and the circuit parameters cannot be estimated for different elements. Physical nature is different for capacitances and inductances. Therefore, probabilistic models have to be evaluated for capacitance parameters and for inductance parameters separately.

The growing technique is used during the evolutionary design. The embryonic circuit is shown in Fig. 1. The evolutionary process is based on the topology of the complete graph. Adjacency matrix of the N-node complete graph is given as follows [22]:

$$A = \begin{cases} a_{ij} = 0 & if \quad i = j \\ a_{ij} = 1 & if \quad i \ne j \end{cases} . \tag{1}$$

First, we specify the boundary limits on the amount of the circuit nodes. Second, each element is determined by its type and parameter. Two bits define the type. Each element can take on 3 different types, namely, switch (0), capacitance (1) and inductance (2). Three bits define a parameter of element. The parameter of switch can take on two fixed values only. We consider that 0 means that a switch is turned off and 1 means that a switch is turned on. Parameters of inductances and capacitances are chosen from the given range. The input of the circuit is always node 1 and the output is node 2. Branch 1 is always series connection of a voltage source and the source resistance. Branch 2 is the load resistance. The circuit being evolved contains (n-2) branches. The amount of the branches is given as follows:

$$n = \sum_{i=0}^{N-1}(N - i). \tag{2}$$

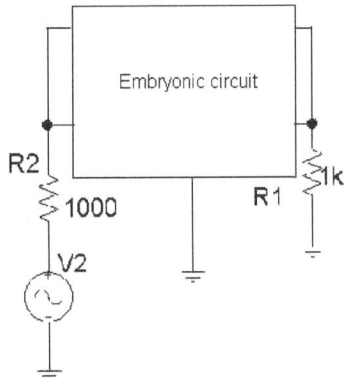

Fig. 1. The embryonic circuit of two-ports

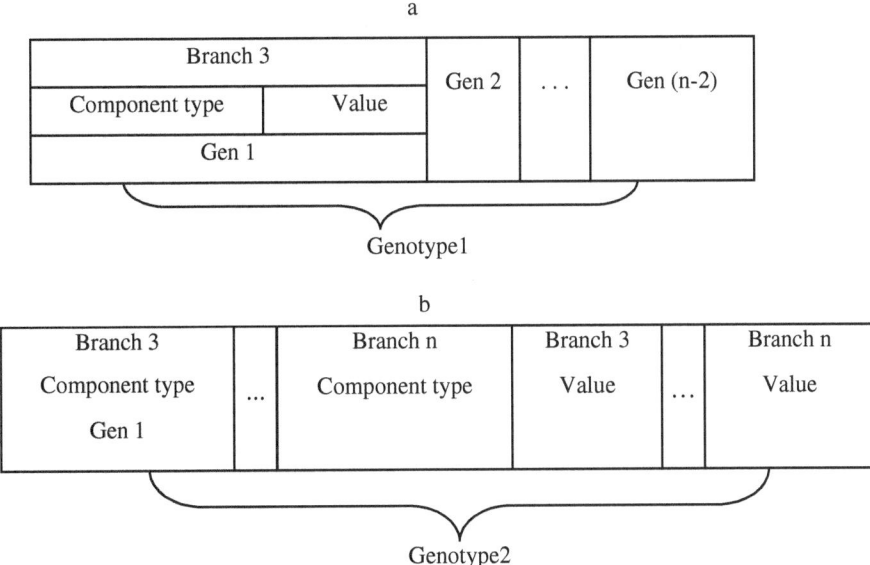

Fig. 2. General structure of the branch circuit representation Genotype1 (a), general structure of the element circuit representation Genotype2 (b)

We propose two circuit representations that use the adjacency matrix (1). Figure 2 shows their general structure. Figure 3 illustrates the case of the 3 node topology. Chromosome length is equal to 23. The best circuit is shown in Fig. 3 (c).

3.1 Branch-List Circuit Representation

The branch-list circuit representation is one of standard circuit representation techniques for evolutionary analog circuit design. A genotype is formed by means of the combination of separate genes for each branch. The circuit representation used is similar to [7, 11, 12, 19]. Briefly, each gene consists of alleles for a type, connecting nodes and value parameters. The genotype [type, connecting nodes, value parameters] [11, 12], the genotype [connecting nodes, value parameters, type] [7], and the hierarchical genotype [type / value parameters] [19] are possible modifications. Fig.2 (a) shows general structure of the branch-list circuit representation (Genotype1). The description of Genotype1 for the chromosome of length 23 is illustrated in Fig. 3 (a). The correspondent adjacency matrix is given as follows:

$$A = \begin{vmatrix} 0 & 1 & 1 & 1 \\ 1 & 0 & 1 & 1 \\ 1 & 1 & 0 & 1 \\ 1 & 1 & 1 & 0 \end{vmatrix}. \tag{3}$$

Amount of nodes	Branch 3		Branch 4	Branch 5	Branch 6
0 1 1	1 0 ↑ Element	0 1 1 ↑ parameter	0 1 1 0 1	0 1 1 0 1	0 0 0 0 1

a

Amount of nodes	Element				Parameter			
	Branch 3	Branch 4	Branch 5	Branch 6	Branch 3	Branch 4	Branch 5	Branch 6
0 1 1	1 0	0 1	0 1	0 0	0 1 1	1 0 1	1 0 1	0 0 1

b

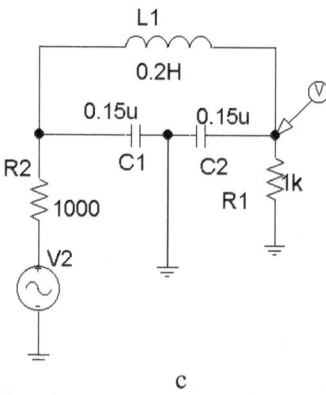

c

Fig. 3. The branch circuit representation Genotype1 (a), the element circuit representation Genotype2 (b), and the best circuit (c)

3.2 Element-List Circuit Representation

The element-list circuit representation is a circuit representation modification. It should be noted that topology information is more important for circuit design. Thus, we suppose that a change of circuit representation could improve the effectiveness of evolutionary search. The topological information is concentrated within the first alleles, whilst the rest of genotype holds alleles for the value parameters. Fig. 2 (b) shows general structure of the element - list circuit representation (Genotype2). The description of Genotype2 for the chromosome of length 23 is illustrated in Fig. 3 (b).

The correspondent adjacency matrix is given by (3). Such representation is effective when the ruggedness of the fitness landscape is reduced. Otherwise it is disadvantageous. Fig.3 indicates that the phenotypes are the same ones, whilst the genotypes are different. Hence, the interactions between the alleles are different as well. Therefore, the efficiency of evolutionary design can be changed. In order to determine the deficiencies of different circuit representation it is useful to compare their behaviour.

4   Experimental Setup

The circuit design benchmark chosen was the lowpass filter. Such circuits have wide applications, although the design theory is based on some simple circuits. The starting point for the design of a filter is the specification of the frequencies for the filter's pass-band and stop-band. The design of filters with high stop-band attenuation and low pass-band attenuation is hard. It is more difficult to design a lowpass filter with small transitional region between two bands. The input specification for the filter design benchmark is summarized in Table 1. The ratio between signal in the pass-band (1 V) and the maximum acceptable signal in the stop-band (1 mV) is 60 decibels, although there is 2-to-1 ratio between the stop-band frequency and the pass-band frequency. The known design methods generate the desired circuit, although there are some fluctuations in the stop-band and in the pass-band.

Table 1. Input specifications for the filter design benchmark

Circuit Specifications	Values
Minimum number of components	3
Maximum number of components	6
Pass-band frequency (Hz)	1000
Stop-band frequency (Hz)	2000
Maximum pass-band gain (dB)	0
Minimum pass-band gain (dB)	-0.26
Maximum stop-band gain (dB)	-60

In order to reduce computational efforts the fitness is evaluated by means of equation-based method [20]. The numerical frequency response evaluation and the modified nodal method are used. The fitness function evaluation technique used is similar to those described in [20, 21]. The fitness is evaluated in terms of the sum of the absolute weighted difference between the actual output voltage and the target one. These error values were summed across evaluation points Nc for the evaluation of the fitness

$$F = \sum_{i=0}^{Nc} W[d(f_i), f_i)] d(f_i) \tag{4}$$

$$d(f) = \left| U_{goal}(f) - U_{out}(f) \right| \tag{5}$$

$$W[d(f), f] = \begin{cases} 1 & \text{if } d(f) \leq 0.03, f \leq 1000 \ Hz, \\ 10 & \text{if } d(f) \geq 0.03, f \leq 1000 \ Hz, \\ 1 & \text{if } d(f) \leq 0.001, f \geq 2000 \ Hz, \\ 10 & \text{if } d(f) \leq 0.001, f \geq 2000 \ Hz. \end{cases} \tag{6}$$

where f is the frequency of the fitness evaluation, $d(f)$ is the absolute value of the difference between the output voltage and the target at the frequency f and $W(d(f), f)$ is the penalty factor for the difference $d(f)$ at the frequency f.

The set of circuits that is generated within the topology of the 3 node complete graph is examined to show how success rate is affected by different circuit representations. The best circuit is given in Fig. 3 (c). The amount of hits is equal to 86. We have found [20] that the effective design is supplied by a big ratio between the population size and the genotype length. It should be noted that it is impossible to use a theoretical framework to determine an effective circuit representation. In order to make a fair comparison, we have to compare actual simulation results.

5   Experimental Results

We examined the behaviour of different circuit representations for the fixed number of bits n=23 with a truncation threshold τ from 0.02 to 0.04. Population size changes from N=300 to N=500. We consider the successful runs after 10% of the population consisted of the best solution only. Figures 4 -7 show how the success rate and the function evaluations are changed when a circuit representation is modified. The results have been obtained under assumption that the amount of iterations is less than the genotype length.

Figures 4, 5 illustrate the case in which the population size varies between N=300 and N=500 and the truncation threshold is equal to τ=0.02. Figures 6, 7 show the case in which the population size is fixed N=300, whilst the truncation threshold is increased to τ=0.04. The selection intensity I depends on a truncation threshold of τ [17]. The most obvious fact is that the success rate is generally enhanced where the element-list circuit representation is used. The branch-list circuit description is less efficient. It is remarkable that the same number of function evaluations is required for two circuit representations approximately. Therefore, the curves have shown that the element-list circuit representation allows us to increase the success rate without the significant computational expenses.

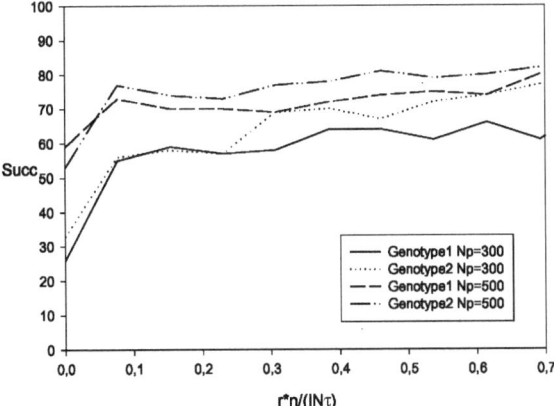

Fig. 4. Performance comparison with varying mutation rate r for different population sizes and truncation selection with $\tau=0.02$. Succ is the number of times, when the best solution was found in 100 runs

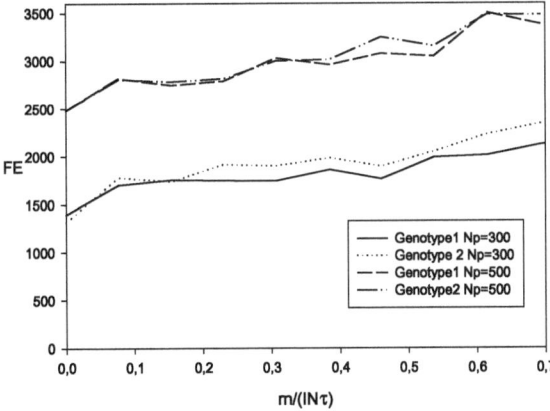

Fig. 5. Performance comparison with varying mutation rate r for different population sizes and truncation selection with $\tau=0.02$. FE is the number of evaluation until 10% of the populations consisted of the best solution

6 Discussions

Two different types of the circuit representation have been presented. The results show that the element-list circuit representation is better than the branch-list circuit representation. We are able to exploit larger population size and truncation selection with smaller τ more effectively using the element-list circuit representation. Therefore, the assignment of topology information and parameters to the correspondent part of the genotype allow us to improve the effectiveness of evolutionary analog circuit design.

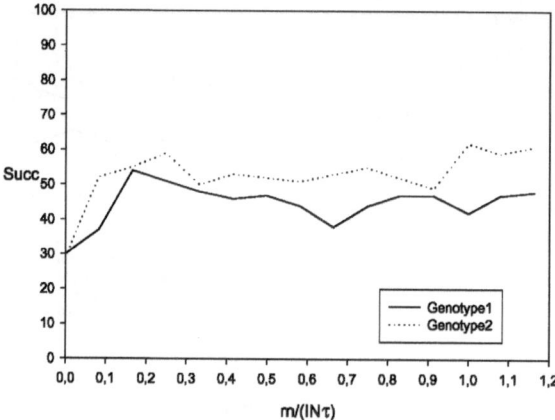

Fig. 6. Performance comparison with varying mutation rate r for the population size N=300 and truncation selection with τ=0.04. Succ is the number of times, when the best solution was found in 100 runs

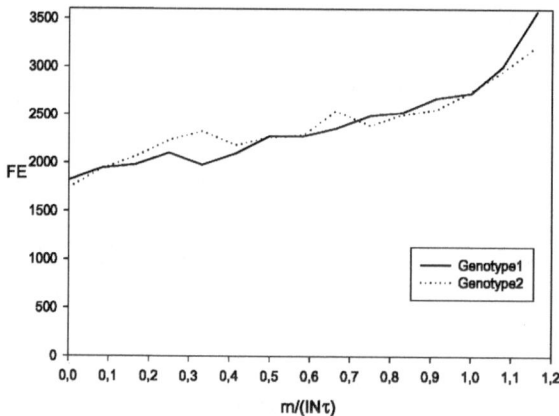

Fig. 7. Performance comparison with varying mutation rate r for the population size N=300 and truncation selection with τ=0.04. FE is the number of evaluation until 10% of the populations consisted of the best solution

Acknowledgments

This work was supported by INTAS (International Association for the promotion of co-operation with scientists from the New Independent States of the former Soviet Union) under grant No YSF 00 - 190. We would like to thank all members of the RWCP Group, FhG - AiS for their support, advice and criticism of this work.
Lyudmilla Zinchenko and Victor Kureichik acknowledge support by Russian Fund of Basic Research (grant 01-01-00044).
The authors also acknowledge the helpful comments of the reviewers.

References

1. Higuchi, T., Iwata, M. etc.: Real World Applications of Analog and Digital Evolvable Hardware. IEEE Transactions on Evolutionary Computation **3** (1999) 220-235
2. Tyrell, A. M., Bradley, D.W.: The Architecture for a Hardware Immune System. In: Proceedings of the Third NASA/DoD Workshop on Evolvable Hardware EH 2001. Computer Press (2001) 193-200
3. Rutenbar, R., Liu, H., Singhee, A., Carley, L. R.: Remembrance of Circuit Past: Macromodeling by Data Mining in Large Analog Design Spaces. In: Proceedings of the ACM/IEEE Design Automation Conference (2002)
4. Lohn, J. D., Colombano, S. P.: A Circuit Representation Technique for Automated Circuit Design. IEEE Transactions on Evolutionary Computation **3** (1999) 205 –219
5. Yao, X.,. Liu, Y.: Getting Most Out of Evolutionary Algorithms. In: Proceedings of the 2002 NASA DoD Conference on Evolvable Hardware EH 2002. Computer Press (2002) 8-14
6. Lohn, J., Laith, G., Colombano, S., Stassinopoulos, D.: A Comparison of Dynamic Fitness Schedules for Evolutionary Design of Amplifiers. In: Proceedings of the First NASA DoD Workshop on Evolvable Hardware EH'99. IEEE Computer Press (1999) 87-92
7. Zebulum, R.S., Pacheco, R.S., Vellasco, M.A.: Analog Circuit Evolution in Extrinsic and Intrinsic Modes. In: Sipper, M., Mange, D., and Perez-Uribe, A. (eds.): Proceedings of the Second International Conference on Evolvable Systems: From Biology to Hardware (ICES98). Lecture Notes in Computer Science, Vol. 1478. Springer-Verlag, Berlin Heidelberg New York (1998) 154-165
8. Koza, J.R., Bennett, F. H., Andre, D., Keane, M. A.: Genetic Programming III: Darwinian Invention and Problem Solving. Morgan Kaufmann, San Francisco (1999)
9. Thomson, A., Layzell, P., Zebulum, R.: Explorations in Design Space: Unconventional Electronics Design Through Artificial Evolution. IEEE Transactions on Evolutionary Computation **3** (1999) 167-196
10. Hartmann, M., Haddow, P., and Eskelund, F.: Evolving Robust Digital Designs. In: Proceedings of the 2002 NASA DoD Conference on Evolvable Hardware EH 2002. Computer Press (2002) 36-45
11. Ando, S., Iba, H.: Analog Circuit Design with a Variable Length Chromosome. In: Proceedings IEEE Int. Congress on Evolutionary Computation. IEEE Press (2000) 994 – 1001
12. Goh, C., Li, Y.: GA Automated Design and Synthesis of Analog Circuits with Practical Constraints. In: Proceedings IEEE Congress on Evolutionary Computation. IEEE Press (2001) 170 -177
13. Goh, C., Chan, L., and Li, Y.: Performance Metrics Assessment for Pareto Fronts with Application to Analog Circuit Evolution. In: Giannakoglou, K.G., Tsahalos, D.T., Periaux, J., Papailiou, K.D., Fogarty, T. (eds.): Evolutionary Methods for Design, Optimization and Control. Theory and Engineering Applications of Computational Methods. CIMNE, Barcelona (2002) 208-213
14. Grimbleby, J.B.: Hybrid Genetic Algorithms for Analogue Network Synthesis. In: Proceedings IEEE Congress on Evolutionary Computation. IEEE Press (1999) 1781 -1787
15. Tamplin, M.R., Hamilton, A.: Ant Circuit World: An Ant Algorithm MATHLAB Toolbox for the Design, Visualization and Analysis of Analogue Circuits. In: Proceedings of the International Conference on Evolvable Systems: From Biology to Hardware. Lecture Notes in Computer Science. Springer-Verlag, Berlin Heidelberg New York (2001)
16. Mühlenbein, H., Paaß, G.: From Recombination of Genes to the Estimation of Distributions. Parallel Problem Solving from Nature PPSN IV. Lecture Notes in Computer Science, Vol. 1141. Springer-Verlag, Berlin Heidelberg New York (1996) 178-187
17. Mühlenbein, H.: The Equation for Response to Selection and its Use for Prediction. Evolutionary Computation **5** (1998) 303-346

18. Mühlenbein, H., Mahnig, T.: Optimal Mutation Rate Using Bayesian Priors for Estimation of Distribution Algorithms, to appear.
19. Mühlenbein, H., Kureichik, V.M., Mahnig, T., Zinchenko, L.A.: Evolutionary Algorithms with Hierarchy and Dynamic Coding in Computer Aided Design. In: Giannakoglou, K.G., Tsahalos, D.T., Periaux, J., Papailiou, K.D., Fogarty, T. (eds.): Evolutionary Methods for Design, Optimization and Control. Theory and Engineering Applications of Computational Methods. CIMNE, Barcelona (2002) 202-207
20. Mühlenbein, H., Kureichik, V.M., Mahnig, T., Zinchenko, L.A.: Application of the Univariate Marginal Distribution Algorithm to Analog Circuit Design. In: Proceedings of the 2002 NASA DoD Conference on Evolvable Hardware EH 2002. Computer Press (2002) 93-101
21. Mühlenbein, H., Kureichik, V.M., Mahnig, T., Zinchenko, L.A.: Effective Mutation Rate of Probabilistic Models for Evolutionary Analog Circuit Design. In: Proceedings of the IEEE ICAIS 2002. Computer Press (2002) 401-406
22. Biggs, N.: Algebraic Graph Theory. Cambridge (1974)

Fault Tolerance via Endocrinologic Based Communication for Multiprocessor Systems

Andrew J. Greensted and Andy M. Tyrrell

Department of Electronics,
Bio-Inspired Research Group,
University Of York, UK
YO10 5DD
{ajg112,amt}@ohm.york.ac.uk
http://www.bioinspired.com

Abstract. The communication mechanism used by the biological cells of higher animals is an integral part of an organisms ability to tolerate cell deficiency or loss. The massive redundancy found at the cellular level is fully taken advantage of by the biological endocrinologic processes. Endocrinology, the study of intercellular communication, involves the mediation of chemical messengers called hormones to stimulate or inhibit intracellular processes.

This paper presents a software model of a multiprocessor system design that uses an interprocessor communication system similar to the endocrine system. The feedback mechanisms that govern the concentration of hormones are mimicked to control data and control packets between processors. The system is able to perform arbitrary dataflow processing. Each processing stage within the system is undertaken by a separate group of microprocessors. The flow of data, and the activation of the next stage within the process is undertaken using the bio-inspired communication technique. The desired result is a system capable of maintained operation despite processor loss. The feasibility of the multiprocessor system is demonstrated by using the model to perform a simple mathematical calculation on a stream of input data.

1 Introduction

Biology provides a diverse source of inspiration that reaches across many fields from both artistic and scientific disciplines. Electronic Engineering is no exception.

One of the strongest impetuses for Bio-Inspired Engineering is the assistance it lends to the development of engineering design, especially where solutions via traditional techniques fall short. Such biological based solutions have led to the creation of artificial learning and pattern matching systems based on neural networks [1,2], as well as self re-configuring systems [3] based on biology's ability to evolve and adapt.

Of more interest to the area of Reliability Engineering is biology's ability to maintain operation in the face of adverse conditions. Biology is able to employ

A.M. Tyrrell, P.C. Haddow, and J. Torresen (Eds.): ICES 2003, LNCS 2606, pp. 24–34, 2003.

automated fault tolerance, detection and recovery characteristics that enable organisms to remain functional despite injury. It is the desire for electronics to also exhibit these three characteristics that has led to an interest in Bio-Inspired Reliability Engineering.

Both Embryonics [4] and Artificial Immune Systems [5,6] mimic aspects of biological reliability systems. The Embryonic Architecture is capable of removing faulty circuit areas. Reconfiguration via the shifting of functionality to healthy circuitry in redundant areas returns the system to full functionality. Whereas Artificial Immune Systems provide a fault detection and removal mechanism based on the biological self, non-self principle [7].

The ability of biological systems to tolerate and recover from a subsystems death is reliant upon the use of redundancy. This is a common feature present on a number of levels within biology's structural hierarchy. Society maintains activity through redundancy in individual organisms, similarly organisms can function without certain organs. However, it is the cellular level that utilizes redundancy to the greatest effect.

Cellular biology of higher animals provides the inspiration for the multiprocessor system presented in this paper. A software model of an inter-processor communication system based on biology's endocrine system is presented, including results that demonstrate the system's fault tolerant characteristics.

The subject of biological cell signalling is discussed in Section 2. How such signalling can be translated into a useful electronic system is presented in Section 3. The model of the resulting multiprocessor system and operation results are described in Sections 4 and 5 respectively. The paper is concluded in Section 6 with suggestion for further work in Section 7.

2   Cell Signalling

Cells are heavily dependant on signalling mechanisms for survival. The ability for cells to influence each other enables a multicellular organism to maintain a level of homeostasis[1]. Even some unicellular organisms utilise signalling to influence proliferation of other like cells [8].

A variety of different communication systems are employed in higher animals. Each system varies in a number of ways, but especially with regard to range and speed. However, in each case communication is achieved via signalling molecules. Each messenger molecule exhibits a biological signature that determines its recognition by other cells. Target cells recognize messenger signatures through receptors. A receptor-messenger match allows molecules to bind to their target and complete the communication process.

The synaptic signalling process shown in Figure 1 is the most directed messenger based communication system. Nerve cells directly steer their messengers via their connected axons. As the neurotransmitter messengers are released so close to their target, affinity between receptor and messenger can be low. In

---

[1] Homeostasis is the process by which organisms or cells maintain a stable internal equilibrium via physiological changes.

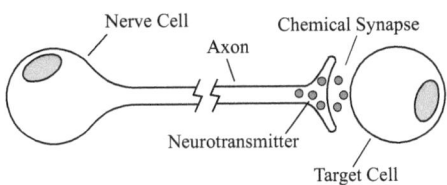

Fig. 1. Synaptic signalling via neurotransmitters.

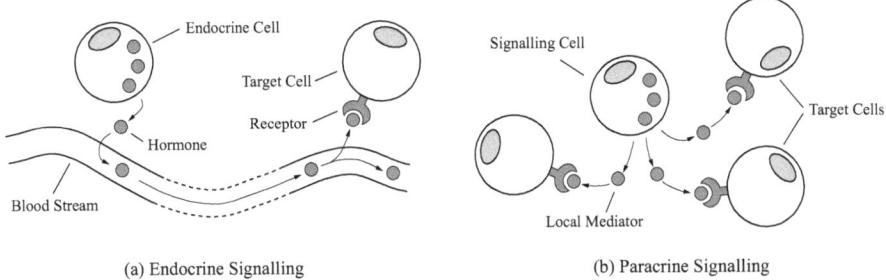

(a) Endocrine Signalling   (b) Paracrine Signalling

Fig. 2. Two types of cell signalling. (a) Endocrine communication, hormones are transported via the blood stream between source and target cells. (b) Paracrine communication, local mediator messengers are restricted to local tissue diffusion to travel from source to target cells.

contrast, messengers used in endocrine and paracrine signalling are not directed straight to their target. Endocrine signalling, as shown in Figure 2a, transport their messengers (called hormones) through the blood stream. Paracrine signalling, Figure 2b, is similar, however the message chemicals are limited to a localized area of tissue.

2.1   Endocrinologic Control

Homeostasis of higher animals is achieved by the body's nervous and endocrine systems [9]. Mediated by their respective communication systems, changes in an organism's internal or external environment invoke responses that stimulate the appropriate physiological changes required for organism adaption.

In the case of the endocrine system, hormones released from special glands are able to stimulate cells, activating a variety of reactions. These reactions may involve the production of other hormones that in turn activate other cells.

It is possible for cells with hormone receptors with different specificities to bind to the same hormone signature. Therefore, the release of a single hormone type is able to initiate a response in more than one type of target tissue. Also, hormones are not limited to stimulatory effects. They may also produce an inhibitory effect on their target. These combined abilities of hormones provide the foundation for hormone based control. Figure 3 depicts two possible scenarios for endocrine control.

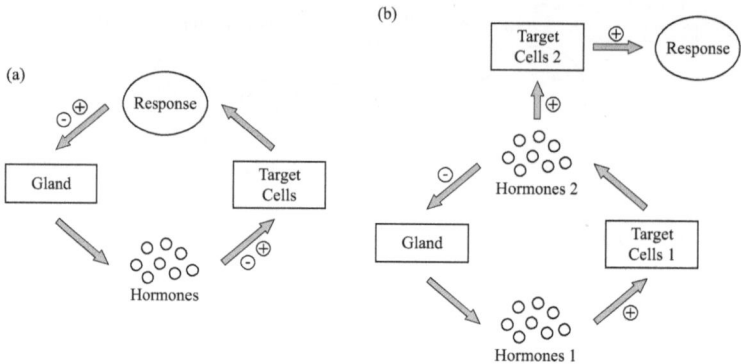

Fig. 3. Two possible scenarios for endocrine based control.

In Figure 3a, the release of hormones by the gland initiates a response by the target cells. The response itself then stimulates or inhibits the original gland to increase or decrease the production and secretion of hormones. The second system, shown in Figure 3b, contains a second stage. The initiating gland produces an initial set of hormones that activate the first group of target cells. The resulting action is the production of a second set of hormones that target both the original gland and a second group of target cells. The gland is inhibited by these hormones to stop its own hormone production, whereas the second cell group is stimulated producing the desired overall response.

3   An Electronic System from Biology

The biological endocrine system provides higher animals with a robust control mechanism. Unfortunately, this is all it is, it does not directly lend itself to use as an information processing or calculation system. However, if the hormone messengers contained data, and target cells performed operations on the data, the endocrinologic paradigm could be used to create such a system.

Figure 4a depicts a system adapted from the biological model shown in Figure 3b. In this case, the system has been extended to contain a number of target groups. Each group, on stimulation, releases messengers that stimulate the next target group and inhibit the previous. The result is a cascade of data processing stages. The equivalent system, in block diagram form is shown in Figure 4b.

3.1   Microcells: Microprocessor Cells

All biological cells develop from a common ancestor type, the stem cell [10]. When new cells are required, they can be developed and put to use. Unfortunately, physical growth of electronic hardware to replace malfunctioning circuitry is not feasible at the moment. Instead, resources have to be made available which can be brought into use, mimicking regrowth.

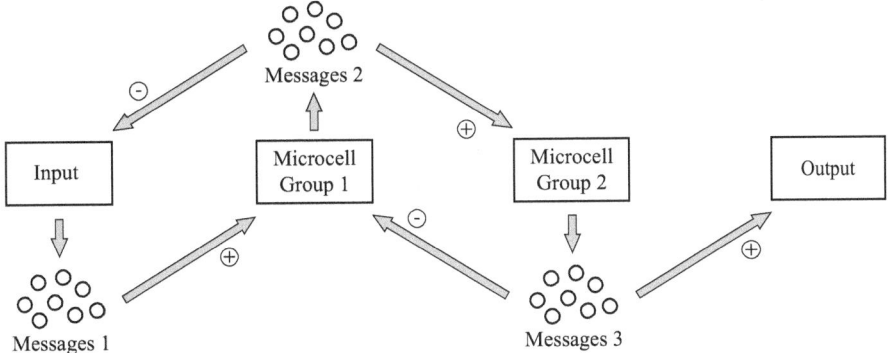

Fig. 4. Adapting the endocrine system to a data processing platform. (a) A succession of messenger activated microcell groups forming a chain of data operations, (b) An equivalent block diagram view of the system.

In the case of the cells in the new system, a hardware unit is required that can assume a variety of functions. Then, in the situation that a cells needs to be replaced, a new generic cell can take its role, configuring itself to perform the required task.

A microprocessor based cell, or microcell, is able to fulfil such a task. Each unit stores all the required cell procedures within memory, and on activation, programs itself with the appropriate section of software. This is an established bio-inspired technique used especially in Embryonics [11,12], and in this case forms a neat parallel with the ability of biological cells to to develop into any number of roles [8].

3.2   Communication Space

Biology exists in a non-rigid framework that allows the movement of matter and the constant reorganization of its constituent parts. In electronics this is simply impossible. The structure of the silicon within a device determines exactly where a component is located, and where it shall stay. This static nature of electronic features make replicating the endocrine communication mechanism difficult. Endocrine hormones must be free to move and co-exist if they are to reach their target effectively.

As it is not possible to create a medium where data packets are able to freely roam between microcells, a more restricted solution is taken in this paper. A fully-connected mesh topology [13,14] would allow free communication between any two microcells, but would result in an unrealistic number of connecting links for any reasonable size system. Instead, a 'closest-neighbour' system is used, where each microcell is connected to a number of its closest neighbouring microcells. Messenger data packets are able to diffuse across the network by passing from one microcell to the next. If the target is reached, the message data is consumed as its travel is complete, if not the message is moved on.

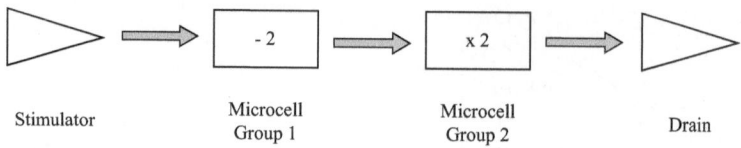

Stimulator   Microcell   Microcell   Drain
                  Group 1   Group 2

Fig. 5. A block diagram of the configuration used to test the multiprocessor system.

3.3   Topology

In a system so dependant on communication, the network topology is an important consideration. To give each microcell an equal communication standing, each cell's view of its IO connections must appear the same. To achieve this, a boundary free topology with an equal quantity of neighbour connections per microcell is used. This takes the form of a toroidal mesh. However, this is an area for further work (see Section 7.1).

3.4   Data Processing

The system resulting from the conversion of the endocrine control system to electronic system is a simple dataflow processor. The system has a single input, the Stimulator and a single output, the Drain as shown in Figure 4b. The microcells that lie in between are grouped depending on the stage they represent in the system. Redundant microcells that do not perform operations on the data are also included in the system. Their role is to simply traffic data without alteration, and when required, assume a data operating function to replace the loss of a faulty microcell.

4   Multiprocessor System Model

The multiprocessor system was simulated using a multi-threaded Java Model [15,16,17]. To demonstrate the operation of the model a simulation of the simple data processing system, shown in Figure 5, has been performed. The results are presented in Section 5.

The network produced by the model is shown in Figure 6. Each sphere represents a single microcell. The connections between the spheres show the connectivity of the network. In this particular network all connections are full-duplex, however half-duplex connections are possible. The system Stimulator and Drain can also be seen.

5   Simulation Results

The key points of the systems operation are shown in the following three Figures 7, 8 & 9 (The data for each graph were taken from the same simulation run). The messaging activity of the system is complex due to the random nature in which microcells pass data. However, a pattern in cell activity levels, with similarities

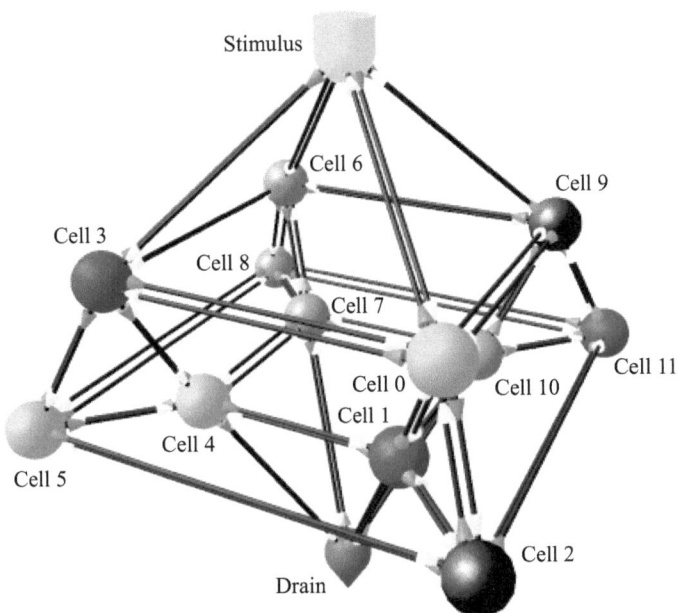

Fig. 6. A representation of the multiprocessor system. Cells 0, 4 & 5 and Cells 1, 3 & 11 perform different processing tasks. Cells 2 and 9 have been killed off to demonstrate fault tolerance. The remaining cells are redundant. The main system input (Stimulator) and output (Drain) are also shown.

to biorhythms [9], can be observed. Figure 7 shows the activity levels of cell 4 (see Figure 6). The stimulation and inhibition components that define the cells activity can be seen.

Each rise and fall in activity is regulated by the initial stimulus of input data. Comparison of this graph and that in Figure 8 show a correlation in the introduction of new data, and subsequent cell activation.

Figure 8 shows the overall system input and output. Each spike of Figure 8a represents the activation of the system Stimulator. It can be seen that these spikes lead to a number of output spikes, Figure 8b. These represent the arrival of final stage messengers at the system Drain. Further study of the graph shows that the correct mathematical operations (see Figure 5) are being performed on the data.

The final graph shown in Figure 9 shows the effects of microcell loss. At two points during the simulation a microcell is killed off, and then replaced by a redundant cell. As Figure 8 proves, operation is maintained despite the loss.

6   Conclusion

So far, the simulation results have been encouraging. Even though the system is in its very early stages of development it is able to perform simple processing

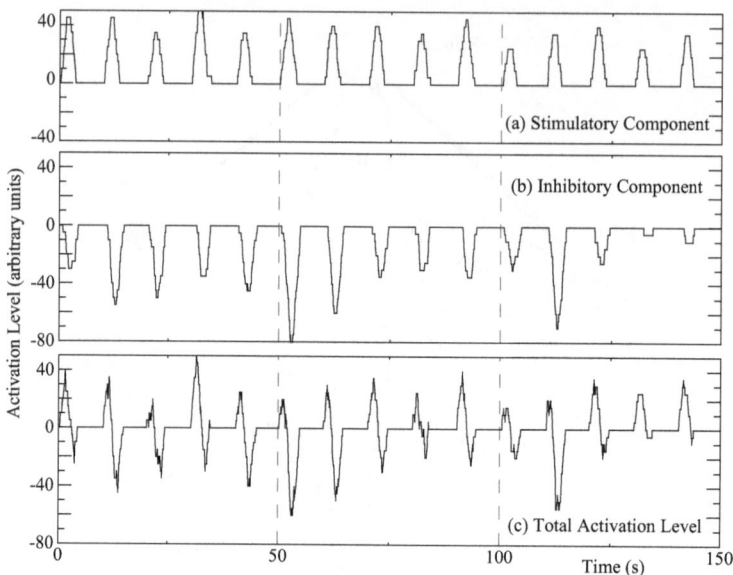

Fig. 7. Graph of the activation levels of microcell 4 (See Figure 6). The total activation level (c) can be seen as the addition of the stimulatory (a) and inhibitory (b) components.

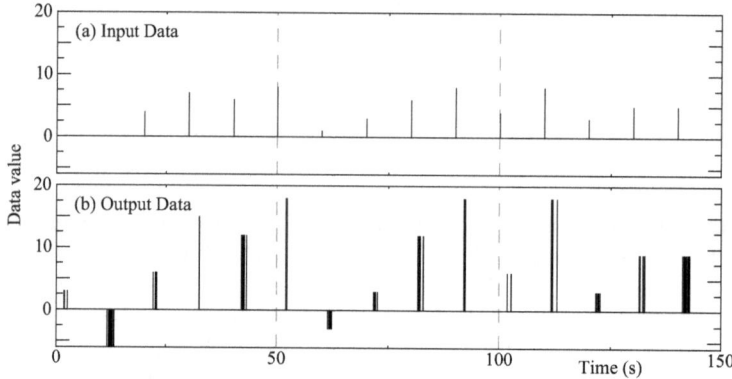

Fig. 8. Graph showing the overall system input (a) and output (b) data streams.

on a flow of data. Microcell loss is also tolerable and the ability to return the system to a full complement of processing units is possible.

Biological Endocrinology has been presented as another source of inspiration for engineering. Furthermore, this paper has made a start to show it is possible to glean some of its reliability properties via its replication as an electronic communication system.

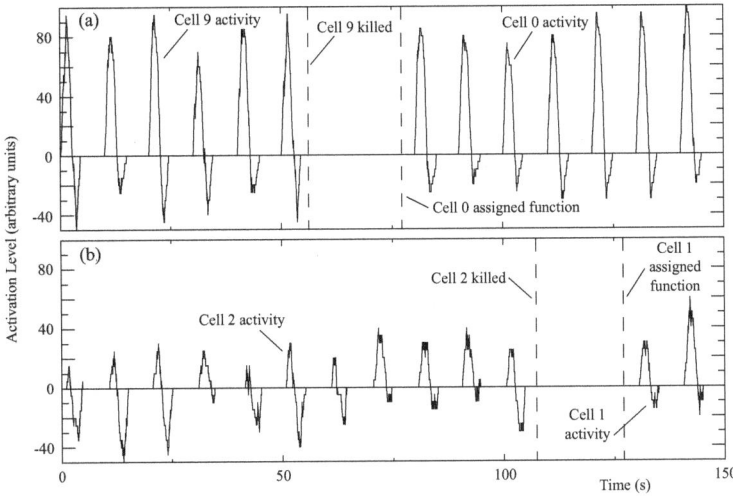

Fig. 9. Graph of activity levels in four different cells as they become activated and deactivated, demonstrating the systems ability to deal with microcell loss.

7 Further Work

The development of the presented system model is currently still within its early stages. Consequently, there remain many features to be included and issues to be dealt with. The rest of this section introduces some of these.

7.1 Different Topologies

At present each microcell is linked to only four other neighbours. This is supported in a toroid topology that enables the removal of boundaries from a standard square grid.

Other network topologies that have more intermicrocell links need to be tested. Extra connections should provide greater freedom of movement for messages. Although this would bring the communication space representation closer to that of biology, there is a cost in extra hardware.

7.2 Pipelining

The system developed so far has a very limited throughput. The activation level of each microcell must always settle to zero between stimulations. This is required to avoid data from different time frames mixing and consequently invalidating the output data stream. Ideally, the system should perform in a similar manner to a pipeline [18,19] with each microcell group performing concurrently on consecutive data samples.

7.3   Message Redirection

The level of activity in each microcell is dependant on its location in the network. This is an undesirable consequence of the random nature in which the choice of next cell is made when passing messages. The flow of blood in a biological organism provides a broad transport medium for hormones. Blood carries hormones past all the cells in the organism, thus averaging the likelihood of cells receiving hormones.

The introduction of a biasing in the redirection algorithm for message transmission would make it possible to create a general direction of message flow within the network. This would lower the likelihood of activity localization within the network.

7.4   Fault Detection and Recovery

Ultimately, a system that has automated fault detection and recovery is desired. Following the design based on endocrine communication has led to a very suitable architecture for implementing such features.

7.5   Real World Application

The need for Reliability Engineering is a consequence of the use of electronics in Real World applications. A feasibility study of the presented system in use within a 'real world' context would be an interesting area for further work. Due to the data flow processing nature of the system, a possible application could be found in the area of Automated Control Systems.

References

1. S.S. Haykin. *Neural Networks: A Comprehensive Foundation*. Prentice Hall, 2nd edition, 1998.
2. A. Pérez-Uribe. *Structure-Adaptable Digital Neural Networks*. PhD. Thesis: Ecole Polytechnique Federale de Lausanne, 1999.
3. A. Thompson. Evolving electronic robot controllers that exploit hardware resources. *Proceedings, The Third European Conference on Artificial Life*, 1995.
4. D. Mange, M. Sipper, A. Stauffer, and G. Tempesti. Toward self-repairing and self-replicating hardware: The embryonics approach. *Proceedings, The Second NASA/DoD Workshop on Evolvable Hardware, IEEE Computer Society, Los Alamitos*, pages 205–214, 2000.
5. D.W. Bradley and A.M. Tyrrell. Immunotronics : Hardware fault tolerance inspired by the immune system. *proceedings of the 3rd International Conference on Evolvable Systems Lecture Notes in Computer Science*, 1801:11–20, 2000.
6. S. Forrest, S. Hofmeyr, and A. Somayaji. Principles of a computer immune system. *New Security Paradigms Workshop*, pages 75–82, 1998.
7. I. Roitt and P.J. Delves. *Essential Immunology*. Blackwell Science, 10th edition, 2001.

8. A. Alberts, D. Bray, J. Lewis, M. Raff, K. Roberts, and J.D. Watson. *Molecular Biology of The Cell*. Garland Publishing, 3rd edition, 1994.

9. G.D.B. Holland and N.J. Marshall. *Essential Endocrinology*. Blackwell Science, 2nd edition, 2001.

10. C.A. Janeway, P. Travers, M. Walport, and J.D. Capra. *Immuno Biology, The Immune System in Health and Disease*. Garland Publishing, 4th edition, 1999.

11. Marchal P. Embryonics: The birth of synthetic life. *LNCS, Towards Evolvable Hardware*, 1062:166–196, 1996.

12. Mange D. Embryonics: A new family of coarse-grained fpga with self-repair and self-reproduction properties. *LNCS, Towards Evolvable Hardware*, 1062:197–220, 1996.

13. J. Duato, S. Yalamanchili, and L. Ni. *Interconnection Networks: An Engineering Approach*. Morgan Kaufmann Publishing, 2002.

14. A. Ferrero. *The Evolving Ethernet*. Addison-Wesley, 1996.

15. J. Shirazi. *Java Performance Timing*. O'Reilly, 2000.

16. S. Oaks and H. Wong. *Java Threads*. O'Reilly, 1999.

17. D. Selman. *Java 3D Programming*. Manning Publications, 2002.

18. F. Hwang, K. Briggs. *Computer Architecture and Parallel Processing*. McGraw-Hill, 1995.

19. I. Englander. *The Architecture of Computer Hardware and System Software: An Information Technology Approach*. John Filey & Sons, 2000.

Using Negative Correlation
to Evolve Fault-Tolerant Circuits

Thorsten Schnier and Xin Yao

School of Computer Science
The University of Birmingham
Edgbaston, Birmingham B15 2TT, UK
{T.Schnier,X.Yao}@cs.bham.ac.uk

Abstract. In this paper, we show how artificial evolution can be used to improve the fault-tolerance of electronic circuits. We show that evolution is able to improve the fault tolerance of a digital circuit, given a known fault model. Evolution is also able to create sets of different circuits that, when combined into an ensemble of circuits, have reduced correlation in their fault pattern, and therefore improved fault tolerance. An important part of the algorithm used to create the circuits is a measure of the correlation between the fault patterns of different circuits. Using this measure in the fitness, the circuits evolve towards different, highly fault-tolerant circuits. The measure also proves very useful for fitness sharing purposes. We have evolved a number of circuits for a simple 2×3 multiplier problem, and use these to demonstrate the performance under different simulated fault models.

1   Introduction

A common technique for improving the fault tolerance of digital circuits is to use redundancy, e.g., multiple copies of the same circuit. Using majority voting within an ensemble of circuits, one or more faults in a single circuit can always be corrected. Faults in different circuits of the ensemble can usually be corrected if and only if they affect different bits in the output. Correlation between the circuits used in the ensemble can degrade the performance, if the same fault is likely to occur in all circuits. Error correction becomes less successful in this case. For this reason, designers generally try to increase the diversity of circuits used in an ensemble (Lala & Harper 1994, Mitra, Saxena & McCluskey 2001, Mitra, Saxena & McCluskey 1999). More advanced word-based error correction algorithms may be able to detect and possibly correct some faults that occur at the same bit (Mitra & McCluskey 2000), but fault correlation still degrades error correction performance. In any case, there does not exist a satisfactory and systematic approach to generating diverse circuits.

Evolutionary algorithms have the advantage that they often are able to create a variety of different circuits that perform the same function. Moreover, given a fault model, correlation between the faults of different circuits can be measured, and used in a fitness function. In this paper, a novel approach to generating

A.M. Tyrrell, P.C. Haddow, and J. Torresen (Eds.): ICES 2003, LNCS 2606, pp. 35–46, 2003.

a population of diverse circuits, based on evolutionary algorithms, is proposed. The approach provide a very different alternative towards fault-tolerant systems from other existing approaches. Our experimental results show that the proposed evolutionary approach can produce highly reliable and fault tolerant circuits.

The rest of this paper is organised as follows. Section 2 explains the ideas behind our evolutionary approach. Section 3 describes the circuit behaviours with and without the presence of faults. Section 4 gives the experimental results. Section 5 discusses related issues to our evolutionary approach. Finally, Section 6 concludes the paper.

2   Evolving Fault-Tolerant Circuits

Redundancy is often necessary in fault-tolerant circuits. However, it is unclear what would be the optimal way to organise redundancy. Diversity has been recognised as a key issue in designing redundant fault-tolerant circuits. However, it is by no means clear how to design an ensemble of diverse fault-tolerant circuits.

Our idea is very simple (Yao & Liu 2002). We use an evolutionary algorithm to evolve digital circuits. The algorithm is different from others in its fitness function which takes into account of the correlation of the error patterns of the circuits. We then combine the best set of circuits from one or more evolutionary runs into an ensemble (of redundant circuits). This ensemble can be tested under different fault models and compared to individual circuits and ensembles of identical circuits. The evolutionary algorithm optimises the fault tolerance of the set of circuits by optimising individual fault tolerance, and by maximising the variety in fault pattern between the circuits. Our approach therefore can be seen as sitting between approaches where individual circuits are directly evolved for fault tolerance (for example Canham & Tyrrell 2002), and approaches where the diversity of the whole population is used (for example Tyrrell, Hollingworth & Smith 2001).

In this section, we first describe our experimental setup, the representation used, and other algorithm details. We then discuss how we calculate the error patterns and their correlation, and how they are introduced into the fitness function.

To test our idea, we used a simple test setup to evolve binary arithmetic circuits. These circuits take two inputs of length a and b bits, and have c output bits. The representation we use is similar to that described in Miller, Job & Vassilev (2000) and V. K. Vassilev (2000). It is based on a K columns by L rows grid of nodes. Each node $N_{k,l}$ is programmed to perform one of a predefined set of binary functions. Each node has two inputs which are connected to other nodes. The genotype encodes, for each node, the function performed and the node coordinates of the two inputs for this node. In order to avoid feedback loops, processing is only allowed in a left-to-right order. Nodes in column C_i can only have inputs from other nodes in column $C_{j<i}$. A special column C_0 contains the inputs to the circuit in a predefined order; this column is not influenced by the genotype. Finally, the rightmost column C_K controls the outputs. Nodes in

this column are treated differently from other columns in that the function is ignored and the output value is directly taken from the first input. The nodes in the last column therefore only define the wiring of the output. In this aspect, our representation differs from Miller et al. (2000), where a special section of the genotype defines the output wiring.

Because the inputs and outputs are in single columns, the grid needs to have at least as many rows as the larger of $a + b$ (the number of inputs) and c (the number of outputs). Nodes in column C_0 that are not connected to inputs are fixed to 0.

For our experiments, we evolve a 2-bit by 3-bit multiplier ($a = 2, b = 3$) with 5-bit output ($c = 5$). We evolve the circuits on a 7x7 grid, where the 7th column is the output column. In order to improve the chances of evolving different circuits, we allow the algorithm to use a large set of different node functions, including and, or, xor, nand, nor. We use uniform crossover and a mutation operator that randomly changes one of the inputs or replaces a node function with a new random one.

3   Computing Error Behaviour

The goal of our algorithm is to create individuals that differ in their error patterns. Hence we need to compute the error patterns. There are two parts to this. The first part is the behaviour of the design in the absence of circuit faults. If we assume that the input-output mapping is completely defined, all correct circuits should produce the same outputs given the same inputs. However, during evolution we generally are not dealing with prefect circuits. Different circuits can differ strongly in the input-output mapping they implement.

The second part is the behaviour of circuits in the presence of faults. Here, the goal is that different circuits will behave differently in response to damaged nodes in the circuit. For both computations, we look at the 'output string' of the circuit. This string contains the signal of all the outputs of the circuit for all possible input combinations. For example, a circuit with 5 inputs and 5 outputs will produce an output string of length $5 * 2^5 = 160$. The 'correct' output string is defined by the function that the circuit implements. For all correct circuits the output string will be identical to the correct string.

3.1   Circuit Behaviour without Faults

The first step in the computation of the fault free behaviour is to convert the genotype into a circuit. The output string can then be calculated by simulating the circuit. By comparing the output string with the correct string, we can get an 'error string' which is '1' wherever the two differ, and '0' otherwise. The number of '1's in the fault string gives a straightforward measure of the quality of the circuit.

3.2   Circuit Behaviour with Faults

The behaviour of a circuit with faults depends on the nature of the fault or faults. Generally, it is not possible to predict which element of a circuit will become faulty, and which fault exactly it will develop. Instead, a probabilistic fault model is used here; which describes the likelihood of a fault in each element of the circuit. For our experiments, we assume a very simple fault model, where each node in the circuit is equally likely to develop a fault where the output is either always 0 ('stuck-at-zero') or always 1 ('stuck-at-one').

With a given phenotype in the node-grid description, it is possible to test the behaviour of the circuit with faults. To do this, we first identify which of the nodes in the circuit contributes to the output. These nodes, which we refer to as active nodes, are found by backward-chaining from the outputs to the inputs in the circuit, and marking the nodes used. We then exhaustively test the fault behaviour by going through all active nodes and switching them to 'stuck-at-one' and 'stuck-at-zero'. A circuit with n active nodes will need $2n$ tests. As in the fault-free case, an error string indicates the difference to the correct circuit; but here the error vector is computed by adding a value of 1 at each position where the signal is incorrect for each of the $2n$ tests.

After all tests, we normalise the values in the error string by dividing it by $2n$. Under the simple fault model described above, the error string gives the probability that for a single, random fault at one of the active nodes in the circuit, a particular bit in the output string is wrong.

3.3   Error Correlation

For the evolutionary algorithm, we are interested in the correlation of the error patterns of different circuits, with and without faults. We therefore use a new error string, created by adding error strings from the fault-free circuit and from the circuit with simulated faults. This new string cannot be interpreted as a string of error probabilities as in the previous section (it will contain values greater than 1.0 for incorrect circuits); but gives a good characterisation of the error behaviour of the circuit. The error correlation between two circuits can then be computed as the scalar product of the error vectors of two circuits. This value will be highest if the two are closely correlated, in other words, if the bits with high fault rates are at the same positions in the error strings.

3.4   Fitness Computation and Selection

For the fitness computation, we compute the average of the error correlation between a candidate circuit and all the circuits in the population. This fitness function rewards circuits that have

- fewer errors when tested without faults,
- fewer errors when tested with faults, and
- an error distribution that differs from that of other circuits in the population.

Parent selection is done using a simple tournament selection with a tournament size of 2. Survival for the next generation is also done using tournament. For each offspring we select the worst circuit from a tournament of 2 or 3 of the current population, and replace the circuit in the population with the offspring if the offspring has an equal or better fitness. Because the population changes and the fitness is based on a comparison between the circuits in the population, the fitness of each circuit in the population has to be updated after each generation. When digital circuits are evolved using the correctness of the output as fitness, 'plateaus' exists in the fitness landscape, and it has been shown that it is important that the population is able to move along these plateaus into different areas of the search space (Vassilev & Miller 2000, T. Yu 2001). Using our fitness function, plateaus do not exist in the fitness landscape. The correlation introduces dynamic gradients within the sets of individuals with identical bit error counts, driving individuals into different areas of the search space.

4   Experiments

To evaluate our idea, we have performed a number of tests. In each test, we have performed one or more runs of the evolutionary algorithm as outlined above. The runs use different setups and termination criteria. At the end of runs, we take the population of one or more runs, selected ensembles, and tested their performance. The following section describes how we test for fault tolerance. The remainder of this section describes the individual tests and results.

4.1   Testing Fault Behaviour

As described in Section 1, we are not so much interested in the behaviour of single circuits, but that of ensembles. We select ensembles of three circuits and use a simple bitwise majority voter to select the outcome of the computation. We then compare this to the outcome of the simulation of three identical circuits with a majority voter.

In order to test the evolved circuits for fault tolerance, we are using a number of tests with different fault models. In each case, we only test correct individuals, i.e. individuals that produce a correct result in the fault free case (see also Section 5.2).

1. Two faults with equal fault distribution. In this simulation, we randomly pick a circuit and introduce a single fault, and then repeat this for a second fault. Each fault will be either a stuck-at-one or a stuck-at-zero fault and will hit any of the active nodes with equal probability. We perform this test 100,000 times and count the number of bit faults in the output string of the ensemble.

   For correct circuits, the outcome of these tests can be calculated from the error strings of the circuits. The ensemble will return a wrong value if two of the circuits return a wrong value.

If we define p_n as the probability that a particular bit is wrong given a single fault in circuit n, we can compute the fault probability for an ensemble of three circuits as

$$\frac{1}{3}p_1*\frac{1}{3}p_2+\frac{1}{3}p_1*\frac{1}{3}p_3+\frac{1}{3}p_2*\frac{1}{3}p_3+\frac{1}{3}p_2*\frac{1}{3}p_1+\frac{1}{3}p_3*\frac{1}{3}p_1 = \frac{2}{9}(p_1p_2+p_1p_3+p_2p_3). \tag{1}$$

For an ensemble of 3 identical circuits, this produces $\frac{2}{3}(p1p1)$. If we repeat this for all bits in the output string, we find that the expected number of bit-errors is 2/9 of the sum of the cross-correlations.

2. $N(N > 2)$ faults with equal fault distribution. For this test, we randomly pick one of the three circuits and introduce a random fault into the active nodes of this circuit. We repeat this until N faults have been introduced, taken care not to introduce a fault into the same node more than once. Again, we perform this test 100,000 times, and count the total number of bit-faults. For this case, the outcome cannot be calculated from the fault strings of the individual circuits, as they are only computed for the single-fault case.

3. $N(N > 2)$ faults with unequal distributions. In real world circuit implementation, it is often observed that the active nodes in a circuit are not all equally likely to develop a fault, and that the likelihood of different types of fault may also vary for the different nodes. Reasons for the variations could be proximity to sources of heat or radiation, proximity to other signals, difference in the quality of power and clock signals, etc. For identical circuits, it is possible that the fault distributions are correlated. In other words, if a certain node is more likely to develop a fault in one instance of the circuit, this would also be the case in the other two instances.

This simulation assumes the extreme case, where the fault distribution is identical for identical circuits. For each test, we first generate between one and three fault distributions. For a circuit with n active nodes, we generate a set of fault probabilities

$$p_1 = \frac{1}{c},\ldots,p_k = \frac{1+(k-1)*\frac{s-1}{n-1}}{c},\ldots,p_n = \frac{s}{c},$$

where c is the normalisation factor and s is a scaling factor which allows us to set the largest value to s times the smallest value. We then associate these probabilities to the active nodes in the circuit in a random order. These values represent the probability that a particular node develops a fault. In each test they will be distributed differently over the circuit, but the set of probabilities will be the same. If we test an ensemble with identical circuits, the order will be the same for all the circuits. If we test different circuits, a separate set (depending on the number of active nodes in the circuit) is used for each, and the distribution is different for each.

4.2   First Perfect Individuals

In this test, we have performed a number of runs with a population size of 200, stopping the runs after the first correct individual appeared. We then load the correct individual of 20 different runs into a program and compute the correlations between the error strings of the individuals. Finally, we pick the ensemble of 3 individuals with the lowest sum of cross-correlation between them. The results of this test are shown in Table 1.

Table 1. Correct Circuits from 10 runs.

Number	Active Nodes	Error	Auto-correlation
0	23	13.0	1.41
1	22	11.66	1.08
2	20	11.52	1.13
3	25	10.66	0.94
4	19	10.63	0.90
5	17	12.38	1.30
6	15	14.17	1.77
7	17	11.53	1.12
8	26	10.71	0.96

The first two columns of Table 1 show the circuit identification number and the number of active nodes used in the circuit. The number of active nodes alone indicates that the circuits are very different from each other. They use between 15 and 26 nodes. The following column reports the sum of the values in the error string of the circuit, it gives the expected number of bit errors for a single, random fault in the circuit. This column shows that the circuits vary strongly in their sensitivity to faults, circuit 6 showing more than 30% more errors than circuits 3 and 4. Finally, the last column shows the auto-correlation of the error string (the correlation with itself). Not surprisingly, it is closely related to the error count, but not entirely: circuit 2 has a lower error count, but a higher auto-correlation, than circuit 4.

The node number indicates that the circuits are different, but this does not guarantee that the error strings are different. For this, we compare the error strings, and calculate the cross-correlations between the circuits. Figure 1 shows the error strings for two circuits, circuit 4 and 8. As can be seen, they are fairly similar to each other for some bits, and vary quite strongly for others. Comparing the patterns with other circuits, it becomes apparent that certain patterns appear in all circuits. For example, the leftmost 32 bits are the lowest bit of the output, in this case the product of the inputs. In a correct circuit, this bit is the result of an 'and' operation of the lowest two inputs. It could be implemented with 2 or more different gates, but there is little evolution can vary to improve fault behaviour. For this reason, there is a limit as to how far the cross-correlation of circuits can be improved.

Table 2 shows the cross-correlations for 9 circuits. The diagonal again shows the auto-correlation. The other fields show the cross-correlation between two circuits. The numbers show that the value of the cross-correlation very much depends on the quality of the individual circuits: circuits with lower auto-correlation

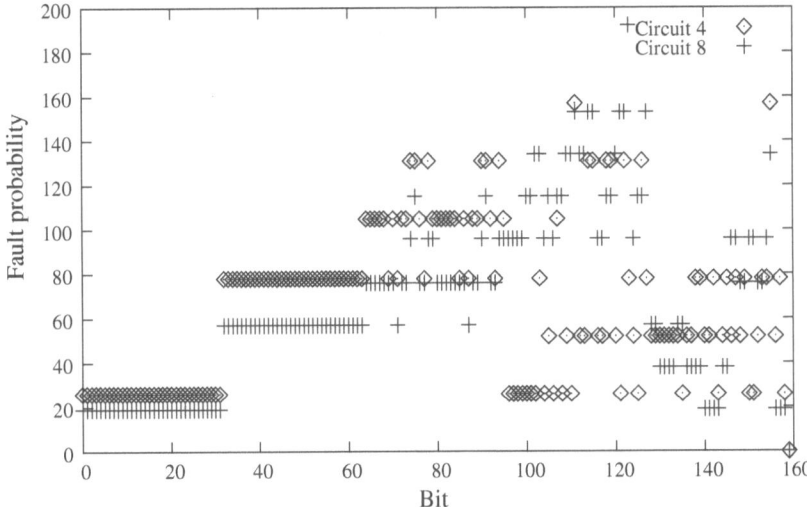

Fig. 1. Fault probabilities for bits in the output string for circuits 4 and 8.

Table 2. Cross-correlations between circuits.

	0	1	2	3	4	5	6	7	8
0	1.41	0.94	1.04	0.88	0.94	1.07	1.19	1.01	0.89
1	-	1.08	1.00	0.94	0.91	1.12	1.26	1.03	0.94
2	-	-	1.13	0.97	0.97	1.17	1.35	1.09	0.90
3	-	-	-	0.95	0.87	1.03	1.22	0.95	0.89
4	-	-	-	-	0.90	1.04	1.24	0.96	0.81
5	-	-	-	-	-	1.31	1.43	1.19	0.99
6	-	-	-	-	-	-	1.77	1.33	1.13
7	-	-	-	-	-	-	-	1.12	0.91
8	-	-	-	-	-	-	-	-	0.96

generally have better cross-correlations, too. Combining different circuits does bring improvements, but only if the auto-correlation of the circuits are similar in value. The best cross-correlation is 0.81, between circuits 4 and 8. Compared to the best auto-correlation (0.90 for circuit 4). This is an improvement of 10%. The importance of using **different** circuits is therefore small compared to using **good** circuits.

Determining the best ensemble of three circuits produces an interesting result: it consists of two copies of circuit 4 and one of circuit 8. The cross-correlation between circuits 4 and 8 is noticeably lower than all the cross-correlations. The auto-correlation of circuit 4 is not much larger than other cross-correlations. It therefore turns out to be beneficial to only use these two circuits.

Test Results. With the ensemble of circuits 4, 4, and 8, we have tested the performance when 2 or more faults are introduced into the circuits. We only

Table 3. Total bit-error count of ensembles, for 100.000 tests.

Ensemble	2 faults	3 faults	4 faults
4-4-4	60,613	168,891	312,264
8-8-8	64,881	175,855	323,992
4-4-8	56,364	156,999	292,672

report the test results (Table 3) with equal fault probabilities for all nodes here (tests 1 and 2 in Section 4.1). Results of tests with unequal fault probabilities are similar to those described in the next section.

The results of the tests confirm that the ensemble of circuits 4 and 8 performs about 10% better than an ensemble of circuit 4 only. It also shows that this improvement still holds if more than 2 faults are introduced into the circuits. The numbers in the 2 fault case tally well with the numbers expected from Equation 1: for the ensemble with circuit 4 only, the expected number of faults is 60000, for the ensemble with circuit 8 only it is 64000. For the mixed ensemble, the expected fault number is 56444.

4.3   Continued Evolution

The previous section has shown that combining different individuals can improve the result, but the individual quality of the circuits is more important. It therefore suggests itself to try and improve fault tolerance of the individual circuits through evolution. We have done another set of runs, this time with 400 individuals, which continued to evolve after the first correct individual had been found. These runs were only stopped after the cross-correlation fitness of the best individual had dropped below 0.5. From 6 of these runs, we took all correct individuals and compared them as in the previous section. The circuits are generally larger, they use between 26 and 31 nodes (see also Section 5.3). The average number of bit-errors for a single circuit with a single fault lies between 7.2 and 8.5, clearly a lot better than the results with the first correct circuits. Similarly, the auto-correlation lies around 0.5, with the best value of 0.475, 50% better than the results of the previous section. Combining different circuits into an ensemble again has an advantage of about 10%, reducing the sum of the cross-correlations from 1.42 to 1.35.

Test Results.   With the best performing ensemble from the six runs, we have again performed fault simulations. Table 4 shows the results.

The tests were performed using unequal fault probabilities, as described for test 4, Section 4.1. The scaling factor s is set to 1.0, 2.0, 10.0 and 100.0. With $s = 1.0$, all nodes have an equal probability of developing a fault. As can be seen, the ensembles perform much better than those used in the Section 4.2. The mixed ensemble produces the best results, by about 10%. The effect of the correlation introduced by the unevenness in fault distribution can be seen in the other numbers: as expected, it leads to a worsening of the fault tolerance for

Table 4. Bit-faults for different circuits, after 100000 simulations.

circuit	s	2 Faults	3 Faults	4 Faults
1-1-1	1.0	31298	89267	170173
	2.0	32001	90136	170532
	10.0	33366	94497	175307
	100.0	33441	94866	179556
2-2-2	1.0	33064	94998	179187
	2.0	34492	95212	182274
	10.0	36016	102351	190663
	100.0	36926	105852	197402
3-3-3	1.0	33285	90685	175942
	2.0	33094	92568	176384
	10.0	34050	98205	181918
	100.0	36348	98931	185828
1-2-3	1.0	30324	86175	165387
	2.0	30519	85747	163070
	10.0	31319	86074	163841
	100.0	30933	86167	162325

those ensembles with identical circuits. For the ensemble with circuits 1, 2, and 3 on the other hand, the uneven fault distribution does not show any effect on the error numbers.

5   Other Aspects

5.1   Fitness Sharing Performance

As described in Section 3.4, the error string used for the fitness computation combines the errors of the circuit in the no-fault case and in the fault simulation. A circuit that has the same number of bit-errors as other circuits in the population, but at different position in the output string, will have a low cross correlation. Because the fitness is based on the average correlation between circuits, it turns out to be a very effective fitness sharing mechanism.

For a comparison, we performed a set of runs with 200 individuals, using the number of bit errors in the output string as fitness instead. We run these until they either found a correct individual, or reached 1,000,000 generations. Of 50 runs, 18 did not find a correct circuit in this time, the rest needed on average 252775 generations to find a correct circuit. Compared to this, a set of 33 runs using the average cross-correlation all finished in the generation limit, and took on average 14715 generations.

5.2   Evolving Ensembles

In the early stages of the evolutionary run, the no-fault bit-error count of the best ensemble is often noticeably better than that of the best individual. This can be intuitively explained: if random individuals are sampled, we are more likely to sample a good individual the more individuals we sample. With an ensemble, 100 individuals can be combined into nearly 1,000,000 combinations. Not all of these will be different, but the number of samples, and therefore the likelihood of a better result, is increased.

We have tried to make use of this by evolving individuals as part of an ensemble. The hope is that this leads to some kind of implicit 'divide and conquer', where each individual is allowed to ignore parts of the output string. So far, however, this has not lead to a satisfactory behaviour of the evolutionary runs. The reason for this seems to be that if the algorithm finds a new circuits with fewer bit-errors than the currently best, this circuit will not automatically improve the bit-errors of the best ensemble - in fact is has been observed that the individual with the lowest bit-rate quickly got lost again in the population. On the other hand, individuals with very large numbers of bit-errors can often become part of good ensembles. The result of these two effects is that the runs fail to improve the bit-error count of the individuals sufficiently to achieve an ensemble without bit errors. Changes in the way the selection and fitness calculation is done might improve the behaviour. A promising variation we are looking at is using ensemble fitness only once we already have correct individuals. This would allow the circuits to improve the ensemble error count with faults, by reducing the correlation between the error patterns.

5.3 Other Fault Models

Our calculations and tests assume a very simple fault model. With different fault models, it is likely that some of the results are different. One of the major assumptions in our model that is unlikely to be observed in the real world is that the likelihood of a fault is independent of the number of nodes in a circuit. When we calculate the fault string, we normalise it by the number of tests (and therefore nodes); and when we test circuits, circuits will get single faults independent of the number of nodes. One result of our experiments was that the individuals evolved to have a low average error correlation (Section 4.3) were generally larger than the first correct circuits (Section 4.2). However, this was not always the case, as both experiments had cases where smaller circuits performed better. Most likely, if the fault likelihood was to be made proportional to the number of nodes, the number of nodes would not grow, but an improvement of the error string correlation would still be possible.

6   Conclusions

Our experiments show that it is possible to use artificial evolution to create digital circuits that have different error patterns when faults are introduced, and that this is beneficial for fault tolerance. The experiments also show that the sensitivity to faults of individual circuits has a much larger influence on the fault tolerance of the ensemble. Continuing the evolution after the first correct circuit is found and using a fitness function that is based on the correlation of fault patterns allow us to breed circuits that have a low sensitivity to faults as well as high diversity. The variation in the population generated by this fitness function also improves the performance of the evolutionary algorithm. The implicit fitness sharing scheme we used was shown to reduce the number of generations required to find a correct individual by more than a factor of 10. The next step of our study is to further research our approach and investigate its real use to large and practical circuits.

References

Canham, R. O. & Tyrrell, A. M. (2002). Evolved fault tolerance in evolvable hardware, *Congress on Evolutionary Computation CEC 2002*, Hawaii, pp. 1267–1272.

Lala, J. H. & Harper, R. E. (1994). Architectural principles for safety critical real-time applications, *Proceedings of the IEEE* **82**(1).

Miller, J. F., Job, D. & Vassilev, V. K. (2000). Principles in the evolutionary design of digital circuits – part i, *Journal of Genetic Programming and Evolvable Machines* **1**(1): 8–35.

Mitra, S. & McCluskey, E. (2000). Word-voter: A new voter design for triple modular redundant systems, *18th IEEE VLSI Test Symposium*, Montreal, Canada, pp. 465–470.

Mitra, S., Saxena, N. & McCluskey, E. (1999). A design diversity metric and reliability analysis for redundant systems, *Proc. 1999 Int. Test Conf*, Atlantic City, NJ, pp. 662–671.

Mitra, S., Saxena, N. & McCluskey, E. (2001). Techniques for estimation of design diversity for combinational logic circuit, *nt. Conf. on Dependable Systems and Networks (DSN'01)*, Goteborg, Sweden, pp. 25–34.

T. Yu, J. F. M. (2001). Neutrality and evolvability of a boolean function landscape, *in* J. F. Miller, M. Tomassini, P. L. Lanzi, C. Ryan & W. Langdon (eds), *Procedings of the 4th European Conference on Genetic Programming (EuroGP2001)*, Springer-Verlag, pp. 204–217.

Tyrrell, A., Hollingworth, G. & Smith, S. (2001). Evolutionary strategies and intirinsic fault tolerance, *in* D. Keymeulen, J. Lohn & R. Zebulum (eds), *The Third NASA/DoD Workshop on Evolvable Hardware*, Long Beach, CA, pp. 98–106.

V. K. Vassilev, Dominic Job, J. F. M. (2000). Towards the automatic design of more efficient digital circuits, *in* J. Lohn, A. Stoica, D. Keymeulen & S. Colombano (eds), *Proceedings of the 2nd NASA/DOD Workshop on Evolvable Hardware*, IEEE Computer Society, Los Alamitos, CA, pp. 151–160.

Vassilev, V. & Miller, J. F. (2000). The advantages of landscape neutrality in digital circuit evolution, *in* J. F. Miller, A. Thompson, P. Thomson & T. Fogarty (eds), *Third International Conference on Evolvable Systems: From Biology to Hardware*, Springer-Verlag, Edinburgh, pp. 252–263.

Yao, X. & Liu, Y. (2002). Getting most of evolutionary approaches, *in* A. Stoica, J. Lohn, R. Katz, D. Keymeulen & R. Zebulum (eds), *2002 NASA/DoD Conference on Evolvable Hardware (EH'02)*, IEEE Computer Society, Alexandria, Virginia, pp. 8–14.

A Genetic Representation for Evolutionary Fault Recovery in Virtex FPGAs

Jason Lohn[1], Greg Larchev[1], and Ronald DeMara[2]

[1] Computational Sciences Division, NASA Ames Research Center,
Mail Stop 269-1, Moffett Field, CA 94035-1000, USA
{jlohn,glarchev}@email.arc.nasa.gov
[2] School of Electrical Engineering and Computer Science, University of Central
Florida, Orlando, FL 32816-2450
demara@mail.ucf.edu

Abstract. Most evolutionary approaches to fault recovery in FPGAs focus on evolving alternative logic configurations as opposed to evolving the intra-cell routing. Since the majority of transistors in a typical FPGA are dedicated to interconnect, nearly 80% according to one estimate, evolutionary fault-recovery systems should benefit by accommodating routing. In this paper, we propose an evolutionary fault-recovery system employing a genetic representation that takes into account both logic and routing configurations. Experiments were run using a software model of the Xilinx Virtex FPGA. We report that using four Virtex combinational logic blocks, we were able to evolve a 100% accurate quadrature decoder finite state machine in the presence of a stuck-at-zero fault.

1 Introduction

Numerous advantages of Field Programmable Gate Arrays (FPGAs) in space-borne electronics have been identified in recent research publications [3,15] and manufacturers' literature [1,16]. Benefits include reconfiguration capability to support multiple missions, the ability to correct latent design errors after launch, and the potential to accommodate on-chip and off-chip failures. Ground Support Equipment (GSE) based FPGA applications primarily employ reprogrammable devices as a means of amortizing development costs over multiple missions. In GSE-enabled applications such as Reusable Launch Vehicles (RLVs), FPGAs are configured or replaced between missions rather than being reprogrammed during flight. For applications such as RLVs, comparatively short mission durations and low levels of ionizing radiation are involved. Hence for many ground reconfigurable applications, conventional Triple Modular Redundancy (TMR) techniques often provide sufficient fault handling coverage.

On the other hand, in-mission reconfigurable FPGAs are advantageous for deep space probes, satellites, and extraterrestrial rovers. In these applications, the radiation exposures, mission durations, and repair complexities are significantly greater. The need for adequate fault coverage during these missions has become further intensified by the increasing number of FPGAs being deployed. For instance, NASA's Stardust probe contains over 100 FPGA devices. Although

A.M. Tyrrell, P.C. Haddow, and J. Torresen (Eds.): ICES 2003, LNCS 2606, pp. 47–56, 2003.

the Stardust's FPGAs are based on a non-reprogrammable antifuse-based technology, a more recent space-qualified SRAM-based technology has become commercially available.

In SRAM-based devices, the number of programming cycles is unlimited. Hence new techniques become feasible for active recovery through reconfiguration of a compromised FPGA. The approach developed here concentrates on autonomous reconfiguration of SRAM-based devices while in-flight. The experiments conducted involve Xilinx's SRAM-based Virtex parts from the same device family as the space-qualified QPRO radiation-hardened series.

Permanent Single-Event Latchup (SEL) failures may impact CLBs and/or programmable interconnections within the FPGA itself. They may also involve other supporting devices that the FPGA interfaces with or processes data from. These failure modes also suggest that the ability to derive an alternative FPGA configuration in-situ would be beneficial. Likewise, SEL exposures exist with regards to the data processing path within the FPGA that is not involved with the device's programmable configuration. In the above cases, the FPGA configuration derived at design time will no longer provide the required functionality for the damaged part. Traditionally, redundant spares have been utilized to replace the damaged device.

Autonomous repair can work in concert with or provide an alternative to device redundancy. While redundant spares exist only in limited quantities, evolutionary recovery methods attempt to facilitate repair through reuse of damaged parts. Hence the potential benefits are two-fold. First, one or more failures might be accommodated by reconfiguring the failed part without incurring the increased weight, size, or power traditionally associated with providing redundant spares. Second, the characteristics of the failure need not be precisely diagnosed in order to be repaired. Here the repair is performed in-situ via intrinsic evaluation of the device's remaining functionality. This implies that any residual functionality, including the electrical characteristics of both the damaged device and its interaction with any supporting devices, is taken into account when realizing the repair. After isolating the fault to a size that is manageable for the evolutionary algorithm, alternate solutions are refined though iterative selection. This can be carried out without detailed knowledge of the underlying failure mechanism itself.

The approach developed here attempts to regain lost functionality due to a fault by evolving a new configuration on the defective FPGA. We assume a dual-redundant FPGA system whereby the faulty FPGA undergoes evolution to recover its functionality while the redundant FPGA maintains proper functionality during evolution on the faulty FPGA. Thus after a fault is detected, redundancy is lost for a short period of time and then restored. Application functionality is maintained throughout this process under the assumption that only one of the FPGAs fails. Our results are that the evolutionary methods are able to fully recover from a simulated stuck-at-zero fault in the input of a state machine implementing a quadrature decoder. Several research challenges remain and they are also discussed.

2   Related Work

Recently, various evolutionary algorithm approaches have been proposed for fault-recovery of FPGAs. Some previous work applies evolutionary algorithms prior to the occurrence of the fault while other approaches attempt to repair the fault after its occurrence. Some techniques involve intrinsic evolution using the failed part itself. Others rely on extrinsic evolution of an abstracted model of the devices.

Three examples of recent work that apply evolutionary algorithms to realize fault-tolerant designs include [11], [4], and [13]. In [11], Miller examined properties of messy gates whereby evolved logic functions inherently contain redundant terms as their functional boundaries change and overlap. In [4], Canham and Tyrrell compare the fault tolerance of oscillators evolved by including a range of fault conditions within the fitness measure during the evolutionary process. A population-based approach scores evolved designs using a fitness function corresponding to desired operation based on the absence of faults. When evolution is complete, an additional pass evaluates the ability of the evolved individuals to tolerate a range of faults, and the most fault-tolerant individuals are retained. In [13], the evolution of designs containing redundant capabilities without the designer having to explicitly specify the redundant parts themselves was investigated. To achieve this, a range of fault cases was introduced throughout the evolution process. This allowed individuals to exploit whatever component behaviors exist, even behaviors known to be faulty.

An evolutionary fault-recovery approach is described by Vigander [14]. He develops a genetic algorithm to restore functionality after random faults are injected into a 4-bit by 4-bit multiplier using standard genetic operators. He simulated the repair of the prior-designed multiplier that consisted of feed-forward interconnection of hypothetical FPGA cells capable of 8 different logic functions. He used as his fitness function the number of correct input-output mappings from the 256 possible input combinations that could be applied to the multiplier. He demonstrated that while it is not exceedingly difficult to derive a solution that can produce a nearly correct repair, completely correct repairs present a challenging problem. To remedy this, he demonstrated that a voting system with as few as three alternatively evolved repaired circuits was capable of producing a majority output that was completely correct.

3   Representation and Operators

Several goals were taken into account while designing the representation scheme. Amenability to recombination is of course a primary concern. After that, our priorities were to let the GA work in the largest, most flexible design space as possible: we wanted to allow all possible LUT configurations and allow the maximum number of CLB interconnections given the constraints of hardware routing support (we will say more about the routing at the end of this section). We also wanted to disallow illegal configurations and to minimize non-coding alleles (introns).

Bitstring representations are a natural choice for FPGA applications, and many times the raw configuration string can be used as the representation. In our case, we chose a bitstring representation mainly out of convenience in programming. Since we knew that only a handful of CLBs would be evolved, our bitstrings would be at most 1000 bits long. We acknowledge that this approach would likely suffer as more CLBs were utilized and the corresponding bitstring enlarged to thousands of bits.

The representation is shown in Figure 1. This scheme is comprised of multiple 128-bit fields, one for each CLB. Within each CLB field are a number of subfields that specify each of the LUT bits and remote connections. There are 16 bits that specify the contents of each LUT. Each LUT has four inputs, and since each of these inputs can be connected to other LUT outputs, the remote CLB/LUT requires addressing bits. Since our system will be comprised of four CLBs, we need only two bits to specify the remote CLB, and another two bits to specify the particular LUT within the CLB. This pattern of sub-fields continues for each LUT until all the LUTs in the CLB are accounted for. An illustration of the CLBs, LUTs and sample routing is shown in Figure 2.

The operators employed were crossover and mutation. Two-point crossover was implemented using cut points allowed between bits. Mutation was applied on individual bits.

Regarding the routing, it is chosen automatically by the JBits software. Our circuits are sufficiently small that we have never experienced a situation where a route could not be found. Successful routes have been found routing 1 LUT output to 48 different inputs (the maximum number of inputs available in a 2-by-2 circuit, where 1 CLB is dedicated to external inputs). It is theoretically

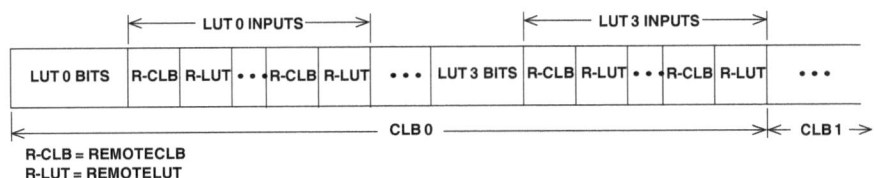

Fig. 1. Genetic representation used showing logic fields and routing fields.

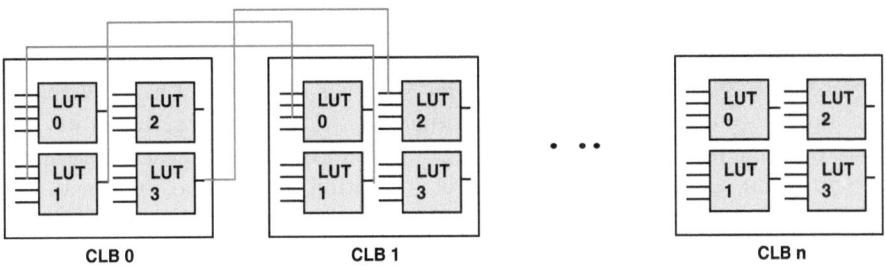

Fig. 2. Example of routing among CLBs.

possible that in some larger designs the routing will become so dense that some routes would not be found. If such a case were to occur, the specific route will simply not get connected. Such individual would then most likely receive a low fitness score, and be automatically eliminated from the gene pool.

4   Fault Recovery of Quadrature Decoder

The quadrature decoder [2] was selected as an initial case study for testing and refinement of our evolutionary recovery strategy. It represents a NASA application of manageable size that is appropriate for tuning of the GA. Quadrature decoders provide a means of counting objects passed back and forth through two beams of light, or alternatively determining the angular displacement and direction of rotation of an encoder wheel turning about its axis. A quadrature decoder that determines the direction of rotation of a shaft is shown in Figure 3.

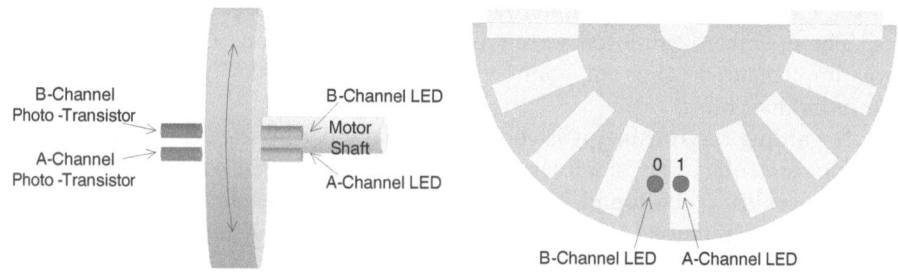

Fig. 3. Rotating shaft application for a quadrature decoder.

The concept of operation for the quadrature decoder is that the objects, or opaque arcs on the rotating wheel, to be counted will first obscure and then move past the two light beams in succession. The order in which the beams are cleared can be used to ascertain the direction of rotation. The use of two beams acts to preclude false counts due to jitter or bounce resulting from multiple phantom reads. For example, to have a valid increment in the rotational count, both beams must be cleared in succession.

To implement the encoder, it is possible to employ a state machine that keeps track of the beam activity. The state machine accepts two single-bit inputs which are asserted only when the corresponding sensor is obscured. When a change of the inputs occurs, the state machine transitions to its next internal state. The state machine is asynchronous and outputs a zero bit if the wheel is rotating in one direction, and a one bit if the wheel is rotating in the opposite direction. If the wheel is not rotating, the output is the same as it previously was. The finite state machine for the quadrature decoder is shown in Figure 4.

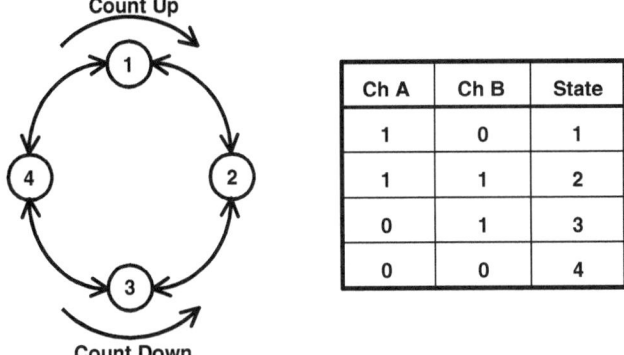

Ch A	Ch B	State
1	0	1
1	1	2
0	1	3
0	0	4

Fig. 4. Quadrature decoder finite state machine.

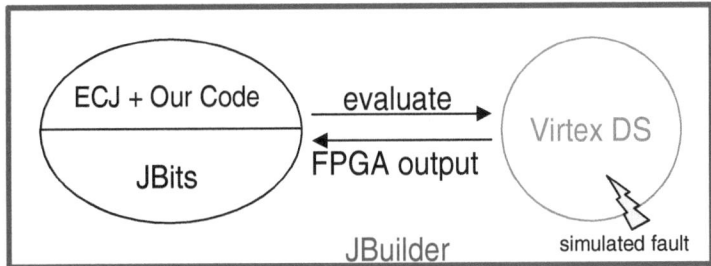

Fig. 5. Software system.

5 Experimental Setup and Results

The software system used is depicted in Figure 5. The entire system is implemented in software. The GA software is ECJ, a Java-based evolutionary computation and genetic programming system by Sean Luke of George Mason University. ECJ is augmented by our code for tasks like decoding individuals and calculating fitness. The GA sits on top of Xilinx Corporation's JBits software [5,8], a set of Java classes which provide an Application Programming Interface to access the Xilinx FPGA bitstream. Xilinx's Virtex DS software, which simulates the operation of Virtex devices, is used to test candidate solutions. Borland's JBuilder Java environment is used for development and to run the system, though Sun Microsystem's Java virtual machine is used beneath JBuilder.

To evaluate the fitness of an individual, an input stream of 500 bit pairs is used. These inputs attempt to fully exercise the evolving finite state machines. The quad decoder inputs are supplied at specified clock intervals, which have nothing to do with how often the finite state machine changes state. The quad decoder inputs can change at any frequency below the frequency of the decoder clock – they might never change (if the wheel is standing still), or they might change at every clock cycle (if the wheel is rotating at the maximum allowed velocity).

The output stream consists of 510 bits sampled across all four CLBs. Ten bits have been added to allow for the delay (in clock cycles) between the time the decoder inputs are fed into the decoder, and the time the output of the decoder is read. Such arrangement allows us to read the output of the decoder from 1 to 11 clock cycles after the inputs have been fed into the circuit. Adding ten bits gives ten output stream windows of length 500, with each output stream shifted by 1-bit from the next. Sampling across all the CLBs allows the GA to maximum flexibility in building the FSM. Thus, fitness is expressed as:

$$F = \max_{i=1,4;j=0,9}(\mathrm{CLB}_i^j)$$

where CLB_i^j represents the number of correct output bits from the ith CLB shifted by j clock ticks. The fitness is simply the highest number of correct output bits seen across all of the CLBs and across the ten output windows. The best score is 500, and the worst score is 0.

The genetic algorithm was set up as shown in Table 1. Small population sizes were necessary since an unfixablememory leak was present in one of the pre-compiled modules.

Table 1. GA parameters.

Number of generations	1000
Population size	40
Tournament Size	4
Elitist Individuals	2
Gen 0 Seeding	20 individuals
Crossover rate	0.8
Mutation rate	0.002 per bit

Approximately 10 experimental runs were conducted using smaller input bit-streams of 100 bit pairs. These were found to evolve finite state machines that were tuned to the test cases, but not robust when interrogated with out of sample input test streams. Two runs were conducted using 500 bit pairs and one these runs was able to evolve a 100% accurate quadrature decoder finite state machine in the presence of an induced fault. The location of the fault was chosen at random, although we made sure that it would adversely affect the functionality of the seeded circuits. Once the fault is present, we assume that it does not get removed (however, if it does, our algorithm can start evolving the circuit configuration again). We assumed that the circuit is operating properly prior to the fault, and the evolution is started once the fault is detected.

The best evolved configuration was found in generation 623 and is shown in Figure 6. Two of the 16 LUTs went unused which is not surprising given that the FSM can be implemented with about 10 LUTs. The GA exploits the induced

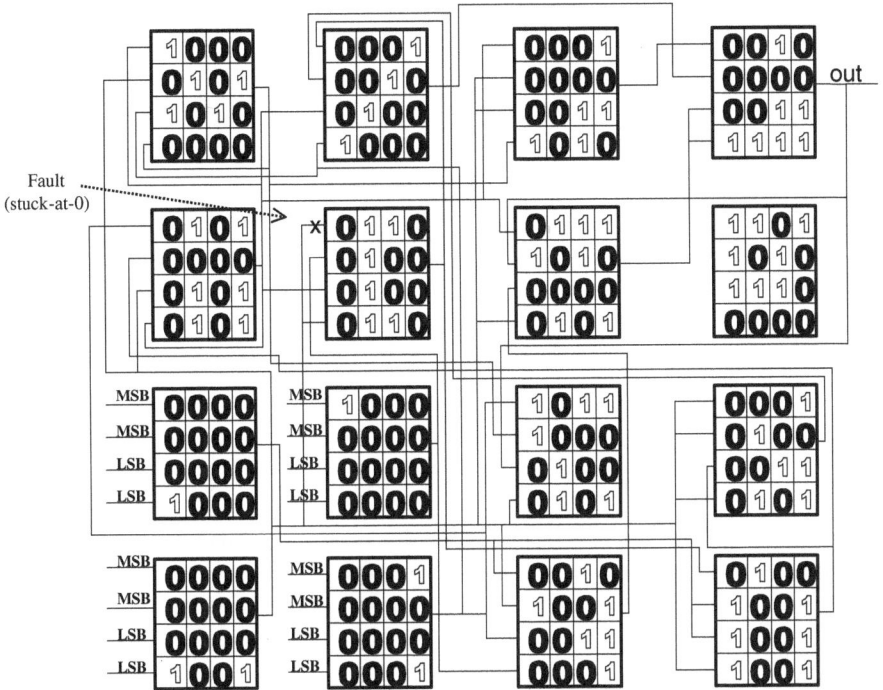

Fig. 6. Evolved configuration showing routing, LUT contents, and simulated fault. Inputs are on the lines labeled MSB and LSB, referring to the least/most significant bit of the input (channel A and B inputs). Wires that are shown crossing perpendicularly (eg, +) are unconnected – only wires that have ⊤ junctions are connected.

fault to its advantage because if you remove the fault in the evolved solution, it no longer functions correctly – it achieves an accuracy of only 93.8%. Also, note that the input LUTs had mostly zeros in their tables. This is because we fix most of those bits to zero in the genome since they do not affect the LUT's function. However, the "corner" bits of each of those input LUTs are involved in processing the input, and therefore, are evolved.

The GA performance curve for this run is shown in Figure 7. The run ramps up quickly showing that useful search is underway, however, the average fitness is stagnant for about 300 generations, which is not encouraging. The runs are quite slow to execute on a 2 GHz Pentium 4 PC. Runtimes were about 45 hours since each evaluation takes approximately 6 seconds.

6 Discussion

Evolutionary systems for fault recovery on FPGAs may be an important tool in the quest for ever-higher levels of fault tolerance in NASA missions and other applications. We have demonstrated a system that is able to evolve a realistic

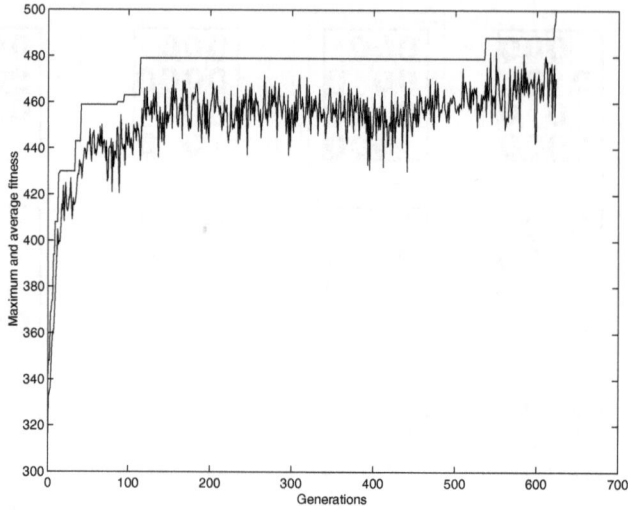

Fig. 7. GA performance curve. The top curve is the best individual's fitness at each generation and the bottom curve is the average fitness.

spacecraft control function in the presence of a permanent stuck-at fault. Using a software simulation of an FPGA, we constructed a genetic representation that included both logic and routing information, and ran a genetic algorithm to evolve a quadrature decoder. As is typical in evolutionary algorithm applications, the evolved solution exploits its resources in unexpected ways. In our case, the algorithm made use of the fault itself in constructing its solution. If there is economy to be gained by exploiting damaged resources, that is certainly a benefit largely unique to evolutionary search.

Potential advantages of this approach are handling a wider range of errors, and relaxing the requirement of fault location/isolation. An autonomous fault recovery system would be possible if the evolution could be done at sub-second speeds. Future work includes investigation of scalability to more complex logic functions and systems that have multiple induced faults. Speeding up the evaluation cycle by doing evolution directly in hardware is our next line of research.

Acknowledgments

The authors would like to thank David Gwaltney of NASA Marshall Space Flight Center for suggesting the quad decoder application, and Delon Levi of Xilinx, Inc. for many helpful discussions. The research described in this paper was performed at NASA Ames Research Center, and was sponsored by NASA's Computing, Information, Communications, and Technology Program.

References

1. Actel Corporation, "Actel FPGAs Make Significant Contribution To Global Space Exploration," Press Release, August 30, 1999. available at: http://www.actel.com/company/press/1999pr/SpaceContribution.html
2. Agilent Technologies, Inc., Quadrature Decoder/Counter Interface ICs, Data Sheet HCTL-2020PLC.
3. N. W. Bergmann and P. R. Sutton, "A High-Performance Computing Module for a Low Earth Orbit Satellite using Reconfigurable Logic," in Proceedings of Military and Aerospace Applications of Programmable Devices and Technologies Conference, September 15-16, 1998, Greenbelt, MD.
4. R. O. Canham and A. M. Tyrrell, "Evolved Fault Tolerance in Evolvable Hardware," in Proceedings of IEEE Congress on Evolutionary Computation, 2002, Honolulu, HI.
5. S. Guccione, D. Levi, P. Sundararajan, "JBits: A Java-based Interface for Reconfigurable Computing," 2nd Annual Military and Aerospace Applications of Programmable Devices and Technologies Conference (MAPLD).
6. P. Haddow and G. Tufte, "Bridging the Genotype-Phenotype Mapping for Digital FPGAs," The Third NASA/Dod Workshop on Evolvable Hardware, pp. 109-115
7. D. Keymeulen, A. Stoica, R. Zebulum, "Fault-Tolerant Evolvable Hardware using Field Programmable Transistor Arrays," IEEE Transactions on Reliability, Special Issue on Fault-Tolerant VLSI Systems, Vol. 49, No. 3, September 2000, pp. 305-316.
8. D. Levi and S. Guccione, "GeneticFPGA: Evolving Stable Circuits on Mainstream FPGAs," In Adrian Stoica, Didier Keymeulen, and Jason Lohn, editors, Proceedings of the First NASA/DOD Workshop on Evolvable Hardware, pp. 12-17, IEEE Computer Society Press, Los Alamitos, CA, July 1999.
9. J.D. Lohn, G.L. Haith, S.P. Colombano, D. Stassinopoulos, "A Comparison of Dynamic Fitness Schedules for Evolutionary Design of Amplifiers," in Proceedings of the First NASA/DoD Workshop on Evolvable Hardware, Pasadena, CA, IEEE Computer Society Press, 1999, pp. 87-92.
10. D.C. Mayer, R. B. Katz, J. V. Osborn, J. M. Soden, "Report of the Odyssey FPGA Independent Assessment Team," NASA/JPL, 2001.
11. J. F. Miller and M. Hartmann, "Evolving messy gates for fault tolerance: some preliminary findings," in Proceedings of the Third NASA/DoD Workshop on Evolvable Hardware, July 12-14, 2001, Long Beach, CA.
12. M. Tahoori, S. Mitra, S. Toutounchi, E. McCluskey, "Fault Grading FPGA Interconnect Test Configuration," in Proceedings of Intl Test Conference, 2002.
13. A. Thompson, "Evolving Fault Tolerant Systems," in Proceedings of 1st IEE/IEEE Intl Conference on Genetic Algorithms in Engineering Systems, IEE Conf. Pub. No 414, pp 524-529, TBD Date, TBD Place.
14. S. Vigander, Evolutionary Fault Repair of Electronics in Space Applications, Dissertation, Norwegian University of Science and Technology, Trondheim, Norway, February 28, 2001.
15. E. B. Wells and S. M. Loo, "On the Use of Distributed Reconfigurable Hardware in Launch Control Avionics," in Proceedings of Digital Avionics Systems Conference, TBD day/month, 2001, TBD location.
16. Xilinx Inc., "Xilinx Radiation Hardened Virtex FPGAs Shipping To JPL Mars Mission And Other Space Programs," Press Release, May 15, 2001.

Biologically Inspired Evolutionary Development

Sanjeev Kumar and Peter J. Bentley

Department of Computer Science
University College London
Gower Street, London WC1E 6BT, UK
{S.Kumar, P.Bentley}@cs.ucl.ac.uk

Abstract. We describe the combination of a novel, biologically plausible model of development with a genetic algorithm. The Evolutionary Developmental System is an object-oriented model comprising proteins, genes and cells. The system permits intricate genomic regulatory networks to form and can evolve spherical embryos constructed from balls of cells. By attempting to duplicate many of the intricacies of natural development, and through experiments such as the ones outlined here, we anticipate that we will help to discover the key components of development and their potential for computer science.

1 Introduction

Talk to any evolutionary biologist and they'll tell you that the standard genetic algorithm (GA) does not resemble natural evolution very closely. While our GAs may evolve their binary genes, most biologists would be horrified to discover that concepts such as genotype and phenotype are so blurred in evolutionary computation that some researchers make no distinction between the two. Should you have the courage to go and talk to a developmental biologist, you'll have an even worse ear-bashing. You'll be told that development is the key to complex life. Without a developmental stage from genotype to phenotype, all you have is a big DNA or RNA molecule. With development you can have layer upon layer of complexity, from cells to organs to organisms to societies.

Of course our motivations in computer science are often very different from the motivations of biologists. Nevertheless, it has long been the goal of evolutionary computationists to evolve complex solutions to problems without needing to program-in most of the solution first. The dream of complex technology that can design itself requires rejection of the idea of knowledge-rich systems where human designers dictate what should and should not be possible. In their place we need systems capable of building up complexity from a set of low-level components. Such systems need to be able to learn and adapt in order to discover the most effective ways of assembling components into novel solutions. And this is exactly what developmental processes in biology do, to great effect.

In this paper we present, for the first time, an overview of a novel biologically plausible model of development for evolutionary design. This system is intended to model biological development very closely in order to discover the key components of development and their potential for computer science. The paper is divided into

A.M. Tyrrell, P.C. Haddow, and J. Torresen (Eds.): ICES 2003, LNCS 2606, pp. 57–68, 2003.
© Springer-Verlag Berlin Heidelberg 2003

sub-sections covering different aspects of the Evolutionary Developmental System (EDS). It begins with an overview of the entire system, followed by sections detailing individual components in isolation. These individual components are then drawn together, and how they work as part of the overall developmental system is detailed as well as the role of evolution and how the genetic algorithm is wrapped around the developmental core. Finally we present some examples of results generated during on-going experiments.

2   Background

Development is the set of processes that lead from egg to embryo to adult. Instead of using a gene for a parameter value as we do in standard EC (i.e., a gene for long legs), natural development uses genes to define proteins. If expressed, every gene generates a specific protein. This protein might activate or suppress other genes, might be used for signalling amongst other cells, or might modify the function of the cell it lies within. The result is an emergent "computer program" made from dynamically forming gene regulatory networks (GRNs) that control all cell growth, position and behaviour in a developing creature [Bentley, 2002].

The field of Computational Development has matured steadily over the past decade or so, with work touching upon a wide range of aspects of development ranging from its use for the construction of neural net robot controllers [Jakobi, 1996], to the large scale modelling of morphogenesis [Fleischer, 1993].

Recently a resurgence of interest into computational development has fuelled much research. Problems of scalability, adaptability and evolvability have led many researchers to attempt to include processes such as growth, morphogenesis or differentiation in their evolutionary systems [Eggenberger, 1996; Haddow et al, 2001; Bongard, 2002; Miller 2002]. For reviews see [Kodjabachian and Meyer, 1994; Kumar and Bentley, 2002].

3   The Evolutionary Developmental System (EDS)

In nature, development begins with a single cell: the fertilised egg, or zygote. In addition to receiving genetic material from its two parents, the zygote is seeded with a set of proteins – the so-called 'maternal factors' deposited in the egg by the mother [Wolpert, 1998]. The maternal factors trigger development causing the zygote to cleave (fast cell division with no cell growth). After cleavage, normal cell division begins; as cells divide they inherit the state of their parents. To ensure the embryo is not homogenous, one or two asymmetric divisions occur, resulting in an unequal distribution of factors to the daughter cells. In doing so, cells become different from one another.

Development is controlled by our DNA. In response to proteins, genes will be expressed or repressed, resulting in the production of more (or fewer) proteins. The chain-reaction of activation and suppression both within the cell and within other nearby cells through signaling proteins and cell receptors, causes the complex processes of cellular differentiation, pattern formation, morphogenesis and growth.

The Evolutionary Developmental System is an attempt to encapsulate many of these processes within a computer model. At the heart of the EDS lies the developmental core. This implements concepts such as embryos, cells, cell cytoplasm, cell wall, proteins, receptors, transcription factors (TFs), genes, and cis-regulatory regions (see figure 1 for a graphical view of the EDS). Genes and proteins form the atomic elements of the system. A cell stores proteins within its cytoplasm and its genome (which comprises rules that collectively define the developmental program) in the nucleus. The overall embryo is the entire collection of cells (and proteins emitted by them) in some final conformation attained after a period of development. A genetic algorithm is wrapped around the developmental core. This provides the system with the ability to evolve genomes for the developmental machinery to execute.

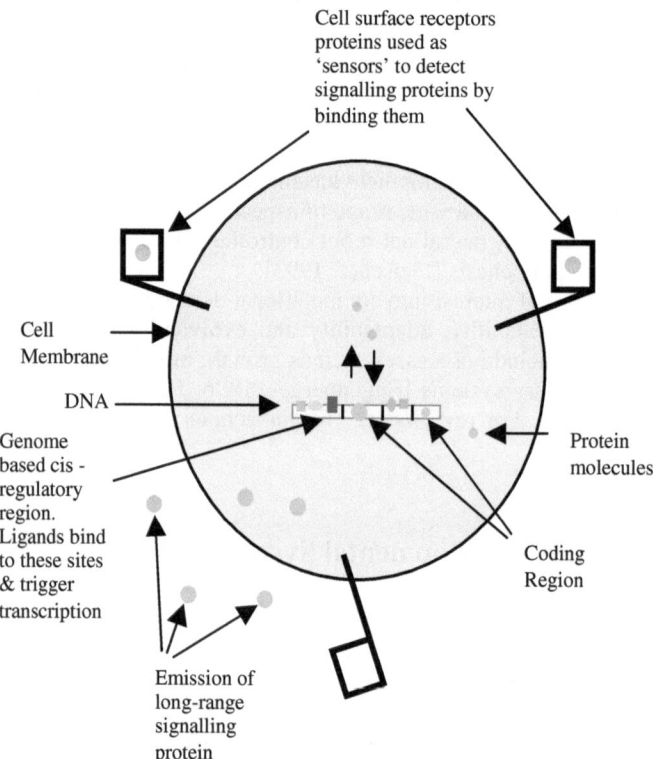

Fig. 1. A single cell in the Evolutionary Developmental System

4   Components of the EDS

The following sections describe the main components of the developmental model: proteins, genes and cells.

4.1 Proteins

In nature, proteins are the driving force of development. They are macromolecules, long chains of amino acids that assemble at protein production sites known as ribosomes. The only function of genes is to specify proteins.

The EDS captures the concept of a protein as an object. Each protein has an ID tag, which is simply an integer number. The EDS uses eight proteins (although number of proteins used is a user defined variable in the system). Protein objects contain both a current and a new state object (at the end of each developmental cycle all protein new states are swapped with current states to provide "parallel" protein behaviour). These protein state objects house important protein-specific information, for example, the protein diffusion co-efficient.

Protein Creation, Initialisation, and Destruction. In the EDS, proteins do not exist in isolation; they are created and owned by cells. Thus, during protein construction each protein is allocated spatial co-ordinates inherited from the cell creating the protein. Handling protein co-ordinate initialisation using this method overcomes the problem of knowing which cell created which proteins.

A protein lookup table (extracted from the genome, see next section) holds details about all proteins and is used to initialise each protein upon creation. It has the following details for each protein:

Rate of Synthesis	amount by which the protein is synthesised
Rate of Decay	amount by which the protein decays
Diffusion coefficient	amount by which the protein diffuses
Interaction strength	strength of protein interaction, i.e., activation or inhibition
Protein Type	ID tag, e.g., long-range hormone, or short-range receptors

Additionally, each protein keeps the following variables:

Bound?   whether or not a receptor protein is currently bound[1]
Protein Source Concentration   the current concentration of the protein
Spatial coordinates   the position of the source of the protein

Protein destruction in the EDS is implemented by simply setting the protein's source concentration to zero: if the concentration is zero there can be no diffusion, unless more of the protein is synthesised.

Protein Diffusion. Diffusion is the process by which molecules spread or wander due to thermal motions [Alberts et al., 1994]. When molecules in liquids collide, the result is random movement. Protein molecules are no different: they diffuse.

The average distance that a molecule travels from its starting point is proportional to the square root of the time taken to do so. For example, if a molecule takes on average 1 second to move 1 μm, it will take 4 seconds to move 2 μm, 9 seconds to move

---

[1]The bound variable is only operational in receptor proteins.

3 μm, and 100 seconds to move 10 μm. Diffusion represents an efficient method for molecules to move short distances, but an inefficient method to move over large distances. Generally, small molecules move faster than large molecules (Alberts et al., 1994).

Protein diffusion in the EDS models this behaviour. Diffusion is implemented by using a Gaussian function centred on the protein source. The use of the Gaussian assumes proteins diffuse equally in all directions from the cell.

In more detail: the source concentration records the amount of the current protein. Every iteration, its value is decremented by the corresponding 'rate of decay' parameter. If expressed by a gene, its value is also incremented by the corresponding 'rate of synthesis' parameter. To calculate the concentration of a protein at a distance x from the protein source:

$$concentration = s \times e^{\frac{-x^2}{2d^2}}$$

Where: d is the diffusion coefficient of the current protein.
x is distance from protein source to current point
s is the current protein source concentration.

Figure 2 illustrates the way protein concentration changes according to the three variables: distance, diffusion coefficient and source concentration.

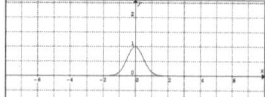 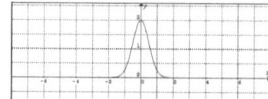 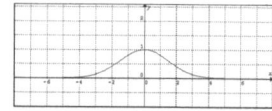

Fig. 2. Plot of protein concentration against distance from source, where: $d = 0.5$ and $s = 1.0$ (left), $d = 0.5$ and $s = 2.0$ (middle), and $d = 1.5$ and $s = 1.0$ (right).

4.2 Genes

The EDS employs two genomes. The first contains protein specific values (e.g., synthesis, decay, diffusion rates, see above). These are encoded as real floating-point numbers. The second describes the architecture of the genome to be used for development; it describes which proteins are to play a part in the regulation of different genes. It is this second genome that is employed by each cell for development; the information evolved on the first genome is only needed to initialise proteins with their respective properties.

In Nature, genes can be viewed as comprising two main regions: the cis-regulatory region [Davidson, 2001] and the coding region. Cis-regulatory regions are located just before (upstream of) their associated coding regions and effectively serve as switches that integrate signals received (in the form of proteins) from both the extra-cellular environment and the cytoplasm. Coding regions specify a protein to be transcribed upon successful occupation of the cis-regulatory region by assembling transcription machinery. Currently, the EDS's underlying genetic model assumes a "one gene, one protein" simplification rule (despite biology's ability to construct multiple proteins); this aids in the analysis of resulting genetic regulatory networks. To this end, the activation of a single gene in the EDS results in the transcription of a single protein.

This is currently ensured by imposing the following structure over genes: each gene comprises both a cis-regulatory region and a consequent protein-coding region.

A novel genome representation (based on eukaryotic genetics) was devised for development in the EDS. This genome is represented as an array of Gene objects (fig. 3). Genes are objects containing two members: a cis-regulatory region and a protein-coding region. The cis-regulatory region contains an array of TF target sites; these sites bind TFs in order to regulate the activity of the gene.

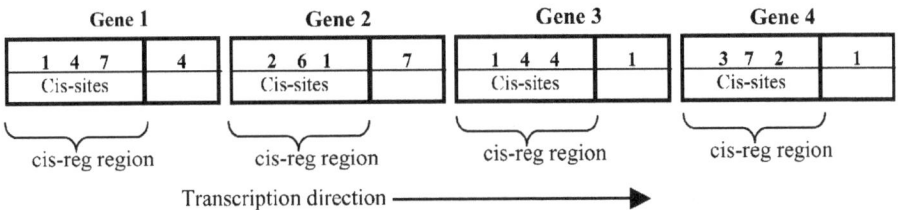

Fig. 3. An arbitrary genome created by hand. Genes consist of two objects: a cis-regulatory region and a coding region. Cis-regulatory regions consist of transcription factor target sites that bind TFs, triggering transcription of the coding region. Each number denotes a protein.

The gene then integrates these TFs and either switches the gene 'on' or 'off'. Integration is performed by summing the products of the concentration and interaction strength (weight) of each TF, to find the total activity of all TFs occupying a single gene's cis-regulatory region:

$$a = \sum_{i=1}^{d} conc_i * int\,eraction\_strength_i$$

where: a is the total activity, i is the current TF,
          d is the total number of TF proteins visible to the current gene,
          $conc_i$ is the concentration of i at the centre of the current cell,
          $interaction\_strength_i$ is the strength of protein interaction for the current TF
(see previous section).

This sum provides the input to a logistic sigmoid threshold function (a hyperbolic tangent function), which yields a value between –1 and 1. Negative values denote gene repression and positive values denote gene activation:

$$g(a) \equiv \tanh(a) \equiv \frac{e^a - e^{-a}}{e^a + e^{-a}}$$

Figure 4 illustrates this sigmoid calculation used to determine whether a gene is activated and produces its corresponding transcription factor or not.

4.3 Cells

Cells can be viewed as autonomous agents. These agents have sensors in the form of surface receptors able to detect the presence of certain molecules within the environment. Additionally, the cell has effectors in the form of hundreds and thousands of protein molecules transcribed from a single chromosome able to affect other genes in other cells. Cells resemble multitasking agents, able to carry out a range of behaviours. For example, cells are able to multiply, differentiate, and die.

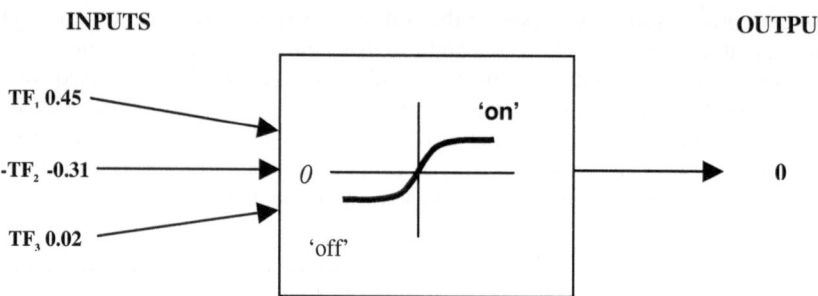

Fig. 4. A gene showing the various positive and negative inputs received in the form of transcription factors, with their respective affinities (weights), and concentrations of 0.24, 0.87, and 0.11 respectively. Internally, the gene integrates these TFs and decides whether or not to switch the gene 'on' or 'off'. TF_1 and TF_3 are both activators, whereas TF_2 is a repressor, denoted by a '-' symbol.

Like protein objects, cell objects in the EDS have two states: *current* and *new*. During development, the system examines the current state of each cell, depositing the results of the protein interactions on the cell's genome in that time step into the new state of the cell. After each developmental cycle, the current and new state of each cell is swapped ready for the next cycle.

The EDS supports a range of different cell behaviours, triggered by the expression of certain genes. These are currently: division (when an existing cell "divides", a new cell object is created and placed in a neighbouring position), differentiation (where the function of a cell is fixed, e.g., colour = "red" or colour = "blue"), and apoptosis (programmed cell death).

The EDS uses an *n-ary* tree data structure to store the cells of the embryo, the root of which is the zygote. As development proceeds, cell multiplication occurs. The resulting cells are stored as child nodes of the root in the tree. Proteins are stored within each cell. When a cell needs to examine its local environment to determine which signals it is receiving, it traverses the tree, checks the state of the proteins in each cell against its own and integrates the information.

4.4 Evolution

A genetic algorithm (GA) is "wrapped around" the developmental model. The GA represents the driving force of the system. Its main roles are to:

1. provide genotypes for development;
2. provide a task or function, and hence a measure of success and failure; and
3. search the space of genotypes that give rise to developmental programs capable of specifying embryos, correctly and accurately according to the task or function.

Individuals within the population of the genetic algorithm comprise a genotype, a phenotype (in the form of an embryo object), and a fitness score. After the population is created, each individual has its fitness assessed through the process of development. Each individual is permitted to execute its developmental program according to the instructions in the genome. After development has ended a fitness score is assigned to the individual based upon the desired objective function.

The EDS uses a generational GA with tournament selection (typically using ¼ of population size), and real coding. Crossover is applied with 100% probability. Creep mutation is applied with a Gaussian distribution (small changes more likely than large changes), with probability between 0.01 and 0.001 per gene.

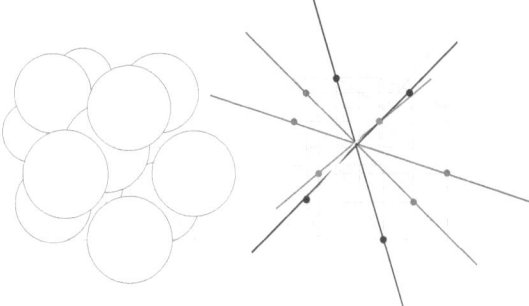

Fig. 5. Isopatial coordinates permit twelve equidistant neighbours for each cell (left) and are plotted using six axis (right).

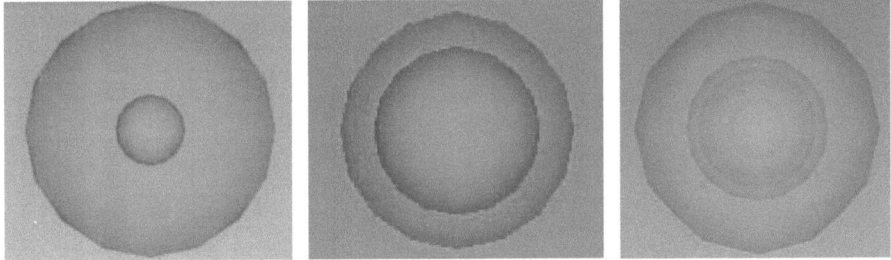

Fig. 6. Examples of proteins with their associated cell (at centre). Left: single cell emitting a long-range hormone-type protein. Middle: single cell emitting a short range (local) protein. Right: single cell emitting four proteins of various spread, reflected by the radius of each protein sphere.

4.5 Coordinates and Visualisation

The underlying co-ordinate system used by the EDS is isospatial. All coordinate systems have inherent biases towards different morphologies; the isospatial system is no different. However, the isospatial system bias results in what can only be described as more natural (biologic) morphologies than its Cartesian counterpart [Frazer, 1995]. Isospatial co-ordinates permit a single cell to have up to twelve equidistant neighbours defined by 6 axis (fig. 5), Cartesian co-ordinates only permit 6.

The EDS automatically writes VRML files of developed embryos, enabling three-dimensional rendered cells and proteins to be visualised. Cells are represented by spheres of fixed radius; proteins are shown as translucent spheres of radius equal to the extent of their diffusion from their source cells. In order to place a cell in VRML its Cartesian co-ordinates need to be defined: to this end, isospatial co-ordinates are converted to Cartesian. Figure 6 illustrates how cells and proteins appear when rendered.

5 Experiments

Because of the complexity of the system, numerous experiments can be performed to assess behaviour and capabilities. Here (for reasons of space) we briefly outline two:

1. The ability of genes and proteins to interact and form genomic regulatory networks within a single cell.
2. The evolution of a 3D multi-cellular embryo with form as close to a prespecified shape as possible.

5.1 Genetic Regulatory Networks

In order to assess the natural capability of the EDS to form GRNs independently of evolution, genomes of five random genes were created and allowed to develop in the system for ten developmental steps. The cell was seeded with a random set of eight proteins (maternal factors).

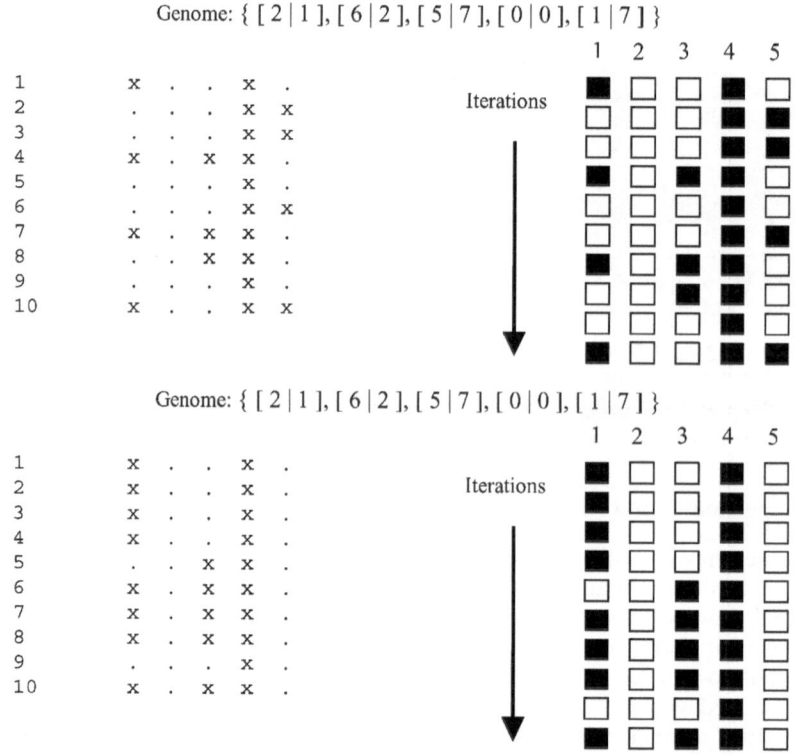

Fig. 7. Gene expression patterns for a run of a randomly created genome seeded with a random subset of proteins. The left side shows the raw output from the system where an 'x' means the gene in that column is 'on' and '.' means the gene is 'off'. The right side depicts this text pattern as a graphical output viewed as a 1D CA iterated over ten time-steps. Note, gene 4, i.e., [0 | 0] is autocatalytic.

Figure 7 (top) shows an example of the results of this experiment. The pattern shows gene four exhibiting autocatalytic behaviour having initially bound to protein zero. (Gene four is activated when in the presence of protein zero, and produces protein zero when activated.)

Figure 7 (bottom) shows an example of the pattern that results when the initial random proteins (initial conditions) are varied very slightly, but the genome is kept constant. Again, gene four shows the same autocatalytic behaviour, but now the GRN has found an alternative pattern of activation. These two runs illustrate the difference the initial proteins can make on the resulting GRN.

5.2 Morphogenesis: Evolving a Spherical Embryo

In addition to GRNs, the other important capability of the EDS is cellular behaviour. The second experiment focuses on morphogenesis, i.e., the generation of an embryo with specific form, constructed through appropriate cellular division and placement, from an initial single zygote. For this experiment, the genetic algorithm was set up as described previously, with the fitness function providing selection pressure towards spherical embryos of radius 2 (cells have a radius of 0.5).

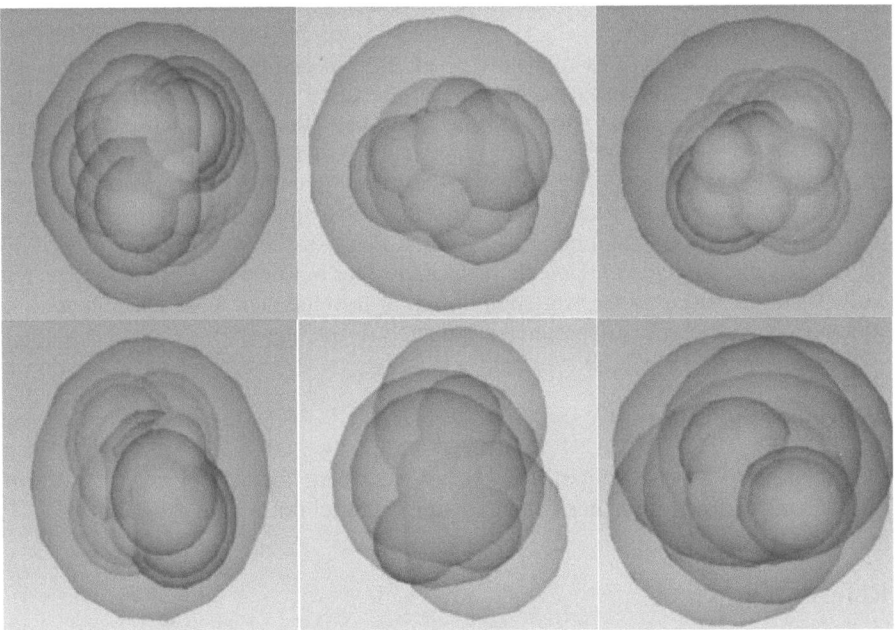

Fig 8. Six random initial embryos.

Figure 8 shows examples of the initially random embryos with their corresponding proteins produced by the GRNs. Figure 9 shows two examples of final "spherical" embryos. As well as having appropriate forms, it is clear that the use of proteins has been reduced by evolution. Interestingly, analysis indicates that evolution did not require complex GRNs to produce such shapes. It seems likely that it is the natural tendency of the EDS to produce near-spherical balls of cells, hence evolution simply did not need to evolve intricate GRNs for this task. Further experiments to evolve more complex morphologies are under way.

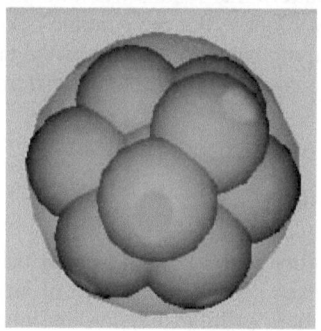 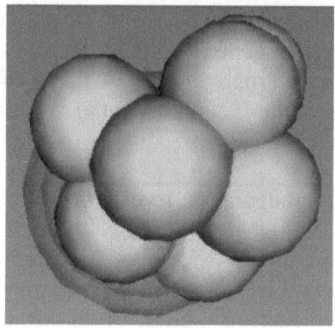

Fig. 9. Two "spherical" embryos. Using the equation of a sphere as a fitness function with sphere of radius 2.0.

6 Summary

The staggering complexities of nature result from a combination of evolution and development. This work has described the combination of a novel, biologically plausible model of development with a genetic algorithm. We have shown how an Evolutionary Developmental System can be constructed based on an object-oriented model of proteins, genes and cells. We have also described how this system permits intricate genomic regulatory networks to form and can evolve spherical embryos constructed from balls of cells. By attempting to duplicate many of the intricacies of natural development, and through experiments such as the ones outlined here, we anticipate that we will help to discover the key components of development and their potential for computer science. Further experiments and analysis are ongoing.

Acknowledgements

Many thanks to Lewis Wolpert and Michel Kerszberg for helpful advice and criticism. Thanks also to Tom Quick and Piet van Remortel for helpful suggestions.

References

1. Alberts et al. (1994) *Molecular Biology of the Cell*. 3rd edition. Garland Publishing.
2. P. J. Bentley (2002) *Digital Biology*. Simon and Schuster, New York.
3. J. Bongard (2002) Evolving Modular Genetic Regulatory Networks. In *Proceedings of the IEEE 2002 Congress on Evolutionary Computation*.
4. E.H. Davidson (2001). *Genomic Regulatory Systems*. Academic Press.

5. P. Eggenberger (1996) Cell interactions as a control tool of developmental processes for evolutionary robotics. In Maes, P. et al.(Eds) *From Animals to Animats 4*. Cambridge, MA: MIT Press.
6. K. Fleischer and A. Barr (1993). A simulation testbed for the study of multicellular development: The multiple mechanisms of morphogenesis. In C. Langton, editor, *Artificial life III*, pages 389--416. Addison-Wesley, 1993.
7. J. Kodjabachian and J.-A. Meyer (1994). Development, learning and evolution in animats. In *Perception To Action Conference Proceedings*, P. Gaussier and J.-D. Nicoud, Eds. 1994, pp. 96--109, IEEE Computer Society Press.
8. J. Frazer (1995). *An Evolutionary Architecture*. Architecture Association, London.
9. P. C. Haddow, G. Tufte, and P. van Remortel (2001) "Shrinking the Genotype: L-Systems for EHW?" In Proc. Of 4th Int. Conf. On Evolvable Systems: From Biology to Hardware, Tokyo, Japan.
10. N. Jakobi (1996c). Harnessing morphogenesis. In Proceedings of *the International Conference on Information Processing in Cell and Tissue*.
11. S. Kumar and P. J. Bentley (2002). Computational Embryology: Past, Present and Future. Invited chapter in Ghosh and Tsutsui (Eds) Theory and Application of Evolutionary Computation: Recent Trends. Springer Verlag (UK).
12. J. F. Miller (2002) "What is a Good Genotype-Phenotype Mapping for the Evolution of Computer Programs?" Presented at the *Software Evolution and Evolutionary Computation Symposium*, EPSRC Network on Evolvability in Biology & Software Systems, University of Hertfordshire, Hatfield, U.K.7-8 February 2002.
13. L. Wolpert (1998), *Principles of Development*, Oxford University Press.

Building Knowledge into Developmental Rules for Circuit Design

Gunnar Tufte and Pauline C. Haddow

The Norwegian University of Science and Technology
Department of Computer and Information Science
Sem Selandsvei 7–9, 7491 Trondheim, Norway
{gunnart,pauline}@idi.ntnu.no

Abstract. Inspired by biological development, we wish to introduce a circuit-DNA that may be developed to a given circuit design organism. This organism is a member of a Virtual EHW FPGA species that may be mapped onto a physical FPGA. A rule-based circuit-DNA, as used herein, provides a challenge to find suitable rules for artificial development of such an organism. Approaching this challenge, the work herein may be said to be of an investigative nature, to explore for developmental rules for developing even the simplest organism.

The artificial developmental process introduced herein, uses a knowledge rich representation including both knowledge of the circuit's (organism's) building blocks and local knowledge about neighbouring cells. Initial experiments for knowledge rich development on our virtual technology platform, are presented.

1 Introduction

Generally in EHW, a one-to-one mapping i.e. a direct mapping, has been chosen for the genotype-phenotype transition. In a one-to-one mapping the genotype is a complete blueprint of the phenotype, including all information of the assembled phenotype. In the case of an FPGA, the genotype holds the configuration data for the FPGA. The larger the circuit to be evolved, the more logic and routing information (configuration data) is required in the genotype. As the complexity of the genotype increases, so increases the computational and storage requirements of the evolutionary process. These high resource requirements limit the complexity, especially functional complexity, of circuits that can realistically be evolved.

Moving away from a direct mapping to an indirect mapping we have the possibility of introducing a smaller genotype for a large phenotype. Turning to nature, we see that DNA is indirectly mapped to the developed organism. DNA is a building plan of how to assemble the organism rather than a blueprint of it. Following nature's example, we could achieve a smaller genotype in the form of a building plan for a given complex circuit and use artificial development to develop our organism.

A.M. Tyrrell, P.C. Haddow, and J. Torresen (Eds.): ICES 2003, LNCS 2606, pp. 69–80, 2003.

By simplifying or rather shrinking the size of the genotype we effectively move the complexity problem over to the genotype-phenotype mapping thus increasing the complexity of the mapping. In this work, artificial development is this mapping and, as such, we need a development process that can handle such a complex mapping. To simplify this problem, we have chosen to split the mapping into two simpler mappings by introducing a virtual technology between the genotype and phenotype. In addition, the development process is now developing towards a phenotype platform more attuned to developmental design than today's technologies.

The mapping from the virtual technology, our Virtual Evolvable Hardware (EHW) FPGA, to a physical FPGA is a relatively simple mapping as both architectures are based on FPGA principles. This mapping is described in [1]. The development process itself is the first stage of the mapping. Instead of developing to our complex organism (phenotype) we are developing to a simpler organism the intertype i.e. the configuration data for our Virtual EHW FPGA.

This work is a continuation of earlier work into L-systems for development of electronic circuits [1]. L-systems is a rule based artificial development system (see section 2). Designing rules that are suitable for such a purpose and that will in fact give rise to the desired phenotype is quite a challenge, even with the help of evolution. To understand why evolution has a problem meeting this challenge we need to look again at the genotype representation and fitness evaluation.

The main goal, herein, is to consider the genotype representation i.e. the developmental rules. We wish to introduce more knowledge into developmental rules so as to tune development towards a given organism i.e. electronic circuit, on a Virtual EHW FPGA platform.

The article is laid out as follows. Section 2 introduces the L-systems concept which is the basis for our developmental rules and section 3 describes our technology platform, how this platform may be seen to be "developmental friendly" and its computational and communication properties as an organism. Section 4 describes our reasoning for introducing knowledge into our developmental rules and in section 5 we describe the way in which we have introduced knowledge. Section 6 describes the experimental work of this paper. Section 7 summarises some of the relevant work in the area of artificial development.

2   L-Systems

An L-system is a mathematical formalism used in the study of biological development. One of the main application areas is the study of plant morphology. An L-system is made up of an alphabet, a number of ranked rules and a start string or axiom. The rules are built up of two types of rules: change and growth rules. Change rules change the content of the target and growth rules expand the target. Applying a rule means finding targets within the search string which match the rule condition. This condition is a pattern on the left hand side (LHS) of the rule. This condition is also a string and the string is made up of elements from the alphabet. Firing the rule means replacing the targets, where possible, with the result of the rule i.e. the right hand side (RHS) of the rule. Firing of the rules continues until there are no targets found for any rule or until the process is interrupted.

Previous work into L-systems for circuit design [1], was a first attempt at using L-systems in combination with a genetic algorithm to achieve development on a digital platform. The rules were given no chosen meaning with respect to improving connectivity or logic. Also the rules were non context-sensitive i.e. did not rely on neighbouring Sblock information. As such, these may be said to be knowledge-poor [2]. To begin with, the rules in the L-system experiments were generated randomly, evolved, fine-tuned to a certain extent through experimental results and then re-evolved. However, with little knowledge in the rules other than random strings it was difficult to interpret individual rule effects on the final phenotype.

3   Decreasing Development Complexity – Introducing an Intertype

The Virtual EHW FPGA contains blocks — Sblocks, laid out as a symmetric grid where neighbouring blocks touch one another [3]. There is no external routing between these blocks except for the global clock lines. Input and output ports enable communication between touching neighbours. Each Sblock neighbours onto Sblocks on its 4 sides.

Each Sblock consists of both a simple logic/memory component and routing resources. The Sblock may either be configured as a logic or memory element with direct connections to its 4 neighbours or it may be configured as a routing element to connect one or more neighbours to non-local nodes. By connecting together several nodes as routing elements, longer connections may be realised.

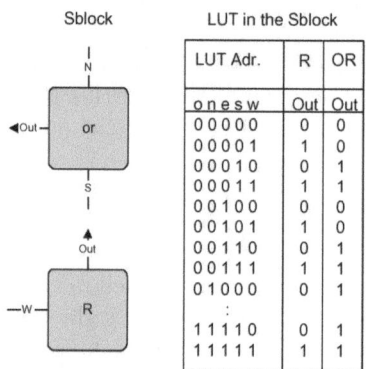

Fig. 1. Sblock as a Routing Module

Figure 1 illustrates a routing Sblock (R) which takes a west input and routes it to the North. The logic block (OR) then reads its south input as well as its

north input and forwards its result to its west output. If we look at the LUT
output for the routing module (R) we see that the output is identical to the west
input (w). As such, the other inputs are treated as don't cares. The LUT for the
logic module (OR) will most likely have different input values, but for simplicity
we consider the same LUT inputs and consider the output values for an OR
module. The OR column shows that in this case the north and the south inputs
have been Or-ed and the other inputs are treated as don't cares. It is worth
noting, however, that for clarity we have chosen to highlight only the relevant
outputs in this figure. The single output from an Sblock is, of course, available
in all directions.

3.1   Virtual EHW FPGA: Development Friendly Features

Artificial development could, as in nature, use: pattern formation, morphogenesis, cell differentiation and growth as interrelated processes. Biologically we are
looking at a cell based organism where the maximum size of our organism is the
maximum grid size of the technology we are using. Each Sblock is a cell in the
organism and can communicate with other cells through local communication —
although limited to 4 neighbours due to the grid structure. An organism (circuit)
should grow from a single cell (an Sblock) by programming a neighbour cell as
part of the organism. Growth should only effect neighbours i.e. growth causes
our growing organism to expand cell-wise within the grid of our technology. In
addition, our development algorithm will need to allow for cell death, which
would mean deactivating an Sblock by reprogramming it to the default value of
non-organism cells.

The development algorithm should control pattern formation within the cells
based on the cell types. However, what is a cell's type with respect to Sblocks? If
as in biology, the type is the same functionality e.g. a muscle cell, then perhaps
we refer to the functionality of the given block e.g. a 3-input AND gate. The
type will then describe how the cell functions i.e. how it reacts to various input
signals.

Cell differentiation may be achieved by reprogramming a given cell. Varying
the structure of an sblock through differentiation would mean reprogramming the
Sblock to connect to or disconnect to individual neighbours. Since the internal
structure of the sblock is fixed no other structural variations are possible. Varying
the functionality of the Sblock through differentiation involves programming
the cell's internal functionality i.e. its look-up table. It should be noted that
programming the cell's connectivity and functionality are intertwined through
the look up table, as described earlier.

The development algorithm should control pattern formation within the cells
based on the cell types. However, what is a cell's type with respect to Sblocks? If
as in biology, the type is the same functionality e.g. a muscle cell, then perhaps
we refer to the functionality of the given block e.g. a 3-input AND gate. The
type will then describe how the cell functions i.e. how it reacts to various input
signals.

Morphogenesis may be defined in terms of the development of form and
structure in an organism. One of the key ways that this is achieved in biology
is through change of cell shape. If we regard single Sblocks as cells then morphogenesis will not be achieved in this way, as each cell has a fixed structure
and thus size with respect to a given technology. However, it is possible to abstract slightly from the technology and not only view the grid as an organism

but perhaps view the cell itself as an expanding body, consisting of one or more cells.

As such, we see features in the Virtual EHW FPGA which may be termed "development friendly". However, the challenge is to find an artificial developmental algorithm which will take some form of electronic circuit representation and develop it to a functionally correct electronic circuit on this platform.

3.2   Artificially Developing Sblock-Based Circuits

Our organism, with respect to development, is the circuit design on our Virtual EHW FPGA described by its configuration data. So the building plan described by the genotype needs to build a circuit description with respect to the components of the virtual EHW FPGA. During the development phase, at a given point in time, the structure and functionality of the developing organism may differ from that of the final organism but the computational and communication properties of the units under development should be static. That is, the blue print includes a description of the computational and communication properties of the organism that will remain unchanged during the development process.

To explain this biologically, if a given DNA can develop into a given organism then our building plan needs to develop to a given type of circuit. To describe the circuit in its loosest terms, we can describe the circuit as an interconnection of Sblocks, where each Sblock has certain computational and communication properties. As such, our building plan has to be limited to the computational and communication properties of our phenotype i.e. our organism.

Considering the description of the virtual EHW FPGA given, communication available within the organism is limited to neighbour communication with up to 4 neighbours. Computationally, an Sblock can compute any logic function with up to 5 inputs. If the full functionality of the Sblock structure is to be used in the organism then the genotype should reflect this in its building plan. In the L-systems work described in [1], the building plan enabled communication with 4 neighbours and any type of Sblock i.e. any 32 bit configuration.

4   Introducing Knowledge to Developmental Rules

In earlier work, lack of knowledge within the rules meant that it was hard to gain any realistic understanding of good or bad rules with respect to development of a given organism. In this section, the introduction of knowledge into developmental rules is considered so as to aid our understanding of these rules.

4.1   Genotype Representation

In the work presented in [1], evolution was used to try to refine the developmental rules. However, this proved to be quite a challenge for evolution. One of the key problems lay in the genotype representation.

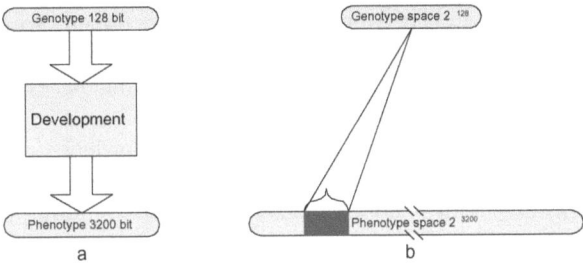

Fig. 2. Genotype-Phenotype mapping and their Respective Solution Space

Figure 2 provides an example of an indirect mapping where a 128 bit geno-type is developed to a 3200 bit phenotype. A 128 bit genotype can search a solution space of 2^{128} different phenotypes of length 128 bit, if we assume a direct mapping. If we assume a deterministic development mapping, as shown, then we still search 2^{128} different phenotypes but this time of length 3200 bit as development has extended the phenotype to 3200 bit. As such, the search space for the 128 bit genotype is only a very small part of the phenotype search space.

How do we represent our 128 bit genotype such that it may be developed to a 3200 bit phenotype? Using development, the genotype is a building plan of how to build the functionality and structure that we wish to achieve in the phenotype. However, how do we express this?

In the work presented in [1], an axiom and a set of change and growth rules was the genotype and both the axiom and the rules themselves were based on random strings. The genotype was 440 bits long and the phenotype, a 16x16 grid of Sblock routers was therefore 16x16x32 bits i.e. 8192 bits. To a certain extent the communication and computational properties of the phenotype were built into the building plan. Growth was restricted to complete Sblocks i.e. one complete 32 bit cell at a time and a given cell was allowed to communicate with up to 4 neighbours. Also the computation properties of the Sblock grid were retained since a random 32 bit string could produce any type of Sblock. However, the building plan did not direct the developing genotype to the actual phenotype required i.e. routing Sblocks in the experiments presented. Of the 2^{32} possible configurations of an Sblock only 4 configurations describe routing blocks. As such, development was guiding us towards any phenotype of the set of the phenotypes available on a 16x16 grid of Sblocks. We were then using evolution to try to refine the rules to rules that might develop to the given phenotype.

4.2   Introducing Restricted Sblock Types

In our former L-system implementation, no knowledge as to the circuit goal was built into our rules, only knowledge about the virtual technology platform. However, if more knowledge is built into the rules of the genotype, one might expect that we achieve a more focussed search for the phenotype sought. In addition, more knowledge in the rules will allow us to study the development process itself to understand the workings of development somewhat better. That is, instead of leaving evolution to wildly explore for solutions in a very large search space, through a better understanding of development, evolution can then be used to refine the rules through exploitation.

We, therefore, need a building plan that has knowledge as to the actual phenotype to be built. That is, instead of the rules being based on random strings, more specific strings should be included in the building plan. This may be seen as a form of Sblock type reduction where the variety of Sblock types allowed for within the building plan is restricted, from the 2^{32} types available in our former work, to a subset of these types attuned to the organism to be grown. That is, an Sblock may be restricted to one of the following types: a 3-input NAND gate; a 2-input NOR gate: a west routing Sblock and a North routing Sblock. That is only 4 types are allowed. This means that both the axiom and the rules in the genotype are limited to strings of these types. Also that the communication and computation model of the developed organism is restricted to that attainable using these types.

4.3   Introducing Local Knowledge to the Development Process

The communication and computational model for the Virtual EHW FPGA are intertwined. A description of the computation of an Sblock i.e. its configuration includes communication with its neighbours. That is its configuration describes its functionality in terms of which neighbours it reacts with and what function is computed. As such, an Sblocks functionality is based on local knowledge i.e. its state and the state of its connected neighbours under a given configuration.

A further improvement on the development process may be achieved by building local knowledge into the development process itself. To include this feature of natural development what we are looking for is that a cell in the developing organism is not just developing through a set of rules applying to its own state, but to the state of its neighbours as well i.e. induction. In the case of development towards the Virtual EHW FPGA, there are 4 neighbours and therefore induction will be similar to that seen in cellular automata. Another possibility is to free the neighbourhood from the restriction of 4 neighbours but update a given cell's state based on a dynamic neighbourhood where the number of neighbours varies over time [4]. This is of course not attuned to today's technology but may be an interesting approach to development of circuits on a more futuristic technology.

5   Knowledge Rich Development

We wish to adapt our rule system for the experiments herein to a more knowledge rich system in two ways. The first being the introduction of limited Sblock typing to guide development towards a given phenotype and the second being to introduce neighbour information to the rules to add local knowledge to the development process. Further, a feature of biological development not present in our previous L-system work is introduced: that of cell death.

In the experiments herein, we limit our Sblock types to one of 5 types. These are don't care **DC**, empty **Z**, north router **NR**, west router **WR** and XOR function **XOR**. That is, the organisms (circuits) that may be developed are limited to those including North or west routers and XOR gates. Empty Sblocks signify those that are not configured as part of the circuit. They may also be Sblocks that during the development process have been configured and then killed i.e. empty. Don't care types are used to provide more flexibility to the rules. They may be used in the condition for a rule (LHS). However, they are not permitted as the result of à rule so as to retain determinism in the development process.

Our rules are based on two types of rules (as in L-systems) i.e. change and growth rules. These rules are, of course, restricted to expressions consisting of up to the 5 Sblock types. In addition, they are context sensitive, relying on neighbouring information. As such, both rule types are expressed in terms of the type and state of a given Sblock and those of its 4 neighbours to the West, North, East and South. It should be noted that the state of a given Sblock is its output value i.e. 0 or 1.

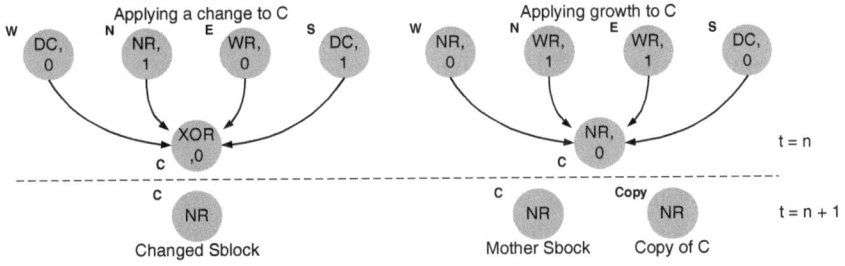

Fig. 3. Applying Change and Growth Rules

Figure 3 illustrates the workings of the change and growth rules. As shown, each of the neighbouring Sblocks (**W, N, E** and **S**) and the target Sblock itself (**C**) have a type and state. The result of the rule is the effect on the target Sblock. As shown, this effect is described in terms of the change of type, for the change rule and a copying of the type to a new Sblock for the growth rule. It should be noted that the resulting state of the target, and its copy where relevant, are not given in the rule. The resulting states are calculated during the development process.

When a growth rule triggers i.e. it finds a match for its LHS in the Sblock array, it grows a new Sblock that is a copy of the target Sblock i.e. **C**. As this is growth, we need to expand the organism i.e. we need to configure an empty Sblock. Growth is directed to the direction of most empty Sblocks. Where there is an equal amount of empty blocks in any direction, growth is prioritised in the following order: **W - N - E - S**. Change rules on the other hand only effect the target Sblock.

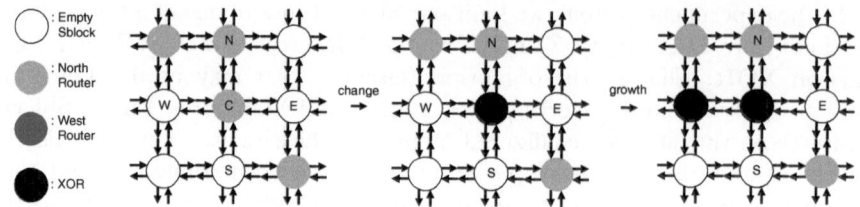

Fig. 4. Change and Growth on an Sblock Grid

Figure 4 illustrates the process of applying change and growth rules. A change rule has targeted C. This means that the condition of this change rule, including types and states of the neighbours and the target match. The change rule result is that the target Sblock **C** is changed from a north router Sblock to an XOR Sblock. A growth rule then targets the new **C**. Again its condition matches the types and states of the neighbouring blocks and the new type and state of the target **C**. The growth rule is applied by copying the target Sblock into the empty Sblock to the west.

6   Experiments and Results

The purpose of these experiments was to investigate if the added knowledge built into the rules could develop to specific although simple organisms. As such, these experiments may be said to be of a more preliminary nature. The experimental platform is a software simulation of a 3x3 Virtual EHW FPGA for growth and differentiation and a 4x4 Virtual EHW FPGA for pattern formation. The genome was chosen to include an axiom — the type and state of an Sblock, together with 3 rules. A rule may be a change or a growth rule and of the format given in section 5.

First, a set **R** of all possible rules was generated. That is all possible combinations of type and state for the rule elements: **C, W, N, E, S** and the result. From this set R, all possible 3 rule combinations were extracted to provide our rule combinations for our set of genomes. Each genome was then developed for 6 development steps. At each step, up to 3 rules may be active. The resulting organism and its genome was stored for each development run. The colour codes for the different Sblocks is given in Figure 4.

Achieving Growth

Our first investigation was to find out if rule combinations could be found that indicated reasonable growth. Reasonable growth was expressed as growing to an organism where more than 4 Sblocks were configured i.e. non empty, within the 6 development steps. This result was not hard to find since many rule combinations could in fact lead to growth across the complete grid. Further, we wanted to see

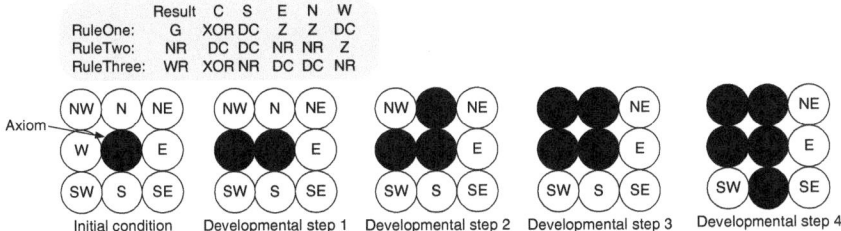

Fig. 5. Development Focused on Growth, XOR axiom shown in center in the initial condition

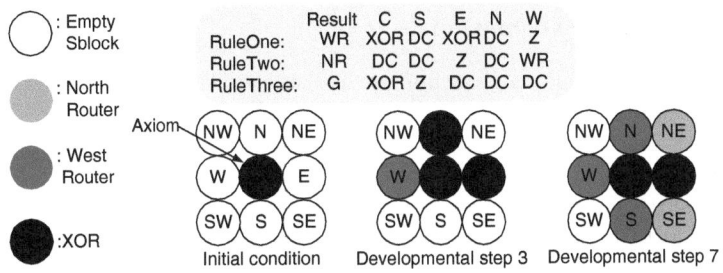

Fig. 6. Development Focused on Differentiation

if we could find rules which would grow in a specific way i.e. grow a specific type of Sblock.

In Figure 5, a genotype consisting of three rules and the Sblock array, is shown, with an XOR axiom in the centre. The development of the genome is shown for the first four development steps. After step four no more rules are activated so the growth stops. The shown rule set illustrates a rule combination where the type of the axiom itself is grown. That is, it is possible to find a rule combination where a more specific growth is achieved.

Achieving Differentiation

To investigate differentiation (change) we again began with a more general goal. That is, to find developed organisms where more than 4 Sblocks were configured. However since change rules may not be applied to empty cells, more than 4 configured cells says more about growth than specialisation. However, since growth only copies the type of the mother cell, then lack of specialisation would mean that all cells configured would hold the type of the mother cell. As such, our criteria for specialisation was set to finding genomes where both more than 4 cells were configured and there existed 3 cell types in addition to possible empty types. Figure 6 illustrates a rule combination together with an XOR axiom which achieves this criteria. In other words we can achieve differentiation.

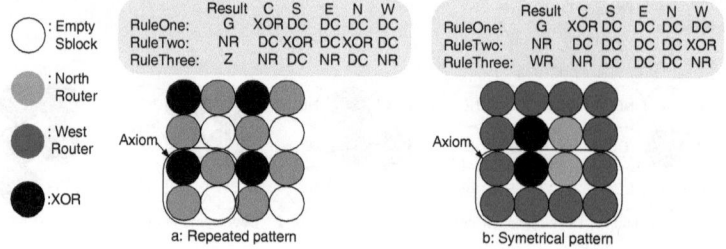

Fig. 7. Development Focused on pattern formation

Achieving Pattern Formation

Having found rules able to grow and specialise, we want to use thise properties towards development of an organism consisting of patterns of connected Sblocks. Pattern may be expressed as collections of Sblocks with given qualities. The two qualities we are looking for in these experiments are a repeated pattern and symmetry. Developing repeated patterns in an organism may be related to connecting logic gates to obtain a certain logical function. This connection of gates, i.e. Sblocks, may be reused in circuit design as a component. The ability to divide an organism in two or more symmetrical patterns may be related to partitioning of recourses in a circuit design into regions.

In Figure 7 a, the shown rule set develops from an XOR axiom to an organism consisting of a pattern repeated four times. Each pattern consists of one XOR and two North Routers. The organism shown uses fourteen development steps to completion. One of the four patterns is marked together with the placement of the axiom. The results is found by storing candidate rules able to make repeated 2x2 structures of Sblocks in the 4x4 Sblock array.

The organism in Figure 7 b is developed from an XOR axiom to an symmetrical organism, the lower and upper half of the organism is symmetrical. The development of this organism took eight development steps. The symmetrical sub organism is marked together with the placement of the axiom. The shown organism is found by storing rules producing symetrical 4x2 organisms.

7   Related Work

The work carried out by Sipper on Quasi-Uniform Cellular Automata [5] addresses the issue of universal computation, the rules in this work may be related to possible Sblocks in our development rules. Inspired by biological development before differentiation, a new family of fault tolerant FPGAs are being developed at York [6]. Cellular encoding [7] is a well-defined formal approach based on biological development. This methodology is a variation of genetic programming [8]. Similar to L-systems, a grammar-based approach may be used to develop neural network topologies. The process is used in experiments to generate large repetitive structural systems [9].

References

1. P.C. Haddow, G Tufte, and P. van Remortel. Shrinking the genotype: L-systems for ehw? In *4th International Conference on Evolvable Systems (ICES01)*, Lecture Notes in Computer Science, pages 128–139. Springer, 2001.
2. P.J. Bentley. Exploring component-based representations - the secret of creativity by evolution. In *Fourth Intern. Conf on Adaptive Computing in Design and Manufacture*, 2000.
3. P. C. Haddow and G. Tufte. Bridging the genotype-phenotype mapping for digital FPGAs. In *the 3rd NASA/DoD Workshop on Evolvable Hardware*, pages 109–115, 2001.
4. P. van Remortel, T. Lenaerts, and B. Manderick. Lineage and induction in the Development of evolved genotypes for non-uniform 2d cas. In *proceedings of the 15th Australian Joint Conference on Artificial Intelligence 2002*, 2002.
5. M. Sipper. *Evolution of Parallel Cellular Machines The Cellular Programming Approach*. Springer-Verlag, 1997.
6. C. Ortega and A. Tyrell. A hardware implementation of an embyonic architecture using virtex FPGAs. In *Evolvable Systems: from Biology to Hardware, ICES*, Lecture Notes in Computer Science, pages 155–164. Springer, 2000.
7. F. Gruau and D. Whitley. Adding learning to the cellular development of neural networks: Evolution and the baldwin effect. *Journal of Evolutionary Computation*, 1993.
8. J. Koza. *Genetic Programming*. The MIT Press, 1993.
9. H. Kitano. Building complex systems using development process: An engineering approach. In *Evolvable Systems: from Biology to Hardware, ICES*, Lecture Notes in Computer Science, pages 218–229. Springer, 1998.

Evolving Fractal Proteins

Peter J. Bentley

Department of Computer Science
University College London
Gower Street, London WC1E 6BT, UK
P.Bentley@cs.ucl.ac.uk

Abstract The fractal protein is a new concept for improving evolvability, scalability, exploitability and providing a rich medium for evolution. Here the idea of fractal proteins is introduced, and a series of experiments showing how evolution can design and exploit them within gene regulatory networks is described.

1 Introduction

Evolutionary computation does it wrong.

That's not to say that an evolutionary algorithm doesn't evolve solutions. Using principles of selection, reproduction with inheritance, and variability, we evolve good solutions to many problems all the time. The trouble is, to make an evolutionary algorithm evolve, you often need to be trained in the black arts of genetic representation wrangling and fitness function fiddling.

Along the way you'll get a few glimpses of the creativity of evolution through the bugs in your code: the little loopholes that are ruthlessly exploited by evolution to produce unwanted and invalid solutions. Music with notes inaudible to the human ear, designs with components floating in mid-air, fraud detectors that detect everything and yet find nothing useful, virtual creatures that sidestep inertia and flick themselves along in bizarre ways. Each result fascinating, and each prevented by the addition of another constraint by the developer. The bugs are never reported in any publication, and yet they point to the true capabilities of evolution.

These capabilities become even more apparent as evolution is embodied in physical media. Researchers such as Thompson [Woolf and Thompson, 2002] or Linden [2002] have shown how exotic designs of electronic circuits or antenna that exploit the physical properties of different media can be readily evolved. Such solutions could equally have been called bugs, for the unconventional nature of their designs made them difficult to use or even understand. But in truth, these results throw light on the nature of evolution.

Evolution is not an optimiser. It is a satisficer, exploiting every conceivable (and inconceivable) way to make the minimum necessary adaptation. Evolution's greatest and unrivalled skill is its ability to find and use every avenue to achieve almost nothing. In nature it is able to do this trick often enough to generate ingenuity and complexity beyond our comprehension.

In nature it can achieve this. But in our computers, suppressed by constraints, limited by stifling representations and severe operators, evolution can barely move through a search space. And even when evolution is able to evolve effectively, the

A.M. Tyrrell, P.C. Haddow, and J. Torresen (Eds.): ICES 2003, LNCS 2606, pp. 81–92, 2003.

problem it solves is usually so simplified and isolated from the true real-world problem that the evolved solution is irrelevant.

Work at UCL on a new project called MOBIUS (modelling biology using smart
materials) is focusing on these issues. The aims of this research are to free evolution
from the traditional constraints imposed by evolutionary computation (EC). Smart
materials will be used to provide a rich environment in which evolution can be embodied [Quick et al, 1999], enabling the intrinsic evolution of potentially unconventional and diverse morphologies. Evolution will employ new genetic representations
based on natural developmental processes, designed to be evolvable, scalable and free
of constraints. This paper describes the initial research on this topic: how evolved
genes can be expressed into fractal proteins that form themselves into complex and
desirable gene regulatory networks.

2 Background

2.1 Development

Development is the set of processes that lead from egg to embryo to adult. Instead of
using a gene for a parameter value as we do in standard EC (i.e., a gene for long legs),
natural development uses genes to define proteins. If expressed, every gene generates
a specific protein. This protein might activate or suppress other genes, might be used
for signalling amongst other cells, or might modify the function of the cell it lies
within. The result is an emergent "computer program" made from dynamically forming gene regulatory networks (GRNs) that control all cell growth, position and behaviour in a developing creature. (An introduction to biological development can be
found in [Bentley 2002].)

There is currently much research being performed on computational development.
Problems of scalability, adaptability and evolvability have led many researchers to
attempt to include processes such as growth, morphogenesis or differentiation in their
evolutionary systems [Haddow et al, 2001; Miller 2002]. For recent reviews see
[Kumar and Bentley, 2002] and [Gordon and Bentley, 2002].

Research at UCL has been focusing on development for some years. Kumar has
developed extensive modelling software capable of evolving genes that are expressed
into proteins, forming GRNs that control the expression of the genes and the development (growth, placement, function) of cells [Kumar and Bentley, 2003]. Gordon
has been investigating the use of development-inspired evolutionary systems for evolvable hardware, which makes use of ideas such as differentiation and growth
[Gordon and Bentley, 2002].

The research described here complements these systems, by focusing on methods
to enrich the genetic space in which evolution searches. Specifically, this paper introduces the novel idea of *fractal proteins*. Here, a gene defines a protein constructed
from a subset of the Mandelbrot Set.

2.2 Mandelbrot Set

Given the equation $x_{t+1} = x_t^2 + c$ where x_t and c are imaginary numbers, Benoit Mandelbrot wanted to know which values of c would make the length of the imaginary

number stored in x_t stop growing when the equation was applied for an infinite number of times. He discovered that if the length ever went above 2, then it was unbounded – it would grow forever. But for the right imaginary values of c, sometimes the result would simply oscillate between different lengths less than 2.

Mandelbrot used his computer to apply the equation many times for different values of c. For each value of c, the computer would stop early if the length of the imaginary number in x_t was 2 or more. If the computer hadn't stopped early for that value of c, a black dot was drawn. The dot was placed at coordinate (m, n) using the numbers from the value of c: $(m + ni)$ where m was varied from -2.4 to 1.34 and n was varied from 1.4 to -1.4, to fill the computer screen. The result was the infinite complexity of the "squashed bug" shape we know so well today. [Mandelbrot, 1982]

3   Fractal Proteins

One of the factors that limits evolution is its genetic representation. Previous work has shown how component-based representations (knowledge-poor representations with few constraints) enable greater diversity and creativity by evolution [Bentley, 2000]. Here, this idea has been combined with ideas of development and fractals, to produce a representation with as few constraints and as much richness as possible.

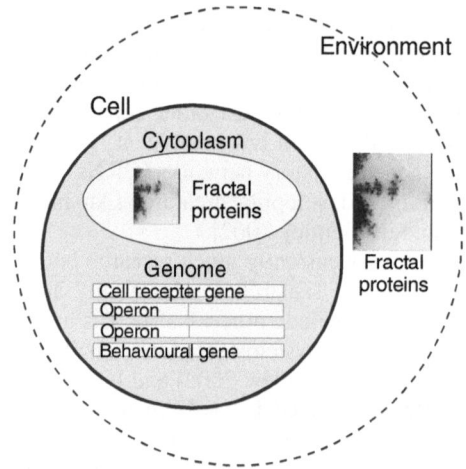

Fig. 1. Representation using fractal proteins.

3.1 Representation

Currently in this representation, there exist:

– *fractal proteins*, defined as subsets of the Mandelbrot set.
– *environment*, which can contain one or more fractal proteins, and one or more *cells*.
– *cell*, which contains a *genome* and *cytoplasm*, and which has some *behaviours*.
– *cytoplasm*, which can contain one or more fractal proteins.
– *genome*, which comprises *structural genes* and *regulatory genes*. In this work, the structural genes are divided into different types: a *cell receptor gene* and one or more *behavioural genes*.
– *cell receptor gene*, a structural gene which acts like a mask, permitting variable portions of the environmental proteins to enter the corresponding cell cytoplasm.
– *behavioural gene*, a structural gene comprising operator (or promoter) region and a cellular behaviour region.
– regulatory gene comprising operator (or promoter) region and coding region.

Figure 1 illustrates the representation.

3.2 Defining a Fractal Protein

In more detail, a fractal protein is a finite square subset of the Mandelbrot set, defined by three codons (x,y,z) that form the coding region of a gene in the genome of a cell. The gene is expressed as a protein by calculating the square fractal subset with centre coordinates (x,y) and sides of length z, see fig. 2 for an example. In this way, it is possible to achieve as much complexity (or more) compared to protein folding in nature.

Fig. 2. Example of a fractal protein defined by $(x = 0.132541887, y = 0.698126164, z = 0.468306528)$

3.3 Fractal Chemistry

Cell cytoplasms and the environment usually contain more than one fractal protein. In an attempt to harness the complexity available from these fractals, multiple proteins are merged. The result is a product of their own "fractal chemistry" which naturally emerges through the fractal interactions.

Fractal proteins are merged (for each point sampled) by iterating through the fractal equation of all proteins in "parallel", and stopping as soon as the length of any is unbounded (i.e. greater than 2). Intuitively, this results in black regions being treated as though they are transparent, and paler regions "winning" over darker regions. See fig 3 for an example.

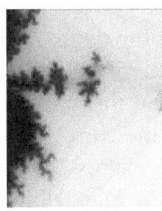

Fig. 3. Two fractal proteins (top) and the resulting merged fractal protein combination (bottom).

3.4 Genes

The cell receptor gene, regulatory, and behavioural genes all contain 7 real-coded values (codons):

xp	yp	zp	thresh	x	y	z

where $(xp,yp,zp,thresh)$ defines the promoter (operator or precondition) for the gene and (x,y,z) defines the coding region of the gene. When *thresh* is a positive value, one or more proteins must match the promoter shape defined by (xp,yp,zp) with a difference equal to or lower than *thresh* for the gene to be activated. When *thresh* is a negative value, one or more proteins must match the promoter shape defined by (xp,yp,zp) with a difference equal to or lower than $|thresh|$ for the gene to be repressed (not activated).

To calculate whether a gene should be activated, all fractal proteins in the cell cytoplasm are merged (including the masked environmental proteins, see below) and the combined fractal mixture is compared to the promoter region of the gene.

The similarity between two fractal proteins (or a fractal protein and a merged fractal protein combination) is calculated by sampling a series of points in each and summing the difference between all the resulting values. (Black regions of fractals are ignored.) Given the similarity matching score between cell cytoplasm fractals and gene promoter, the activation probability of a gene is given by:

activationprob = (1 + tanh((*matchnum* – *thresh* - Ct) / Cs)) / 2
where: *matchnum* is the matching score,
                   thresh is the matching threshold from the gene promoter
                   Ct is a threshold constant (normally set to 50)
                   Cs is a sharpness constant (normally set to 50)

Regulatory gene. Should a regulatory gene be activated by other protein(s) in the cytoplasm matching its promoter region, its corresponding coding region (x,y,z) is expressed (by calculating the subset of the Mandelbrot set). This is added to the cell cytoplasm (at the end of the developmental cycle), which holds all current proteins.

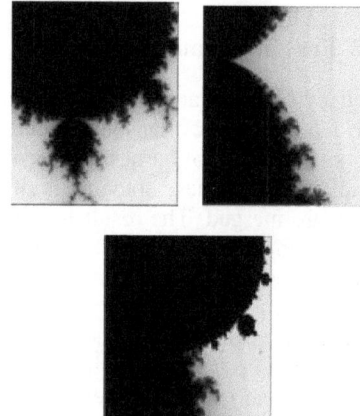

Cell receptor gene. At present, the promoter region of the cell receptor gene is ignored, and this gene is always activated. As usual, the corresponding coding region (x,y,z) is expressed by calculating the subset of the Mandelbrot set. However, the resultant fractal protein is treated as a mask for the environmental proteins, where all black regions of the mask are treated as opaque, and all other regions treated as transparent, see fig. 4.

Fig. 4. Cell receptor protein (top left), environment protein (top right), resulting masked protein to be combined with cytoplasm (bottom).

Behavioural Gene. Should a behavioural gene be activated by other protein(s) in the cytoplasm matching its promoter region, then its cell behavioural region (x,y,z) is expressed. These three real values are decoded to specify a range of different cellular functions, depending on the application. Interestingly, experiments indicate that the matching score of the promoter region should *not* be taken into account when decoding the output behaviours. (If it is, evolution designs generalized proteins that "cheat" and match too many different gene promoter regions).

3.5 Fractal Sampling

Fractal proteins are normally stored in memory as lists of constructor (x,y,z) values rather than as bitmaps. All fractal calculations (masking, merging, comparisons) are performed at the same time, by sampling the fractals at a resolution of 15x15 points. During a comparison, where differences between fractals are found to be larger than a preset threshold, the fractals are sub-sampled in a higher resolution in those regions. Figure 5 illustrates the calculations made by the computer.

3.6 Development

An individual begins life as a single cell in a given environment. To develop the individual from this zygote into the final phenotype, fractal proteins are iteratively calculated and matched against all genes of the genome. Should any genes be activated, the result of their activation (be it a new protein, receptor or cellular behaviour) is generated at the end of the current cycle. Development continues for *d* cycles, where *d* is dependent on the problem. Note that if one of the cellular behaviours includes the creation of new cells, then development will iterate through all genes of the genome in all cells.

Fig. 5. Comparing fractal proteins (top) – what the computer "sees" (bottom).

3.7 Evolution

The genetic algorithm used in this work has been used extensively elsewhere for other applications (including GADES [Bentley 1999]). A dual population structure is employed, where child solutions are maintained and evaluated, and then inserted into a larger adult population, replacing the least fit. The fittest *n* are randomly picked as parents from the adult population. The degree of negative selection pressure can be controlled by modifying the relative sizes of the two populations. Likewise the degree of positive selection pressure is set by varying *n*. When child and adult population sizes are equal, the algorithm resembles a canonical GA. When the child population size is reduced, the algorithm resembles a steady-state GA. Typically the child population size is set to 80% of the adult size and *n* = 40%. (For further details of this GA, refer to [Bentley 1999].)

Unless specified, alleles are initialised randomly, with (*xp,yp,zp*) and (*x,y,z*) values between –1.0 and 1.0 and *thresh* between –10000 and 10000. The ranges and precision of the alleles are limited only by the storage capacity of *double* and *long* 'C' data types – no range constraints were set in the code. Currently genomes of a fixed number of genes are used.

Genetic Operators. Genes are real-coded, so the crossover operator simply performs uniform crossover, randomly picking alleles from the two parents to construct the child genome. Mutation is more interesting, particularly since these genes actually code for proteins in this system. There are three main types of mutation used here:

1. Creep mutation, where (*xp,yp,zp*) and (*x,y,z*) values are incremented or decremented by a random number between 0 and 0.5 and *thresh* is incremented or decremented by a random number between 0 and 16384.

2. Duplication mutation, where a (xp,yp,zp) or (x,y,z) region of one gene randomly replaces a (xp,yp,zp) or (x,y,z) of another gene. (This permits evolution to create matching promoter and coding regions of different genes quickly.)
3. Sign flip mutation, where the sign of *thresh* is reversed.

Crossover is always applied; all mutations occur with probability 0.01 per gene.

3.8 Discussion

The key idea behind this representation is the concept of the fractal protein. Expressing a simple gene as a highly complex fractal form captures some of the richness and diversity seen in natural protein folding. By allowing these fractals to combine and interact with each other, a fractal chemistry as complex as natural chemistry is formed. And because the space of fractal subsets defined by genes is coherent (similar fractals are near to each other), redundant (similar fractals are found in many different places), and infinite (limited only by computer memory), evolvability and exploitability of this representation should be impressive.

Interestingly, the representation flies in the face of conventional wisdom: rather than make the search space as small as possible, the space is made as close to infinite as computers can manage (using the data types described). But despite such vast genetic spaces, it will be shown that evolution is able to find good solutions without difficulty – proving that evolvability, not size of search space is critical in evolution.

4 Evolving Distinct Fractal Proteins

Before attempting to evolve genes that can regulate behaviour via protein interactions, the first experiments simply checked to see if sensible fractal proteins could be evolved at all. In a vast search space, (although highly ordered) it was a vital test for evolution. For this experiment, no developmental cycle was used (i.e., only the coding regions of regulatory genes were evolved; promoters, receptor genes, and the environment were not used). A fitness function was created which produced a selection pressure towards the creation of 10 distinct and identifiable (amongst others) fractal proteins. In more detail, it maximised:

1. the differences between each protein and the other 9.
2. the differences between all 10 merged proteins and all 9 merged proteins (excluding each protein in turn).[1]

The goal of this experiment was to generate 10 fractal proteins that could be merged together and still have part of every protein identifiable in the mixture. Small population sizes of 20 (child) and 25 (adult) were used, for a maximum of 100 generations. 20 runs were performed.

---

[1] This fitness function took a little while to develop. Initial functions that simply maximized the difference between each protein and the other 9 merged proteins, resulted in a little bit of creativity by evolution. Without fail, a single white protein would evolve, with the other 9 being mostly identical fractals. When merged, the white protein dominates or "sits on top" of all the others. As long as none of the other 9 were white, they would be equally different from the merged group, whether identical to each other or not.

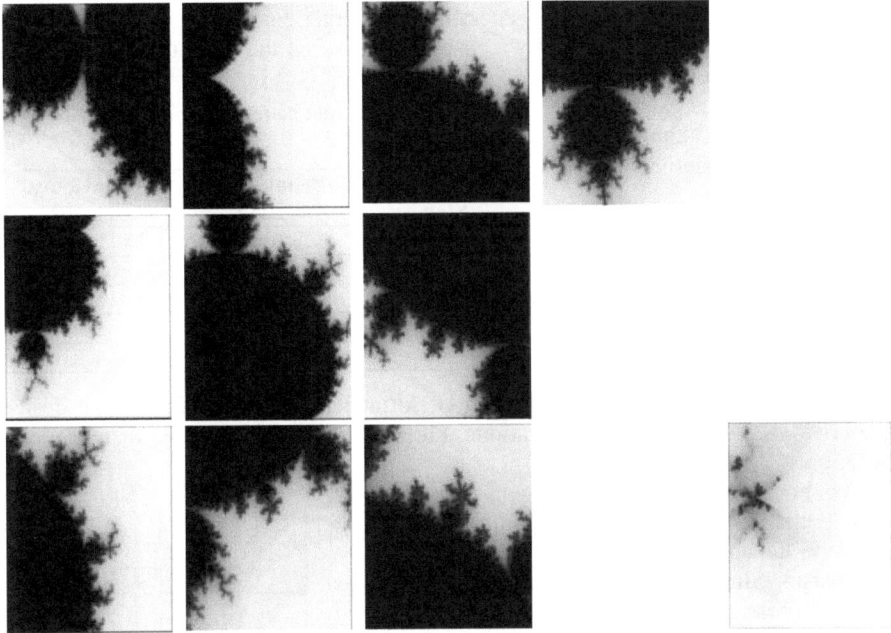

Fig. 6. 10 evolved fractal proteins and the proteins merged together (bottom right).

4.1 Results

Figure 6 shows an example of 10 fractals evolved in one of the runs. Performance was consistently good, and it is clear that despite the somewhat non-traditional nature of the search space, evolution has successfully found 10 distinct and different fractal subsets. Indeed, although results varied, at no time did evolution appear to become "stuck" – a healthy gradual change was always observed. And with an infinite amount of fine-tuning possible to the fractals, this seems very satisfactory.

Figure 6 (right) shows the result of merging all 10 proteins into one. Although difficult to see, it is possible to trace distinct portions of all ten in the merged product. Evolution has produced a complicated jigsaw of interlocking fractal proteins, as desired.

5   Evolving Gene Regulatory Networks

5.1 Experiment 1

Having shown that evolution can create and modify fractal proteins without difficulty, the next step was to attempt the evolution of gene regulatory networks. A single cell with 4 regulatory genes and 2 behavioural genes was used. The genes were randomly initialised with the alleles that defined the protein fractals shown in fig. 6. The environment and cell receptor gene were not used. Five developmental steps were employed, and the system ran for 25 generations. All other parameters were as described above. A simple fitness function was used: it measured the output from the behavioural genes at every developmental step, and calculated how much this output deviated from a predefined pattern, see table 1.

Table 1. Desired output pattern to be created by GRN in developing phenotype

Developmental Step	Behavioural gene 1	Behavioural gene 2
1	Activated	Not activated
2	Not activated	Activated
3	Activated	Not activated
4	Not activated	Activated
5	Activated	Not activated

5.2 Results 1

Although other representations had struggled to achieve any desired GRN pattern, in every run performed for this experiment, the desired pattern was found within 5 generations. Figure 7 shows an example of one typical solution.

As should be clear, all four regulatory genes have evolved to become active in the absence of proteins in the cytoplasm. Since development begins with no proteins (the environment was not used), all four become active on the first step and produce proteins. In the second development step, the presence of proteins regulates the four regulatory genes and they become inactive. In the third step, the absence of proteins makes them active again, and so on. To create the desired pattern, the first behavioural gene also evolved to become active in the absence of proteins, and so oscillates with the regulatory genes. The second behavioural gene evolved to become active in the presence of protein(s) produced by the regulatory genes, and so oscillates against them.

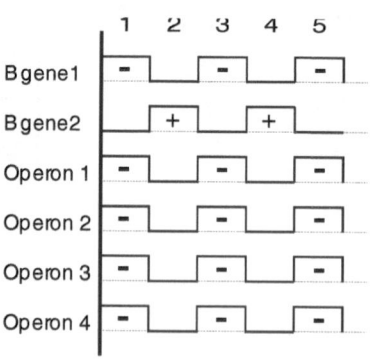

Fig. 7. Simple evolved GRN. A high '-' indicates a gene active because it has not been repressed. A high '+' indicates a gene activated by a match to its promoter. A low indicates the gene is inactive.

This is clearly a very simple GRN that only required one regulatory gene, but it demonstrates an important feature: the evolution of a desired pattern of gene activation during development. Also note how the behaviour of all four regulatory genes has converged (something that consistently occurred in all runs, despite early solutions containing some very intricate GRN patterns). Again, such behaviour makes sense: the best way to produce the desired output pattern in a robust and stable way is to use all four regulatory genes to reinforce the underlying "clock signal". It is a lovely demonstration of evolution building redundancy and graceful degradation into the solution. The desired output is generated even if one or two of the regulatory genes mutate.

5.3 Experiment 2

With evolution clearly able to find a simple GRN pattern, a second experiment was performed, with a rather more complex pattern set as the target, see table 2. Instead of requiring a simple oscillator from each behavioural gene, this target required 'B gene 1' to be active on step 1 and 5, but inactive on steps 2, 3 and 4, and the opposite from 'B gene 2'. Effectively, the target requires a GRN to count to 3 in order to construct such a pattern. All other system settings were as described for the first experiment.

Table 2. Desired output pattern to be created by GRN in developing phenotype

Developmental Step	Behavioural gene 1	Behavioural gene 2
1	Activated	Not activated
2	Not activated	Activated
3	Not Activated	Activated
4	Not activated	Activated
5	Activated	Not activated

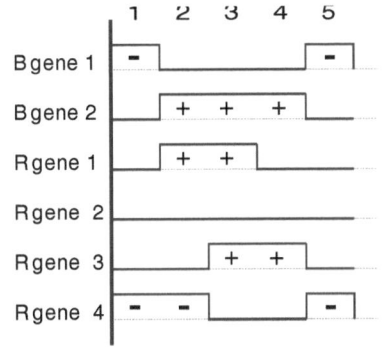

Fig. 8. Evolved GRN showing counting behaviour using regulatory genes 1, 3 and 4.

5.4 Results 2

The results were unexpectedly good. In 13 out of 20 runs, the solutions were two gene activations away from the perfect solution. In 1 run the solution was a single gene activation away from the perfect solution. Perfect solutions were found in 5 out of 20 runs. Given that very small population sizes were used and the GA only ran for 25 generations, these results are considered impressive. Figures 8 and 9 show two examples of a perfect solution.

In the first example (fig. 8) the desired activation patterns from the two behavioural genes were produced by an ingenious gene regulatory network that appears to act like a counter. Regulatory gene 4 triggers regulatory gene 1, which triggers regulatory gene

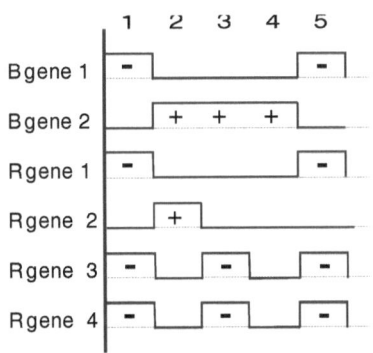

Fig. 9. Evolved GRN showing counting behaviour derived from an evolved "clock signal" using regulatory genes 1, 2 and 3, 4.

3. In the second example (fig 9), the desired patterns appear to result from an evolved "clock signal" provided by regulatory genes 3 and 4. In both examples, if the GRN is permitted to run for more time steps, the same behavioural pattern is repeated indefinitely. Also in both cases, the regulatory genes seem to act like a hidden layer in a neural network, with the behavioural genes somehow "sitting on top" of many diverse patterns and choosing to exploit or extract only those patterns that result in the desired

output behaviour. Many other examples of GRNs were also observed, each using different 'ideas' to achieve the same results.

6 Conclusions

Ideas of evolvability, scalability, exploitability and embodiment in a rich media have led to the development of a novel representation based on *fractal proteins*. This paper has introduced the concept of fractal proteins and shown that genes defining these complex forms can be evolved effectively. Two further experiments demonstrated that evolution can create regulatory genes that harness the complexity of fractal proteins. These regulatory genes were used to form intricate gene regulatory networks, that output desired behavioural patterns.

The work described here is only a few weeks old. There remains considerable research to investigate the evolution of more complex GRNs, the effects of environment and cell receptor genes, cellular behaviours and eventually the evolution of multicellular forms. This research will be linked with work into new smart materials and the creation of novel morphologies and controllers.

Acknowledgments

Thanks to Sanjeev Kumar for his comments and for the gene activation probability function. This material is based upon work supported by the European Office of Aerospace Research and Development (EOARD), Airforce Office of Scientific Research, Airforce Research Laboratory, under Contract No. F61775-02-WE014. Any opinions, findings and conclusions or recommendations expressed in this material are those of the author and do not necessarily reflect the views of EOARD. MOBIUS is an ▨ EMBLEM project.

References

1. P. J. Bentley (2002) *Digital Biology*. Simon and Schuster, New York.
2. P. J. Bentley (2000). "Exploring Component-Based Representations - The Secret of Creativity by Evolution?" In Proc. of the *Fourth International Conference on Adaptive Computing in Design and Manufacture* (ACDM 2000), April 26th - 28th, 2000, University of Plymouth, UK.
3. P. J. Bentley (1999). "From Coffee Tables to Hospitals: Generic Evolutionary Design" Chapter 18 in Bentley, P. J. (Ed) *Evolutionary Design by Computers*. Morgan Kaufmann Pub. San Francisco, pp. 405-423.
4. T. W. Gordon and P. J. Bentley (2002) "Towards Development in Evolvable Hardware." To appear in Proc. of the 2002 NASA/DoD Conference on Evolvable Hardware (EH-2002, Washington D.C., July 15-18, 2002).
5. P. C. Haddow, G. Tufte, and P. van Remortel (2001) "Shrinking the Genotype: L-Systems for EHW?" In Proc. Of 4th Int. Conf. On Evolvable Systems: From Biology to Hardware, Tokyo, Japan.

6. S. Kumar and P. J. Bentley (2002). Computational Embryology: Past, Present and Future. Invited chapter in Ghosh and Tsutsui (Eds) Theory and Application of Evolutionary Computation: Recent Trends. Springer Verlag (UK).

7. S. Kumar and P. J. Bentley (2003). Biologically Plausible Evolutionary Development. In Proc. of ICES 2003.

8. D. Linden (2002) "Innovative Antenna Design using Genetic Algorthms" Chapter 20 in Bentley, P. J. and Corne, D. W. (Eds) *Creative Evolutionary Systems*. Morgan Kaufmann Pub, San Francisco, pp. 487-510.

9. B. Mandelbrot (1982) *The Fractal Geometry of Nature* . W.H. Freeman & Company.

10. J. F. Miller (2002) "What is a Good Genotype-Phenotype Mapping for the Evolution of Computer Programs?" Presented at the *Software Evolution and Evolutionary Computation Symposium*, EPSRC Network on Evolvability in Biology & Software Systems, University of Hertfordshire, Hatfield, U.K.7-8 February 2002.

11. T. Quick, K. Dautenhahn, C. Nehaniv, and G. Roberts (1999) "The Essence of Embodiment: A framework for understanding and exploiting structural coupling between system and environment" In Proc. Of Third Int. Conf. On Computing Anticipatory Systems (CASYS'99), Symposium 4 on Anticipatory, Control and Robotic Systems, Liege, Belgium, pp. 16-17.

12. S. Woolf and A. Thompson (2002) "The Sound Gallery – An Interactive A-Life Artwork". Chapter 8 in Bentley, P. J. and Corne, D. W. (Eds) *Creative Evolutionary Systems*. Morgan Kaufmann Pub, San Francisco, pp. 223-250.

A Developmental Method
for Growing Graphs and Circuits

Julian F. Miller[1] and Peter Thomson[2]

[1] School of Computer Science, The University of Birmingham, UK
j.miller@cs.bham.ac.uk
http://www.cs.bham.ac.uk/~jfm
[2] School of Computing, Napier University, UK
p.thomson@napier.ac.uk
http://www.dcs.napier.ac.uk/~petert

Abstract. A review is given of approaches to growing neural networks and electronic circuits. A new method for growing graphs and circuits using a developmental process is discussed. The method is inspired by the view that the cell is the basic unit of biology. Programs that *construct* circuits are evolved to build a sequence of digital circuits at user specified iterations. The programs can be run for an arbitrary number of iterations so circuits of huge size could be created that could not be evolved. It is shown that the circuit building programs are capable of correctly predicting the next circuit in a sequence of larger even parity functions. The new method however finds building specific circuits more difficult than a non-developmental method.

1 Introduction

Natural evolution builds organisms using a process of biological development. In this a fertilized cell begins a process of replication and differentiation that culminates in an organism made of a huge number of specialist cells. Each cell carries the same genetic information yet somehow this decentralized, distributed community of cells builds a whole organism. Human beings design things in a top-down manner. Electronic circuits are built from high-level specification, to synthesis of symbolic components. These are then converted into transistor-level designs and finally wafers of silicon and wires are created. This method of design will become increasingly untenable as the size of electronic components continues to shrink.

In evolutionary computing a problem of scalability has become evident. The word scalability has a number of meanings: the evolution time increases markedly with problem size, the genotype length is proportional to problem size, the density of solutions in the search space associated with larger problems is a sharply decreasing function, and the time for fitness evaluation increases rapidly with problem size. The problem is particularly evident in the evolution of neural networks, where each link requires a floating-point weight that must be determined. In Genetic Programming researchers have generally concentrated on relatively small and simple problem such

A.M. Tyrrell, P.C. Haddow, and J. Torresen (Eds.): ICES 2003, LNCS 2606, pp. 93–104, 2003.

as parity functions and the Santa Fe Ant Trail. This has partly been motivated by scientific reasons in that these, though small are sufficiently difficult benchmarks for the evaluation of new techniques. Koza has been able to produce large and complex, human competitive analog circuit designs, but only by providing enormous computational power [19]. The problems of scalability have partly motivated a number of attempts to consider the use of a developmental approach to evolutionary design. In the field of evolvable hardware researchers have also tended to concentrate on relatively small circuit design problems. In digital circuit design the scalability problem is particularly acute [26].

2   Lindenmeyer Systems, Graph Re-writing and Developmental Approaches

A number of researchers have studied the potential of Lindenmeyer systems [20] for developing artificial neural networks (ANNs) and generative design. Boers and Kuiper have adapted L-systems to develop the architecture of artificial neural networks (ANNs) (numbers of neurons and their connections) [3]. They used an evolutionary algorithm to evolve the rules of a L-system that generated feed-forward neural networks. Backpropagation was used, and the accuracy of the neural networks on test data was assigned to the fitness of the encoded rules. They found that this method produced more modular neural networks that performed better than networks with a predefined structure. Kitano had developed another method for evolving the architecture of an artificial neural network [17] using a matrix re-writing system that manipulated adjacency matrices. Although Kitano claimed that his method produced superior results to direct methods (i.e. a fixed architecture, directly encoded and evolved), it was later shown in a more careful study that the two approaches were of equal quality [23]. Gruau devised an elegant graph re-writing method called cellular encoding [10]. Cellular encoding is a language for local graph transformations that controls the division of cells that grow into artificial neural networks. This method was shown to be effective at optimizing both the architecture and weights at the same time, and they found that, to achieve as good performance with direct encoding, required the testing of many candidate architectures [11]. Others have successfully employed this approach in the evolution of recurrent neural networks that control the behaviour of simulated insects [18]. Koza has successfully employed a modified cellular encoding technique to allow the evolution of programs that produce human-competitive designs, especially in analog circuit design [19]. Recently Hornby and Pollack have also evolved context free L-systems to define three dimensional objects (table designs) [14]. They found that their generative system could produce designs with higher fitness and faster, than direct methods.

Jacobi created an impressive artificial genomic regulatory network, where genes code for proteins and proteins activate (or suppress) genes [16]. He used the proteins to define neurons with excitatory or inhibitory dendrites. This allowed him to define a recurrent ANN that was used to control a simulated Khepera robot for obstacle avoid-

avoidance and corridor following. Nolfi and Parisi evolved encoded neuron position and branching properties of axonal trees that would spread out from the neurons and connect to other neurons [22] and in later work introduced cell division using a grammar [5]. Astor and Adami have created a developmental model of the evolution of ANN that utilizes an artificial chemistry [1].

Eggenberger suggests that the complex genotype-phenotype mappings typically employed in developmental models allow the reduction of genetic information without losing the complex behaviour. He stresses the importance of the fact that the genotype will not necessarily grow as the number of cells, thus he feels that developmental approaches will scale better on complex problems [8]. Sims evolved the morphology and behaviour of simulated agents   [24]. Bongard and Pfeifer have evolved genotypes that encode a gene expression method to develop the morphology and neural control of multi-articulated simulated agents [4].

Bentley and Kumar examined a number of genotype-phenotype mappings on a problem of creating a tessellating tile pattern [2]. They found that the indirect developmental mapping (that they referred to as an implicit embryogeny) could evolve the tiling patterns much quicker, and further, that they could be subsequently grown (iterated) to much larger sized patterns. One drawback that they reported, was that the implicit embryogeny tended to produce the same types of patterns.

Hemmi, Mizoguchi and Shimohara evolved the rules of a rewriting system to define HDL programs [12]. Recently interest in developmental approaches in Evolvable Hardware has begun to increase. Haddow, Tufte and van Remortel have considered the use of Lindenmeyer systems for digital circuit design [13]. Their "cells" are like configurable logic blocks in FPGAs and cell interactions are defined by neighbouring cells in a two-dimensional grid system. They note that there are many epistatic effects that make evolution of the genotype difficult. Gordon and Bentley have compared a developmental evolutionary approach with a non-developmental encoding. They found that the could evolve these direct mappings more easily [9]. Edwards also examined developmental genotypes and possible physical implementations [7].

3 Cartesian Genetic Programming for Circuits

Cartesian Genetic Programming was developed from methods developed for the automatic evolution of digital circuits [21]. Since electronic circuits can be represented by graphs (often referred to as netlists) it was natural to generalize it to the general problem of evolving computer programs. CGP represents a program or circuit as a list of integers that encode the connections and functions. The representation is readily understood from a small example. Consider a one bit binary adder circuit. This has three inputs that represent the two bits to be summed and the carry-in bit. It has two outputs: sum and carry-out. A possible implementation of this is shown below:

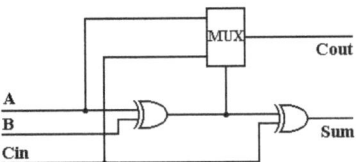

Fig. 3.1. One-bit adder circuit.

CGP employs an indexed list of functions that represent in this example various two input logic gates and also three input multiplexers. Suppose that XOR is function 10 and MUX is function 16. The three inputs A, B, Cin are labeled 0, 1, 2. The output of the left (right) XOR gate is labeled 3 (5). The output of the MUX gate is labeled 6. In CGP, this circuit could have the following genotype representation:

$$0\ 1\ 0\ \mathbf{10}\qquad 2\ 0\ 2\ 6\qquad 0\ 2\ 3\ \mathbf{16}\qquad 3\ 2\ 2\ \mathbf{10}$$

The integers in bold represent the functions, and the others represent the connections between gates, however, if it happens to be a two input gate then the third input is ignored. It is assumed that the circuit outputs are taken from the last two nodes. The second group of four integers (shown in grey) represent an AND gate that is not part of the circuit phenotype (hence doesn't appear in Fig 3.1). Since we are only interested in feedforward circuits here, it is important to note that the connections to any gate can only refer to gates that appear on its left. Typically in CGP we use point mutation (that is constrained to respect the feedforward nature of the circuit). Suppose that the first input of the MUX gate (0) was changed to 4. This would connect the AND gate into the circuit. Similarly, a point mutation might disconnect gates. Thus it is clear that in CGP we use a many to one genotype-phenotype mapping, as redundant nodes may be changed in any way and the genotypes would still be decoded to the same phenotype. We use an (1+4)-ES evolutionary algorithm that uses characteristics of this genotype-phenotype mapping to great advantage (i.e. genetic drift). This is given below:

1. Generate 5 chromosomes randomly to form the population
2. Evaluate the fitness of all the chromosomes in the population
3. Determine the best chromosome (called it *current_best*)
4. Generate 4 more chromosomes (offspring) by mutating the *current_best*
5. The *current_best* and the four offspring become the new population
6. Unless stopping criterion reached return to 2

Step 3 is a crucial step in this algorithm: if more than one chromosome is equally good then the algorithm always chooses the chromosome that is not the *current_best* (i.e. equally fit but genetically different). In a number of studies this step has been proved to allow a genetic drift process that turns out be of great benefit [25][29]. In later sections we refer to the mutation rate, this is the percentage of each chromosome that is mutated in step 4.

4   Developmental Cartesian Genetic Programming

The philosophical standpoint for the form of development described here is that *the cell is the basic unit of biology*. We see evolution as a process of evolving a cell. The cell is a very clever piece of machinery that can in co-operation with an environment self-replicate and differentiate to form a whole organism. Thus in developmental Cartesian GP (DCGP) we attempt to evolve a cell that can construct a larger program by iteration of the cell's program in its environment[1]. To elaborate this further, the idea is to define a program (encoded as a CGP graph) that is run inside the nodes (cells) of another graph (that is the phenotype or organism, or in this case the final digital circuit). It is important to note that in the context of the organism being a digital circuit, the cell is identified with a logic gate. All the gates of the final circuit are produced by running the same program inside all the nodes (gates) of the circuit graph for a given number of iterations. In the version of developmental CGP described here, we define the "environment" of a given cell to be the position of the cell and the integer labels of the cells that are connected to that cell. The analogues between the abstract graph re-writing rules and real biological entities are given below:

DNA bases: **Cartesian Cell Genotype**; *Translation*: **Genotype-Phenotype decoding**
Positional information: **Node position**; *Cell division*: **Node replication**
Cell type: **node function type (gate)**; *Zygote*: **seed node**
Neighbouring cell signals: **Node inputs (gate inputs)**

Of course, this is a highly idealized environment and further investigations are planned that will take into account the functions of the cells that are connected to the cell in question (this will allow semantic information to affect cell function). Imagine that we start with a 'seed' cell, this can only be connected to the program inputs. Here is one possible seed (and the one used for all the experiments described in this paper):

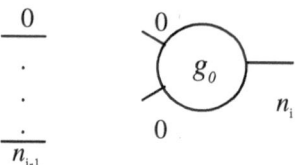

Fig. 4.1. Seed cell with n_i program inputs (0- n_{i-1}), g_0 denotes the first gate type in a list of possible gates.

Now we want to run a program inside the cell that uses information about the cell and its environment to construct a new one, indeed, we also want it to be able to replicate itself, so that we can grow a larger program. The cell needs four pieces of information, its two connections, its function and its position and the program inside

---

[1] The technique described here is much more biologically motivated than either L-systems or cellular encoding in addition the re-writing rules have the an enormously richer space of transformations that are possible with either, and with a much simpler conceptual apparatus.

the cell needs to take this information and create a new node with connections and a function and whether the cell should be duplicated. So the situation is now

Fig. 4.2. Cell with program inputs and outputs.

The cell program is a mapping from the integers that define the cell's connections, function and position to a new set of integers defining its new connections, function and whether it will replicate itself. Since we are using this program to construct a circuit that is feed-forward , the integers that come out of the node program must take the correct values (i.e. the connections must be to nodes on the left of our current position, the functions must be ones on the list of valid functions, and finally the divide must be 0 or 1). Unfortunately it is very difficult to construct programs that will automatically respect the feed-forward requirement. After we have run the cell program we will need to carry out some sort of operation to bring the numbers into the correct ranges. A simple way to do this is to apply a modulo operation (note for the new function F' we use the $X \bmod N_F$ as the *address* in the function look up table). Finally, we are left with deciding how to allow the evolution of the program inside the cell. This can be accomplished using CGP with primitive functions that manipulate integers.

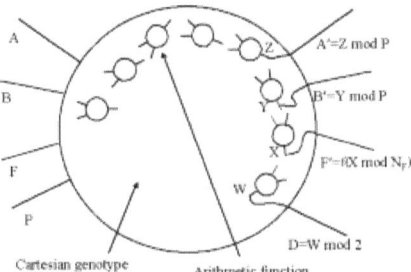

Fig. 4.3. Inside a cell is a Cartesian program defined using arithmetic functions. The phenotype nodes are valid feed-forward nodes due to the modulo operations.

To construct a circuit we run a single Cartesian program in each cell of a developing circuit (organism). This is called an iteration. Then at each iteration, we execute the same program. After a user defined number of iterations we test the developed circuit against the specification for the circuit (i.e. we count how many bits are correct). To clarify this still further let us consider an example of an evolved cell program that constructs the even-parity function with four inputs. Suppose that there are two primitive arithmetic functions f_0 and f_s of arity two, defined as follows:

$$f_0 = f_0(x, y) = x + y \quad , \quad f_5 = f_5(x, y) = \begin{cases} 1 & x \ge y \\ 0 & x < y \end{cases}. \tag{4.1}$$

Consider the following genotype

node labels	4	5	6	7	8	9	10	11
Cartesian genotype	1 2 **5**	4 4 **0**	3 2 **5**	2 0 **5**	2 1 **0**	8 7 **0**	4 3 **0**	1 5 **5**

There are four inputs: connection1 (A), connection2 (B), function (F), position (P) with labels 0, 1, 2, 3. The outputs of the program are taken from nodes 8, 9, 10, 11 and these outputs are denoted Z, Y, X, W (as seen in Fig. 4.3). Note that in this genotype node 6 is not referenced by nodes on its right, so is redundant. We can decode the genotype into a set of arithmetic rules dependent on the four inputs.

$$Z = F + B \ , \quad Y = Z + f_5(F, A) \ , \tag{4.2}$$
$$X = P + f_5(B, F) \ , \quad W = f_5(B, 2f_5(B, F))$$

To clarify the origin of these rules, consider the expression for W, this is the output of node 11, the first input is 1 (B), the second input refers to node 5. This is the addition of its two inputs, both of which refer to node 4. Node 4 is function 5 acting on it two inputs 1 and 2, that is, B and F (hence $2f_5(B,F)$ is the second argument of f_s in W). Armed with the rules of equation 4.2 we can apply them to the seed cell and carry out the first iteration. Note that in this example we are using two functions in the phenotype, XOR (labelled as decimal 10) and XNOR (labelled as decimal 11). We are considering even-4 parity which has four inputs (labelled 0, 1, 2, 3). We found that the program worked much better when we allowed the right daughter cell to differentiate again (disallowing further replication)[2]. Thus, we define the word "iteration " to mean running the cell program in each cell and running the cell program (ignoring replication) in the right daughter cell. The reason why this helps markedly is still under investigation. The steps required for one iteration are shown in table 4.1. In the first "interpretation" line in the table a 10 appears (asterisk) because the cell program (after modulo operation) is 0, this refers to the zeroth function, which in this case is 10 denoting the XOR operation. After one iteration the circuit is represented by the encoded phenotype: 2 3 **10**   3 4 **11**   which is the circuit shown in Fig. 4.4.

In the next iteration the rules defined in equation 4.2 are then applied to *each* cell of the phenotype. Applying the rules to the cell defined by 2 3 **10** we obtain 1 2 **10** that is replicated (since D=1) giving 1 2 **10** 1 2 **10** 3 4 **11** and when the program is run with 1 2 **10** (node 5) we obtain 2 3 **11** (note D is ignored on the second differentiating step - see earlier). Finally applying the rules to the cell defined by 3 4

---

[2] This step was found to allow more diversity in nodes in the phenotype, without it there tended to occur repeat node groups.

11 we find that if replicates to 0 1 **11** 0 1 **11** and on differentiating the last cell we obtain the new encoded phenotype   1 2 10   2 3 **11**   0 1 **11**   5 6 **11**. This is even-four parity circuit below shown on the right in Fig. 4.4.

Table 4.1. Applying the cell program to the seed cell (first iteration)

Seed cell						Cell program outputs (see equation 4.2)			
A	B	F	P	f_s(F, A)	f_s(B, F)	Z	Y	X	W
0	0	10	4	1	0	10	11	4	1
						Zmod 4	Ymod4	Xmod2	Wmod2
New left cell						2	3	0	1
Interpretation						**2**	**3**	**10***	replicate
Right cell									
2	3	10	5	1	0	13	14	5	1
						Zmod 5	Ymod5	Xmod2	Wmod2
New right cell						3	4	1	1
Interpretation						**3**	**4**	**11**	ignore

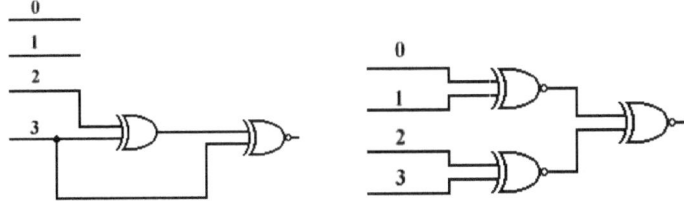

Fig. 4.4. The developed circuit (one iteration left), two iterations (right).

The idea of growing graphs or circuits leads to some very interesting consequences. It is possible for us to provide rewards during the growth process. It also allows us to consider trying to evolve programs that construct classes of circuits, by requiring that at particular points in time the circuit have a particular behaviour. In the next section we describe experiments and results for the problem of growing the class of even-parity functions.

5 Experiments and Results

In the experiments reported here we were able to choose from a set of seven primitive arithmetic functions to build the cell program. It is extremely difficult to know what would be a good choice for such functions[3]. The functions chosen are listed below:

---

[3] This is akin to trying to define a sequence of DNA base pairs that will give rise to a specific phenotypic trait. The function set chosen (equation 5.2) so as to provide a "complete" set of arithmetic operations with the addition of comparisons (f_5) and controlled swapping (f_6).

$$f_0 = f_0(x, y) = x + y \quad, \quad f_1 = f_1(x, y) = x - y$$
$$f_2 = f_2(x, y) = x \bmod (y + 1) \quad, \quad f_3 = f_3(x, y) = xy \quad, \quad f_4 = f_4(x, y) = x / y$$
$$f_5 = f_5(x, y) = \begin{cases} 1 & x \geq y \\ 0 & x < y \end{cases} \quad, \quad f_6 = f_6(x, y, z) = \begin{cases} x & z = 0 \\ y & z \neq 0 \end{cases} \tag{5.2}$$

Where x, y, z are the integer-value inputs to any node in the cell program. Note that in f_4 the operator is integer division. The circuit phenotype is built from the two binary input (a and b) functions $p(a, b)$ that are defined in table 5.2:

Table 5.2. Possible circuit functions

Integer code	$p\ (a, b)$	Integer code	$p\ (a, b)$
0	0	8	AND(NOT(a),b)
1	1	9	NOR(a, b)
2	a	10	XOR(a, b)
3	b	11	XNOR(a, b)
4	NOT(a)	12	OR(a, b)
5	NOT(b)	13	OR(a, NOT(b))
6	AND(a, b)	14	OR(NOT(a), b)
7	AND(a, NOT(b))	15	NAND(a, b)

In the first experiment we wanted to test the feasibility of the new approach, so we tried a simple problem of evolving the program that would construct a binary adder with the function defined in section 3. The experimental parameters are seen in table 5.3. We obtained three perfect adders designs. When observing the runs we observed that the algorithm reached fitness 14 (two bits incorrect) quite quickly but was often unable to improve. The average fitness obtained was 14.18. This does not compare well with a directly (non-developmental) evolved circuit (where the genotype directly encoded the circuit using the same representation as the phenotype) with the same parameters. We obtained 100 perfect adder circuits in that experiment. In the second experiment we tried to evolve a program that would build a series of even parity functions. The experimental parameters are shown in table 5.4. The circuit functions (phenotype) chosen for this experiment allow even parity functions to be built easily. The fitness of the cell program was defined as the sum of the correct truth table bits for all the target functions. We obtained 60 cell programs that could correctly build the even parity functions at the desired iterations. All these programs were subsequently tested at iteration four to see whether any of them could correctly build even-6 parity, and one program succeeded.

Unfortunately it proved not to be a general solution to the even-parity problem as it was found that it did not correctly build even-7 parity at the fifth iteration. Other researchers have successfully evolved completely general solutions to even-parity problems. However they all provide bit strings serially and work at a much higher level of abstraction.: machine code [15], recursive functions[27], and lambda abstractions [28].

Table 5.3. Parameters for experiment 1

Parameter	value
Population size	5
Mutation rate	8%
Number of generations	20,000
Number of runs	100
Number of circuits	1
Desired circuit	One-bit adder
Iteration at which circuit required	4
Maximum number of genotype nodes	20
Genotype node functions used	0 - 6
Phenotype node functions	6,7,10

Table 5.4 Parameters for experiment 2

Parameter	value
Population size	5
Mutation rate	3%
Number of generations	100,000
Number of runs	100
Number of circuits	3
Desired circuits	even-3, even-4, even5 parity
Iteration at which circuit required	1, 2, 3
Maximum number of genotype nodes	40
Genotype node functions used	0 - 6
Phenotype node functions	10,11

6   Conclusions

We have presented a new way to evolve digital circuits that uses a developmental mapping that is based on the idea of evolving the program for a cell. It is capable of constructing a number of Boolean circuits at user specified points in time. The circuit construction program can in principle build circuits of arbitrary size. However, in practice the genotypes are not as evolvable as direct encodings. Designing an effective developmental mapping is a very difficult problem. The mapping we have presented is very general and flexible. However there are many aspects that require further investigation. In our representation the "neighbourhood" of a cell are the *positions* of the cells that are connected to it. Intuitively it seems more natural to take into account the *functions* of the neighbours. We have used a form of Cartesian GP that is linear in nature. In would be interesting to investigate the behaviour of the developmental algorithm if the circuit was defined in terms of a two-dimensional grid. One could evolve a circuit in which gates were arranged in columns and one could define the neighbours of a gate to be some gates that were *physically* close. When the behaviour of the program that developed correct even parity functions was examined it was found that the circuits at successive iterations were generally quite different. One could potentially inherit more structure from previous circuits if the cell program was not run in every cell but only in the *last* cell. This of course, would depart from the biological inspired form of development. The form of development described here works at a quite abstract level in that the integers that form the input to the cell program are labels and addresses. It may be that a more direct representation would be more effective, for instance, one in which wires are particular types of cells. The problems we have studied here are all relatively simple and it may be that developmental approaches are more valuable when the circuits being grown are enormous. Many authors tend to believe that the shorter developmental genotypes will be more evolvable, however, this work suggests that the issue is much more complex.

References

1. Astor J. C., and Adami C. (2000), "A Development Model for the Evolution of Artificial Neural Networks", Artificial Life, Vol. 6, pp. 189-218.
2. Bentley P., and Kumar S. (1999), "Three ways to grow designs: A comparison of embryogenies for an Evolutionary Design Problem", in Proceedings of the Congress on Evolutionary Computation, IEEE Press, pp. 35-43.
3. Boers, E. J. W., and Kuiper, H. (1992), "Biological metaphors and the design of modular neural networks", Masters thesis, Department of Computer Science and Department of Experimental and Theoretical Psychology, Leiden University.
4. Bongard J. C. and Pfeifer R. (2001), "Repeated Structure and Dissociation of Genotypic and Phenotypic Complexity in Artificial Ontogeny", in Spector L. et al. (eds.) Proceedings of the Genetic and Evolutionary Computation Conference, Morgan-Kaufmann, pp. 829-836.
5. Cangelosi, A., Parisi, D., and Nolfi, S. (1993), "Cell Division and Migration in a 'Genotype' for Neural Networks", Technical report PCIA-93, Institute of Psychology, CNR, Rome.
6. Dellaert, F. (1995), "Toward a Biologically Defensible Model of Development", Masters thesis, Department of Computer Engineering and Science, Case Western Reserve University.
7. Edwards R. T (2003), "Circuit Morphologies and Ontogenies", in NASA/DOD Conference on Evolvable Hardware, IEEE Computer Society, pp. 251-260.
8. Eggenberger P. (1997), "Evolving morphologies of simulated 3D organisms based on differential gene expression", in Proceedings of 4th European Conference on Artificial Life, pp. 205-213.
9. Gordon T., and Bentley P. J. (2003) "Towards Development in Evolvable Hardware", in NASA/DOD Conference on Evolvable Hardware, , IEEE Computer Society, pp. 241-250.
10. Gruau, F. (1994), "Neural Network Synthesis using Cellular Encoding and the Genetic Algorithm", PhD thesis, Ecole Normale Supérieure de Lyon.
11. Gruau, F., Whitley, D., and Pyeatt, L. (1996) "A Comparison between Cellular Encoding and Direct Encoding for Genetic Neural Networks", in Proceedings of the 1st Annual Conference on Genetic Programming, Stanford.
12. Hemmi H., Mizoguchi, J., and Shimohara K. (1994), "Development and Evolution of Hardware Behaviors", in Proceedings of Artificial Life IV, pp. 371-376.
13. Haddow, P. C., Tufte G., and van Remortel P. (2001) "Shrinking the genotype: L-systems for evolvable hardware", in 4th International Conference on Evolvable Systems: From Biology to Hardware, Lecture Notes in Computer Science, Vol. 2210, Springer-Verlag, pp. 128-139.

14. Hornby G. S., and Pollack J. B. (2001), "The Advantages of Generative Grammatical En-codings for Physical Design", in Proceedings of the Congress on Evolutionary Computa-tion, IEEE Press, pp. 600-607.
15. Huelsbergen L. (1998), "Finding general Solutions to the Parity Problem bu Evolving Ma-chine-Language Representations", in Proc. Conf. on Genetic Programming, pp. 158-166.
16. Jacobi, N. (1995), "Harnessing Morphogenesis", Cognitive Science Research Paper 423, COGS, University of Sussex.
17. Kitano, H. (1990), "Designing neural networks using genetic algorithms with graph gen-eration system", Complex Systems, Vol. 4, pp. 461-476.
18. Kodjabachian, J. and Meyer, J-A. (1998), "Evolution and Development of Neural Controleres for Locomotion, Gradient-Following and Obstacle-Avoidance in Artificial Insects", IEEE Transactions on Neural Networks, Vol. 9, pp. 796-812.
19. Koza, J., Bennett III, F. H., Andre, D., and Keane, M. A. (1999). Genetic Programming III. Darwinian Invention and Problem Solving. Morgan Kaufmann.
20. Lindenmeyer, A. (1968), "Mathematical models for cellular interaction in development, parts I and II", Journal of Theoretical Biology, Vol. 18, pp. 280-315.
21. Miller, J. F. and Thomson, P. (2000), "Cartesian genetic programming", in Proceedings of the 3rd European Conference on Genetic Programming. LNCS, Vol. 1802, pp.121-132.
22. Nolfi, S., and Parisi, D. (1991) "Growing neural networks", Technical report PCIA-91-15, Institute of Psychology, CNR, Rome.
23. Siddiqi, A. A., and Lucas S. M. (1998), "A comparison of matrix rewriting versus direct encoding for evolving neural networks", in Proceedings of the 1998 IEEE International Conference on Evolutionary Computation, IEEE Press, pp. 392-397.
24. Sims K. (1994), "Evolving 3D morphology and behaviour by competition", in Proceedings of Artificial Life IV, pp. 28-39.
25. Vassilev V. K., and Miller J. F. (2000), "The Advantages of Landscape Neutrality in Digi-tal Circuit Evolution", 3rd International Conference on Evolvable Systems: From Biology to Hardware, Lecture Notes in Computer Science, Vol. 1801, Springer-Verlag, pp. 252-263.
26. Vassilev V. K. and Miller J. F., (2000), "Scalability Problems of Digital Circuit Evolu-tion", in Proceedings of the 2nd NASA/DOD Workshop on Evolvable Hardware, IEEE Computer Society, pp. 55-64
27. Wong M. L., and Leung K. S. (1996), " Evolving Recursive Functions for the Even-Parity Problem Using Genetic Programming", in Advances in GP, MIT Press, Vol. 2, pp. 221-240.
28. Yu, T. (1999), "An analysis of the Impact of Functional Programming Techniques on Ge-netic Programming, Ph. D. Thesis, University College London, UK.
29. Yu, T. and Miller, J. (2001), "Neutrality and the evolvability of Boolean function land-scape", in Proceedings of the 4th European Conference on Genetic Programming, Springer-Verlag, pp. 204-217.

Developmental Models
for Emergent Computation

Keith L. Downing

The Norwegian University of Science and Technology
Trondheim, Norway
keithd@idi.ntnu.no
http://www.idi.ntnu.no/~keithd

Abstract. The developmental metaphor has clear advantages for the design of physically-realizable artifacts, particularly when coupled with evolutionary algorithms. However, the embodiment of a developmental process in a purely computational system appears much more problematic, largely because embryogenesis evolved for the purpose of synthesizing 3-dimensional structure from a linear code, not for growing Universal Turing Machines.

This research considers possible models of computational problem-solving based on the 5 primary developmental stages: cleavage division, patterning, cell differentiation, morphogenesis, and growth. A specific developmental approach to the NP-Complete problem, vertex cover (VC), is discussed, as well as a general model of developmental computation based on a multicellular extension of PUSH [12], a new stack-based language designed specifically for evolutionary computation.

1 Introduction

While applied computer scientists hunt feverishly for the next *killera app*(application), basic researchers search for a new *killer an* (analogy). Biology has provided a steady stream of exceptionally useful metaphors, resulting in technologies such as evolutionary computation (EC)[6,8], artificial neural networks (ANN) [9], ant-colony optimization (ACO) [2], and artificial immune systems (AIS)[5]. One of the newest hot possibilities is natural development: the process by which a single cell gives rise to a fully-functioning multicellular organism.

The potential advantages abound for incorporating a developmental process into tasks such as automated design, particularly when an evolutionary approach is employed. The most convincing argument lies in the simple fact that natural genomes encode recipes for development, not blueprints for the 3-d layout of the final organism. Hence, the genome is a very compact description of a very complex product. Automated evolutionary design systems using a blueprint approach must resort to huge genomes for describing complex artifacts, whether hardware circuits, computer programs, simulated organisms, airplanes or even furniture. However, in many cases, these products have an organized hierarchical structure that is often replete with symmetries, redundancies and recursion. The difficulty of evolving such patterns in a monolithic blueprint genome scales

A.M. Tyrrell, P.C. Haddow, and J. Torresen (Eds.): ICES 2003, LNCS 2606, pp. 105–116, 2003.

exponentially with the size of the desired artifact, but concise recipe genomes can generate the appropriate structure with only a constant scaling factor, often zero, relating the sizes of genotypes to phenotypes.

However, the complexity of the natural developmental process is daunting, particularly to a computer scientist. One may argue that so too is evolution, and Holland [6] and others have masterfully tamed it for conventional use. But, the founding fathers of EC had a beautifully simple foundational abstraction, laid down by Darwin himself, on which to build their fortress: the triad of variation, heritability and selection (survival of the fittest). By instantiating these three factors in a myriad ways, EC has become an omnipotent tool for difficult search problems.

Development does boil down to one simple abstraction, that of the genome as a recipe for phenotype generation, which has already been widely exploited in EC. But delving a bit deeper in search of a slightly more detailed set of useful, transferable principles, the going gets much tougher. There are a few basic stages of development, as described below, but the utility of their analogs in artifact design tasks is far from obvious. Many basic developmental mechanisms, such as cell duplication, differentiation and migration, are exemplary tools for generating a structured, heterogeneous 3-dimensional object in vivo, but it is unclear whether they have legitimate utility in solving problems in silico.

2   Five Basic Stages of Development

Natural development can be characterized by five distinct (but temporally and spatially overlapping) processes: 1) cleavage division, 2) pattern formation, 3) morphogenesis, 4) differentiation, and 5) growth. The following is a brief summary of these processes, as detailed in [16].

In cleavage division, the original egg cell divides repeatedly, with each daughter cell receiving a copy of the mother cell's DNA plus half its biomass. The daughter cells do not grow before themselves dividing. Hence, after n rounds of division, an egg cell of mass M is replaced by 2^n cells of mass $M \div 2^n$. The function of this process thus appears to be the copying and distribution of the genome into many (nearly) identical cells, which can later differentiate, migrate and grow to form the heterogeneous phenotype. The embryo thus becomes a ball of cells called the *blastula*.

Although even a fertilized egg has slight asymmetries, the second process, pattern formation involves critical symmetry breaking to form the anterior-posterior (head to tail) and dorsal-ventral (back to front) axes. Chemical gradients created by certain signalling-center cells or *organizers* form these axes, an orthogonal set of which creates a coordinate system. Regional organizers and their gradients can also provide some of the finer asymmetries and details of the developing embryo, such as the formation of the different fingers of a hand. The three germ layers (*ectoderm, mesoderm and endoderm*), which are initially concentric, are also created during pattern formation.

Morphogenesis involves major changes in 3-dimensional shape of the embryo. In one morphogenic process, gastrulation, the three germ layers split and fold to form the mouth and gut cavities, thus transforming the blastula into a *gastrula*. In neurulation, the ectoderm folds to create the neural crest and tube, precursors to the vertebrae, limbs and spinal cord. Cell migration underlies most morphogenetic processes. This results from cellular changes in gene expression, which can, among other things, a) alter the surface chemistry of the cell, changing its repellant/attractive tendencies toward other cells, and b) modify the elastic properties of the cell membrane, making the cell more or less likely to change shape and/or move when pressed upon by other cells.

Cell differentiation is the process by which cells achieve different patterns of gene expression within the embryo. Some of these activity changes are reversible, but many are not. As development proceeds, more and more cells become *determined*: essentially stuck in an activity pattern. These attractors in activity space eventually gives rise to one of the many morphologically distinct cell types, such as a pancreatic cell or a pyramidal neuron.

There are at least 3 key mechanisms for differentiation: 1) induction, 2) asymmetric cell division, and 3) timing. Induction involves intercellular signalling, achieved by either direct contact between cells or via diffusion of chemical messengers into the extracellular fluid. In asymmetric cell divisions, the two daughter cells do not get exactly the same chemical constituents from the mother cell and hence may subsequently follow different paths of development. Finally, cells may have built-in chemical clocks that trigger changes in gene activity.

The final main developmental process, growth, is just an increase in size of the organism, with only minor changes in shape. This may result from the growth of individual cells, the replication of cells (with daughter cells growing to adult size before dividing) or the accumulation of extracellular material such as fluids or bone matrix. Induction and chemical timing mechanisms play key roles in controlling growth.

The sequencing, overlap and interaction between these processes varies across species, although the order of their initiation roughly follows the enumeration above. That is, the egg cell begins development with many cleavage divisions, followed by overlapping pattern formation, morphogenesis and differentiation. This results in a miniature organism which then grows to newborn size with visible external changes but only minor internal and external rearrangements.

3   Utility of Development for Evolutionary Design

From an artificial life (ALife) perspective, each developmental process is interesting and well worth simulating. For example, Japanese researchers [3,15] have achieved the emergence of several developmental stages from the primitive behaviors of individual cells: their metabolisms and adhesive properties.

However, from the perspective of automated evolutionary design, the elements of natural development stimulate interesting, but largely unjustified, analogies. For example, although there have been attempts to grow artificial neural network topologies using everything from simple, evolvable rules [7] to detailed neurogenetic-like processes [1], these methods cannot match the performance of less biological approaches to ANN evolution such as SANE [11] and NEAT [13].

The basic concept of simulated evolutionary developmental design (SEDD) is straightforward. Assume a generic evolutionary algorithm, in which the genome embodies a recipe for creation of the phenotype. The SEDD system will begin with a single module/cell containing the genome plus some other cell-specific properties, e.g., a vector of state variables such as size, location, activity level, age, etc. By executing the genome, the cell may alter its own state, send signals to other cells, reproduce (by splitting), or die. Over time, the SEDD will create a collection of cells that can be interpreted (by, for example, integrating information from each cell's state) as a testable phenotype, whether an ANN, an arithmetic circuit, a computer program, or a coffee table.

4   A Specific Problem: Vertex Cover

Consider the classic NP-complete problem of vertex cover (VC) [4]: given a constant integer k and an undirected graph consisting of a vertex set, V, and an edge set E, find a set of vertices, $V_s \subseteq V$, such that $size(V_s) \leq k$ and each edge in E has at least one of its two vertices in V_s.

To grow a VC solution, begin with a single cell whose internal state represents $v_i \in V$. The propensity for a cell to divide is proportional to the size of the edge set of v_i, so the seed cell should represent a well-connected vertex. These cleavage divisions would quickly lead to many v_i cells.

Inductive signalling from these cells could then cause problem-related differentiation by a simple procedure: v_i cells would emit inhibitor signals for any vertices in N_i, the immediate neighbors of v_i. Cells would then read gradients of inhibitors (and activators) and possibly change their own vertex as a weighted deterministic function of the signals. Cells could also have a *not-vertex* state, such as $\sim v_j$, and such cells would emit activation signals for any vertices in N_j.

Finally, cells could migrate up and down signal gradients for their vertex type. For instance, a cell expressing v_i ($\sim v_i$) would migrate toward increasing (decreasing) v_i activator and away from increasing (decreasing) v_i inhibitor, using a deterministic function based on weighted contributions from the two gradients. Over time, this might lead to relatively stable configurations with few conflicts between the ambient signals and the local cell types.

After a certain number of developmental steps, or after a certain degree of stability were attained, a potential solution, V_p could be read from the set of all cell vertices, with each $\sim v_i$ cell cancelling out one v_i cell.

To put this process under evolutionary control, the genome could encode an initial seed cluster of cells, a set of parameters for the weighted state-change and movement routines, or even a GP program with primitives for expressing activators/inhibitors, for migrating, etc.

Although the odds of this working successfully are slight, it illustrates that even a conventional problem could be attacked from a developmental perspective, in this case, one involving a) cleavage division, b) differentiation by induction, and c) morphogenesis by migration.

5   Multicellular Programs

Although tailoring a developmental process for a particular problem might prove fruitful for certain types of well-structured problems, a higher-level metaphor that links development to a general computational device would provide a flexible platform for embryonic problem-solving.

To fully exploit the developmental metaphor in a computational setting, we need a meaningful correspondence between living cells and computational units, as found in the Biowatch project [14]. Fortunately, the recent development of PUSH [12], a stack-based language designed for genetic programming [8], provides an ideal substrate for this analogy.

In PUSH, an (evolvable) genome encodes a list of primitive operations (in postfix form) that can act on a variety of data types. So a single program may include standard arithmetic, boolean, list and array operators. While standard GP must handle type mixing with strongly-typed programs [10] and their ensuing computational overhead, PUSH very elegantly solves the problem with multiple data stacks, one for each data type. Thus, when an operator needs arguments of particular types, PUSH simply pops them from the appropriate data stacks; the operator's result is then pushed onto the correct stack as well. PUSH even provides code stacks, so that programs can build and run code chunks on the fly.

PUSH is implemented in LISP, a language that treats data and programs equally in the sense that programs can be a) passed as arguments to functions, just like data, and b) constructed by standard data-manipulation routines from bits of numeric and symbolic data, just like any other data structure. So data can be processed directly or incorporated into other programs that process data, which is analogoues to living cells, where proteins and other compunds are both metabolic resources AND components of the machinery that performs metabolism, mitosis, RNA translation, etc. Hence, by running Lisp-like code in a population of weakly-communicating modules, one can achieve a degree of abstract biological realism.

5.1   Cellular PUSH Programs

A Cellular PUSH Program (CEPP) is a module that runs pieces of PUSH-like code and resides in a network containing other CEPPs. It communicates by pushing data values onto the stacks of neighboring CEPPs, migrates by swapping places with a neighbor, and reproduces by copying its genotype (a PUSH-like code chunk) into a newly-formed neighbor cell. The current CEPP prototype, running in Common Lisp, provides a basic feel for the language, as illustrated by the detailed examples below.

As in PUSH, CEPP relies on a type hierarchy to control the inputs and outputs to typed functions, making sure that they are popped from and pushed onto the proper stacks. This hierarchy has two main branches: *atom* and *structure*. Atoms are all single items such as booleans, numbers or symbols (with numbers being further partitioned into integers and reals). A structure is any data structure, such as an expression (a possibly-nested list) or an array. The two main subtypes of expression are *code* and *data*. A CEPP runs by popping and executing items from its code stack. Only the leaf types, i.e. most specific types, of the hierarchy have corresponding stacks.

CEPP also has a type stack, which represents a prioritized list of types used to disambiguate general function calls. For example, the function + takes two input arguments of type number and returns a number. When a + is invoked, CEPP checks the type stack to find the first occurence of a leaf subtype of number, whether integer or real. It then pops from and pushes to the appropriate stack when performing the addition. Unlike the other stacks, the type stack is never popped.

A CEPP colony consists of one or more CEPPs, all sharing the same genotypic code chunk. Simulation of the colony involves the repeated running of individual CEPPs for a fixed number of *execution* steps. The colony begins as a single seed cell, which may or may not multiply to form an interacting cell population.

Consider the following CEPP genotype:

```
( quote (integer 1 - real dup)
  quote (real +)
  integer dup 1 > if))
```

To test this CEPP, push the integer 8 and the real 13.0 onto their respective stacks. The genotype is then pushed onto the code stack, and CEPP's *do* command is invoked. This executes the top item on the code stack and pops it AFTERWARDS, thus facilitating self-modifying code.

CEPP uses postfix notation, but the quote command is an important exception. When reading the symbol *quote*, the interpreter sets a flag, telling it to push the next item onto an expression stack. The default type priority at the bottom of the type stack lists *code* as the highest priority expression. In this case, the next item is a list, so the sequence *(integer 1 - real dup)* is pushed onto the code stack. The second quote then forces *(real +)* to be pushed onto the code stack, which now contains:

```
(real +)
(integer 1 - real dup)
(quote (integer 1 - real dup) quote (real +) integer dup 1 > if)
```

The symbol *integer* is then pushed onto the *type* stack, and this instructs the next command, *dup*, to duplicate the top of the integer stack, which then consists of (8, 8), where the top item is leftmost. *dup* is useful for preserving the top of a stack in situations where an upcoming operator (in this case, *if*) will pop it.

Next, the integer 1 is pushed, and then 8 > 1 is computed, giving *true*, which is pushed onto the boolean stack. The *if* operation pops the code stack twice and the boolean stack once. Since the boolean is true, the second code chunk, (integer 1 - real dup), is executed. This dictates the subtraction of 1 from the top of the integer stack and then a duplication of the top of the real stack, which then reads (13.0, 13.0).

During the CEPP's alloted execution steps, if the code stack becomes empty, the genotype is re-pushed and executed again. Also, the CEPP's stack state is preserved between *time* steps, with interrupted code chunks being resumed on the next time step.

Hence, in the current example, the genotype is re-pushed and executed again, but this time with an integer stack of (7) and a real stack of (13.0 13.0). This process of decrementing the top integer and adding another 13.0 to the real stack continues until the top integer is 1 and the real stack contains eight 13.0's. Since $1 > 1$ is now false, each execution of the *if* in the genotype will cause the *(real +)* code chunk to be executed, eventually resulting in 104.0 (8 x 13.0) atop the real stack.

5.2   Computational Resources in CEPP

CEPP adds the concept of resources into the PUSH paradigm, such that every item that is pushed onto a stack needs to be created from the available resource pool, and every popped item replenishes the pool. Resources are of different types, corresponding to the leaf nodes on the CEPP type tree. For each descendent leaf subtype of the class *atom*, every instance of that leaf type requires one unit of the specific resource. For example, pushing a boolean onto the boolean stack requires one boolean unit of resource.

The leaf descendents of the class *structure* demand greater resources, essentially one unit for each atomic item and one unit of *glue* to bind it to the data structure. For example, with lists, the glue is similar to a LISP cons cells. Hence, the previous code chunk, *(integer 1 - real dup)*, is a structure that requires 1 integer resource unit, 2 type units (for *integer* and *real*), two symbol units (for − and *dup*) and 5 units of code glue.

Most primitive CEPP functions, such as the arithmetic, logical and list operations, pop input items from stacks and push their results onto the same or other stacks. In many cases, such as arithmetic, the resources freed up by the input pops are more than sufficient to cover the resource needs of the ensuing push. However, some operators such as the boolean comparators, require a resource that is not provided by the inputs. Hence, some operations can be blocked by unavailable resources, despite the presence of all necessary input arguments.

Also, any data item appearing in a code chunk requires additional resource in order to push it onto a stack. Thus, while executing (3 4 9.2 integer +), CEPP will require an additional 2 integer units, one real unit and one type unit in order to push the first 4 items onto their respective stacks. In short, when the contents of a code expression become active (via their presence on stacks), they drain resources. Also, the code chunk itself drains resources, since it too resides on a stack. Thus, although a recursive genotype can easily copy several versions of itself onto the code stack, using a *(code dup)* sequence, each copy requires resources, whose availability can thereby limit recursive depth.

This resource abstraction can be viewed as a computer memory in which certain finite areas are statically allocated for the different data types. Data on stacks represents biomass, in the sense that the complete stack state is a structured product of the ongoing computation, just as real biomass is the product of life. Similarly, code mirrors genetic material (DNA, RNA, etc.) in that it serves as a recipe for building this computational edifice.

5.3   Cell Division, Migration and Communication

To reproduce, a CEPP genotype must include the *repro* symbol. When executed, *repro* attempts to copy the genotype, using the CEPP's resources. If successful, a new CEPP is created and given a share of the mother's resources, where the amount is determined by the ratio of the top two elements on the integer stack. The mother and daughter then become neighbors in the network/colony.

The seed CEPP cell begins with resources that constitute the entire pool for the CEPP colony; no resources are ever created, only redistributed among stacks and CEPPs. This puts a natural restriction on the number of possible cell divisions and hence the colony's size.

The CEPP colony has a simple topology: each cell has a neighborhood, initially composed of its parent cell and all of its own children. However, cell migration, via the *mig* primitive, can quickly alter this arrangement. It causes a CEPP, C, to pop the integer stack and use that value as an index (modulo the neighborhood size) into its neighbor list. The chosen neighbor, N, and C then swap neighborhoods.

To exchange information, C can push items onto any of its neighbors' stacks by using the *pipe* operator, which commands the CEPP to pop the integer stack, again using the value to select a neighbor, N, which becomes the *piped-cepp* for C. Now, all future operations performed by C will take inputs from C's stacks and will use C's resources to push outputs onto N's stacks. Even operations like *dup* will duplicate the top element of a C stack and then push the copy onto the corresponding N stack. The *cut* operator signals a closing of the pipe and a return to normal internal processing. Also, the *pipar* operator assigns the parent as the piped-cepp.

5.4   A Developmental CEPP Example

Consider the following genotypic code string:

```
(cut code quote (real 1.0 - 10 1 repro integer 1 + 0 pipe dup 0.0
                          code quote (integer dup * data pipar conv))
             real dup 3.0 > if
cut   code quote (data pop mig real 2.0 *)
             data append dup length 4 > if
```

Skipping the deepest details, a functional overview suffices to illustrate the developmental process. The genotype essentially encodes two conditional actions:

reproduction and migration. A CEPP can reproduce only if a) $X > 3$, where X is the top element of its *real* stack, and b) it has the available resources to copy its genotype. By pushing 10 and 1 onto the integer stack prior to reproduction, the mother insures that she will retain 10/11ths of her resources after splitting. When the CEPP reproduces, it uses the sequence *(0 pipe)* to open a pipe to its daughter cell. It then copies the code sequence *(integer dup * data pipar conv)* onto the top of the daughter's code stack, along with a copy of N, the top element of the mother's integer stack. When the daughter begins running, it first executes this sequence, causing it to compute N^2. The *pipar* command then opens a pipe back to the parent, which the daughter uses to copy N^2 onto the top of the mother's *data* stack.

In short, the mother uses its children as integer-squaring subroutines. However, each also receives a copy of the mother's genome, so after squaring an integer, each daughter will push the genome onto its code stack and behave similarly to the mother. However, note that the mother pushes a 0.0 onto the daughter's *real* stack. This helps prevent the daughter's from satisfying the criteria for reproduction ($X > 3$), although the complexities of resource-restricted computation and CEPP interactions often lead to the pushing of positive reals onto the stacks of many different CEPPS. (Accurately predicting the behavior of these colonies is very difficult). In the simulation below, the seed CEPP begins with a 6.0 atop its real stack, giving it the opportunity to act as a structural organizer of the colony.

The other main activity, migration, occurs only if the CEPP has a list containing 5 or more elements (i.e., squared integers) on top of its *data* stack. The list is created by the append operation near the end of the genotype, which gradually compresses the elements of the data stack into a single list.

In a nutshell, this genome was handcrafted to allow the seed CEPP to create daughters, use them to produce squared integers, and then migrate to a new neighborhood (by swapping places with one of its daughters) for another round of reproduction.

Figure 1 shows the basic processor topology generated by a run of this genotype. The seed CEPP, cell 0, generates the network's central nodes (1, 2, 3, 12 and 25), which gradually unfold into the horizontal spine as the seed migrates from left to right. Static analysis of the genotype indicates that the seed should reproduce 5 times, thus receiving 5 squared integers, before migrating. This clearly occurs in the region of cell 12, where all 6 cells (10, 11, 12, 13, 18 and 19) are daughters of the seed. However, further to the left, the pattern indicates that a) the seed produced only a few offspring before migrating, and b) non-seed cells also managed to reproduce (although none migrated).

That the emergent pattern conflicts with the intended design of the genotype, which was meant to produce several groups of 5 non-reproducing cells, indicates the complexity of CEPP-colonies. Two factors confound our predictions: 1) resource-limited computation, and 2) intercellular stack pushing.

In resource-limited computation, a series of actions may be only partially carried out. For example, the sequence *(real dup 0.0 >)* is intended to test

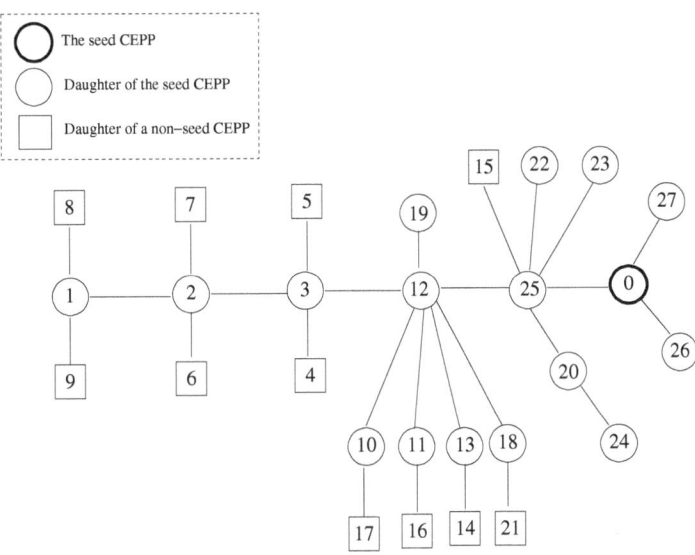

Fig. 1. CEPP topology generated largely by the reproduction and migration of the seed CEPP. Numbers indicate order of birth, while links connect neighboring cells in the emerging computational network.

whether the top real number, X, is positive. After the test, the real stack will appear as before, with X on top; and the boolean stack will hold the result of the test. However, if the CEPP lacks any boolean resource, then it will duplicate the X, push a 0, and then fail to perform the greater-than test. Thus, it fails to consume the two new stack items, and the real stack then contains (0, X, X).

The results of inter-cellular signalling can be even more severe, as whole new expressions can be pushed onto neighboring stacks, thus interrupting their planned action sequences. Like PUSH, CEPP is designed for use with evolutionary algorithms, so it will be interesting to see in future work whether evolving CEPPs include robust genotypes that can handle these unexpected stack changes, or whether the main prospects for robustness lie in the distributed colony itself.

6   Discussion

This paper shows that the developmental process can conceivably map to both specific problem solvers and more general computational systems, although neither has yet to be proven useful.

Celluar PUSH programs enable the emergence of a computational colony from a single program, thereby providing a developmental basis for problem solving. To assess the quality of this metaphor, return to the 5 basic stages of development.

1. Cleavage division occurs when a CEPP repeatedly copies its genotype, without performing too many other intermediate computations. This often occurs early in a run, when cells tend to have a lot of free resources.
2. Pattern formation can easily occur in CEPP, since cells are mobile and can communicate. In the previous example, the seed CEPP governs patterning by producing children and then swapping places with them (i.e., migrating) to form the horizontal spine.
3. Morphogenesis is clearly embodied in CEPP migration.
4. Differentiation occurs through the execution of both a) a CEPPs own code, which may force changes to its state and/or its code, and b) piped code from a neighbor. Also, a neighbor can alter behavior by piping atoms, such as the 0.0 sent by the above seed CEPP to its children to (help) prevent them from reproducing. Piping thus enables cell induction. The other two mechanisms for differentiation, asymmetric division and timing, are also coverd by CEPP, since a) cell splitting need not evenly distribute resources between mother and daughter, and b) any stack value can serve as a timing variable.
5. Growth is somewhat tricky: the concept of size in an ongoing computation is difficult to assess. One feasible alternative is the amount of data on the stacks, since this is information structured by the computation for the computation, in much the same way that biomass is structured material for supporting life.

Although the accuracy of the analogy is an interesting academic exercise, the important question is whether this type of development can improve problem solving. Our ongoing efforts to evolve CEPPs will hopefully shed more light on this issue.

One key factor, redundancy, separates software from the physical realities of both biology and hardware. In nature, redundancy accounts for large degrees of both homogeneity and heterogeneity at the cellular level. Similarly, physical devices, such as circuit boards, are replete with redundant components. In software, however, redundancy is preferably avoided, with single subroutines being created for common activities. Thus, evolving developmental computation may not provide a good piece of working software, but rather a computational topology that can later be translated into efficient software by modularization of redundant CEPPs.

In general, the proof is in the pudding, but we are still shopping for ingredients. Still, the fascinating capabilities of natural development inspires our further investigation, in the hopes that this, like evolution itself, holds a little something for everyone.

References

1. J. Astor and C. Adami, *A developmental model for the evolution of artificial neural networks*, Artificial Life, 6 (2000), pp. 189–218.
2. M. Dorigo, V. Maniezzo, and A. Colorni, *Ant system: Optimization by a colony of cooperating agents*, IEEE Trans. on Systems, Man, and Cybernetics–Part B, 26 (1996), pp. 29–41.

3. C. FURUSAWA AND K. KANEKO, *Complex organization in multicellularity as a necessity in evolution*, Artificial Life, 6 (2000), pp. 265–282.
4. M. GAREY AND D. JOHNSON, *Computers and Intractability: A Guide to the Theory of NP-Completeness*, W.H. Freeman and Company, San Francisco, 1979.
5. S. HOFMEYR AND S. FORREST, *Architecture for an artificial immune system*, Evolutionary Computation, 7 (2000), pp. 1289–1296.
6. J. H. HOLLAND, *Adaptation in Natural and Artificial Systems*, The MIT Press, Cambridge, MA, 2 ed., 1992.
7. H. KITANO, *Designing neural networks using genetic algorithms with graph generation system*, Complex Systems, 4 (1990), pp. 461–476.
8. J. R. KOZA, *Genetic Programming: On the Programming of Computers by Natural Selection*, MIT Press, Cambridge, MA, 1992.
9. W. MCCULLOCH AND W. PITTS, *A logical calculus of the ideas immanent in nervous activity*, Bulletin of Mathematical Biophysics, 5 (1943), pp. 115–133.
10. D. J. MONTANA, *Strongly typed genetic programming*, Evolutionary Computation, 3 (1995), pp. 199–230.
11. D. E. MORIARTY AND R. MIIKKULAINEN, *Forming neural networks through efficient and adaptive coevolution*, Evolutionary Computation, 5 (1997), pp. 373–399.
12. L. SPECTOR AND A. ROBINSON, *Genetic programming and autoconstructive evolution with the push programming language*, Genetic Programming and Evolvable Machines, 3 (2002), pp. 7–40.
13. K. STANLEY AND R. MIIKKULAINEN, *Evolving neural networks through augmenting topologies*, Evolutionary Computation, 10 (2002), pp. 99–127.
14. A. STAUFFER, D. MANGE, G. TEMPESTI, AND C. TEUSCHER, *A self-repairing and self-healing electronic watch: The biowatch*, in Lecture Notes in Computer Science, Y. Liu, K. Tanaka, M. Iwata, T. Higuchi, and M. Yasunaga, eds., vol. 2210, Berlin, 1998, Springer-Verlag, pp. 112–127.
15. H. TAKAGI, K. KANEKO, AND T. YOMO, *Evolution of genetic codes through isologous diversification of cellular states*, Artificial Life, 6 (2000), pp. 283–306.
16. L. WOLPERT, *Principles of Development*, Oxford University Press, New York, 2002.

Developmental Effects
on Tuneable Fitness Landscapes

Piet van Remortel, Johan Ceuppens, Anne Defaweux,
Tom Lenaerts, and Bernard Manderick

COMO – Department of Computer Science
Vrije Universiteit Brussel, Pleinlaan 2, 1050 Brussels, Belgium
{pvremort,jceuppen,adefaweu,tlenaert,bmanderi}@vub.ac.be

Abstract. Due to the scalability issue in genetic algorithms there is a
growing interest in adopting development as a genotype-phenotype mapping. This raises a number of questions related to the evolutionary and
developmental properties of the genotypes in this context. This paper
introduces the NK-development (NKd) class of tuneable fitness landscapes as a variant of NK landscapes. In a first part the assumptions
and choices made in defining a simplified model of development genomes
are discussed. In a second part we present results of the comparison of
NK and two variants of NKd landscapes. The statistical properties of
the landscapes are analysed, and the performance of a standard GA on
the different landscapes is compared. The analysis is aimed at identifying
the influence of the properties by which the landscapes differ. The results and their implications for the design of computational development
models are discussed.

1 Introduction

In the context of the evolution of complex phenotypes, such as electronic circuits
in Evolvable Hardware (EHW) or the evolution of neural networks, the scalability problem of genetic algorithms (GAs) has kept applications in a toy-problem
phase. A promising suggestion is to take the analogy with biology one step further and adopt a bio-inspired development mapping between the genotype and
phenotype [Kit98,BK99,HTvR,vRLM02]. This holds the promise of improved
modularity of the phenotype, relatively shorter genotypes for complex phenotypes, remapping of the search space to improve search performance (e.g. by
neutrality) etc. An obvious disadvantage is the low degree of evolvability of developmental representations: by their very nature, developmental genes depend
on high epistatic interaction in constructing the phenotype, hence rendering the
genome less modular and hard to evolve. Nothing restricts engineers of computational developmental genomes to construct genomes in such a way that these
effects can be avoided, or at least minimised. In order to do this, more indicative
knowledge about the fitness landscapes of developmental genomes is required.

A useful tool in studying the general properties of fitness landscapes are
Kauffman's NK landscapes [Kau93]. In this paper we propose NKd landscapes

A.M. Tyrrell, P.C. Haddow, and J. Torresen (Eds.): ICES 2003, LNCS 2606, pp. 117–128, 2003.
© Springer-Verlag Berlin Heidelberg 2003

as adapted NK landscapes to model basic principles of biological development. The properties of these landscapes will be analysed and compared to standard NK landscapes.

2 Simple Models for Developmental Genomes

2.1 Basic Modelling Assumptions

For computationally complex mechanisms such as multi-cellular development, only very simple models allow analysis of macroscopic properties such as fitness landscape structure. From the biological reality, as well as from known (more complex) computational models derived from it, a number of basic assumptions can be made. Although contestable in a strictly biological context, they will serve as basic assumptions for the landscape models investigated later on.

Assumption 1: The first assumption is the presence of *time* in the interpretation of the genome. This means that the action of reading and interpreting a gene is to be associated with a moment in time, rather than with the information content of the gene only. This induces an order of interpretation in the genome.
Assumption 2: The *order of interpretation* induced by the first assumption is *crucial*. In other words, genes expressed later in time exert their influence on a situation which is the result of genes activated earlier on during development. Due to the increasing amount of cells during development (due to cell division), this implies that 'early' genes have a more profound effect on the phenotype structure than 'late' genes.

Having derived the basic assumptions for our model, we will discuss these further in the context of the NK model for tuneable fitness landscapes.

2.2 NK Landscapes

In order to continue the discussion, we will first briefly introduce the basics of Kauffman's NK landscapes [Kau93]. NK landscapes are a class of tuneable fitness landscapes, whose two main parameters are N and K (hence the name NK). A genotype consists of N genes, where each gene's fitness contribution is a factor of it's own allele, and that of K others in the genome. These K influencing loci can be chosen according to the *nearest neighbours* scheme ($K/2$ left and right of the focal gene) or the *random neighbours* scheme. For the first the genome has periodic boundaries, for the latter it doesn't. The values of the fitness contributions of a gene is randomly assigned and fixed (between 0 and 1) for every combination of its value in the context of all the possible 2^K allele combinations of its influencing genes. The fitness of the total genome is the average fitness contribution of its genes. For details about fitness calculations and examples, we refer to the work of Kauffman[Kau93] or Altenberg[Alt97].
The parameter K tunes the ruggedness (or multi-modality) of the fitness landscape. In general this parameter is the main influence of the landscape's properties, while the neighbour scheme tends to be of minor importance.

The parameter K and the scheme used for choosing which loci provide the epistatic inputs per gene (nearest neighbours or random neighbours), decides on a combination of incoming and outgoing epistatic connections per gene. For NK landscapes, each gene has K incoming connections, and on average K outgoing connections. For future reference we define $I(i)$ and $O(i)$ as the functions designating the number of incoming and outgoing connections of gene i.

2.3   Development in NK Landscapes: NKd

The discussion will now focus on the implementation of the two assumptions mentioned above:

Implementation of Assumption 1 (order of interpretation due to time): We will arbitrarily consider this order to be from left to right in the genome. Genes on the left of the genome will now be considered 'early', while genes on the right are 'late'.

Implementation of Assumption 2 (increasing influence of earlier genes): This can be implemented in two ways:

a. by keeping K constant for all genes, but giving early genes more epistatic outputs (or *pleiotropy*) then late genes
b. by varying K along the genome, so that early genes have low K and late genes have higher K.

NKd(Pl) landscapes are inspired by option (a) above. The principles of this NK landscape variant was introduced by Kauffman in the context of a phylogenic model for studying generative entrenchment [Kau93]. In that context the model was called the 'hierarchical NK model'. In the context of development and the alternative model proposed below, we will adopt the term 'NKd(Pl)' here, indicating a NK-development landscape, where the pleiotropy of loci follows a designated non-uniform distribution, while the K value is constant.

NKd(K) landscapes are based on option (b) above. NKd(K) landscapes model the fact that early loci are hardly affected by any others, while late loci are affected by (almost) all previous loci in time. Exactly which loci affect a focal gene will be defined by the neighbour schemes used. More details on the neigbouring schemes will be discussed in section 3.

Summarising, both NKd variants are based on introducing an unbalance in the epistatic interactions of genes, based on their position on the genome. This is done in such a way that 'early' genes influence more other genes then 'late' genes. In NKd(Pl) landscapes, this unbalance is introduced by dictating the distribution of outgoing interactions, while NKd(K) landscapes are based on dictating the distribution of incoming interactions.

3   Implementation Details of NKd Landscapes

We will now take a closer look at the details of the NK and NKd landscapes that will be analysed in the next section. Experiments were conducted on NK,

NKd(Pl) and NKd(K) landscapes. Throughout all experiments N was fixed at 32. For the NK landscape experiments we can be brief: standard NK landscapes with K loci chosen as nearest or random neighbours for the interactions were used. We will now focus the discussion on the NKd landscapes used in the experiments.

3.1   NKd(K)

An NKd(K) landscape is specified by its $I(i)$ distribution (indicating K for every gene), and a scheme for choosing the K inputs per gene. NKd(K) landscapes can be compared using the *average* K value which results from the $I(i)$ distribution.

The different types of NKd(K) landscapes examined ranged over $I(i)$ distributions *exponential increasing* (model distribution: $I(i) = e^{0.1i}$), *linear increasing* (model distribution $I(i) = i$) and *root increasing* (model distribution $I(i) = \sqrt{i}$). The distributions are then scaled to obtain average K values of $\{0, 1, 2, 3, 4, 8, 15\}$. In deciding which genes influence which, four neighbour schemes were tested:

1. *random neighbours:* $I(i)$ influencing genes are chosen at random for gene i.
2. *nearest neighbours:* $\frac{I(i)}{2}$ loci are chosen left and right of gene i. The genome has periodic boundaries.
3. *left nearest neighbours:* $I(i)$ neighbours are chosen in an adjacent complete block to the left of gene i. The genome has periodic boundaries.
4. *development neighbours:* In this scheme the influencing genes are deterministically chosen as far left (or 'early') on the genome as possible, using up the maximum of $N - 1$ outgoing influences of the genes. Assignment of the incoming influences is done in a left-to-right manner on the genome.

3.2   NKd(Pl)

An NKd(Pl) landscape is specified by its value of K ($I(i) = K$ for all genes), and the shape of the pleiotropy-function $O(i)$. The different types of NKd(Pl) landscapes examined ranged over K values in $\{0, 1, 2, 3, 4, 8, 15\}$. The shapes of $O(i)$ adopted in the tests were *exponential decreasing* (model distribution: $O(i) = e^{-0.1i}$)), *linear decreasing* (model distribution $O(i) = -i + 31$) and *root decreasing* (model distribution $O(i) = \sqrt{31 - i}$). These basic shapes were scaled in order to obtain a correct match between the total amount of pleiotropic (sum of the values dictated by the distribution) and epistatic interactions ($N \times K$). A second requirement is that when scaling a distribution, the maximum value allowed for any gene's pleiotropy is $N - 1$. In deciding which genes influence which, two neighbour schemes were tested (nearest and leftnearest schemes are incompatible with imposing a $O(i)$ distribution):

1. *random neighbours:* $I(i)$ influencing genes are chosen at random for gene i, while obeying the maximum pleiotropy dictated by the $O(i)$ distribution of the influencing genes.
2. *development neighbours:* Similar as described above, but now the maximum pleiotropy of a gene is dictated by the $O(i)$ distribution.

4   Analysis of NKd Landscapes

4.1   Overview of Experiments

The three types of landscapes within the scope of this paper differ by tree parameters:

1. global landscape type (NK, NKd(Pl) and NKd(K))
2. neighbour or interaction schemes (random, nearest, left nearest and development)
3. interaction distributions (constant, linear, exponential and root shaped)

The goal of the experiments reported is to gain knowledge of the influence of these three global parameters on both the statistical properties of the landscapes, and the performance of a standard GA on them. The statistical properties of the landscapes were extracted from the data gained from two types of walks on the landscape:

1. *greedy walks:* Starting from a random starting genotype, continuously move to a random fitter neighbour at Hamming distance 1, until a local optimum is attained. The goal of the experiment is threefold: first of all, statistical information is gathered of fundamental properties of the landscape such as peaks and fitness values. Second, knowledge can be gained of the overall organisation of the landscape such as relative positions of the optima, basins sizes etc. In the experiments 100 greedy walks were performed on 100 random instances of every landscape type and variant.
2. *random walks:* Starting from a random genotype, continuously move to a random neighbour at Hamming distance 1, until a predefined number of steps is taken. The goal of this experiment is to measure the correlation length of the landscape [MdWS91,Wei90]. In the experiments 50 random walks of 1000 steps were performed on 50 random version of landscapes of each type.

The evolutionary experiments on the landscapes consisted of running a standard genetic algorithm over a fixed amount of generations and observing the results. The population size was fixed at 30, letting the GA run for 100 generations. Bit-mutation probability was 3% per bit and one-point crossover probability 30%. Roulette-wheel selection was used without elitism. For each landscape type, 30 GA runs were performed.

4.2   Classification of Obtained Information

From all these experiments the following data was obtained:

1. *Mean walk length to a local optimum* ($\overline{WL}$): Rugged landscapes tend to have a local optimum nearby, while smoother landscapes allow longer walks of continuous improvement.

2. *Mean and maximum fitness of optima encountered* ($\overline{F}$ and Max F): This data will be used to evaluate the performance of the GA. The mean fitness of local optima in highly rugged NK landscapes dwindles to 0.5 as a result of the 'complexity catastrophe' [Kau93].
3. *Massif Central effect* (MC): This refers to the effect that high peaks in the landscape tend to be close to each other, forming a 'Massif Central'. This effect is favourable, since it indicates that in the surroundings of a high optimum, other high optima can be found. This effect can be observed from the relation between the distance of encountered optima to the highest optimum and their relative fitness fitness. If the effect is present, the distribution of optima according to their distance to the global optimum will be sloped as a result of the unbalance in fitness. The effect can then be numerically observed from the slope of a linear fit to the dataset. The steeper the slope of the regression curve, the more prominent the effect. Figure 1(a) gives an example of a fitted plot that exhibits the 'Massif Central' effect.
4. *Unequal Distribution of basin sizes of optima* (UBS): The optima of smoother NK landscapes show a tendency to have large basins. This can be observed from a sloped distribution of optima fitness scores vs. the number of times the optimum is reached during the greedy walks. This effect is favourable, since optima with large basins of attraction are reached more easily by an optimisation algorithm. An indication of the trend of the effect can again be obtained by fitting the data and observing the slope of the regression curve. Figure 1(b) shows an example. On the plot, fitness values relative to the fitness of the best optimum found per landscape (vertical axis) are plotted against the number of times the optimum was reached during the greedy walks.
5. *Correlation length of the landscape* (CorrL) [MdWS91,Wei90]: The correlation length of a fitness landscape is the distance h from a genotype where the autocorrelation function $\rho(h) = 1/2$. The autocorrelation function is defined as:

$$\rho(h) = \frac{1}{\sigma^2} R(h) = \frac{R(h)}{R(0)} = \frac{E[(F_h - \mu_F)(F_0 - \mu_F)]}{E[(F_0 - \mu_F)(F_0 - \mu_F)]}$$

where σ^2 is the variance and $R(h)$ the auto-covariance of the stationary stochastic process. $E[x]$ is the expected value or mean, F_h the fitness at distance h from the observation point, and μ_F the mean fitness. The correlation length is an overall measure of ruggedness of the landscape.
6. *Relative fitness reached by a standard GA* (Evol. $\overline{F}$ and Ratio): The highest fitness obtained from performing a series of GA runs on a landscape was compared to the highest fitness ever encountered on the same type of landscape (but other random instances) during the greedy walks. This value can be used to compare the fitness that can be achieved by a GA in comparison to the (limited) knowledge available of the fitness achievable on that type of landscape.
7. *Number of evaluations to reach maximum fitness by a standard GA* ($\overline{nrEvals}$): The number of evaluations needed to reach the maximum score in a GA runs was recorded and the results averaged. This can give an indication of

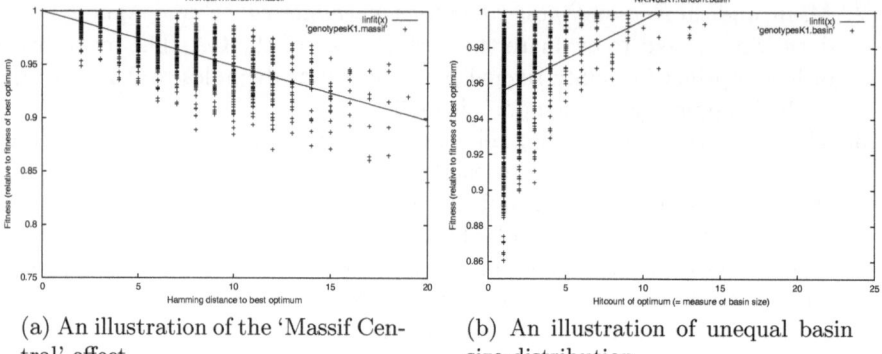

(a) An illustration of the 'Massif Central' effect

(b) An illustration of unequal basin size distribution

Fig. 1.

the number of generations a GA can increase the maximum fitness before converging. This can then be compared to the quality of the optimum (see previous item) and the overall properties of the landscape obtained from the walks.

4.3 Results and Observations

The experiments conducted in the scope of this paper generate huge amounts of data, that simply cannot be listed in a single paper. We will limit the discussion to a comparing low ($\overline{K} = 1$), medium ($\overline{K} = 3$) and higher ($\overline{K} = 8$) epistatic landscapes. The upper limit of $\overline{K} = 8$ also allows comparison of all variants of landscapes, since for example it is impossible to scale an exponential K distribution in such a way to obey all restrictions discussed before. In order to discuss the influence of landscape type, interaction distribution and neighbour scheme, we will list a selected number of results from landscapes that differ only by the property under investigation.

The results for the 'Massif Central' (MC) effect and the unequal basin size distributions (UBS) will be summarised by the slope a of the linear fit to the data defined as: $linfit(x) = ax + b$. See Figure 1 for an illustration. In contrast to the results for MC, the fitting slope of the UBS is only indicative, and not intended for detailed comparisons, due to the natural trend of the data. The x-range over which the data for the UBS measurements is defined, differs with the ruggedness of the landscape (the smoother the landscape, the less optima, and the broader the x-range). For this reason, the data for UBS was xx-scaled to a range of 100, before the fitting slope was calculated. If the effect was virtually non-existent from the experiments performed, the result was marked between brackets. The results of the experiments are shown in Tables 1, 2 and 3.

4.4 Comparing Landscape Types

In order to make a basic comparison between NK, NKd(Pl) and NKd(K) landscapes, we select the random interaction scheme for all landscape types and

Table 1. Results of comparison of landscape types.

Landscape	$\overline{K}$	I-Scheme	I-Distr.	WL	$\overline{F}$	UBS ($\times 10^{-3}$)	MC ($\times 10^{-3}$)	CorrL
NK	1	random	-	15.9 (3.5)	0.696 (0.035)	1.10	-5.08	8.32 (2.01)
	3	random	-	14.9 (4.2)	0.712 (0.032)	0.861	-3.83	5.14 (0.91)
	8	random	-	10.4 (3.6)	0.699 (0.029)	(0)	-2.28	2.21 (0.26)
NKd(Pl)	1	random	linear	15.3 (3.5)	0.692 (0.038)	0.689	-4.53	8.21 (1.85)
	3	random	linear	14.9 (4.2)	0.712 (0.031)	0.750	-3.46	5.33 (0.80)
	8	random	linear	10.2 (3.5)	0.700 (0.030)	(0)	-2.46	2.22 (0.27)
NKd(K)	1	random	linear	15.7 (3.6)	0.686 (0.038)	0.933	-4.52	8.26 (1.86)
	3	random	linear	14.3 (4.5)	0.701 (0.036)	1.41	-4.41	5.01 (1.05)
	8	random	linear	9.3 (3.6)	0.678 (0.034)	(0)	-3.27	2.29 (0.35)

Landscape	$\overline{K}$	I-Scheme	I-Distr.	Max F	Evol. $\overline{F}$	Ratio	$nrEvals$
NK	1	random	-	0.788	0.752 (0.014)	0.952	1558.8 (510.9)
	3	random	-	0.799	0.746 (0.031)	0.933	1702.8 (676.8)
	8	random	-	0.795	0.735 (0.016)	0.919	1386 (560.1)
NKd(Pl)	1	random	linear	0.813	0.719 (0.004)	0.884	1455 (567)
	3	random	linear	0.802	0.731 (0.017)	0.911	1254 (567)
	8	random	linear	0.805	0.727 (0.019)	0.903	996 (282)
NKd(K)	1	random	linear	0.799	0.732 (0.009)	0.916	1381.2 (451.8)
	3	random	linear	0.823	0.745 (0.015)	0.905	1531.2 (649.2)
	8	random	linear	0.811	0.731 (0.025)	0.901	1124.4 (483)

OBSERVATIONS (from Table 1):

1.1 Generally the results for all landscapes are comparable. This implies that keeping K constant is equivalent to using a skewed distribution of K that amount to the same average K (this describes the difference between NK and NKd(K)). Also, keeping K fixed for all genes, but enforcing an skewed distribution of pleiotropy values per gene is of no visible influence on the landscape (this describes the difference between NK and NKd(Pl)).

1.2 The NKd(K) landscape shows a slightly more prominent 'Massif Central' effect for higher (>1) $\overline{K}$.

linear the interaction distribution for both NKd landscapes. The results for $\overline{K}$ = 1,3 and 8 are shown in Table 1.

4.5   Comparing Interaction Schemes

In order to make a basic comparison between the influence of the different interaction schemes, we compare the different landscapes with identical settings, only changing the interaction scheme. We choose the linear interaction distribution for both NKd landscapes. The 'random' and 'development' interaction scheme were compared, since these are the only two available for the NKd(Pl) landscape. The results for $\overline{K}$ = 1,3 and 8 are shown in Table 2.

4.6   Comparing Interaction Distributions

In order to make a basic comparison between the influence of the different interaction distributions, we compare the different landscapes with identical settings,

Table 2. Results of comparison of interaction schemes.

Landscape	$\overline{K}$	I-Scheme	I-Distr.	WL	$\overline{F}$	UBS ($\times 10^{-3}$)	MC ($\times 10^{-3}$)	CorrL
NKd(Pl)	1	random	linear	15.3 (3.5)	0.692 (0.038)	0.689	-4.53	8.21 (1.85)
	3	random	linear	14.9 (4.2)	0.712 (0.031)	0.750	-3.46	5.33 (0.80)
	8	random	linear	10.2 (3.5)	0.700 (0.030)	(0)	-2.46	2.22 (0.27)
	1	dev	linear	15.9 (3.6)	0.698 (0.040)	0.592	-4.11	8.38 (1.78)
	3	dev	linear	14.3 (4.1)	0.703 (0.033)	0.798	-3.60	5.45 (1.15)
	8	dev	linear	9.5 (3.4)	0.675 (0.029)	(0)	-2.32	2.25 (0.31)
NKd(K)	1	random	linear	15.7 (3.6)	0.686 (0.038)	0.933	-4.52	8.26 (1.86)
	3	random	linear	14.3 (4.5)	0.701 (0.036)	1.41	-4.41	5.01 (1.05)
	8	random	linear	9.3 (3.6)	0.678 (0.034)	(0)	-3.27	2.29 (0.35)
	1	dev	linear	15.7 (3.2)	0.667 (0.046)	1.36	-3.36	7.98 (1.80)
	3	dev	linear	15 (3.8)	0.667 (0.039)	1.85	-4.81	4.91 (1.13)
	8	dev	linear	11.8 (3.8)	0.648 (0.037)	(0.967)	-3.73	2.26 (0.42)

Landscape	$\overline{K}$	I-Scheme	I-Distr.	Max F	Evol. $\overline{F}$	Ratio	$nrEvals$
NKd(Pl)	1	random	linear	0.813	0.719 (0.004)	0.884	1455 (567)
	3	random	linear	0.802	0.731 (0.017)	0.911	1254 (567)
	8	random	linear	0.805	0.727 (0.019)	0.903	996 (282)
	1	dev	linear	0.803	0.740 (0.007)	0.922	1401 (366)
	3	dev	linear	0.823	0.719 (0.010)	0.874	1590 (789)
	8	dev	linear	0.783	0.716 (0.021)	0.914	1419 (669)
NKd(K)	1	random	linear	0.799	0.732 (0.009)	0.916	1381.2 (451.8)
	3	random	linear	0.823	0.745 (0.015)	0.905	1531.2 (649.2)
	8	random	linear	0.811	0.731 (0.025)	0.901	1124.4 (483)
	1	dev	linear	0.764	0.632 (0.007)	0.827	1302 (309)
	3	dev	linear	0.774	0.675 (0.020)	0.872	1341 (336)
	8	dev	linear	0.773	0.700 (0.024)	0.906	1461 (516)

OBSERVATIONS (from Table 2):

2.1 In neither of the two tested NKd landscapes does the changing of the interaction scheme provide any substantial change in the landscape properties.

2.2 The NKd(K) landscape with 'development' interaction scheme and 'linear' distribution shows slightly longer average walk lengths to an optimum and relatively strong 'Massif Central' effect for higher (>1) $\overline{K}$. The effect for $\overline{K} = 1$ is less prominent then for the other landscapes examined.

only changing the interaction distribution. We choose the random interaction scheme for all landscapes. The 'exponential' and 'square root' interaction scheme were compared, since their shape differs the most. The results for $\overline{K} = 1, 3$ and 8 are shown in Table 3.

5   Discussion

General Experimental Conclusions: The general conclusion from the numerous experiments in the scope of this paper is that enforcing an skewed distribution on the incoming or outgoing epistatic interactions on NK-like landscapes is of hardly any influence to the overall landscape properties, and the performance of a standard GA on them.

Table 3. Results of comparison of interaction distributions.

Landscape	$\overline{K}$	I-Scheme	I-Distr.	WL	$\overline{F}$	UBS ($\times 10^{-3}$)	MC ($\times 10^{-3}$)	CorrL
NKd(Pl)	1	random	exp	15.9 (3.6)	0.693 (0.040)	1.125	-4.51	8.71 (1.95)
	3	random	exp	15.0 (4.2)	0.714 (0.033)	1.345	-4.21	5.22 (0.97)
	8	random	exp	10.5 (3.6)	0.700 (0.030)	(0.734)	-2.34	2.24 (0.26)
	1	random	sqrt	15.8 (3.5)	0.692 (0.031)	0.867	-4.44	8.11 (1.63)
	3	random	sqrt	14.7 (4.3)	0.712(0.032)	0.944	-3.52	5.02 (0.94)
	8	random	sqrt	10.3 (3.7)	0.699 (0.030)	(1.754)	-2.75	2.41 (0.29)
NKd(K)	1	random	exp	15.7 (3.7)	0.692 (0.035)	0.934	-4.29	8.08 (1.55)
	3	random	exp	14.2 (4.4)	0.694 (0.038)	1.372	-4.84	4.92 (0.97)
	8	random	exp	9.0 (3.7)	0.671 (0.038)	(0)	-3.94	2.57 (0.49)
	1	random	sqrt	15.8 (3.6)	0.693 (0.030)	0.821	-4.21	8.12 (1.72)
	3	random	sqrt	14.3 (4.1)	0.708 (0.031)	1.176	-3.55	5.03 (0.98)
	8	random	sqrt	10.0 (2.6)	0.691 (0.033)	(0)	-2.83	2.32 (0.32)

Landscape	$\overline{K}$	I-Scheme	I-Distr.	Max F	Evol. $\overline{F}$	Ratio	*nrEvals*
NKd(Pl)	1	random	exp	0.781	0.631 (0.013)	0.808	1479 (531)
	3	random	exp	0.813	0.734 (0.011)	0.903	1254 (513)
	8	random	exp	0.792	0.732 (0.026)	0.924	1368 (588)
	1	random	sqrt	0.783	0.734 (0.009)	0.937	1548 (531)
	3	random	sqrt	0.813	0.745 (0.027)	0.916	1728 (717)
	8	random	sqrt	0.793	0.726 (0.022)	0.916	1503 (663)
NKd(K)	1	random	exp	0.782	0.776 (0.017)	0.992	1494 (498)
	3	random	exp	0.809	0.731 (0.014)	0.904	1314 (447)
	8	random	exp	0.786	0.721 (0.015)	0.917	1308 (762)
	1	random	sqrt	0.767	0.709 (0.008)	0.924	1299 (327)
	3	random	sqrt	0.800	0.736 (0.021)	0.920	1407 (597)
	8	random	sqrt	0.826	0.736 (0.027)	0.891	1206 (636)

OBSERVATIONS (from Table 3):

3.1 Changing the interaction curve in any of the two NKd landscapes does not cause a considerable change in the landscape properties.

3.2 The NKd(K) landscape with 'random' interaction scheme and 'exponential' interaction distribution shows a relatively strong 'Massif Central' effect for higher (>1) $\overline{K}$.

Keeping the above in mind as a general conclusion, in some cases the NKd(K) landscape showed to have somewhat deviating properties from NK and NKd(Pl) landscapes. In a few cases the 'Massif Central' effect was slightly more prominent, and the mean distance to a local optimum was slightly higher then for other landscapes with comparable settings. Bear in mind that NKd(K) landscapes are the only landscape variant where the value of K per bit is not constant. Apart from that, the maximum pleiotropy per gene is not limited by an induced distribution (as in NKd(Pl)), but mere limited to $N-1$. This leads us to conclude that forcing a part of the genotype to lower K values then the average, renders the landscape somewhat smoother and more structured, while doing the same for the pleiotropy values of genes showed to have no effect.

Relation to Development in EC: Interpreting the results in a broader perspective we can conclude that the properties of fitness landscapes that were investigated depend on the average value of K. They generally do not depend on the way the values of K per gene are distributed on the genotype (the interaction distribution from the experiments), neither on the way the K influencing genes are chosen from the $N-1$ possible genes. For NK landscapes this conclusion has already been drawn Kauffman [Kau93].

Computational models of development mappings between genotype and phenotype often consist of rule-type genes. In order to develop a given genotype (which is a collection of rules), the rules are sequentially applied to a starting structure (equivalent to a zygote) to obtain a final phenotype (for an example see [vRLM02]). In this context our results imply that:

1. (NKd(Pl) $\equiv$ NK) If the fitness contribution of a rule-gene depends on the activation of a constant amount of other rule-genes, then the position on the genotype of the influencing rule-genes is of no influence on the evolutionary properties of the genotype. In the scope of the definition of our model this actually means that from an evolutionary point of view it is of no influence whether the influencing rule-genes have already been activated, or still have to be activated in the future.

2. (NKd(K) $\equiv$ NK) A genotype in which the fitness contributions of a rule-gene depend on a skewed distribution of on average $\overline{K}$ other rule-genes, then this genotype is evolutionary equivalent to a genotype in which each rule-gene depends on exactly K others.

Of course, these conclusions depend on the correctness of the proposed model. We conclude the discussion with some critical remarks to the proposed model, which will be the basis of future work: First of all, the fitness contributions of all genes are between 0 and 1, which is based on the NK model. In a developmental model this is unlikely to be the case, which suggests extending the model to genes with different maximum fitness contributions. Secondly, although it is interesting to use a model where it is possible to reason based on the fitness contribution of individual genes as a starting point, this is much harder in a computational development model, since the phenotype is built using a complex system of interacting rules. Hence numerically valuing the fitness contribution becomes virtually impossible. And as a third and final remark, it is an open question under which conditions the dependencies between genes follow a certain type of distribution (exponential, linear, ...). Possibly there is a link with phenotype modularity, but this is still under investigation and at this point highly speculative.

References

Alt97.   Lee Altenberg. NK landscapes. In Thomas Bäck, David B. Fogel, and Zbigniew Michalewicz, editors, *Handbook of Evolutionary Computation*, pages B2.7:5–10. Institute of Physics Publishing and Oxford University Press, Bristol, New York, 1997.

BK99.   P. J. Bentley and S. Kumar. Three ways to grow designs: A compari-
           son of embryogenies for an evolutionary design problem. In *Genetic and
           Evolutionary Computation Conference (GECCO '99)*, pages 35–43, 1999.
HTvR.   Pauline Haddow, Gunnar Tufte, and Piet van Remortel. Shrinking the
           Genotype: L-systems for Evolvable Hardware. In M. Iwata T. Higuchi M.
           Yasunaga Y. Liu, K. Tanaka, editor, *International Conference on Evolvable
           Systems 2001 (Tokyo) (LNCS 2210)*, Lecture Notes in Computer Science,
           pages 128–139. Springer Verlag.
Kau93.   Stuart A. Kauffman. *The Origins of Order*. Oxford University Press,
           Oxford, 1993.
Kit98.   Hiroaki Kitano. Building complex systems using developmental process: An
           engineering approach. *Lecture Notes in Computer Science*, 1478:218–229,
           1998.
MdWS91.   Bernard Manderick, Mark de Weger, and Piet Spiessens. The genetic al-
           gorithm and the structure of the fitness landscape. In *Proceedings of the
           4th International Conference on Genetic Algorithms*, pages 143–150, San
           Diego, CA, July 1991. Morgan Kaufmann.
vRLM02.   Piet van Remortel, Tom Lenaerts, and Bernard Manderick. Lineage and
           Induction in the Development of Evolved Genotypes for Non-Uniform 2D
           CAs. In *proceedings of the 15th Australian Joint Conference on Artificial
           Intelligence 2002*, Canberra, Australia, 2002.
Wei90.   Edward D. Weinberger. Correlated and Uncorrelated Fitness Landscapes
           and How to Tell the Difference. *Biological Cybernetics*, 63:325–336, 1990.

POEtic Tissue: An Integrated Architecture for Bio-inspired Hardware

Andy M. Tyrrell[1], Eduardo Sanchez[2], Dario Floreano[2], Gianluca Tempesti[2],
Daniel Mange[2], Juan-Manuel Moreno[3], Jay Rosenberg[4], and Alessandro E.P. Villa[5]

[1] University of York, York, UK,
Corresponding author: amt@ohm.york.ac.uk
[2] Swiss Federal Institute of Technology at Lausanne, Lausanne, Switzerland
[3] Technical University of Catalunya, Barcelona, Spain
[4] University of Glasgow, UK
[5] University of Lausanne, Lausanne, Switzerland

Abstract. It is clear to all, after a moments thought, that nature has much we might be inspired by when designing our systems, for example: robustness, adaptability and complexity, to name a few. The implementation of bio-inspired systems in hardware has however been limited, and more often than not been more a matter of artistry than engineering. The reasons for this are many, but one of the main problems has always been the lack of a universal platform, and of a proper methodology for the implementation of such systems. The ideas presented in this paper are early results of a new research project, "Reconfigurable POEtic Tissue". The goal of the project is the development of a hardware platform capable of implementing systems inspired by all the three major axes (phylogenesis, ontogenesis, and epigenesis) of bio-inspiration, in digital hardware.

1 Introduction

The implementation of bio-inspired systems in silicon is quite difficult, due to the sheer number and complexity of the biological mechanisms involved. Conventional approaches exploit a very limited set of biologically-plausible mechanisms to solve a given problem, but often cannot be generalized because of the lack of a *methodology* in the design of bio-inspired computing machines. This lack is due to the heterogeneity of the hardware solutions adopted for bio-inspired systems, which is itself due to the lack of *architectures* capable of implementing a wide range of bio-inspired mechanisms.

The aim of this paper is to present some of the preliminary ideas developed in the framework of a new research project, called "Reconfigurable POEtic Tissue" (or "POEtic" for short), recently started under the aegis of the European Community. After a short introduction to the POE model for the design of bio-inspired hardware, the paper will present an outline of the POEtic project and of the main ideas behind the development of the tissue, to finish with some of the preliminary results in the form of an overall architecture designed for the implementation of bio-inspired systems.

A.M. Tyrrell, P.C. Haddow, and J. Torresen (Eds.): ICES 2003, LNCS 2606, pp. 129–140, 2003.

2 The POE Model

The goal of the POEtic project is the

> "... *development of a flexible computational substrate inspired by the evolution-ary, developmental and learning phases in biological systems.*"

Biological inspiration in the design of computing machines finds its source in es-sentially three biological models [1]: *phylogenesis* (P), the history of the evolution of the species, *ontogenesis* (O), the development of an individual as orchestrated by his genetic code, and *epigenesis* (E), the development of an individual through learning processes (nervous system, immune system) influenced both by the genetic code (the innate) and by the environment (the acquired). These three models share a common basis: the genome.

What follows is a very concise and simplified description of the three models and of their application to the conception of electronic hardware. A much more detailed analysis of the three models and of existing hardware implementations, in the form of internal project reports, is available online on the POEtic project website [2].

2.1 Phylogenesis

On the phylogenetic axis we find systems inspired by the processes involved in the evolution of a species through time, i.e. the evolution of the genome. The process of evolution is based on alterations to the genetic information of a species, occurring through three basic mechanisms: *selective reproduction, crossover* and *mutation*.

These mechanisms are, by nature, non-deterministic. This represents both their strength and their weakness, when applied to the world of electronics. It is a strength, because they are fundamentally different from traditional algorithms and thus are potentially capable of solving problems which are intractable by deterministic ap-proaches. It is a weakness, because computers are inherently deterministic (it is very difficult, for example, to generate a truly random number, a basic requirement for non-deterministic computation, in a computer).

Even with this disadvantage, algorithms that exploit phylogenetic mechanisms are carving themselves a niche in the world of computing. These algorithms, commonly referred to as evolutionary algorithms (a label that regroups domains such as genetic algorithms [3], evolutionary programming [4], and genetic programming [5]), are usually applied to problems which are either too ill-defined or intractable by determi-nistic approaches, and whose solutions can be represented as a finite string of symbols (which thus becomes the equivalent of the biological genome). An initial, random population of individuals (i.e., of genomes), each representing a possible solution to the problem, is iteratively "evolved" through the application of mutation and cross-over. The resulting sequences are then evaluated on the basis of their efficiency in solving the given problem (fitness function) and the best solutions are in turn evolved. This approach, is not guaranteed to find the best possible solution to a given problem, but can often find an "acceptable" solution more efficiently than deterministic ap-proaches.

It appears, then, that the phylogenetic axis has already provided a considerable amount of inspiration to the development of computer systems. To date, however, its impact has been felt mostly in the development of software algorithms, and only marginally in the conception of digital hardware. Koza et al. pioneered the attempt to apply evolutionary strategies to the synthesis of electronic circuits when they applied genetic algorithms to the evolution of a three-variable multiplexer and of a two-bit adder [5]. Also, evolutionary strategies have been applied to the development of the control circuits for autonomous robots [6], [7]. Other research groups are active in this domain [8], [9] however, technical issues have posed severe obstacles to the development of evolvable electronic hardware [10]. A possible solution for the POEtic project is given in [11].

2.2 Ontogenesis

The ontogenetic model concerns the development of a single multi-cellular biological organism. This process exploits a set of very specific mechanisms to define the *growth* of the organism, i.e. its development from a single mother cell (the *zygote*) to the adult phase. The zygote divides, each offspring containing a copy of the genome (cellular division). This process continues (each new cell divides, creating new offspring, and so on), and each newly formed cell acquires a functionality (e.g., liver cell, or epidermal cell) depending on its surroundings, i.e., its position in relation to its neighbors (cellular differentiation).

Cellular division is therefore a key mechanism in the growth of multi-cellular organisms, impressive examples of massively parallel systems: the 6×10^{13} cells of a human body, each a relatively simple elements, work in parallel to accomplish extremely complex tasks (the most outstanding being, of course, intelligence). If we consider the difficulty of programming massively parallel computers, biological inspiration could provide some relevant insights on how to handle massive parallelism in silicon.

A fundamental feature of biological organisms is that each cell contains the blueprint for the entire organism (the genome), and thus can potentially replace any other cell [12]: no single cell is indispensable to the organism. In fact, cells are ceaselessly being created and destroyed in an organism, a mechanism at the base of one of the most interesting properties of multi-cellular organisms: healing.

The ontogenetic axis has been almost completely ignored by computer scientists, despite a promising start in the fifties with the work of John von Neumann, who developed a theoretical model of a universal constructor, a machine capable of constructing any other machine, given its description [13]. Given a description of itself, the universal constructor can then self-replicate, a process not unlike cellular division.

Unfortunately, electronic circuits in the 1950s were too primitive to allow von Neumann's machine to be realized, and the concept of self-replicating machines was thus set aside for a long time. Probably the main obstacle to the development of self-replicating machines was the impossibility of physically creating self-replicating hardware. In fact, such a process requires a means to transforming information (i.e. the description of a machine) into hardware, and such a means was unpractical until the introduction of programmable logic circuits (FPGAs), which provide a partial solution to the problem.

The difficulty of implementing processes such as growth in silicon has limited research on this model. The Logic Systems Laboratory at EPFL began to investigate ontogenesis of electronic circuits with its Embryonics project [14]. Another aspect of ontogenesis, morphogenesis (that is, the generation of the *shape* of an organism) has been the subject of some research [15], [16], and recently the issue of *development* is attracting growing attention as a source of inspiration for artificial systems [17]. More details for this project can be found in [18].

2.3 Epigenesis

The human genome contains approximately 3×10^9 bases. An adult human body contains around 6×10^{13} cells, of which approximately 10^{10} are neurons, with 10^{14} connections. In this case, and in many complex organisms, the genome cannot contain enough information to completely describe all the cells and synaptic connections of an adult organism.

There must therefore exist a process that allows the organism to increase in complexity as it develops. This process can be labeled *epigenesis*, and includes the development of the nervous, immune, and endocrine systems.

Epigenetic mechanisms have already had considerable impact on computer science, and particularly on software design, notably through the concept of learning. The parallel between a computer and a human brain dates to the very earliest days of the development of computing machines, and led to the development of the related fields of artificial intelligence [19] and artificial neural networks.

All living organisms interact with their environment and are capable of responding to sensory inputs. In many cases this behaviour is learnt over a period of time, after which a specific stimulus will trigger the same response each time, sometimes the response may depend on the context of the stimulus. It is now recognised that such behaviour is controlled by neurones or nerve cells and their interactions. The majority of neurones underlying learning are spiking neurones – they communicate through neuronal spike trains. In this project we will focus on the use of spiking (or pulsed) neuronal models to implement learning [20]. These devices are based on leaky integrate-to-threshold and fire neurone models, we will adapt these for implementation in hardware. Learning in spiking neural networks usually takes inspiration from Hebb's ideas, some recent work has suggested that learning in spiking neural networks may be possible using very coarse grained synaptic weightings [21], which are particularly suited to hardware implementation. A view on the way forward for the POEtic project is given in [22].

A more recent addition in the general area of hardware systems and epigenetic processes are artificial immune systems [23]. Here the sophisticated mechanisms associated with "fault tolerance" in nature have been adapted for electronic hardware system designs.

3 The POEtic Project

While each of the three models, taken separately, has to a greater or lesser extent been used as a source of inspiration for the development of computing machines, their amalgamation into a single system is a challenge yet to be met. The aim of the POEtic project [2] is the development of a computational substrate optimized for the implementation of digital systems inspired by all of the three above-mentioned models.

The POEtic tissue is a multi-cellular, self-contained, flexible, and physical substrate designed to interact with the environment, to develop and dynamically adapt its functionality through a process of evolution, growth, and learning to a dynamic and partially unpredictable environment, and to self-repair parts damaged by aging or environmental factors in order to remain viable and perform similar functionalities.

Following the three models of bio-inspiration, the POEtic tissue will be designed as a three-layer structure (figure 1):

- The phylogenetic model acts on the genetic material of a cell. Each cell can contain the entire genome of the tissue. Typically, in the architecture defined above, it could be used to find and select the genes of the cells for the *genotype layer*.
- The ontogenetic model concerns the development of the individual. It should act mostly on the *mapping or configuration layer* of the cell, implementing cellular differentiation and growth. In addition, ontogenesis will have an impact on the overall architecture of the cells where self-repair (healing) is concerned.
- The epigenetic model modifies the behavior of the organism during its operation, and is therefore best applied to the *phenotype layer*.

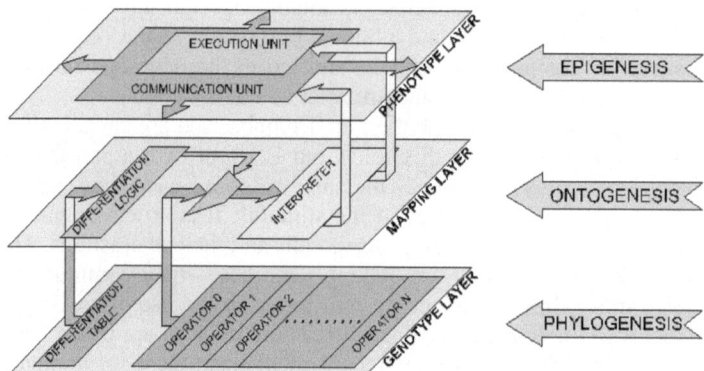

Fig. 1. The three organizational layers of the POEtic project.

Defining separate layers for each model has a number of advantages, as it allows the user to decide whether to implement any or all models for a given problem, and lets the structure of each layer to be adapted to the model. This adaptability, which will be detailed in the next section, is achieved by implementing the cells on a *molecular substrate*, in practice a surface of programmable logic.

Beside this combination of the three main axes of biological self-organization, the proposed tissue presents two more innovative hardware-oriented aspects:

- a layered hardware structure that matches the three axes of biological organization;
- an input/output interface with the external world that allows each cells to perceive and modify its environment when and where necessary.

4   POE Integration

There are at least two advantages in a circuit that integrates phylogenesis, ontogenesis, and epigenesis. The first is that each of the corresponding algorithms can be optimised for a particular class of problems. The second advantage comes from the combination of the different self-organization modalities. For example, instead of combining evolutionary algorithms with off-the-shelf learning algorithms, it will be possible to genetically encode and evolve the rules by which the phenotype plane of the tissue can dynamically adapt to external or internal perturbations without need of additional evolution. It has been shown that in the case of simulated neural networks, by evolving the adaptation mechanisms instead of a fixed configuration pattern, the evolutionary circuits cannot capitalize upon genetically determined solutions that fit a particular environment, but are rather forced to evolve the ability to develop on the fly the structure required by their specific environment [24]. This is quite interesting because it potentially offers a solution to the strong dependency of conventional evolved circuits to the evolutionary environment.

4.1 Cellular Structure

In contrast to conventional cellular electronic devices (such as Field Programmable Gate Arrays), each cell in the POEtic tissue contains the entire genome for the whole tissue. This solution, which is also closer to biological reality, increases the robustness of the system, allowing individual cells to reconfigure their function for growth, self-replication, and self-repair at a local level.

The ontogenetic concept of growth has been largely ignored in the past in the development of bio-inspired systems. The reason is obvious: growth implies the "creation" of new material during the lifetime of an organism (and therefore during the operational life of a machine), a feature which remains impossible to this day in the world of electronic circuits.

However, while the silicon substrate of an electronic circuit cannot be modified, a reconfigurable circuit like the POEtic tissue allows its function to be defined after fabrication, and thus potentially modified at runtime. The tissue allows the realization of a growth process based on information rather than matter.

The process of growth is based on cellular division and on cellular differentiation. From the standpoint of both biological organisms and electronic circuits, these two processes are made possible by the existence of the same genome inside each cell. Despite their structural similarity, each cell can specialize to perform a unique function by "executing" only a specific part of the genetic program depending on the cell's location in the organism. This differentiation allows multi-cellular organisms, including the POEtic tissue, to display behaviors of a greater complexity than those of single-cell organisms.

Finally, biological multi-cellular organisms are capable of surviving an astounding amount of damage. The "fault tolerance" of biological organisms is based on a set of extremely complex systems ranging from the organism-wide immune system to single-cell error correction. The ontogenetic features of the tissue will include a hierarchy of fault-tolerance mechanisms, acting "behind the scenes" with little or no effect on the operation of the machine.

4.2 Layered Organization

The structure in figure 1, naturally accommodates the building blocks and processes that regulate biological life. The genetic material specifying the functionality of the circuit is stored in the genotype layer, the rules and mechanisms of gene expression that map the genotype into a phenotype are stored and executed in the configuration layer, and the resulting function performed by each cell is implemented and executed in the phenotype layer.

The explicit distinction in three layers allows the study and implementation of complex genotype-to-phenotype mappings that can exploit growth and reconfiguration properties of the tissue, reduce the genome length, and eventually improve the evolvability of the tissue.

The mapping between genotypes and phenotypes is mainly a one-to-one function in artificial evolutionary systems. This relatively straightforward strategy does not capture the complex, but highly regulated, mechanisms of gene expression that characterise the development of biological organisms and barely supports the creation of genetic building blocks that could be exploited by crossover operators. At the same time, since in one-to-one mappings the length of the genotype is proportional to the complexity of the phenotype, large structures such as neuronal circuits are difficult to evolve. POEtic tissue will provide a substrate that naturally supports the implementation in the configuration layer of smarter and more biologically inspired mechanisms of gene expression. This structure, together with the cellular organization of the tissue, requires the evolutionary mechanisms that go beyond conventional genetic algorithms in that they capitalize and leave more space to ontogenetic and epigenetic mechanisms of dynamic self-organization of the tissue.

Finally, the presence of three distinct hardware layers allows the implementation of different mechanisms of self-organization in isolation without necessarily integrating them all. For example, one may decide to implement a given neuronal structure and its learning mechanisms on the circuit without going into evolution and gene expression. Similarly, another may wish to implement conventional electronic functions and exploit only the self-configuration and self-repair properties of the circuit without resorting to evolution and learning. This flexibility of the proposed circuit will indeed be exploited in the project to study the different aspects of self-organization and gradually integrate them all in the final POEtic tissue.

5 A Bio-inspired Architecture

Adaptation and specialization are key features of all biological organisms. As a consequence, no bio-inspired architecture can be rigid, but must rather be able to adapt and specialize depending on the desired application. The POEtic tissue will exploit reconfigurable logic to the fullest extent to allow for these characteristics: the bio-inspired machines will consist of two-dimensional arrays of *cells*, where each cell is a small, fully reconfigurable, processing element.

The role of the tissue is to provide a *molecular substrate* for the implementation of the cells: as organic cells are constituted by molecules, so artificial cells will be constituted by the programmable logic elements of the POEtic circuit. The three layers of the POEtic systems (figure 1) will all be implemented on this substrate.

Integrating reprogrammability in an architecture, however, is far from trivial. The heterogeneity of the applications that can potentially be implemented poses considerable design problems, particularly for the control of the processing elements. A novel kind of architecture needs to be developed to fully exploit the potential of reconfigurable logic for bio-inspired systems.

The three-layer architecture proposed in the POEtic tissue, together with its implementation on the molecular substrate, allows the hardware structure of each layer
to be adapted to the bio-inspired mechanism that it implements. What follows is an
outline of a plausible structure for each of the three layers, keeping in mind that the
reprogrammability of the tissue will allow any or all of these layers to be essentially
silent, should the bio-inspired mechanism concerned not be required.

5.1 The Genotype Layer

The genotype layer (figure 2) contains the genetic information of the organism, which
consists essentially of:
- A *set of operators*, which define the set of all possible functions for the cell. The
  quantity and functionality of the operators depends essentially on the application.
  For example, if the phenotype layer implements an array of neurons, one could
  imagine that the operators will consist of the internal parameters for the neuron, together with a set of initial connection weights. Instead, if the phenotype implements, say, a sound processing surface, the operators will consist of different filters
  and sound generation functions.

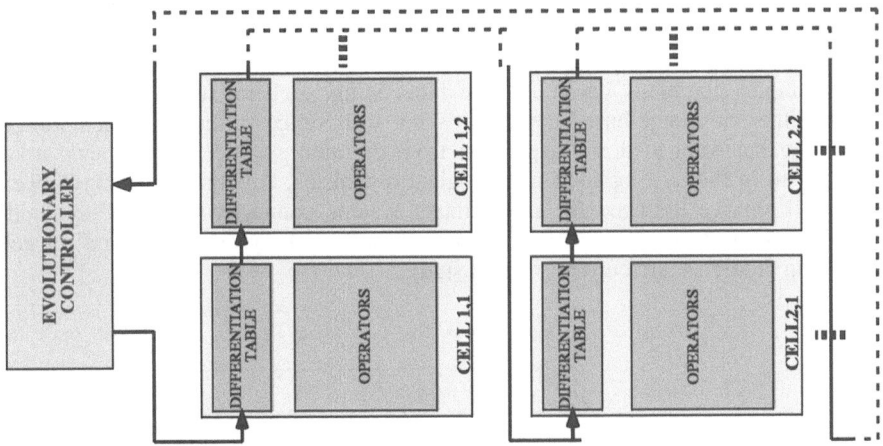

Fig. 2. The genotype layer in an array of cells and the external controller for evolution.

- A *differentiation table*, used to determine which operators will be implemented in
  which cell. The table contains a compact, perhaps redundant, representation of the
  operators to be implemented by the cells. The main advantage of maintaining a
  separate table storing this information is its *size*: the table will represent an extremely compact representation of the structure of the system, and evolutionary
  mechanisms in hardware will be applied to the table, rather than to the operators
  themselves. The differentiation table also plays an important role in the mapping
  layer, as described below.

The key feature of the genotype layer lies in the presence of a set of *dedicated connections*, used to provide user access to the differentiation tables of all the cells in the

system. The advantage of this kind of access lies in the presence of an *on-chip micro-controller*, whose function is to access the tables and to manipulate their contents through a set of user-defined evolutionary mechanisms.

5.2 The Configuration or Mapping Layer

The configuration layer of the POEtic systems is designed to implement the processes of cellular differentiation and cellular division (growth), the bases of the ontogenetic model (the fault-tolerance aspect of the model is also partially present in this layer, but is in general distributed throughout the layers of the array).

In its simplest conception (figure 3), the function of the configuration layer is to select which operator will be implemented by the cell. Each cell must be able to identify its position within the array. A set of coordinates, incremented in each cell and transmitted along the horizontal and vertical axes, is one approach. Once the cell has established its position, it can use this to select from the differentiation table which operator to express. The operator will then be interpreted to generate the appropriate control signals for the phenotype layer.

However, this approach is only a very basic solution to achieve cellular differentiation, and does not exploit many of the most interesting features of the development process.

For example, the solution of figure 3 is limited to a *one-to-one mapping* between genotype and phenotype. From an evolutionary standpoint, however, this approach imposes considerable disadvantages in the application of phylogenetic mechanisms. More complex mappings, based on redundant coding, are being researched.

Also, the coordinate-based approach is not well suited to the implementation of processes that want to draw inspiration from the natural process of *growth*, a key mechanism in the development of biological organisms. Automatic routing between cells will remove the need for a coordinate system, connectivity will be symbolic. More complex differentiation mechanisms, based on L-systems [16] and on cell-signalling gradients, are currently under study.

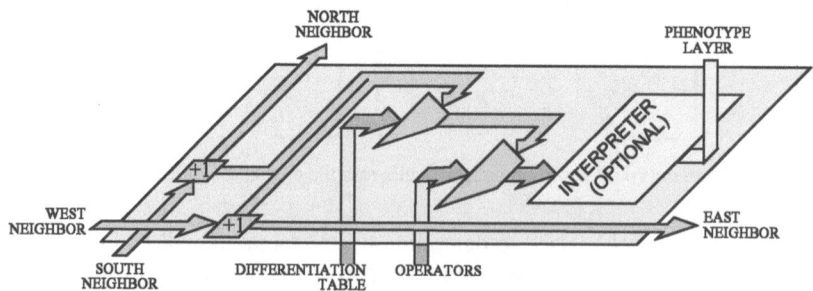

Fig. 3. A possible structure for the configuration or mapping layer.

5.3 The Phenotype Layer

The phenotype layer is probably the most application-dependent layer of the three. If the final application is a "conventional" neural network, the phenotype layer of the cell will simply consist of an artificial neuron. The architecture of the neuron is a choice left to the user. The POEtic project will concentrate on *spiking neurons* but this choice does not imply that the tissue will be limited to such a model.

However, since the POEtic tissue is meant to allow the implementation of bio-inspired systems that do not necessarily involve exclusively neural-like cells, a more general-purpose phenotype architecture will be required.

A simplified layout of a candidate architecture of the type that is being investigated in the POEtic project is shown in figure 4. It consists of two main units:

- An execution unit, consisting of a set of application-dependent *resources* (e.g., adders, and counters). The resources are defined by the user at design time, and are accessed through a set of input and output ports.
- In view of the massive amounts of connectivity of bio-inspired systems, the communication unit will probably have to rely on *serial communication*. The communication unit is seen by the execution unit as just another resource, with an input and output port.

This kind of architecture has two main advantages: it can support specialisation and adaptation through the modification of the resources depending on the application and it can be integrated in a design environment easily.

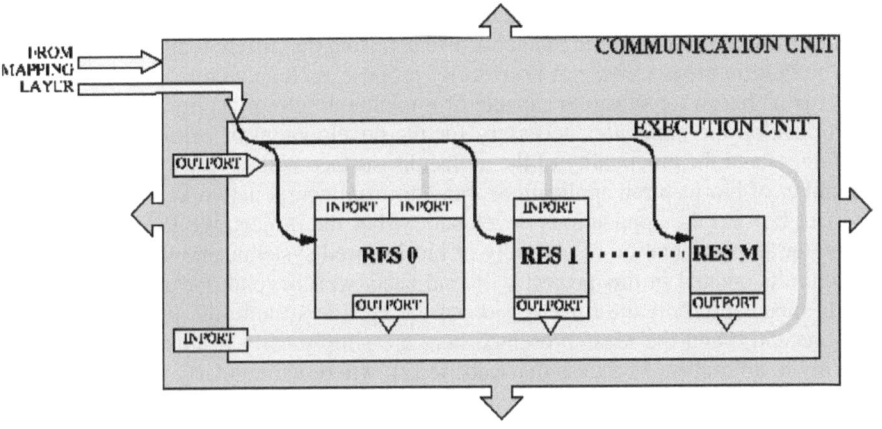

Fig. 4. Outline of a possible architecture for the phenotype layer.

5.4 A Design Methodology

The definition of three separate layers for the different models of bio-inspiration has the advantage of being able to select which of the models are to be used for a given problem. This feature is made possible by the complete programmability of the systems: all the three layers are implemented on the same substrate, a programmable digital logic circuit that incorporates dedicated features for all three models. The implementation of the layers in programmable logic adds the possibility of adapting the architecture of the system to the application, a flexibility that will allow the definition of an integrated design environment. Such an environment should include a *customizable design flow* allowing the user to concentrate on the desired approaches. A typical application-driven design flow for a bio-inspired system could then be:

- Based on the application, the user defines which models will be implemented in the system. The user then accesses a *library of operators* and selects those that will be required by the application. The phylogenetic model can be applied at this stage both to the selection of the functions and to the evolution of novel functions to be added to the library. The presence of the epigenetic model is defined at this stage by selecting functions capable of implementing learning processes.
- The environment will then integrate the desired functions into a *totipotent cell*, that is, a processing element capable of realizing any of the functions. This step represents the implementation of the ontogenetic model. Other optional ontogenetic features, such as growth and self-repair, can be added automatically at this stage.
- Finally, the design environment assembles the cells into a two-dimensional array, integrates the user-specified I/O modules, and generates a configuration for the programmable logic array that will implement the system.

6 Conclusions

Systems inspired by biological mechanisms are carving themselves a niche in several areas of computation, but the solutions adopted often cannot be generalised for the lack of a universal architecture capable of integrating the different approaches.

The POEtic project does not necessarily propose such an architecture: what it proposes is an hardware substrate capable of implementing bio-inspired systems. Its aim is to develop the *molecules* necessary for the development of cellular systems. The usefulness and the practicality of the molecular surface will obviously be validated on a number of bio-inspired applications, ranging from neural networks to music synthesis. But beyond the applications developed within the project, the POEtic tissue will allow the implementation of a variety of bio-inspired systems, and it is our hope that the circuits created in this project will find users well beyond our group. This paper by no means presents the final version of the POEtic system: the project has started only recently, and the ideas presented here are simply early results of our research. We invite all readers to access our website [2], where we regularly post the progress reports and deliverables of the project.

Acknowledgements

The authors would like to acknowledge the contribution of all the other members of the POEtic project, who all contributed to the material presented herein. This project is funded by the Future and Emerging Technologies programme (IST-FET) of the European Community, under grant IST-2000-28027 (POETIC). The information provided is the sole responsibility of the authors and does not reflect the Community's opinion. The Community is not responsible for any use that might be made of data appearing in this publication. The Swiss participants to this project are supported under grant 00.0529-1 by the Swiss government.

References

1. E. Sanchez, D. Mange, M. Sipper, M. Tomassini, A. Perez-Uribe, A. Stauffer. "Phylogeny, Ontogeny, and Epigenesis: Three Sources of Biological Inspiration for Softening Hardware". Lecture Notes in Computer Science, vol. 1259, Springer-Verlag, Berlin, 1997, pp. 35-54.
2. http://www.poetictissue.org
3. M. Mitchell. *An Introduction to Genetic Algorithms*. MIT Press, Cambridge, MA, 1996.
4. D.B. Fogel. *Evolutionary Computation: Toward a New Philosophy of Machine Intelligence*. IEEE Press, Piscataway, NJ, 1995.
5. J.R. Koza. *Genetic Programming*. The MIT Press, Cambridge, MA, 1992.
6. D. Floreano, J. Urzelai.. "Evolutionary Robots with on-line self-organization and behavioral fitness". *Neural Networks*, 13, 431-443.
7. Krohling, R., Zhou, Y. and Tyrrell, A.M. 'Evolving FPGA-based robot controllers using an evolutionary algorithm', 1st International Conference on Artificial Immune Systems, Canterbury, September 2002.
8. T. Higuchi, M. Iwata, I. Kajitani, M. Murakawa, S. Yoshizawa. "Hardware evolution at gate and function level". *Proc. of Biologically Inspired Autonomous Systems: Computation, Cognition, and Action*, Durham, NC, March 1996.
9. A. Thompson, I. Harvey, and P. Husbands. "Unconstrained evolution and hard consequences", *Towards Evolvable Hardware*, Springer, 1996, pp.136-165.
10. Canham, R.O. and Tyrrell, A.M. 'Evolved Fault Tolerance in Evolvable Hardware' Congress on Evolutionary Computation, Hawaii, pp 1267-1272, May 2002.
11. Roggen, D., Floreano, D., Mattiussi, C. 'A Morphogenetic System as the Phylogenetic Mechanism of the POEtic Tissue. Elsewhere in this volume.
12. C. Ortega, A. Tyrrell, "MUXTREE revisited: Embryonics as a Reconfiguration Strategy in Fault-Tolerant Processor Arrays", Lecture Notes in Computer Science, Vol. 1478, Springer-Verlag, Berlin, 1998, pp. 206-217.
13. J. von Neumann. *The Theory of Self-Reproducing Automata*. A. W. Burks, ed. University of Illinois Press, Urbana, IL, 1966
14. D. Mange, M. Sipper, A. Stauffer, G. Tempesti. "Towards Robust Integrated Circuits: The Embryonics Approach". *Proceedings of the IEEE*, vol. 88, no. 4, April 2000, pp. 516-541.
15. F. Gruau. "Genetic systems of boolean neural networks with a cell rewriting developmental process", *Combination of Genetic Algorithms and Neural Networks*. IEEE Press, Los Alamitos, CA, 1992.
16. A. Lindenmayer. "Mathematical models for cellular interaction in development, parts I and II". *Journal of Theoretical Biology*, 18:280-315, 1968.
17. H. Kitano. "Building complex systems using developmental process: An engineering approach", Lecture Notes in Computer Science, vol. 1478, Springer-Verlag, Berlin, 1998. pp. 218-229.
18. G Tempesti, D. Roggen, E Sanchez, Y Thoma, R. Canham, A.M. Tyrrell. 'Ontogrnetic Development and Fault Tolerance in the POEtic Tissue', Elsewhere in this volume.
19. P.H. Winston. *Artificial Intelligence*. Addison-Wesley, Reading, MA, 3rd edition, 1992.
20. Maass, W and Bishop C.M. *"Pulsed Neural Networks"*, MIT Press, 1999.
21. Fusi, S. "Long term memory: Encoding and storing strategies of the brain", Neurocomputing, 38-40, 1223-1228.
22. Eriksson, J., Torres, O., Mitchell, A., Tucker, G., Lindsay, K., Halliday, D., Rosenberg, J., Moreno, J., Villa, A. 'Spiking Neural Networks for Reconfigurable POEtic Tissue. Elsewhere in this volume.
23. Bradley, D.W. and Tyrrell, A.M. 'Immunotronics: Novel Finite State Machine Architectures with Built in Self Test using Self-Nonself Differentiation', IEEE Transactions on Evolutionary Computation, Vol 6, No 3, pp 227-238, June 2002.
24. Urzelai, J. and Floreano, D. (2001) Evolution of Adaptive Synapses: Robots with Fast Adaptive Behavior in New Environments. *Evolutionary Computation*, **9**, 495-524.

Ontogenetic Development and Fault Tolerance in the POEtic Tissue

Gianluca Tempesti[1], Daniel Roggen[1], Eduardo Sanchez[1], Yann Thoma[1], Richard Canham[2], and Andy M. Tyrrell[2]

[1] Swiss Federal Institute of Technology in Lausanne, Switzerland
[2] University of York, England
gianluca.tempesti@epfl.ch

Abstract. In this article, we introduce the approach to the realization of ontogenetic development and fault tolerance that will be implemented in the POEtic tissue, a novel reconfigurable digital circuit dedicated to the realization of bio-inspired systems.

The modelization in electronic hardware of the developmental process of multi-cellular biological organisms is an approach that could become extremely useful in the implementation of highly complex systems, where concepts such as self-organization and fault tolerance are key issues.

The concepts presented in this article represent an attempt at finding a useful set of mechanisms to allow the implementation in digital hardware of a bio-inspired developmental process with a reasonable overhead.

1 Introduction

The POEtic tissue is a self-contained digital integrated circuit aimed at the implementation of bio-inspired systems being developed in the framework of a new three-year research project, called "Reconfigurable POEtic Tissue" or "POEtic" for short (see the project's website at http://www.poetictissue.org for more details), recently started under the aegis of the Information Society Technologies (IST) program of the European Community, involving the Swiss Federal Institute of Technology in Lausanne (EPFL, Switzerland), the University of York (England), the Technical University of Catalunya (UPC, Spain), the University of Glasgow (Scotland), and the University of Lausanne (Switzerland).

POEtic tissues are designed to implement a vast range of bio-inspired systems, covering all the major axes of research in the domain (Phylogenesis, Ontogenesis, and Epigenesis) [7,16,17,19,20]. In this article, we will present the approaches we have selected to realize one of the three biological models that inspire circuit design: ontogenesis, that is, the development and growth of multi-cellular organisms, along with the fault tolerance that such structures imply.

In the next section, we will introduce the main features and the global architecture of POEtic systems as developed in our project, with particular emphasis on those aspects that are directly involved in the ontogenetic axis. We will then examine in some detail the two main areas of research currently under study for this part of the project, namely development and fault tolerance, and present our solutions for their implementation in silicon.

A.M. Tyrrell, P.C. Haddow, and J. Torresen (Eds.): ICES 2003, LNCS 2606, pp. 141–152, 2003.

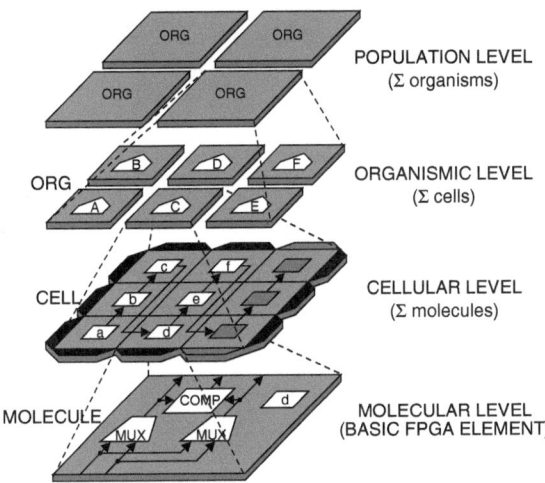

Fig. 1. The organizational layers of an electronic organism.

2   POEtic Hardware

As the general features of POEtic systems have been detailed elsewhere [19,20], in this section we will concentrate on those aspects of the architecture most relevant to the topics of the article, i.e., ontogenetic development and fault tolerance.

POEtic systems draw inspiration from the multi-cellular structure of complex biological organisms. As in biology, at the heart of our systems is a *hierarchical* structure (Fig. 1) ranging from the molecular to the population levels. In particular, the subjects of this article relate mainly to the cellular and organismic levels, with a brief foray into the molecular level for fault tolerance.

It is important to note, in this context, that the only hardware (and hence fixed) level of the POEtic systems is the molecular level, implemented as a novel digital FPGA. All other levels are *configurations* for this FPGA, and thus can be seen as *configware*, that is, entities that can be structurally programmed by the user. As a consequence, most of the mechanisms described in this article, and notably ontogenetic development and cellular fault tolerance, can be altered (or removed) to fit a particular application or a novel research approach.

The current (being configware, it could be modified in the future as new solutions are researched) architecture of the POEtic cells is described in Fig. 2. It consists of three logical layers (physically, of course, the three layers are "flattened" onto the molecular FPGA), each with a specific function within the domain of bio-inspired systems. In this article, we shall concentrate on the *mapping layer*, as, together with the *differentiation table*, it is the layer directly concerned by the ontogenetic development of our machines (fault tolerance, as we shall see, is involved in the system at all levels).

To illustrate with a practical example the structure of POEtic circuits, we shall use a very simple application: a *timer* (Fig. 3) capable of counting minutes

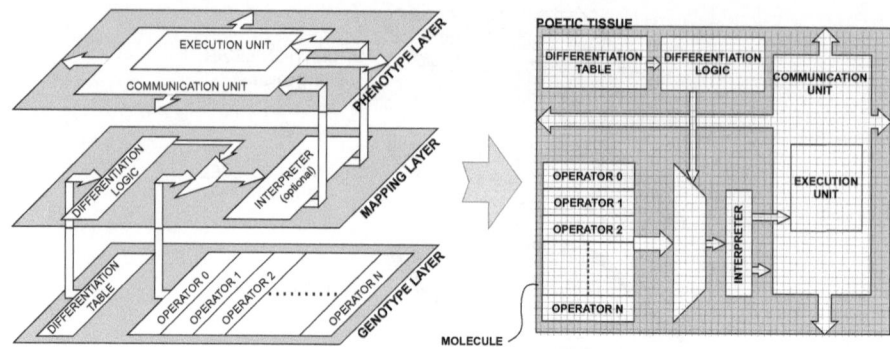

Fig. 2. The three logical layers of an electronic cell, mapped on the molecular tissue. The use of a programmable circuit allows the layers to be adapted to each application.

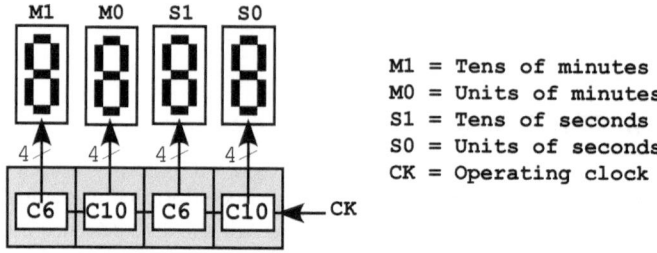

Fig. 3. Cellular implementation of a simple four-digit timer.

and seconds. This architecture has the twin advantage of being conceptually very simple and of being easily broken down into a cellular structure.

The first step in the design of a POEtic application is in fact to identify a logical division into cells. For this example, the division is obvious: tens of minutes, units of minutes, tens of seconds, units of seconds, which incidentally also correspond to the four outputs of the system (while the only input is the 1Hz synchronization clock). In general, the difficulty of this step depends of course on the application, and is probably the main criterion to determine whether an application is amenable to being implemented in a POEtic circuit.

Next, it is necessary to identify, on the basis of the cellular structure of the application, the operators or functions required by the cells of the system. For the timer, a minimum of two operators is required: a counter by six (for the tens of seconds and minutes) and a counter by ten (for the units of seconds and minutes). In addition, the communication pattern of the cells must also be fixed: for the timer of Fig. 3, the output of the counters is sent to the north, while the synchronization signals are sent out to the west.

At this stage, the architecture of the cells can be defined. This step implies essentially a design choice to decide where to place the complexity of the system. In fact, in a typical example of hardware/software co-design, the complexity can reside in the genome and in its interpreter, Embryonics-style [11], or in the phenotype. For example, in the case of the timer, the first approach would require a genome implemented by a structured program, used to control a simple 4-bit register in the phenotype. Or, in the second approach, the genome could well be reduced to a single number, used as a parameter for a programmable counter in the phenotype layer. Probably a more efficient solution in this case, this second approach would practically eliminate the requirement for an interpreter of the genome, and shift all the (in this case limited) complexity to the phenotype layer.

Of course, the set of operators represents only part of the genetic information of the organism, the rest being made up by the differentiation table used to assign operators to the cells in the system. The definition of a mechanism to implement differentiation in a POEtic system is the subject of the next section.

3   Development in the POEtic Tissue

The modelization in electronic hardware of the developmental process of multi-cellular biological organisms is a challenging task, for reasons that should be obvious: on one hand, the sheer complexity of even the simplest multi-cellular biological organisms is orders of magnitude beyond the current capability of electronic circuits, and on the other hand the development of such organisms implies, if not the creation, at least the transformation of matter in ways that cannot yet be realized in digital or analog hardware.

And yet these same considerations make such a process extremely interesting for hardware design: a developmental approach could well provide a solution, via the concept of self-organization, to the design of electronic circuits of a complexity beyond current layout techniques, and introduce into electronics at least some of the astounding fault tolerance of multi-cellular organisms, while an approximation of the growth process of biological entities could be an invaluable tool to introduce adaptability and versatility to the world of electronics.

Setting aside, for the moment, the concept of fault tolerance, in this section we will describe two different approaches to the twin mechanisms of *cellular division* and *cellular differentiation*, mechanisms at the basis of biological development.

This section will introduce two solutions to the implementation of the second mechanism: cellular differentiation, that is, the mechanism that allows a cell to determine its function within the organism (in other words, the architecture of the mapping layer of our cell). As far as cellular division is concerned, we are currently working on the development of some algorithms allowing our circuit to implement a rough approximation of this mechanism (see the "Implementation Issues" subsection for more details).

3.1   Coordinate-Based Development

Probably the simplest approach to cellular differentiation in an architecture such
as that of the POEtic circuits is to assign to each cell in the system a unique
coordinate (in fact, for two-dimensional structures, a set of [X,Y] coordinates).
The cell can then determine its function by selecting an operator depending on
its (unique) coordinates. This is the approach used in the Embryonics project,
and has been described in much detail elsewhere [11,18].

Applied to the timer example (Fig. 4), this approach would require three
different kinds of functional cell (count by 6, count by 10, and pass-through),
and the timer can be realized by four cells (out of the 15 that could fit in the
circuit in this example). The computation of the coordinates is quite simple,
implemented by incrementing the coordinate in each cell. It is then trivial to
define a differentiation table to assign an operator to each set of coordinates.

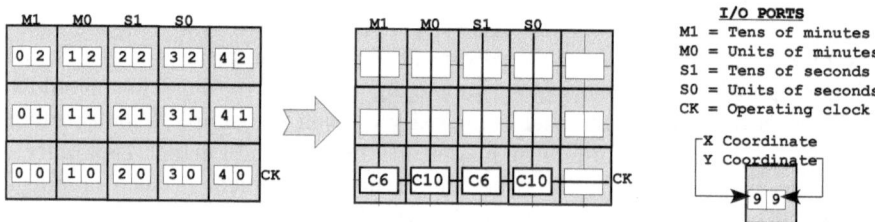

Fig. 4. A possible layout for the timer in a coordinate-based development system. In
this system, the position of each cell in the array is determined at design time.

It might be worthwhile to mention that in the Embryonics approach the
complexity of the system resides mostly in the genome, implemented as an exe-
cutable program. As a result of this design choice, the interpreter for the genome
is the most complex part of the cell, while the phenotype layer is limited to a single register plus some rudimentary connection network. The POEtic approach,
insofar as a coordinate-based development is concerned, will investigate a wide
range of tradeoffs between the complexity of the genome and the complexity of
the phenotype, as described above for the timer example.

3.2   Gradient-Based Development

The coordinate-based approach described above has a number of advantages, and
notably its simplicity on the one hand, and the fact that it has been extensively
used and tested in the Embryonics project over many years on the other. In a
way, as long as the systems to be implemented are completely pre-determined
(i.e., as long as the entire development of the organism is fully specified in the
genome, as is the case, for example, of the *caenorhabditis elegans* [15]), it is also
the most efficient and complete, and in fact any other approach can be reduced
to a coordinate-based approach during development.

However, the growth and development of complex multi-cellular organisms in nature is influenced by the environment from the earliest stages. While a problem from the point of view of an electronic implementation, this feature can lead to extremely interesting systems capable of adapting to environments unknown at design time.

In order to illustrate how such adaptation could be applied to digital circuits, we shall exploit, for clarity's sake, the same example used above for the coordinate-based approach, that is, the timer. It should be noted, however, that such a simplistic example is far from the ideal candidate for an approach designed to address the issue of adaptation: a much more interesting field of application are neural networks, systems that are inherently capable of exploiting rich interactions with the environment.

To come back to the simple example of the 4-digit timer, let us then assume that the circuit has to operate in an *a priori* unknown environment. The environment of a timer is, of course, extremely limited, and could be seen, in the most complex case, as consisting of one input port for the operating frequency of the timer, and of four output ports for the four digits. An unknown environment could then be an environment where the exact position of these input and output ports is not known at design time (once again, this is an extremely unlikely scenario in the design of a timer, but makes much more sense from the point of view of, for example, the control of a robot). A coordinate-based system is not capable of handling such a changing environment, since coordinates are defined as a function of the position of the cells *within* the circuit, independently of the external conditions. A more versatile approach is then required, capable of assigning a function to a cell depending not only on the internal configuration of the circuit, but also of the environment in which the circuit operates. Again drawing inspiration from biology, our choice has fallen on an approach based on the protein gradients that direct the development of organisms in nature.

The concept of protein gradients has been examined in detail in many publications, for biological [4,14] as well as electronic [5,6,9] organisms. Essentially, a gradient-based system consists of a set of *diffusers* that release a given protein into the system. The concentration of the protein in the system is highest at the diffuser, and decreases with the distance. Cells are assigned a functionality depending on the proteins' concentration, and hence on the distance from the proteins' diffusers, implementing a cellular differentiation based on their position *relative* to the diffusers rather than on their coordinates within the system.

For the timer example, we can consider that each of the five I/O ports mentioned above is a diffuser for a different kind of protein (again, we chose this solution for the sake of clarity, as in reality there is a tradeoff between the number of different proteins and the complexity of the mapping circuit). It is then relatively simple to design a cell that selects its function depending on the concentration of the appropriate protein. For example (Fig. 5), a simple algorithm could place the timer so that the rightmost cell is contiguous to the input port for the operating frequency and use the other cells in the circuit to create paths from the four cells to the four output ports, following the corresponding gradients.

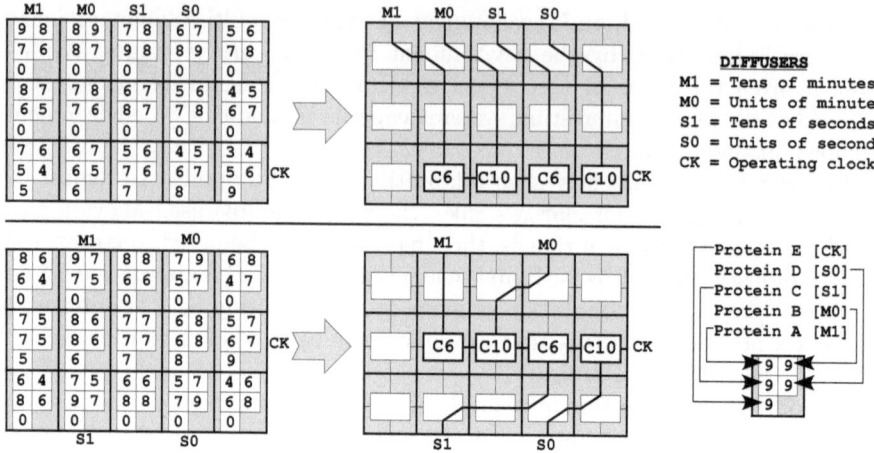

Fig. 5. Two possible layouts for the timer in a gradient-based development system. The position of each cell in the array is determined by the system during configuration.

Even in such a simplistic example, the versatility of a gradient-based system should be obvious. This kind of approach can also exploit the other axes of bio-inspiration to increase the adaptability of multi-cellular systems (aside from the obvious interest for adaptive systems such as neural networks, evolution can be used, for example, to define the position within the system of some emitters, allowing the genome to partially direct the development of the organism).

3.3   Implementation Issues

We are actively pursuing research on both of the developmental approaches described above. As we mentioned, for systems that can be completely defined in the genome at design time, a coordinate-based system would be more efficient. On the other hand, for adaptive systems where the environments influences development, a gradient-based approach provides a much more versatile tool.

In parallel, we are also investigating more complex, richer mappings between the genotype and the phenotype. The Embryonics approach, for example, requires a rigid one-to-one mapping from coordinates to function. A much more interesting approach exploits the possibility of using partially-defined differentiation tables that are accessed not through a one-to-one mapping, but rather through more complex algorithms (e.g., by computing the Hamming distance between the coordinates or the protein concentrations and the table entries).

The use of partially-defined differentiation tables, coupled with the possibility of environment-directed development (e.g., through protein diffusers), also opens the possibility of mechanisms that approximate the growth process in biological organisms. If the organism's structure is only partially defined in the genome, and if the environment is used to define the cells' function, then an informational (rather than physical, as in nature) growth process can be realized.

4   Fault Tolerance in the POEtic Tissue

Even if the complexity of digital hardware remains orders of magnitude behind that of the more developed biological organisms, some of the issues that nature had to confront during the millions of years required for the evolution of such organisms are gaining interest for electronics. For example, as we mentioned earlier, silicon might well be replaced, in a not-so-distant future,by molecular-size components with densities beyond what could be handled through current layout techniques. Among the many challenges represented by such novel hardware, one in particular stands out: these incredibly small components will never be "perfect". In other words, circuits will inevitably contain faulty elements.

Nature, faced with this problem, has developed extremely efficient solutions. Fault tolerance is definitely one domain where nature is vastly more advanced that electronics, and one of our goals is to draw inspiration from biological healing and repair mechanisms to achieve stronger fault tolerance in digital hardware.

From a "practical" standpoint, it is important to note that fault tolerance implies in fact two different processes: self-test, to detect that a fault has occurred within the circuit, and self-repair, to somehow allow the circuit to keep operating in the presence of one or more faults. The two processes are in fact very distinct, both conceptually and in practice. In particular, self-test implies that the circuit be capable of detecting incorrect operation and, since the fault will eventually have to be neutralized, the exact site where the incorrect operation occurs. Self-repair, on the other hand, implies that all parts of the circuit must be replaceable, physically or at least logically, via some sort of reconfiguration.

There exist, for both processes, many "conventional" approaches [1,10,12], but no really efficient off-the-shelf solution. Drawing once again inspiration from nature, where "fault tolerance" is achieved through not one, but a set of mechanisms ranging from molecular repair to complex organism-level immune systems, we are developing a *hierarchical* approach to fault tolerance, spanning all the levels (Fig. 1) of our POEtic systems.

4.1   Molecular-Level Fault Tolerance

Molecular-level fault tolerance is probably the most sensitive area of our hierarchical approach, as the molecules (and hence their test and repair mechanisms) represent the hardware of our system, and thus will necessarily have to be fixed at design time (rather than being modifiable by configware as the other layers).

However, the hardware layer of the POEtic tissue has not yet been fully designed (it represents the main research effort under way at the moment within the project) and, by their nature, logic level test and repair mechanisms require a completed layout to be efficiently implemented.

In general, the fault tolerance mechanisms implemented for the POEtic tissue will bear a loose resemblance to those used for Embryonics designs [11,18], at least in that they will rely on a hybrid approach mixing mechanisms such as duplication (a mechanism found in nature in the DNA's double helix and in its error-correcting processes) and memory testing [10].

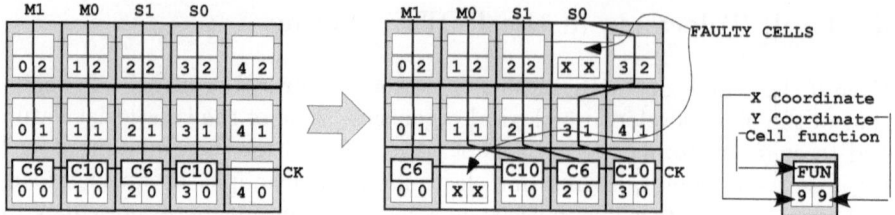

Fig. 6. Self-repair through reconfiguration in a coordinate-based system. As each cell is totipotent, reconfiguration consists simply of a re-computation of the cells' coordinates.

4.2   Cellular-Level Fault Tolerance

The coarser grain of cells with respect to molecules can be exploited to introduce more complex test and repair patterns. Fault tolerance at the processor level (cells can be viewed, for testing purposes, as small processing units) is usually implemented through techniques radically different from the logic level, and more complex reconfiguration patterns can be realized by exploiting the development approaches implemented in the POEtic circuits.

For example, testing at the processor level can take advantage of the presence of multi-bit busses to test the correct operation of the machine through techniques such as parity-checking. As far as self-repair is concerned, reconfiguration mechanisms can exploit the features of the POEtic architecture to simplify the rearrangement of tasks within the array: the presence of *the full genome in each cell*, coupled with the developmental mechanism that assigns the function to the cells depending on local conditions, can for example allow a relatively simple implementation of on-line self-repair via reconfiguration (Fig. 6).

Of course, it should also be noted that many of the systems that POEtic architectures are meant for, and notably neural networks, are *inherently* fault tolerant at the cellular level: the death of a neuron within the system will force the learning algorithms to automatically avoid the faulty neuron and to find a solution that can perform the desired computational task in a reduced network.

4.3   Organismic-Level Fault Tolerance

A simplistic vision of organismic-level fault tolerance consists of exploiting the developmental approach of POEtic systems to create multiple redundant copies of an organism during configuration. This approach, used in the Embryonics systems [11], introduces improved fault tolerance through a simple comparison of the outputs of multiple identical circuits (the biological analogy would reside in the survival of a population even if individuals die).

This form of fault-tolerance, while relatively simple to realize (through cycles in the coordinates, as in Fig. 7, or by using multiple diffusers of the same kind in a gradient-based system), requires material *outside* of the POEtic circuits to identify (and possibly kill) faulty organisms.

A more interesting approach (and one of the research axes of the POEtic project) tries to exploit mechanisms inspired by biological immune systems [2,3].

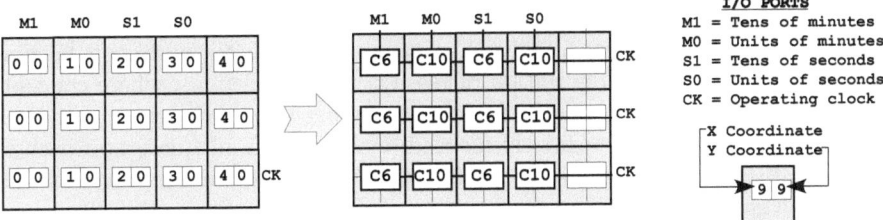

Fig. 7. Multiple redundant copies of an organism through coordinate cycling.

In higher organisms, these systems rely on a multi-layered, distributed approach that is robust and capable of identifying numerous pathogens and other causes of illness, and is undoubtedly an interesting source of inspiration for the development of reliability mechanisms in silicon.

The approach relies on a negative selection algorithm [8] coupled with a more "conventional" mechanism that could be based on roving self-test areas (STAR) [1]. In practice, the approach merges a fixed testing mechanism (the *innate* part of the immune system) with a *learning* negative selection mechanism (the *acquired* part), working at a system or subsystem level to monitor the state of the circuit and identify and kill cells that develop faults at runtime.

4.4 Implementation Issues

The final goal of the mechanisms we introduced is to illustrate how a hierarchical approach to fault tolerance, along the same general lines as the one present in complex biological organisms, is a very efficient solution from the point of view of assuring the correct operation of the circuit in the presence of faults. By adopting a hierarchy of mechanisms, we can exploit the best features of each level of our systems, and thus limit the overhead required to obtain an "acceptable" level of reliability for the system as a whole.

In reality, the only relevant overhead generated by introducing fault tolerance in our POEtic circuits resides in the molecular level: since all other levels are implemented in configware, reliability (and its associated overhead) becomes an option, and can be removed should fault tolerance not be a priority.

The important implementation issues for fault tolerance thus concern essentially the test and repair mechanisms at the molecular level. As we mentioned, however, the design of the molecule is currently under way, and all the important parameters required to introduce reliability to the molecular FPGA (connection network, degree of homogeneity, configuration mechanism, etc.) have yet to be fixed. A complete description of the implementation of fault tolerance in the POEtic tissue will thus be the subject of a future article.

5   Conclusions

The concepts presented in this article represent an attempt at finding a useful set of mechanisms to allow the implementation in digital hardware of a bio-inspired developmental process with a reasonable overhead. The modelization of multi-cellular organisms in silicon is an approach that could soon become extremely useful for the realization of highly complex systems, where concepts such as self-organization and fault tolerance are becoming key issues.

Much of our effort is currently focused on finding an adequate approach to implement development in the POEtic tissue. We are examining processes that are at the same time efficient from the point of view of the required hardware resources and scalable across one or even multiple chips. Notably, we hope to develop a mechanism versatile enough to allow on-line addition of new chips (that is, the user will be able to add new hardware to the system without disrupting its operation), thus achieving something closer to the *physical* growth of biological organisms.

The two developmental models described in this article fit remarkably well our requirements (scalability, robustness, efficiency, etc.), and complement each other remarkably well: the coordinate-based approach is most efficient for "conventional" systems where the layout of the circuit is known in advance, while the gradient-based approach provides an increased versatility invaluable for the realization of adaptive systems such as neural networks. Moreover, the architecture of our POEtic machines places the developmental mechanism where it can be modified by configware, allowing in the future the implementation of different approaches.

It is definitely too soon to draw any conclusions on the topic of fault tolerance in the POEtic tissue: until the hardware layer is fixed, the application of self-test and self-repair techniques is impossible. One aspect that is already apparent, however, is that our research is focusing on a *hierarchical* set of mechanisms meant to work together to achieve a level of reliability that would not be possible for a single mechanism.

Finally, we wish to reiterate the *reconfigurable* aspects of the architecture we described: the molecular tissue is meant to be a *universal* tissue for the implementation of bio-inspired systems. The architectures and mechanisms described in this article represent only some of the ideas that will be pursued in the POEtic project. The circuit, once developed, will be made publicly available and should prove a useful tool for research in the domain beyond the participants to the project.

Acknowledgments

This project is funded by the Future and Emerging Technologies programme (IST-FET) of the European Community, under grant IST-2000-28027 (POETIC). The information provided is the sole responsibility of the authors and does not reflect the Community's opinion. The Community is not responsible for any use that might be made of data appearing in this publication. The Swiss participants to this project are supported under grant 00.0529-1 by the Swiss government.

References

1. Abramovici, M., Stroud, C.: Complete Testing and Diagnosis of FPGA Logic Blocks. IEEE Trans. on VLSI Systems **9:1** (2001).
2. Bradley, D.W., Tyrrell, A.: Multi-Layered Defence Mechanisms: Architecture, Implementation, and Demonstration of a Hardware Immune System. Proc. 4th Int. Conf. on Evolvable Systems, LNCS **2210** (2001), 140–150.
3. Canham, R.O., Tyrrell, A.M.: A Multilayered Immune System for Hardware Fault Tolerance within an Embryonic Array. Proc. 1st Int. Conf. on Artificial Immune Systems (ICARIS 2002), Canterbury, UK, September 2002.
4. Coen, E.: The Art of Genes. Oxford University Press (1999), New York.
5. Edwards, R.T.: Circuit Morphologies and Ontogenies. Proc. 2002 NASA/DoD Conf. on Evolvable Hardware, IEEE Computer Society Press (2002), 251–260.
6. Eggenberger, P.: Cell Interactions as a Control Tool of Developmental Processes for Evolutionary Robotics. From Animals to Animats 4: Proc. 4th Int. Conf. on Simulation of Adaptive Behavior, MIT Press - Bradford Books (1996), 440–448.
7. Eriksson, J., Torres, O., Mitchell, A., Tucker, G., Lindsay, K., Halliday, D., Rosenberg, J., Moreno, J.-M., Villa, A.E.P.: Spiking Neural Networks for Reconfigurable POEtic Tissue. Elsewhere in this volume.
8. Forrest, A., Perelson, A.S., Allen, L., Cherukuri, R.: Self-Nonself Discrimination in a Computer. Proc. 1994 IEEE Symposium on Research in Security and Privacy, IEEE Computer Society Press (1994).
9. Gordon, T.G.W., Bentley, P.J.: Towards Development in Evolvable Hardware. Proc. 2002 NASA/DoD Conf. on Evolvable Hardware, IEEE Computer Society Press (2002), 241–250.
10. Lala, P.K.: Digital Circuit Testing and Testability. Academic Press (1997).
11. Mange, D., Sipper, M., Stauffer, A., Tempesti, G.: Towards Robust Integrated Circuits: The Embryonics Approach. Proc. of the IEEE **88:4** (2000), 516–541.
12. Negrini, R., Sami, M.G., Stefanelli, R.: Fault Tolerance through Reconfiguration in VLSI and WSI Arrays. The MIT Press, Cambridge, MA (1989).
13. Ortega, C., Tyrrell, A.: MUXTREE revisited: Embryonics as a Reconfiguration Strategy in Fault-Tolerant Processor Arrays. LNCS **1478**, Springer-Verlag, Berlin (1998), 206–217.
14. Raven, P., Johnson, G.: Biology. McGraw-Hill (2001), 6th edition.
15. Riddle, D.L., Blumenthal, T., Meyer, B.J., Priess, J.R., eds.: C. Elegans II. Cold Spring Harbor Laboratory Press (1997).
16. Roggen, D., Floreano, D., Mattiussi, C.: A Morphogenetic System as the Phylogenetic Mechanism of the POEtic Tissue. Elsewhere in this volume.
17. Sipper, M., Sanchez, E., Mange, D., Tomassini, M., Pérez-Uribe, A., and Stauffer, A.: A Phylogenetic, Ontogenetic, and Epigenetic View of Bio-Inspired Hardware Systems. IEEE Transactions on Evolutionary Computation, **1:1** (1997) 83–97.
18. Tempesti, G.: A Self-Repairing Multiplexer-Based FPGA Inspired by Biological Processes. Ph.D. Thesis No. **1827** (1998), EPFL, Lausanne, Switzerland.
19. Tempesti, G., Roggen, D., Sanchez, E., Thoma, Y., Canham, R., Tyrrell, A., Moreno, J.-M.: A POEtic Architecture for Bio-Inspired Systems. Proc. 8th Int. Conf. on the Simulation and Synthesis of Living Systems (Artificial Life VIII), Sydney, Australia, December 2002.
20. Tyrrell, A.M., Sanchez, E., Floreano, F., Tempesti, G., Mange, D., Moreno, J.-M., Rosenberg, J., Villa, A.E.P.: POEtic Tissue: An Integrated Architecture for Bio-Inspired Hardware. Elsewhere in this volume.

A Morphogenetic Evolutionary System: Phylogenesis of the POEtic Circuit

Daniel Roggen, Dario Floreano, and Claudio Mattiussi

Autonomous Systems Laboratory, Institute of Systems Engineering, EPFL,
Lausanne, Switzerland
http://asl.epfl.ch
name.surname@epfl.ch

Abstract. This paper describes a new evolutionary mechanism developed specifically for cellular circuits. Called morphogenetic system, it is inspired by the mechanisms of gene expression and cell differentiation found in living organisms. It will be used as the phylogenetic (evolutionary) mechanism in the POEtic project. The POEtic project will deliver an electronic circuit, called the POEtic circuit, with the capability to evolve (Phylogenesis), self-repair and grow (Ontogenesis) and learn (Epigenesis). The morphogenetic system is applied to the generation of patterns and to the evolution of spiking neural networks, with experiments of pattern recognition and obstacle avoidance with robots. Experimental results show that the morphogenetic system outperforms a direct genetic coding in several experiments.

1 Introduction

The POEtic circuit is a multi-cellular electronic circuit composed of a regular 2D array of cells (i.e. functional units), which has capabilities to evolve (Phylogenesis), self-repair and grow (Ontogenesis) and learn (Epigenesis). The POEtic circuit is reprogrammable and its functionality comes from its configuration bits or genotype, which will be evolved using the phylogenetic mechanism described herein. To implement self-repair and growth, each cell may contain the genome of the whole circuit. In addition, each cell has virtual input/output connections to sensors and actuators. The POEtic circuit will be used in a wide range of applications, including autonomous mobile robotics. A more detailed description of the project is given in [10].

This paper is the first accounting of the morphogenetic evolutionary system which will be implemented in the POEtic circuit. This system is designed to be computationally simple (hardware implementation) and suited to multi-cellular circuits like POEtic. The morphogenetic system relies on a simple signalling mechanism and expression rules to encode the functionality of a cellular system which can be a neural network or any other cellular electronic circuit (e.g. the BioWall [9]). Its development is motivated by the dynamical reconfiguration needs of the POEtic circuit, which cannot be easily handled by direct genetic codings with one-to-one mappings between the genotype and the phenotype [11].

A.M. Tyrrell, P.C. Haddow, and J. Torresen (Eds.): ICES 2003, LNCS 2606, pp. 153–164, 2003.

In this paper we describe the morphogenetic system and compare it with a direct genetic encoding in experiments of pattern coverage and in the evolution of spiking neural networks for pattern recognition and obstacle avoidance using real miniature robots. Results show that the morphogenetic system outperforms direct genetic encoding by finding individuals of better fitness in less generations. Analysis suggests that the morphogenetic system generates more individuals of higher fitness than the direct genetic encoding, thus easing the evolutionary process.

2   Morphogenetic Coding

Introduction. The morphogenetic system is motivated by the requirement of the POEtic circuit to have a phylogenetic mechanism which is suited to its cellular structure and its capacity to dynamically reconfigure, e.g. when errors are detected, when the environment changes or when the circuit is expanded with new cells. Direct genetic encodings are not suited for that kind of architecture because the number of elements in the system must be known in advance and cannot change throughout the life of the system.

The morphogenetic system assigns a functionality to each cell of the circuit from a set of predefined functionalities (something akin to skin, muscle, neuron cells, etc. in living organisms). The process works in two phases: first a signalling phase then an expression phase. The signalling phase relies on the ability of the cellular circuit to exchange signals among adjacent cells to implement a diffusion process. The second phase, expression, finds the functionality to be expressed in each cell by matching the signal intensities in each cell with a corresponding functionality stored in an expression table. The genetic code contains the position of diffusing cells (diffusers) and the content of the expression table, which are both evolved using a genetic algorithm. The morphogenetic system is inspired by the mechanisms of gene expression and cell differentiation of living organism [2], notably by the fact that concentrations of proteins and inter-cellular chemical signalling regulate the functionality of cells. Other works related to this morphogenetic systems include L-Systems [7], use of developmental processes [3,6], and also approximate representations [1] seeking to compress the genotype.

Description. The morphogenetic system relies on a set of predefined functionalities which is called a family of functions. The family of functions must offer a rich enough repertoire of functionalities to realize the desired circuit. The family which is used when evolving neural networks on the POEtic circuit is composed of spiking neurons with different connectivity patterns, and either excitatory or inhibitory characteristics (fig. 1). Spiking neurons are used because they display rich non-linear dynamics and because they are well suited for implementation in digital circuits [5].

The cellular circuit allows for signals to be exchanged between adjacent cells. A signal is a simple numerical value (the signal intensity) that the cell owns, and that adjacent cells are able to read (the intensity of signal s in cell i is

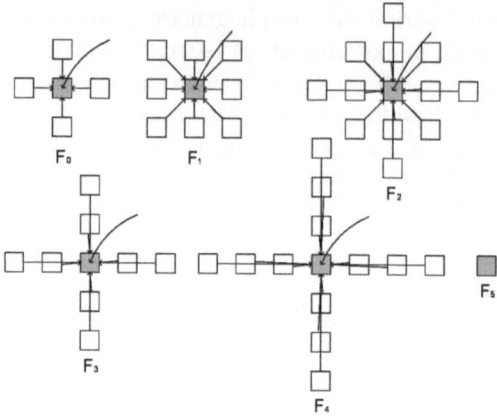

Fig. 1. In the current implementation, a family composed of 12 functions is used when evolving neural networks. The functions are spiking neurons with different connectivities (the 6 types of connectivity shown in the figure) and with either excitatory or inhibitory characteristics. Each cell of the multi-cellular circuit implements a single neuron, shown in gray. It receives inputs from neighboring neurons (outlined), which are implemented in neighbouring cells. Each neuron has an extra external input (curved arrow). Neuron F_5 is equivalent to a void cell. At the boundary of the cellular array the connectivity is truncated (no periodic boundary condition).

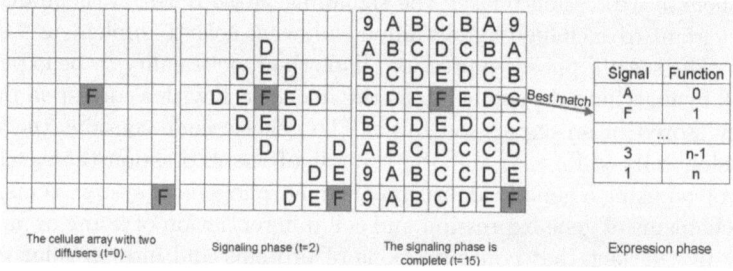

Fig. 2. The three arrays on the left are snapshots of the signalling phase with one type of signal and two diffusers (gray cells) at the start of the signalling phase (left array), after two time steps (middle) and when the signalling is complete (right). The number inside the cells indicates the intensity of the signal in hexadecimal. The expression table used in the expression phase is shown on the right. In this example the signal D matches the second entry of the table with signal F (smalleing distance), thus expressing function F_1.

noted by C_s^i). Special cells, called *diffusers*, own a signal of maximum intensity. The signal intensity in the neighbouring cells then decreases linearly with the Manhattan distance to the diffuser. In the current implementation four type of signals (each represented by a 4-bit number) are used. They are diffused independently, without interaction among them. Figure 2 shows an example of the signalling phase in the case of a single type of signal, with two diffusers placed in the cellular circuit.

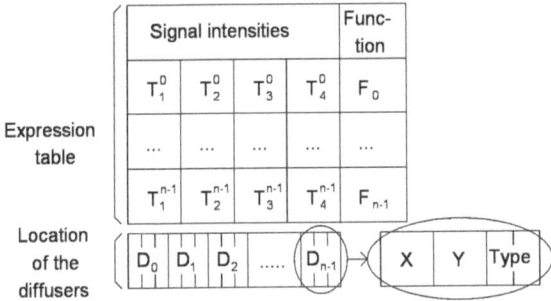

Fig. 3. The genetic code contains two parts. The first is the expression table T, here with n entries. Each entry is 16 bits long. The second part contains the location of the diffusers and their type. The number of bits for the X and Y coordinates depends on the size of the network. The type of the diffuser is encoded on 2 bits.

The expression phase assigns a function to each cell by matching the signal intensities inside that cell with the entries of an expression table T (fig. 2) stored in the genetic code. Each entry of the table contains the intensities of the four signals and the function to express in case of match. The intensity of signal s in the entry j of the table is noted by T_s^j. A cell i is said to match an entry j of the expression table when the distance $d = \sum_{s=1}^{4} DOp(C_s^i, T_s^j)$ is minimum. The distance operator DOp is the Hamming distance.

The genetic code contains two parts. The first part is the expression table T and the second is the location of the diffusers (fig. 3). The expression table contains 16 bits for each entry (4 signals coded on 4 bits each). Each entry corresponds to a predefined function. The functions are not encoded and evolved in these experiments. The location of the diffuser is stored as a pair of X,Y coordinates, plus two bits indicating the type of the diffuser (i.e. $2^2 = 4$ types of diffuser).

A genetic algorithm is used to evolve the morphogenetic coding. In all of the experiments presented here the population is composed of 50 individuals, selection consists of rank selection of the 15 best individuals, the mutation rate is 1%, one-point crossover rate is 20% and elitism is used by copying the 5 best individuals without modifications in the new generation.

3   Pattern Coverage Experiment

This experiment is aimed at testing whether the morphogenetic system can generate circuits with diversified structures (without any specific functionality). It consists of covering an array of 8x8 cells with a specific binary pattern. Four different patterns are tested (fig. 4). A family of two cell functions (black or white cell) is used. The morphogenetic system uses two entries in the expression table and 16 diffusers. Preliminary tests with numbers of diffusers in the range of 2 to 64 have shown that a too low or a too high number of diffusers hinder the performance of the morphogenetic system with some patterns. Values in the

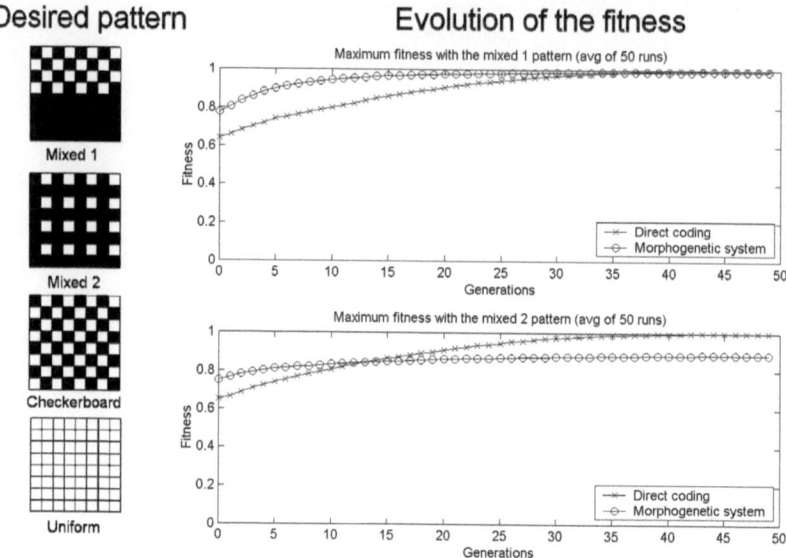

Fig. 4. The pattern coverage experiment consists in covering an array of 8x8 cells with a binary pattern. The left column shows the four desired pattern (*mixed1, mixed2, checkerboard* and *uniform*). The right column shows the evolution of the maximum fitness for the *mixed1* and *mixed2* patterns.

range of 5 to 20 showed quite similar performance. As a good compromise, 16 diffusers have been chosen. The size of the morphogenetic coding is 160 bits: 2 entries in the expression table * 16 bits + 16 diffusers * 8 bits (6 bits to store the coordinates and 2 bits to store the type of the diffuser). Also, a direct genetic encoding is used for comparison, which consists of a binary string with 1 bit per cell (indicating whether the cell is black or white), resulting in a code of 64 bits.

The figure 4 shows the evolution of the maximum normalized fitness, which is proportional to the number of cells covered correctly, for the *mixed1* and *mixed2* patterns, averaged over 50 runs. The evolution of the fitness for the *uniform* and *checkerboard* patterns is very similar to the *mixed1* pattern. In the first few generations, the maximum fitness increases notably faster with the morphogenetic system than with the direct coding with all four patterns. With the *uniform, checkerboard* and *mixed1* pattern the morphogenetic system covers the patterns almost three times faster than the direct coding in terms of generation number (about 10 generations instead of 30). The morphogenetic system cannot fully cover the *mixed2* pattern, but it comes very close. This seems to indicate that some type of regularity is necessary for the morphogenetic system to perform well, or that the system has some biases toward some types of pattern. Further investigation are necessary to clarify this point. However, in the context of the POEtic circuit this may not be an issue, as the epigenetic (learning) mechanism may deal with structures at a finer granularity level, while the phylogenetic mechanism deals with structures at a coarser level. Further experiments may also be performed to assess how the morphogenetic system scales with larger arrays.

4   Spiking Neuron Model

The spiking neuron model used in the following experiments is a discrete-time, integrate-and-fire model with leakage and a refractory period. Each neuron has weighted inputs (+2 or -2 depending on whether the presynaptic neuron is excitatory or inhibitory) from connected neurons. Each cell can express one neuron type, from a family of 12 (6 connectivity patterns times 2 sign types) shown in fig 1. It has one more connection from an external input, e.g. to connect from a sensor, with fixed weight of +10. The neuron integrates the incoming spikes in the membrane potential, according to the weights of the connections. Once the membrane potential reaches a threshold (fixed to 4), the neuron fires (emits a spike), resets its membrane potential to 0 and enters a refractory period where it does not integrate incoming spikes for one time step. After the integration phase and if the neuron has not fired, leakage is applied by decrementing membrane potential by 1 (or incrementing it if the potential is below 0), so that the potential tends to 0.

5   Pattern Recognition Experiment

A circuit of 8x8 cells (fig. 5) was evolved to recognize characters (any other pattern could be used) using two training sets: one set contains corrupted versions of two characters, the other set contains random patterns. The circuit must indicate whether the current pattern is one of both characters or not. The fitness of the network is evaluated by presenting successively all the patterns of a training set, and it is equal to the number of times it correctly classifies the input pattern. Fig. 5 shows the training set for the recognition of characters A and C (noted as A+C). The upper line shows the subset of patterns to recognize. The second line contains random patterns that must be rejected. The maximum fitness achievable is 20 (20 patterns in the training set). The experiments have been performed with four different training sets, for the recognition of characters A+B, A+C, A+D and A+E.

Figure 5 shows the array of cells in the circuit. The input pattern is applied on a subset of cells through the external input of the cells. Each cell receives one pixel of the pattern: if the pixel is black, then it receives a spike every two time steps, otherwise it receives no spike. The network is run for 100 time steps with the input applied to it, after which the activity of the output neuron on the right of the network is read. The activity level (number of spikes) of that neuron indicates whether the pattern is recognized or not (threshold=50% of maximum spike number).

A morphogenetic system with 16 diffusers and 12 entries in the expression table (one for each type of neuron) is compared to the direct coding. The size of the morphogenetic coding is 320 bits: 12 entries in the expression table * 16

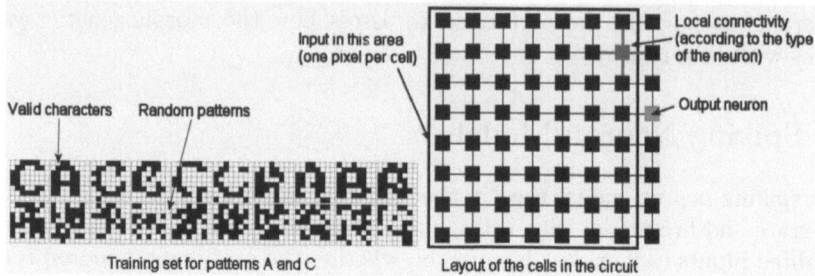

Fig. 5. Left: training set for the recognition of the patterns A and C. It is composed of 20 patterns in two subsets of 10 patterns. The upper line contains the patterns to recognize which are the letters A and C. The second line contains the random patterns that the network must reject. Right: layout of the cells in the circuit. The input pattern is applied on an array of 7x8 cells on the left, with each cell receiving the input from one pixel of the pattern. The output of a single cell at the right column is used to indicate whether the character has been recognized or not.

Table 1. Number of runs, out of 100 performed for each training set, reaching the maximum fitness.

	8x8 network	
Training set	Direct	Morph.
A+B	18	27
A+C	8	34
A+D	19	20
A+E	14	54
Total (max is 400):	59	135

bits + 16 diffusers * 8 bits (6 bits for the coordinates and 2 bits for the type of the diffuser). The size of the direct coding is 256 bits (12 type of neurons, thus 4 bits/cells * 64 cells).

The circuit has been evolved one hundred times for each of the four training sets (fig. 6). The morphogenetic system outperforms the direct coding, both when comparing the maximum fitness reached at a given generation and when comparing the number of runs that have reached the maximum fitness after 50 generations. Table 1 reports the number of runs (on the 100 runs performed) where the maximum fitness of 20 is reached. Averaged over the four training sets, runs finding a maximum fitness using the morphogenetic coding are more than twice as frequent than when using the direct coding.

6   Mobile Robot Controller

A spiking circuit is evolved as a controller for a Khepera miniature autonomous robot. The objective is perform obstacle avoidance using the sensory informations coming from the proximity sensors of the robot (fig. 7). Four sensory groups

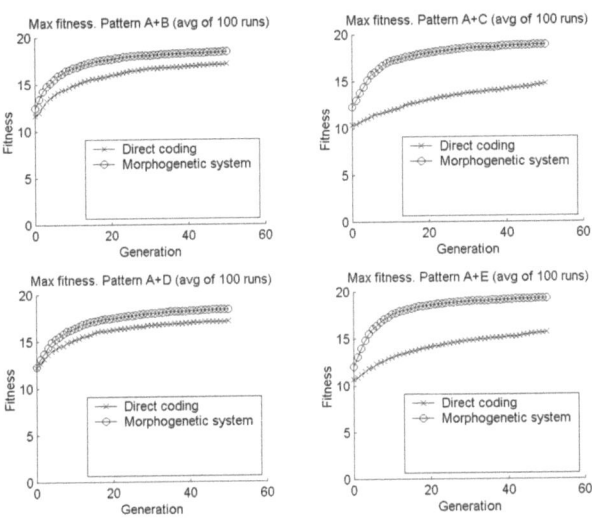

Fig. 6. Evolution of maximum fitness for pattern recognition with the four training sets.

of two cells are connected to the infrared sensors. The activity of two other cells are used to set the speeds of the wheels.

The robot has a sensory-motor period of 100ms. During that period, the neurons on the circuit are updated 20 times and, according to the distance to the obstacles, either 0, 1 (the "low" activity), or both input neurons ("low" and "high" activity) of each sensory group receive a spike train of period 2 (one spike every two time steps). At the end of the sensory-motor period, the speeds of the wheels are updated and the proximity sensors are read to compute spike trains for the next sensory-motor cycle. Some amount of noise is introduced in the spike trains. The cells ML and MR control the speed of the wheels in a way which is inversely proportional to their activity. This allows the robot to move forward when no obstacles are sensed and thus when there is potentially no activity in the network. A minimum activity of the neuron sets the speed of the wheel to +80 mm/s. A maximum activity sets the speed to -80 mm/s. The speed of the wheels scales linearly in between.

The spiking neuron model used here is the same as used in the pattern recognition experiment (section 4), with the addition of random variations in the threshold value to avoid locked oscillations during the 100 ms sensory-motor period. For each neuron and at each time step the threshold value is incremented or decremented by 1 with a probability of 5%.

The fitness of the robot is measured on two tests of 30 seconds in a rectangular arena (40x65 cm). It is the average of the fitness computed at each sensory-motor step using the following equation [4]: $f = \bar{v} \cdot (1 - \Delta v) \cdot (1 - p)$, where $\bar{v}$ is the average speed of the two motors, Δv is the absolute value of the difference of speed of the motors, and p is the activity of the most active sensor ($\bar{v}$, Δv and

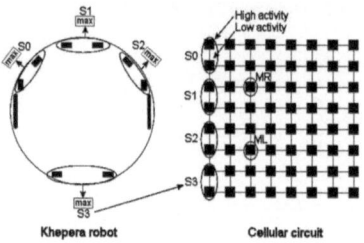

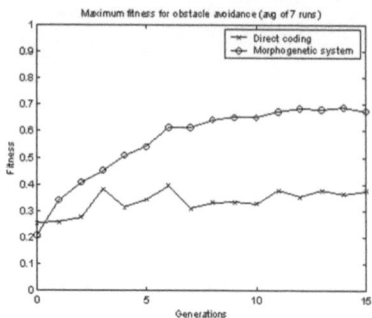

Fig. 7. The Khepera robot (left) and the neural controller (right). The Khepera has 8 proximity sensors. They are grouped by two, taking value of the most active sensors, to have 4 sensory inputs S1 to S4. The circuit receives S1 to S4 as sensory inputs. Each input is coded on two neurons. The cells ML and MR control the speed of the wheels according to their activity.

Fig. 8. Evolution of the maximum fitness in the obstacle avoidance experiment (average of 7 runs on a physical robot).

p are in the range $[0;1]$). The three parts of this function aim to 1) maximize the speed of the robot, 2) minimize the rotation and 3) maximize the distance to the obstacles.

Seven runs were performed to compare the same morphogenetic system and direct coding as those used in the pattern recognition experiments. The morphogenetic system performs better than the direct coding (fig. 8): it clearly achieves a higher maximum fitness than the direct coding. Moreover, after 15 generations, only three runs managed to find individuals displaying obstacle avoidance behavior with direct coding, whereas with the morphogenetic system individuals were found displaying this behavior in all seven runs. It is thought that the difference in performance may come from some characteristics that individuals generated with the morphogenetic system possess but which are harder to get with the direct coding. A preliminary analysis is discussed below.

7   Analysis

The better performance of the morphogenetic system over the direct genetic encoding may be due to differences in characteristics of the fitness landscape. Ruggedness is often linked to the difficulty of search when genetic algorithms are used [8]. To investigate the level of ruggedness, random walks using the mutation operator have been performed in the pattern recognition experiment. For 600 individuals of maximum fitness, 5 random walks of 10 steps (10 consecutive applications of the mutation operator) have been performed on circuits evolved using the morphogenetic system and the direct genetic encoding. The fitness averaged on the 3000 random walks is plotted against the number of application of the mutation operator in fig 9. The fitness drops faster with the morphogenetic

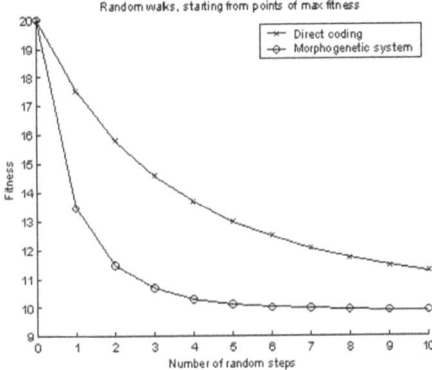

Fig. 9. Comparison of the random walks. The fitness drops faster with the morphogenetic system when going farther away from points of maximum fitness. This seems to indicate that the fitness landscape is more rugged with the morphogenetic system.

Table 2. Maximum fitness and standard deviation of the fitness in a sample of 5000 randomly generated individuals.

Training set	Direct		Morph.	
	Max	Std dev	Max	Std dev
A+B	11.7	0.77	11.92	0.88
A+C	10.38	0.74	12.14	1.08
A+D	12.22	0.64	12.1	0.85
A+E	10.58	0.95	12.28	1.03

system when moving away from a point of maximum fitness, which may imply a more rugged fitness landscape. The better performance of the morphogenetic system thus cannot be explained by having a smoother fitness landscape.

We also measured the fitness values of 5000 randomly generated individuals using the morphogenetic system and the direct genetic encoding. The statistics of table 2 for the pattern recognition task show that the morphogenetic system tends to display higher values and higher standard deviation of the fitness distribution. Although the differences are small and may not be significative, the higher fitness variability may explain why the morphogenetic system can generate individuals of the same fitness faster than direct genetic encoding. The reason why morphogenetic systems also generate individuals with higher fitness is still not clear. We speculate that the difference can be explained by structural properties of the circuits generated by the two genetic encodings. For example, in the case of the robotic experiments, we noticed that the morphogenetic system generated very easily large patches of interconnected excitatory neurons that link sensors to motor neurons, causing a reversal of wheel speeds when the robot approaches an obstacle. Instead, direct genetic encodings take more time to set all the redundant cells to suitable values. However, these data are only preliminary require further statistical investigations.

8   Conclusions

We have presented a simple morphogenetic system and compared it to a direct genotype-to-phenotype mapping. The morphogenetic system uses a simple signalling and expression mechanism, which is inspired by the mechanisms of gene expression and cell differentiation of living beings. Several experiments were performed and the morphogenetic system outperformed a direct coding, both when comparing the maximum fitness reached at a given generation and when comparing the number of runs reaching the maximum fitness. The morphogenetic system, designed to be simple (hardware implementation), and its good performance make it well suited for use in the POEtic circuit. Further experiments need to be done to see how the morphogenetic system scales to networks of larger size. Also, further analysis are needed to explain the better performance of the morphogenetic system. As the mechanism is inspired upon biology, it may be that knowledge from that field will help to shed some light on this topic.

However, the real strength of the morphogenetic system is yet to be realized. It resides in its capacity to handle run-time dynamically reconfigurable circuits, e.g. by adding or removing diffusers, or by expressing the functionality of newly inserted cells in the circuit. Those aspects must still be explored, however they may pave the way toward dynamically reconfigurable electronic circuits that reorganize when a sensor fails or when new sensors or actuators are added to the system. In the future we also plan to extend the system by allowing environmental interactions while the circuit evolves, as well as including the evolution of functions that each cell can take, which, for the sake of simplicity, here have been limited to a predefined set.

Acknowledgments

The authors thank Jesper Blynel, Eduardo Sanchez, Gianluca Tempesti and Yann Thoma for their ideas and comments. This project is funded by the Future and Emerging Technologies programme (IST-FET) of the European Community, under grant IST-2000-28027 (POETIC). The information provided is the sole responsibility of the authors and does not reflect the Community's opinion. The Community is not responsible for any use that might be made of data appearing in this publication. The Swiss participants to this project are funded by the Swiss government grant 00.0529-1.

References

1. S. Boshy and E. Ruppin. Small is beautiful: Near minimal evolutionary neurocontrolllers obtained with self-organizing compressed encoding. In B. Hallam, D. Floreano, J. Hallam, G. Hayes, and J.-A. Meyer, editors, *Proceedings of the Seventh International Conference on Simulation of Adaptive Behaviour*, pages 345–346, Cambridge, MA, 2002. MIT Press-Bradford Books.
2. E. Coen. *The art of genes*. Oxford University Press, New York, 1999.
3. P. Eggenberger. Cell interactions as a control tool of developmental processes for evolutionary robotics. In P. Maes, M. J. Mataric, J.-A. Meyer, J. Pollack, and S. W. Wilson, editors, *From Animals to Animats 4: Proceedings of the Fourth International Conference on Simulation of Adaptive Behavior*, pages 440–448, Cambridge, MA, 1996. MIT Press-Bradford Books.

4. D. Floreano and F. Mondada. Automatic creation of an autonomous agent: Genetic evolution of a neural-network driven robot. In D. Cliff, P. Husbands, J. Meyer, and S. W. Wilson, editors, *From Animals to Animats III: Proceedings of the Third International Conference on Simulation of Adaptive Behavior*, pages 421–430, Cambridge, MA, 1994. MIT Press-Bradford Books.

5. D. Floreano, N. Schoeni, G. Caprari, and J. Blynel. Evolutionary bits'n'spikes. In *Artificial Life VIII Proceedings*. MIT Press, 2002.

6. F. Gruau. Automatic definition of modular neural networks. *Adaptive Behavior*, 3:151–183, 1994.

7. P. C. Haddow, G. Tufte, and P. van Remortel. Shrinking the Genotype: L-systems for EHW? In Y. Liu, K. Tanaka, M. Iwata, T. Higuchi, and M. Yasunaga, editors, *Evolvable Systems: From Biology to Hardware; Proceedings of the Fourth International Conference on Evolvable Systems (ICES 2001)*, pages 128–139, Berlin, 2001. Springer.

8. T. Smith, P. Husbands, and M. O'Shea. Not measuring evolvability: Initial investigation of an evolutionary robotics search space. In *Congress on Evolutionary Computation 2001*, pages 9–16. IEEE Press, 2001.

9. G. Tempesti, D. Mange, A. Stauffer, and C. Teuscher. The biowall: An electronic tissue for prototyping bio-inspired systems. In A. Stoica, J. Lohn, R. Katz, D. Keymeulen, and R. S. Zebulum, editors, *Proceedings of the 2002 NASA/DoD Conference on Evolvable Hardware*, pages 221–230. IEEE Computer Society, Los Alamitos, CA, 2002.

10. A. M. Tyrrell, E. Sanchez, D. Floreano, G. Tempesti, D. Mange, J.-M. Moreno, J. Rosenberg, and A. Villa. POEtic Tissue: An Integrated Architecture for Bio-Inspired Hardware. In *Evolvable Systems: From Biology to Hardware; Proceedings of the Fifth International Conference on Evolvable Systems (ICES 2003)*, Berlin, 2003. Springer.

11. X. Yao. A review of evolutionary artificial neural networks. *International Journal of Intelligent Systems*, 4:203–222, 1993.

Spiking Neural Networks
for Reconfigurable POEtic Tissue

Jan Eriksson[4], Oriol Torres[3], Andrew Mitchell[1,*], Gayle Tucker[2], Ken Lindsay[2],
David Halliday[1], Jay Rosenberg[2], Juan-Manuel Moreno[3],
and Alessandro E.P. Villa[4,5]

[1] University of York, York, UK
[2] University of Glasgow, UK
[3] Technical University of Catalunya, Barcelona, Spain
[4] Lab. of Neuroheuristics, University of Lausanne, Lausanne, Switzerland
[5] University Joseph-Fourier, Grenoble, France
* acmll@ohm.york.ac.uk

Abstract. Vertebrate and most invertebrate organisms interact with their environment through processes of adaptation and learning. Such processes are generally controlled by complex networks of nerve cells, or neurons, and their interactions. Neurons are characterized by all-or-none discharges – the spikes – and the time series corresponding to the sequences of the discharges – the spike trains – carry most of the information used for intercellular communication. This paper describes biologically inspired spiking neural network models suitable for digital hardware implementation. We consider bio-realism, hardware friendliness, and performance as factors which influence the ability of these models to integrate into a flexible computational substrate inspired by evolutionary, developmental and learning aspects of living organisms. Both software and hardware simulations have been used to assess and compare the different models to determine the most suitable spiking neural network model.

1   Introduction

Simple spiking neuron models are well suited to hardware implementation. The particular aspects which are hardware friendly include the ability to translate the output of spiking models to a 1-bit digital representation, and the ability to completely specify the models with a few key parameters. Here we describe some recent research into Spiking Neural Networks (SNN) which can be realised using the POEtic tissue: a flexible computational substrate inspired by evolutionary (phylogenesis), developmental (ontogenesis), and learning (epigenesis) aspects of living organisms. The POEtic tissue follows a 2-D cellular structure consisting of three layers: a genotype plane; a configuration plane; and a phenotype plane, whose combination allows multi-cellular electronic organisms to be easily evolved and grown. For the case where the tissue is required to learn, initially the phenotype will be a SNN although other learning organisms may also be developed on the tissue. A more detailed description of the POEtic project can be found in [1].

Several bio-inspired models capable of endowing the tissue with the ability to learn have been studied [2], ranging from a complex biophysical representation based

A.M. Tyrrell, P.C. Haddow, and J. Torresen (Eds.): ICES 2003, LNCS 2606, pp. 165–173, 2003.

on modified Hodgkin & Huxley dynamics to a greatly simplified Integrate to thresh-old & Fire model. To balance the conflicting requirements of hardware resources and bio-realism, we concentrate on an intermediate level of complexity [3].

Biophysical and computational aspects of spiking neuron models have been extensively studied [4]. Many SNN models now also exist which are suitable for implementation in digital hardware, e.g., [5, 6], with hardware platforms continually being developed for their real-time parallel simulation, e.g., [7, 8]. However, to our knowledge, no SNN models exist which have been successfully developed for their implementation on a digital hardware device which are suitable for combination with onto-genetic [9] and phylogenetic [10] methods.

2   Model Description

In this section we describe the three principle aspects of SNN models. These are the method used in modelling the internal membrane dynamics of an individual neuron; the learning dynamic (which in this case is to be built-in to the individual neuron); and the topological rules by which networks of these neuronal elements are connected.

2.1   Membrane Dynamics

The general form of linear differential equation for a spiking neuron is given in equation (1)

$$C_m \frac{dV_i(t)}{dt} = -G_m(V_i(t) - V_{rest}) - \sum_{j=0}^{n} G_j(t)(V_i(t) - V_j) \tag{1}$$

Where $V_i(t)$ is the membrane potential of cell i at time t, C_m and G_m are the membrane capacitance and conductance, respectively, V_{rest} is the resting potential of the cell, and $G_j(t)$ and V_j are the time dependent conductance change and reversal potential for the n synaptic inputs. The membrane potential is compared with a threshold value to determine if the neuron outputs a spike. After each output spike the membrane potential is instantaneously reset to the resting potential, and no further output spikes are generated for a fixed refractory period. Equation 1 can be rearranged to give

$$\frac{dV_i(t)}{dt} = -f_i(t)V_i(t) + h_i(t) \tag{2}$$

where $f_i(t) = (G_m + \sum G_j(t))/C_m$ and $h_i(t) = (G_m V_{rest} + \sum G_j(t)V_j)/C_m$, the solution of which can be approximated, for a time step of Δt as

$$V_i(t + \Delta t) = (V_i(t) - h_i(t)/f_i(t))\exp(-f_i(t)\Delta t) + h_i(t)/f_i(t) \tag{3}$$

The quantity $1/f_i(t)$ can be regarded as the time constant of the cell. Ignoring the effects of the synaptic inputs, this can be approximated by the constant $\tau_m = C_m/G_m$. Thus

$$V_i(t + \Delta t) = V_{rest} + k_m \left(V_i(t) - V_{rest} \right) + \left(1 - k_m \right) \sum I_{ij}(t) \tag{4}$$

where $k_m = \exp(-\Delta t/\tau_m)$ and $\sum I_{ij}(t)$ approximates the change in membrane potential resulting from synaptic (and any other) input currents. We refer to k_m as the kinetic membrane constant. This allows the exponential decay of the membrane to be approximated using a simple algebraic expression. An alternative approach is described in section 4, dealing with hardware aspects. A further simplification results from setting V_{rest} to zero to give the final form of the spiking neuron model. The firing condition is

$$\text{if } Vi(t) \geq \theta \text{ then } S_i(t) = 1$$
$$\text{if } Vi(t) < \theta \text{ then } S_i(t) = 0 \tag{5}$$

where $S_i(t)$ is the binary valued output of the model, and θ is the (fixed) firing threshold of the cell, a value of 1 for the variable S_i signals the presence of an output spike. The update equation used in combination with the firing condition is

$$\text{if } S_i(t) = 1 \text{ then } V_i(t + \Delta t) = 0$$
$$\text{if } S_i(t) = 0 \text{ then } V_i(t + \Delta t) = k_m V_i(t) + \left(1 - k_m \right) \sum I_{ij}(t) \tag{6}$$

2.2   Learning Dynamics

Although several attempts at applying conventional artificial neural network learning algorithms (e.g., backpropagation) to SNNs exist, e.g., [11], for the purposes of the POEtic project we required a simple unsupervised learning method which would allow the adaptation of weights based on pre- and post-synaptic activity. A recent discussion of several such spike-timing dependent synaptic plasticity (STDP) learning rules can be found in [12]. Whilst these rules offer a fast and unsupervised adaptation of synaptic weights, the resulting networks are often prone to memory loss in noisy environments. However, a method recently proposed by Fusi [13, 14] aims to address this problem through discretising the synaptic weight variable. It has also been shown via simulation that this method reduces the computational load per neuron for models using plastic synaptic weights [15, 16]. An added advantage of this technique, from our perspective, is its increased suitability for digital hardware implementation. The learning rule proposed by Fusi uses an internal learning state variable for each synaptic input which functions as a short term memory, the value of this variable is used to discretise the synaptic weight into a predetermined number of states. We have chosen to modify the STDP method used by Fusi for updating the synaptic weight, so that the state variable is increased if the post-synaptic spike occurs after the pre-synaptic spike, and is decreased if the opposite temporal relation holds.

Each synapse has two main variables for the connection from cell j to cell i, these are an activation variable A_{ij} and a learning state L_{ij}. The activation variable depends on the class of cells i and j, i.e. inhibitory or excitatory. The synapses of excitatory to excitatory neurons have activation variables which can vary between a chosen number of discrete states, i.e., [0, 1, 2, 4], all others have a constant $A_{ij}=1$.

The learning state, L_{ij}, is governed by the following equation:

$$L_{ij}(t+1) = k_{act} \times L_{ij}(t) + \left(YD_j(t) \times S_i(t)\right) - \left(YD_i(t) \times S_j(t)\right) \qquad (7)$$

where k_{act} is a kinetic activation constant used to approximate an exponential decay with time constant τ_{act} of $L_{ij}(t)$. The YD variables define a time window (with exponentially decaying effect) after each pre- and post-synaptic spike where modification of the learning state, $L_{ij}(t)$, is possible. For post-synaptic neuron i, YD is modeled as:

$$\begin{array}{ll} \text{if } S_i(t) = 1 & YD_i(t+1) = YD_{MAX} \\ \text{if } S_i(t) = 0 & YD_i(t+1) = k_{learn} \times YD_i(t) \end{array} \qquad (8)$$

where k_{learn} is a kinetic learning constant approximating an exponential decay with time constant τ_{learn}. From equation 7, we can see that in the absence of any spiking activity, the learning state is a continually decreasing function. However, if the pre-synaptic cell fires just before a post-synaptic cell the learning state will increase by $YD_j(t)$, whereas if the pre-synaptic cell fires just after the post-synaptic cell, the learning state will be decreased by $YD_i(t)$. The respective value of the YD variables reflects the level of coincidence of the pre- and post-synaptic firings as defined by equation 8. The learning state is used to determine the discrete level of activation of the synapse $A_{EXC->EXC}$. The number of discrete activation states is a fixed parameter of the model, and may be two [0,1], three [0,1,2], or more. Note that, according to equation 7, if there is no activity at the synapse then L_{ij} drifts back towards 0 (whether it's value is positive or negative). If L_{ij} reaches a positive threshold ($L_{ij} > L_{th}$) and A_{ij} is not already equal to A_{max}, then A_{ij} is incremented and the value of L_{ij} is reset to $L_{ij} - (2 \times L_{th})$. Likewise, when L_{ij} reaches a negative threshold ($L_{ij} < -L_{th}$) and A_{ij} is not already equal to 0, then A_{ij} is decremented and the value of L_{ij} is reset to $L_{ij} + (2 \times L_{th})$. The input I_{ij} in equation 4 is then given by

$$I_{ij} = \left(W_{ij} \times A_{ij}\right) \qquad (9)$$

where W_{ij} is a constant multiplier used for generating the final weight added to the membrane potential on receiving a given input.

2.3 Network Topology

In this section, we consider the method of network creation. While evolutionary techniques [10] may be applied to determine the parameters of any part of this model, ontogenesis [9] is essentially concerned with the topology. For our purposes this process has been automated using a set of topological rules. These rules define the size of the network, the method and depth of connectivity and the neuronal class proportion. In the experiments which follow, the network contained 50 x 50 neurons N=[0,2499], 80% of which were excitatory, leaving 20% inhibitory, in line with the generally accepted ratio of the neo-cortex [17]. The excitatory cells may connect to other cells with differing synaptic weight values (see 2.2 above). In the simulations described below, each cell makes connections to other cells within a 5 x 5 neighbourhood, i.e., 24 other cells.

3   Software Simulation Results

In this section, we explore the potential of the SNN model described previously to mimic experimental results of neuroscientific studies. We first study the behaviour of the network with static synaptic weights, then with the learning activated we look at its ability to perform a visual discrimination task.

The network was first initialised (t=0) where all neuronal states are set to zero $(S_i(0) = 0)$ and all membrane levels are set to zero $(V_i(0) = 0)$. A brief pulse stimuli, with sufficient amplitude to fire the neuron, was then input to all excitatory cells in the network. Sustained activity was then observed within the network, achieving low average network activities (e.g., 1 spike every 96 sampling intervals) without causing global oscillations. A trace recorded from one of the output cells can be seen in figure 1. This simulation without learning demonstrates that a simple network can hold information indefinitely through persistent activity.

Fig. 1. A five second recording of the activity of a single cell with a mean interval of 40 samples.

The next stage is then to allow learning to investigate the possibility of exploiting different activity patterns as memory depending on the input stimuli. We apply the learning rule as described earlier to the same network used in the previous experiment. The network is initialised at t=0 as before. As well as the input stimuli, this time a low-level Gaussian noise input is also applied at each time step to every cell in the network. This stimuli was designed to simulate sinusoidal gratings moving horizontally at different speeds over a visual receptive field. Training of the network then involved moving the stimulus across the network from left to right at speed=1 (i.e., 50 neurons per second) for a period of 10 seconds.

Figure 2 shows the results following training for applied stimulus speeds of 1 and 2, respectively. In addition to the response of a single excitatory neuron a "local" activity was also measured. The local activity corresponds to the average activity in a narrow strip perpendicular to the movement of the grating. These results demonstrate that the network has been successfully trained to discriminate direction of a moving stimulus across the network.

The effect of reducing the strength of the stimulus is illustrated in figure 3. Here only the forward moving stimulus triggered any response, relying on the presence of additional input noise to provoke the discharge of several neurons. This result suggests that the trained network is also sensitive to an input stimulus of varying amplitude.

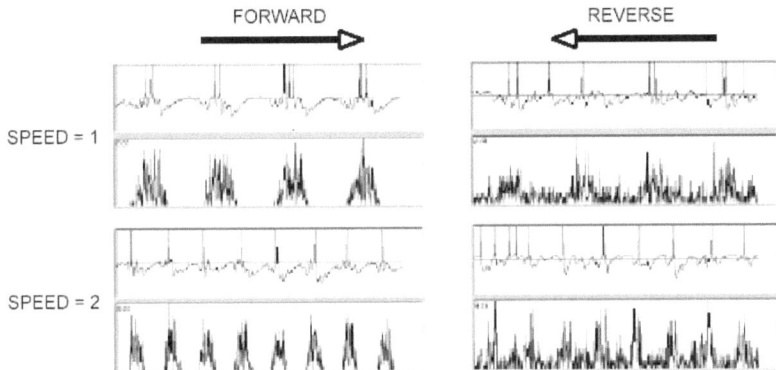

Fig. 2. The response of a single excitatory cell (top trace) and local activity (bottom trace) for input stimuli moving at two speeds, in both forward and reverse directions, across the network.

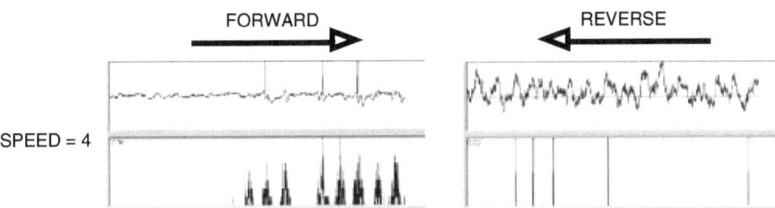

Fig. 3. Recordings from a single cell and associated local activity as a result of applying low level stimulation (original amplitude/4) moving at a faster speed (200 neurons/sec).

4   Digital Hardware Implementation

Several aspects of the previously described model remain under development. We present here the latest implementation of several functional units following the specifications defined by the theoretical model.

A block diagram of a neuron with two inputs is shown in figure 4. From this diagram it can be seen that the inputs are added and the result is stored in a register. At the following clock event this result is added, using the second adder, to the current membrane potential fed back from the decay function block. The result of this addition gives the new membrane potential now found at the output of the decay function block. However, if the result of the first addition is zero ,i.e., no inputs are present, then the membrane potential is not updated and instead the decay function becomes active implementing the leaky membrane.

Equation 4 represents a common approach to approximating exponential decay. However, the decay function currently used is an approach more suited to hardware implementation approximating exponential decay through the control of two parameters M and *STEP*. With the membrane potential held in a shift register, the algorithm is implemented as follows:

```
m=Most_significant_1(membrane)+1                    (i)
m=min(m,M)-1                                        (ii)
delta_mem=m>>membrane                              (iii)
iSTEP=(M-m)*2^STEP                                  (iv)
```

- The first step is to determine the Most Significant Bit (MSB) of the membrane potential which is stored as the variable, m (i).
- If the MSB is set then the membrane potential is decreased by one and steps (ii) and (iii) are bypassed. If the MSB is not set then the minimum of M and the MSB of the membrane potential less one is calculated (ii).
- The shift register holding the membrane potential is then shifted down by m bits to give delta_mem (iii).
- Finally, the time delay (iSTEP) until implementing the new value is calculated (iv).

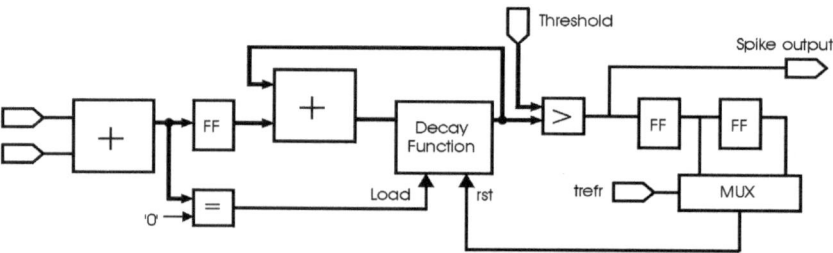

Fig. 4. A block diagram representation of the neuron model implementation prototype for digital hardware

The output of the decay function block is continually compared with the firing threshold. If the firing threshold is surpassed then the output of the comparator goes high, producing an output spike. The output is also fed back resetting the decaying membrane potential to implement an approximate refractory period.

The second part of the implementation consists in the definition of a memory for each synapse which works according to the learning rules explained earlier. Referring to figure 5, the counter holds a variable '*xmem*'. Upon receiving a spike at the synaptic input a new value of '*xmem*' is loaded into the counter. If the membrane potential of the neuron is greater than mem_threshold (note: not the firing threshold), $xmem(t+1)=xmem(t)+\Delta x$, otherwise $xmem(t+1)=xmem(t)-\Delta x$. However, if no spike is present at the synaptic input the counter increases or decreases its value depending upon its current value i.e. if xmem is in the range [0,7] it will decrease to zero otherwise it will be increased to its maximum value of 15.

This implementation represents a simplified version of the methods discussed in section 2.2 where four separate activation states were described. Here, xmem represents the learning variable L_{ij} where $xmem=[0,7]$ and $xmem=[8,15]$ represent the two activation states $A_{00}=0$ and $A_{00}=1$ respectively. Simple methods for implementing further activation states are currently under investigation.

Simulations of this implementation have also been completed showing its correct operation governed by the theoretical model, for more details see [2].

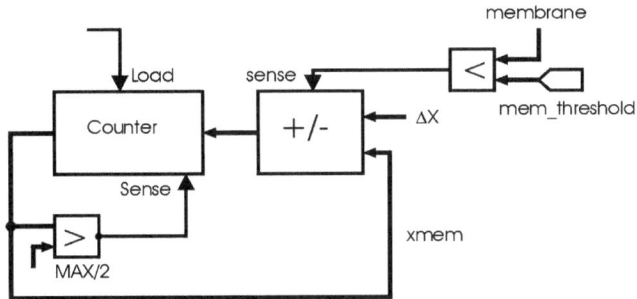

Fig. 5. A block diagram representation of the digital circuit used to implement synaptic weight storage and adaptation

5   Discussion

The model presented has been developed following stringent guidelines, tested first using a software model and finally presented as a hardware implementation.

Neurophysiological studies of neuronal correlates of working (short-term) memory show that the maintenance of a stimulus in working memory is accompanied by persistent (enhanced relative to spontaneous activity) activity in a selective subset of neurons. A long-standing hypothesis has been that persistent activity is a collective property of a network, and that it is maintained by recurrent excitatory feedback [18].

It is difficult to achieve sustained activity without high firing rates. In a network of the type we have envisaged, depending on appropriately selected synaptic strengths, the activity obtained consists generally of synchronous oscillations. Only if external input (noise) is applied other sustained patterns of activity will appear.

The simulations that were made indicate that various patterns of sustained activity without external input can be obtained if some modifications are made to the typical integrate-and-fire model. Our simulation results do not represent an attempt at an exhaustive analysis of the model but an exemplification of some possibilities for uses of the model once combined with the ontogenetic and phylogenetic mechanisms.

The theoretical model and simulation results presented have been used to develop a prototype for digital hardware implementation. This implementation is currently under development and once complete should fully respect the theoretical model.

Acknowledgements

This project is funded by the Future and Emerging Technologies programme (IST-FET) of the European Community, under grant IST-2000-28027 (POETIC). The information provided is the sole responsibility of the authors and does not reflect the Community's opinion. The Community is not responsible for any use that might be made of data appearing in this publication. The Swiss participants to this project are partially supported under grant OFES 00.0529-2 by the Swiss government.

References

1. A. M. Tyrrell, E. Sanchez, D. Floreano, G. Tempesti, D. Mange, J.-M. Moreno, J. Rosenberg, and A. Villa. POEtic Tissue: An Integrated Architecture for Bio-Inspired Hardware, *Submitted to the Fifth International Conference on Evolvable Systems, ICES03.*
2. http://www.poetictissue.org
3. S. L. Hill, A.E.P. Villa, Dynamic transitions in global network activity influenced by the balance of excitation and inhibition, Network, *Comp. Neural Syst.* 8: 165-184, 1997.
4. W. Maass, C. M. Bishop, Pulsed Neural Networks, MIT Press, 1998.
5. C. Christodoulou, G. Bugmann, J. G. Taylor, T. G. Clarkson, An Extension of the Temporal Noisy-Leaky Integrator Neuron and its Potential Applications, *Proceedings of the International Joint Conference on Neural Networks*, Vol.3, pp. 165-170, 1992.
6. S. Maya, R. Reynoso, C. Torres, M. Arias-Estrada, Compact Spiking Neural Network Implementation in FPGA, in R.W. Hartenstein and H. Grünbacher (Eds.): FPL 2000, LNCS 1896, pp. 270–276, 2000.
7. A. Jahnke, U. Roth and H. Klar, A SIMD/Dataflow Architecture for a Neurocomputer for Spike-Processing Neural Networks (NESPINN), *MicroNeuro'96*, pp. 232-237, 1996.
8. G. Hartmann, G. Frank, M. Schafer, C. Wolff, SPIKE 128k - An Accelerator for Dynamic Simulation of Large Pulse-Coded Networks, *Proceedings of the 6th International Conference on Microelectronics for Neural Networks, Evolutionary & Fuzzy Systems*, pp.130-139, 1997
9. G. Tempesti, D. Roggen, E. Sanchez, Y. Thoma, R. Canham, A. Tyrrell, Ontogenetic Development and Fault Tolerance in the POEtic Tissue, *Submitted to the Fifth International Conference on Evolvable Systems, ICES03.*
10. D. Roggen, D. Floreano, C. Mattiusi, A Morphogenetic System as the Phylogenetic Mechanism of the POEtic Tissue, *Submitted to the Fifth International Conference on Evolvable Systems, ICES03.*
11. S. M. Bohte, J. N. Kok, H. La Poutre, SpikeProp: Backpropagation for Networks of Spiking Neurons, proceedings of the European Symposium on Artificial Neural Networks, pp.419-424, 2000.
12. P. D. Roberts, C. C. Bell, Spike-Timing Dependant Synaptic Plasticity: Mechanisms and Implications, *http://www.proberts.net/RESEARCH.HTM, to appear in Biological Cybernetics.*
13. S. Fusi, Long term memory: encoding and storing strategies of the brain, *Neurocomputing* 38-40, pp.1223-1228, 2001.
14. S. Fusi, M. Annunziato, D. Badoni, A. Salamon, D. J. Amit, Spike-Driven Synaptic Plasticity: Theory, Simulation, VLSI Implementation, *Neural Computation* 12, pp.2227–2258, 2000.
15. M. Mattia, P. Del Giudice, Efficient Event-Driven Simulation of Large Networks of Spiking Neurons and Dynamical Synapses, *Neural Computation* 12, pp. 2305-2329, 2000
16. P. Del Giudice, M. Mattia, Long and Short term synaptic plasticity and the formation of working memory: a case study, *Neurocomputing* 38-40, pp.1175-1180, 2001.
17. R. Douglas, K. Martin, Neocortex, Chapter 12 of *The Synaptic Organisation of the Brain* (G. M. Shepherd, Ed.), Oxford University Press, 1998.
18. A. E. P. Villa, Empirical Evidence about Temporal Structure in Multi unit Recordings, in: Time and the Brain (R. Miller, Ed.), Conceptual advances in brain research, vol. 2., Harwood Academic Publishers, pp. 1 51, 2000.

A Learning, Multi-layered, Hardware Artificial Immune System Implemented upon an Embryonic Array

Richard Canham and Andy M. Tyrrell

University of York, Heslington, York. YO10 5DD. UK
roc100@ohm.york.ac.uk
www.bioinspired.com

Abstract. A simple robot control system is used to demonstrate bio-inspired fault tolerance techniques; a multi-layered hardware immune system is used for fault detection and an embryonic array for fault avoidance. The acquired layer of the immune system monitors system behaviour for unusual activity (normal behaviour is learnt during an unsupervised teaching period). An non-learning innate layer is then employed to localise the fault if possible. The embryonic array allows simple, robust reconfiguration to avoid the fault; as many faults as spare cells can be accommodated. The complete system, including the learning algorithm, is implemented on a Virtex FPGA. Results showing the appropriate response to different types of faults are given.

1 Introduction

Through millions of years of refinement biology has produced many living creatures that are remarkably fault tolerant. They can survive injury, damage, wear and tear, and are under continual attack from other living entities in the form of infectious pathogens. They make use of multiple mechanisms, or multiple layers, to generate this fault tolerance. This paper details a demonstration hardware system that is inspired by some of these system to provide fault tolerance. A multi-layered hardware artificial immune system that learns normal behaviour is coupled with an embryonic array to provide fault detection and fault avoidance. A simple control system, to perform object avoidance for robot control, is used to demonstrate the system.

The embryonic array is a homogeneous array of logic units (called cells) that use their location within the array to extract appropriate configuration data. Each cell contains all the configuration details of all cells and hence can perform any cell's function as required.

This is part of the POEtic project which aims to produce a circuit that combines other forms of bio-inspired techniques [26]. Ultimately the POEtic device will be produced in silicon using ASIC (Application Specific Integrated Circuit) fabrication. However, this paper details the initial stages of just the immune system upon an embryonic array which has been implemented on a Xilinx Virtex Field Programmable Gate Array (FPGA).

A.M. Tyrrell, P.C. Haddow, and J. Torresen (Eds.): ICES 2003, LNCS 2606, pp. 174–185, 2003.

The paper provides a background to artificial immune systems, including some details of immunotronics, and embryonics in section 2. The system is described in section 3 which is followed by the results in section 4. Some discussions on further work and comments are given in section 5 before the conclusions are given in section 6.

2   Background

2.1   Artificial Immune Systems

The immune system found in higher organisms is a multi-layered, distributed system that is robust and can identify numerous pathogens and other harmful effects. Many of the properties found in such a system would be most advantageous in many computer and other systems. Artificial immune systems do just this.

One of the most researched algorithms is the negative selection algorithm. Developed by Forrest et al. [18], the negative selection algorithm is based upon the detection of non-self from self, as found within the immune system. Various immune cell types (such as lymphocytes) have receptors that allow them to bind to specific sets of proteins. The maturation of each lymphocyte cell involves the presentation of proteins that are naturally present within the body (self). Lymphocytes that bind with them are destroyed. Hence, when released within the body, binding to a protein indicates it is non-self and may be a harmful pathogen. See [1, 14, 21, 23] for more details of immune systems.

Forrest uses a string to represent the system's state; partial matching of these strings and a detector string is used to distinguish between self and non-self. A set of detector strings is generated such that they do not match with the normally occurring states - they only match with non-self. Hence, a detector match gives an indication that some event has occurred.

Artificial immune systems have been applied to many applications, those that make use of the negative selection algorithm include computer security [18], computer intrusion [16, 17], network security [9, 19, 22], pattern recognition [11, 20], industrial anomaly detection [12, 13] and hardware fault detection (immunotronics) [4–7]. Comprehensive surveys and reviews can be found in [10, 14, 15].

Avizienis [2] discusses, with no specific detail, the concept of using an immune inspired process that surrounds a processor to provide fault tolerance. Separate and distributed hardware is suggested to isolate the fault detection from the processor.

2.2   Immunotronics

All the applications listed are software implementations of an artificial immune system, with the exception of immunotronics (immune + electronics). This was devised by Bradley and Tyrrell [4–7]. It makes use of a negative selection algorithm to identify faults within a hardware circuit, specifically a finite state machine. The current state, next state and the current inputs are used to define the current transition of the machine. Some of the transitions are not present in normal, error free operation and so are considered as non-self. A 4 bit BCD counter implemented upon a Xilinx Virtex FPGA was immunised. The detector set was generated offline and implemented using a content addressable memory (CAM) [31]. Partial matching between the detectors and the current state, based upon the number of contiguous bits, was employed. Experimental results showed that a fault coverage of 93% could be achieved with 103 detectors. This is very impressive but requires a CAM of 103 words, each 10 bits in width. This is large in size compared to the counter that was being immunised. Only fault detection has been considered in current implementations.

The size of the detectors is not dependent on the complexity of the circuit being immunised. It might be considered that the counter is the wrong granularity and would be more appropriate to implement immunotronics at a system, or sub-system level. It would not now be used to identify every error, but unusual behaviours that indicate something erroneous has occurred. If a robot controller is considered, the immune system would provide monitoring for situations that should not occur. This may include situations such as a robot that is heading for an object.

2.3 Embryonics

All multi-cellular organisms start life as a single cell. This cell divides repeatedly to generate numerous identical copies of itself. Embryonics (embryo + electronics) is inspired by the cloning and differentiation of cells within multi-cellular organisms to generate electronic circuits with some of the properties of such organisms [25, 27, 28, 30]. An embryonic array consists of a homogeneous array of cells, each containing the full specification of the device. The coordinate of each cell is calculated dependent upon its neighbours and is used to define its funcionality.

Errors within the array are accommodated by killing the faulty cell. The routing becomes transparent and the coordinate system no longer increments, thus the next cell takes over the functionality of the faulty cell. The array contains spare cells that are not utilised until a fault occurs. To maintain cellular alignment, removing the row or column that contains the error is typical. Fault avoidance has also been performed at sub-cell level. As many faults as there are spare cell columns or rows can be accommodated.

3 System Description

3.1 Overview

The system implemented uses a number of layers to provide fault tolerance using the immune system for inspiration. The top level is a negative selection algorithm that learns and is analogous to the acquired immune system. This can work at either a system or subsystem level and monitors the state of the system for non-self. When non-self is identified the system could be relocated by killing the cells within the embryonic array. For larger systems that use a number of cells, a second layer could be used to localise the fault. This could take many forms: a section of the device could implement a reduced functionality while the rest is tested. The functional section is then relocated to allow complete testing. Alternatively the device is placed in a safe state and taken offline while the testing takes place. Cells that have errors are identified and killed. These tests do not learn and hence have been labelled as the innate immune system which is present in biological systems. However, this is just an identifier and does not reflect the complex processes that occur in a biological innate system.

A simple embryonic array has been implemented to test the immune system. A simple example application is used to demonstrate the system. This takes the form of the simplest of robot controllers which has three single bit inputs and two single bit outputs; the inputs represent the detection of an object to the robot's left, front and right. The outputs represent signals to the robot's left and right motors. Hence a signal on the left motor will turn the device to the right, both motors will cause it to travel forward and a signal on the right motor will turn the robot to the left. This is shown in figure 1. The controller is implemented within a single cell and a fault detected will result in that cell being killed. Hence an innate system cannot localise the fault any further and so is used to test the transparency mechanism. The nature of the controller is not important; in this

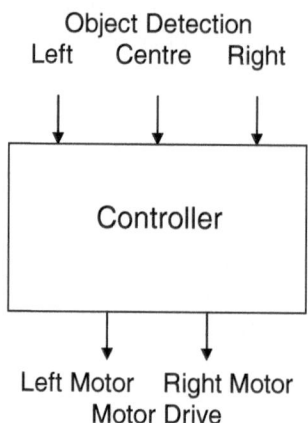

Fig. 1. Test System

demonstration it took the form of a simple lookup table which was driven by a linear feedback shift register (LFSR) to generate a random, complete set of inputs. When the outputs are considered with the inputs there are a number of states that would be considered as normal (self-states) and those that would be considered as abnormal (non-self). An example of non-self would be the detection of an object ahead with both motors driving the robot into it. A period of fault free operation is used as a learning period where normal, self states are presented to the acquired layer.

3.2 The Embryonic Array

The embryonic array is based upon the arrays produced by Ortega and Tyrrell [24, 25]. However, several changes are required. Together with its control circuitry, each cell comprises an array of sub-units (called molecules). The functional unit of each molecule comprises a four input lookup table (LUT) and a D-type flip-flop. A three bit bus connects each molecule to its four nearest neighbours via a switch block. The output of the molecule's LUT or flip-flop can also be connected to any of the switch block's outputs. Each input of the LUT can be connected to any of the switch block's inputs.

A one-to-one genotype (the configuration data) to phenotype (the resultant hardware) mapping is employed. The configuration of each molecule is controlled by a configuration register; the coordinate of the cell selects the appropriate gene from within the genotype to fill this configuration register.

The coordinates for each cell are calculated based upon its neighbours; each cell increments the address in the x and y axis which is then propagated on. Faults are accommodated by typically killing the entire column in which the faulty cell is present. A gene is selected that sets all the switch blocks to be transparent and the cell coordinates are no longer incremented. However, it is also possible to kill a row. This is necessary if the cell will not die appropriately or there is a fault present in the transparency of the cell. This allows a cell to be completely removed, no matter the degree of damage. More details can be found in [8]

3.3 The Artificial Immune System

The fault detection takes the form of a number of immune layers. The top-most layer is the implementation of a negative selection algorithm which is complemented by the innate layer.

The Negative Selection Algorithm. Using the negative selection algorithm to identify unusual situations at system level and just considering the current state reduces the number of error states dramatically. In this example the number is reduced to such an extent that there are more self states than non-self states. This makes partial matching between detectors and the current state no longer possible, since the partial match is more likely to match with another self state than non-self state. The number of holes becomes too large to be practical, even with appropriate changes in the state representations [19]. The nature of the partial match within the negative selection algorithm (and in biological systems) is very powerful and, some may argue, a key component of the system. It allows a greatly reduced set of detectors to identify a wider range of pathogens. However, when implemented in hardware there are a number of constraints and restrictions that prevent full advantage of the partial matching. The luxury of complex software algorithms is not available. The reduced number of non-self states, and the small number of bits in this example allows a complete detector set which is still very small. With a detector of only five bits a complete comparison can be implemented in a simple five input lookup table (implemented in three 4 input LUTs). This compares favourably with the typical CAM implementation used by Bradley [5] which required a LUT for every two bits of a detector and used over 100 detectors. The system still learns and identifies non-self states which is the fundamental aspect of the algorithm.

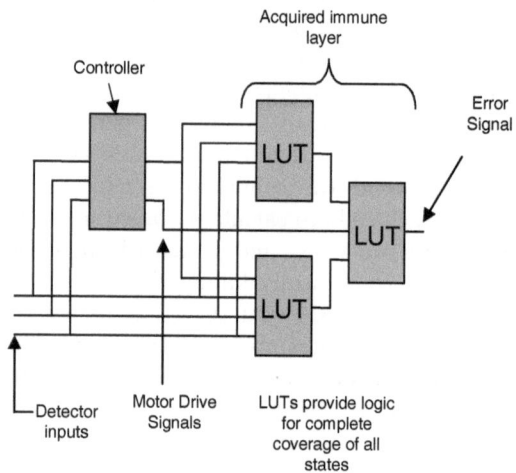

Fig. 2. System Block-Diagram

Figure 2 shows a block diagram of the controller and acquired immune system. The implementation used cells of 10 molecules; the controller, complete with LFSR for simulated object detection, is contained within one cell and the acquired immune system in another.

The acquired layer learns during a teaching period. This is a simple process since all detectors are present. No random number generator or other complex elements are required and so it is practical in hardware. This is different to the Forrest implementation of the negative selection algorithm that first generates a number of detectors that are then destroyed if not appropriate. Here we start with the complete set of detectors that are then reduced. Initially all states are set to indicate a fault. During the learning period states that occur change the appropriate bit within the detector's LUT. Hence the output generated from that state no longer indicates an error; the detector that matched self was destroyed. A *learning line* is fed to the immune cell to prevent it from killing the control cell during the learning process. Since the *normal* genotype is reloaded following the innate testing process (see section 3.3), the changes to the detector set are changed within the genotype to restore them when the genotype is reloaded. This is performed externally to the embryonic array but still in hardware within the FPGA.

The Innate Layer. In this demonstration implementation the innate immune layer is used to confirm that the fault detected by the acquired layer was not such that the cell could not become transparent. In a more complex system tests could be performed to localise the fault. This is achieved by reconfiguring the device and performing test patterns at a low level. The number and type of tests performed can generate an exhaustive test sequence in a manner similar to Sundararajan [29]. A large number of tests would be performed in parallel, with the repetition of the same gene in a number of the cells.

A control process is necessary to carry out the reconfiguration and error checking. On the final POEtic device a microcontroller will be included within the device which will be utilised. However, this generates a large single point of failure. In an ultimate solution the controller would be implemented within the array itself which then offers all the fault tolerant protection provided. It should be noted that unlike many other FPGA fault tolerant techniques that use reconfiguration to locate and/or avoid faults (such as [3]) there is no complex reconfiguration calculation required; this is all achieved by the embryonic array. Hence, an integrated controller would be practical. In this example the control was provided within the FPGA, although not within the embryonic array.

As previously described, this example application implements the robot controller in a single cell, any non-self detected causes that cell to be killed and its column made transparent. However, innate tests are performed to check the correct transparency of the device. Failure of this test results in the appropriate row being made transparent. Two tests are required; the second necessary to test the transparency of the cell used to generate the test signals in the first test. Details can be found in [8].

4   Results and Discussion

Figure 3 shows a simulation of the system. At time 0 the configuration of the array is initiated with the controller and acquired immune cell for the learning period. The system state is denoted as learning_config and is followed by a very brief learning period. Each state generated during this period changes the detector set data appropriately. The *normal* period now commences; first the configuration followed by normal running. An expansion is shown in figure 4. The lower traces show the output of the system; the pseudo-random sequence of the LFSR that simulates the object detection (a time shifted copy of the previous trace) together with the motor drive signals. It can be seen that cell (0,0), that contains the controller, has an OK state and is using gene 0001. At time $600\mu s$ a molecule output within the controller has a stuck at one fault injected. This is detected (the error line goes high); the cell state goes to dead and gene

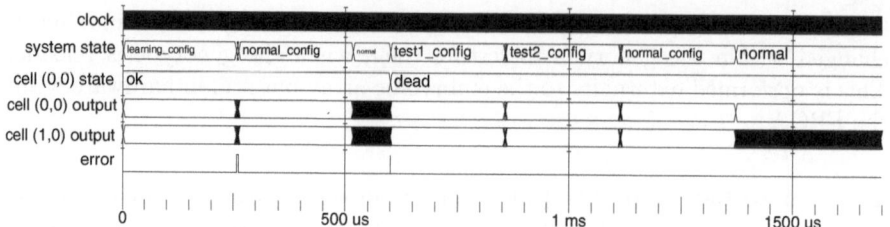

Fig. 3. Simulation of function stuck at fault

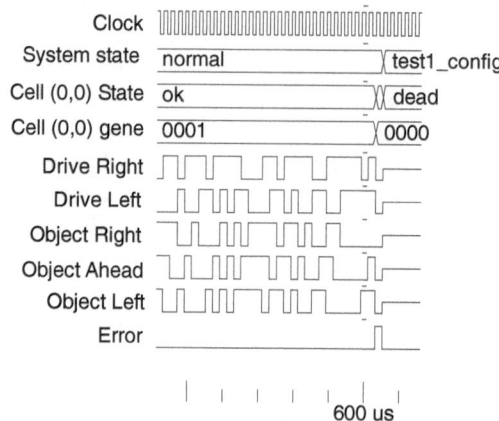

Fig. 4. Expanded section of normal running

0000 (transparent) is selected. Tests 1 and 2 are then performed; this can be more readily seen in figure 3. The combined output of cell (0,0) can be seen on this trace. Once dead the output remains constant and cell (1,0) takes over the controller function. It should be noted that the cell references used here are the physical location, not the address calculated within the embryonic array.

A similar process is shown in figure 5, however, the injected fault is within the routing between molecules and hence will not be avoided by making that column transparent. In this example the transparent cell is unused but the next stage of the work is to inject real inputs from a robot. These would typically enter the array on the left and pass through this cell, hence an error in the routing would block these signals.

As before, cell (0,0) dies when the fault is detected but test 2 detects that the row can no longer become transparent and so the bottom row also becomes transparent (note that cell (1,0) has become transparent). Cell (1,1) now implements the controller. Before any faults the states of the six cells can been seen in figure 6a. The fault injected in the molecule output results in the states shown in 6b. Finally the fault in the routing between molecules results in the states shown in 6c. It should be noted that a dead cell attempts to become transparent to all signals. However, this may not be possible if a fault in the routing is present.

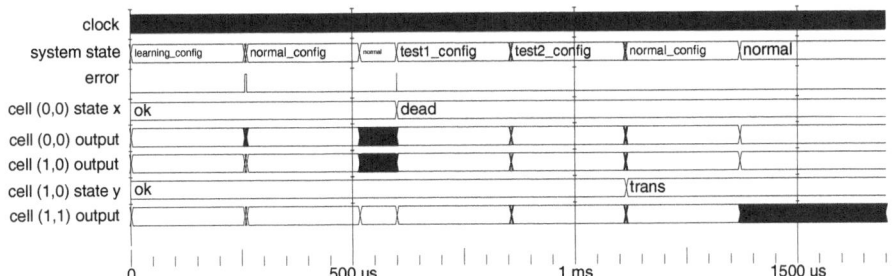

Fig. 5. Simulation of route stuck at fault

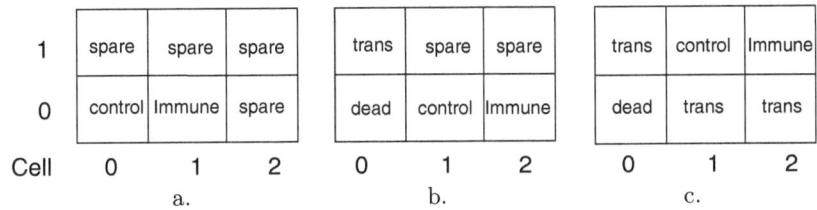

Fig. 6. Cell States

The circuit simulated above was synthesised and implemented upon a Xilinx Virtex 1000. Both routing and non-routing faults were injected and successfully detected and accommodated in a similar manner to the simulations.

5   Further Comments

5.1   Advanced Learning

The current learning requires a period of fault free operation during which all the non-fault states are presented. Although there are applications where this is possible, this can become a non-trivial task in some complex systems. Biological immune systems are presented with a similar problem; not all self proteins are presented to the maturing lymphocytes. However, lymphocytes will only react if the immune system is stimulated by other indicators of a pathogen presence (including detection of damage by the innate system). Hence a learning period could be used to locate most self matching detectors. However, if a fault was repeatedly identified by a given detector but no fault was located by the innate tests then the detector should be removed. Care would have to be taken over temporary faults and design faults. Further work is required here.

5.2   Temporal Characteristics

Faults in the controller example are only identified if they cause the device to approach an obstacle. It could be possible that the fault forces it to follow a circle, or spin, such that it never approaches an obstacle. A second, small detector set could consider the length of time a particular state is held, since the robot's normal behaviour would not include being in a turn state for a protracted length of time.

5.3   Independent Nature

This paper details a hardware immune system that protects a simple hardware robot controller. However, the system is independent of the controller; it makes use of a totally different process than the controller. Although it is implemented within the same device this is not necessary. The controller (or any other suitable application) could be a complex hardware or software device implemented upon separate hardware or microprocessor; the hardware immune system can be added without interference. Replication of the detector set within an embryonic array would provide a robust fault detection system such as that envisaged by Avizienis [2] (see section 2.1).

6   Conclusions

A simple robot controller has been implemented on a commercial FPGA which makes use of biologically inspired *layers* to provide fault tolerance. An immune system that learns normal behaviour is used for fault detection which then makes use of an embryonic architecture to easily avoid the fault (fault recovery).

Using the acquired immune system to monitor at a system level for unusual behaviour (rather than all faults) allows a very small detectors set to be used with an efficient unsupervised learning algorithm. The learning algorithm has been implemented in hardware. This should be compared to implementations of immunotronics by Bradley and Tyrrell which is the only other hardware implementation of a negative selection algorithm. This requires a very large detector set which is generated on a host PC in software. Also their system only implements fault detection. The use of the embryonic array detailed allows as many faults as there are spare cell columns and rows to be accommodated.

The nature of the controller immunised is of little consequence since only the inputs and outputs are considered - the immune system is independent and can be added to any suitable system.

Acknowledgements

The authors would like to thank other members of the POEtic team for their help and ideas. This project is funded by the Future and Emerging Technologies programme (IST-FET) for the European Community, under grant IST-2000-28027 (POETIC). The information provided is the sole responsibility of the authors and does not reflect the Community's opinion. The Community is not responsible for any use that might be made of data appearing in this publication.

References

1. B. Alberts, D. Bray, J. Lewis, M. Raff, K. Roberts, and J. Watson. *Molecular Biology of the Cell.* Garland Publishing, New York, 3 edition, 1994.
2. Algirdas Avizienis. Design diversity and the immune system paradigm: Cornerstones for information system survivability. *Information Survivability Workshop,* 2000.
3. R.D. Blanton, S.C. Goldstein, and H. Schmit. Tunable fault tolerance via test and reconfiguration. *International Fault-Tolerant Computing Symposium,* 1998.
4. D.W. Bradley and A.M. Tyrrell. Hardware fault tolerance: An immunological solution. *Proceedings of IEEE Conference on Systems, Man and Cybernetics,* 1:107–112, 2000.
5. D.W. Bradley and A.M. Tyrrell. Multi-layered defence mechanisms: Architecture, implementation and demonstration of a hardware immune system. *4th International Conference on Evolvable System. Lecture Notes in Computer Science,* 2210:140–150, 2001.
6. D.W. Bradley and A.M. Tyrrell. A hardware immune system for benchmark state machine error detection. *Congress on Evolutionary Computation,* 2002.
7. D.W. Bradley and A.M. Tyrrell. Immunotronics - novel finite-state-machine architectures with built-in self-test using self-nonself differentiation. *IEEE Transactions on Evolutionary Computation,* 6(3):227–38, 2002.
8. R.O. Canham and A.M. Tyrrell. A multilayered immune system for hardware fault tolerance within an embryonic array. *1st International Converence on Artificial Immune Systems,* pages 3–11, 2002.
9. D. Dasgupta. An overview of artificial immune systems and their applications. In *Artificial Immune Systems and Their Applications,* pages 3–21. Springer-Verlag, 1998.
10. D. Dasgupta and N. Attoh-Okine. Immunity-based systems: A survey. *Immunity-Based Systems: A Survey. Proceeding "IEEE" International Conference on Systems, Man and Cybernetics,* 1:369–74, 1997.
11. D. Dasgupta, Y. Cao, and C. Yang. An immunogenetic approach to spectra recognition. In *Proceedings of the Genetic and Evolutionary Computation Conference,* volume 1, pages 149–155, 1999.
12. D. Dasgupta and S. Forrest. Tool breakage detection in milling operations using a negative-selection algorithm. Technical Report CS95-5, Department of Computer Science, University of New Mexico, 1995.
13. D. Dasgupta and S. Forrest. Novelty detection in time series data using ideas from immunology. *Proceedings of The International Conference on Intelligent Systems,* 1999.
14. L. N. de Castro and F. J. Von Zuben. Artificial immune systems: Part1 - basic theory and applications. Technical Report TR DCA 01/99, State University of Campinas, 1999.
15. L. N. de Castro and F. J. Von Zuben. Artificial immune systems: Part2 - a survey of applications. Technical Report DCA-RT 02/00, State University of Campinas, 2000.
16. S. Forrest, S.A. Hofmeyr, and A. Somayaji. Computer immunology. *Communications of the ACM,* 40(10):88–96, 1997.
17. S. Forrest, S.A. Hofmeyr, A. Somayaji, and T. Longstaff. A sense of self for unix processes. *Proc. IEEE Symposium on Research in Security and Privacy,* pages 120–128, 96.

18. S. Forrest, A.S. Perelson, L. Allen, and R. Cherukuri. Self-nonself discrimination. In *Proceedings of the 1994 IEEE Symposium on Research in Security and Privacy*, pages 202–212. IEEE Computer Society Press, 1994.
19. S.A. Hofmeyr and S. Forrest. Architecture for an artificial immune system. *Evolutionary Computation*, 7(1):45–68, 1999.
20. J. E. Hunt and D. E. Cooke. Learning using an artificial immune system. *Journal of Network and Computer Applications*, 19:189–212, 1996.
21. C.A. Janeway, P. Travers, and M. WalPort. *Immuno Biology. The Immune System in Health and Disease.* Current Biology Publications, London, New York, 4 edition, 1999.
22. J. Kim and P. Bentley. Negative selection and niching by an artificial immune system for network intrusion detection. In *Late Breaking Papers at the 1999 Genetic and Evolutionary Computation Conference*, pages 149–158, 13 1999.
23. J.W. Kimball. Biology. *http://www.ultranet.com/˜jkimball/BiologyPages/W/Welcome.html*, 2000.
24. C. Ortega-Sánchez. *Embryonics: A Bio-Inspired Fault-Tolerant Multicellular System.* PhD thesis, The University of York, Heslington, York, UK., 2000.
25. C. Ortega-Sánchez and A.M. Tyrrell. Design of a basic cell to construct embryonic arrays. *IEE Transactions on Computer and Digital Techniques*, 145(3):242–248, 1998.
26. POEtic Project. Poetic web page. *http://POEticTissue.org*, 2002.
27. L. Prodan, G. Tempesti, D. Mange, and A. Stauffer. Embryonics: Artificial cells driven by artificial dna. *4th International Conference "ICES", Lecture Notes in Computer Science*, 2210:100–111, 2001.
28. A. Stauffer, D. Mange, G. Tempesti, and C. Teuscher. BioWatch: A giant electronic bio-inspired watch. *The 3rd NASA/DoD Workshop on Evolvable Hardware*, pages 185–192, 2001.
29. P. Sundararajan, S. McMillan, and S. Guccione. Testing FPGA devices using JBits. *2001 MAPLD*, 2001.
30. G. Tempesti. *A Self-Repairing Multiplexer-Based FPGA Inspired by Biological Processes.* Phd thesis, École Polytechnique Féderale De Lausanne, 1998.
31. Xilinx. An overview of multiple cam designs in virtex family devices. application note 201, 1999.

Virtual Reconfigurable Circuits for Real-World Applications of Evolvable Hardware

Lukáš Sekanina

Faculty of Information Technology, Brno University of Technology
Božetěchova 2, 612 66 Brno, Czech Republic
sekanina@fit.vutbr.cz, phone: +420 541141215

Abstract. The paper introduces a new method for the design of real-world applications of evolvable hardware using common FPGAs (Field Programmable Gate Arrays). In order to avoid "reconfiguration problems" of current FPGAs a new *virtual reconfigurable circuit*, whose granularity and configuration schema exactly fit to requirements of a given application, is designed on the top of an ordinary FPGA. As an example, a virtual reconfigurable circuit is constructed to speed up the software model, which was utilized for the evolutionary design of image operators.

1 Introduction

In recent years, researchers have recognised the essential difference between evolutionary circuit design and evolvable hardware [1].

In case of *evolutionary circuit design* a single circuit is evolved. An evolutionary algorithm is used only during design phase and thus it plays the role of a "designer". While quality and implementation cost of the resulting circuit are main design objectives, the time needed to evolve a satisfactory circuit for a given application may not be critical. One could imagine that it could be economically profitable to evolve a single circuit for one month, provided that it can be reused for the next five years. Hence the design can be approached using a circuit simulator in software (that is, in *extrinsic evolution*).

On the other hand it can be supposed that a noticeable portion of a target (embedded) system will be implemented in hardware in case of real-world applications of *evolvable hardware* (EHW). Such applications typically operate in time-varying environments. Then EHW ensures either adaptation (e.g. in robotics) or high performance computation (e.g. in image compression), which is unreachable using conventional approaches. In contrast to evolutionary circuit design, the evolutionary algorithm is a part of a target system. However its role is rather to provide *sufficient long-time mean performance* (quality) of the evolved circuits than to find "innovative" circuits. Unlike evolutionary circuit design, the amount of resources (gates) needed to realize the final circuit is not usually important because some resources are always devoted for evolutionary purposes and may be used for free.

While there exist some ASIC-based (perhaps already commercialized) real-world applications of EHW [2], implementations of FPGA-based real-world applications of EHW remain practically unexplored. Although FPGAs are widely

A.M. Tyrrell, P.C. Haddow, and J. Torresen (Eds.): ICES 2003, LNCS 2606, pp. 186–197, 2003.

accessible reconfigurable circuits (RC), the following points represent obstacles for their usage in EHW: (1) In current FPGAs the reconfiguration system is not sufficient for EHW because the *partial reconfiguration* is not fully supported (the XC6200 family is off the market now and the usage of JBits is too complicated [5]) and FPGAs can only be reconfigured externally. (2) FPGAs are based on Configuration Logic Blocks (CLBs) whose granularity is too fine for evolution of complex circuits. And these complex circuits are required in real-world applications. Hence in addition to the problems mentioned previously, *scalability* problems of EHW have to be solved in real-world applications [3,4].

In this paper, we present a method for the design of real-world applications of EHW using common FPGAs. A new *virtual* RC, whose granularity and reconfiguration schema exactly fit to the requirements of a given application, will be designed on the top of a common FPGA. It will enable the implementation of a target system in a low-cost, commercial off-the-shelf FPGA and to achieve adaptation to a changing environment. The method will be illustrated in the design of the virtual RC for an evolvable image pre-processing component.

The design will be approached in two major steps: (1) Circuit software simulator will be employed during the evolutionary circuit design of some typical circuits from application domain in order to find "optimal" organization of the virtual RC and evolutionary algorithm. (2) The virtual RC and evolutionary algorithm will be designed and synthesized into an FPGA according to outcomes of step 1. Note that this paper deals only with design of the virtual RC.

Section 2 introduces the problem domain and summarizes the results we obtained from the previous research. The implementation of the virtual RC is described in Section 3. Section 4 discusses the proposed approach. And finally, conclusions are given in Section 5.

2   Evolutionary Image Operator Design

In order to ensure the adaptability of a target embedded system (e.g. to a varying type of noise), we have to find such an organization of the RC and genetic operators, which will enable us to evolve *various* image operators and filters in a reasonable time. Every image operator will be considered as digital circuit of nine 8bit inputs and a single 8bit output, which processes gray-scaled (8bits/pixel) images. As Fig. 1 shows every pixel value of the filtered image is calculated using a corresponding pixel and its eight neighbors of the processed image.

We approached the problem using Cartesian Genetic Programming (CGP) [6]. However after a number of experiments we claim that it is practically impossible to evolve a gate-level image operator of 72 input and 8 output bits, which is able to compete against conventional solutions. In order to allow the design of such complex circuits, we extended CGP to *functional level* [7]. Instead of CLBs and 1 bit connection wires, Configurable Functional Blocks (CFBs) and 8bit datapaths are utilized (see Fig. 1 and Fig. 3).

Similarly to conventional CGP, only very simple variant of the evolutionary algorithm is employed. Population size is 16. The initial population was gener-

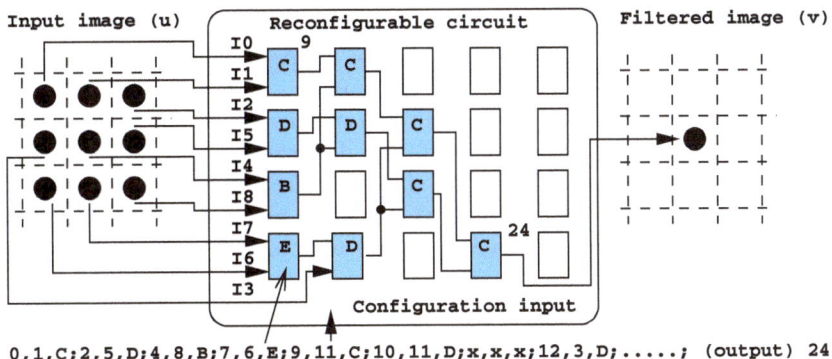

Fig. 1. A typical arrangement of an experiment for evolutionary design of image operators. A sample chromosome (circuit) is uploaded and the active CFBs are marked.

ated randomly, however only function "C" (see Fig. 3) was used in some runs. In case of evolutionary circuit design, the evolution was typically stopped (1) when no improvement of the best fitness value occurs in the last 50000 generations, or (2) after 500000 generations.

Only mutation of two randomly selected active CFBs is applied per circuit. Four individuals with the highest fitness are utilized as parents and their mutated versions build up the new population. In order to evolve a *single* filter (digital circuit), which suppresses a *given* type of noise, we need an original image to measure the fitness of the evolved filter. The generality of the evolved filters (i.e. whether the filter operates sufficiently also for other images of the same type of noise) is tested using a test set.

The design objective is to minimize the difference between the corrupted image and the original image. We chose to measure *mean difference per pixel* since it is easy for hardware implementation. Let u denote a corrupted image and let v denote a filtered image. The original (uncorrupted) version of u will be denoted as w. The image size is $K \times K$ (K=256) pixels but only the area of 254 x 254 pixels is considered because the pixel values at the borders are ignored and thus remain unfiltered. The fitness value of a candidate filter is obtained as follows: (1) the circuit simulator is configured using a candidate chromosome, (2) the circuit created is used to produce pixel values in the image v, and (3) fitness is calculated as

$$fitness = 255.(K-2)^2 - \sum_{i=1}^{K-2}\sum_{j=1}^{K-2} |v(i,j) - w(i,j)|.$$

Now we can summarize CGP parameters that lead to efficient evolutionary design: RC consists of 2-input CFBs placed in array of 10 columns and 4 rows. An input of each CFB may be connected to circuit's inputs or to the output of a CFB, which is placed in the previous two columns (i.e. CGP's L-back parameter is L=2). After statistical analysis of the functions utilized in the evolved circuits, we recognized that CFB should support functions listed in Fig. 3. Let us note that functions "C", "E", and "F" are the most important for successful evolution.

A number of unique image filters for Gaussian, uniform, block-uniform and shot noise was reported [7,9]. These filters typically consist of 10-25 CFBs and show better quality in comparison with typical conventional filters (such as mean and median filters). Surprisingly in some cases the number of equivalent gates needed for implementation in FPGA XC4028XLA was also lower than in case of conventional filters. The very efficient edge detectors were also evolved using the same system. Furthermore, the system was able to produce sequences of very good filters in simulated time varying environment (changing types of noise) where only 20000 populations were generated per circuit [10].

3 Virtual RC for Evolution of Image Operators

The previous section dealt with an evolutionary image operator design problem in which it was not any problem to spend 1 day with a single filter design on common PC. However if an application is operating in a dynamic environment (e.g. an embedded system) it should be considered the software approach becomes slow. The speed at which the approach is able to evolve the efficient image operators automatically, determines a class of applications for which the approach is suitable for. The evolutionary image operator design has clearly demonstrated that very time consuming fitness calculation is the most painful problem of EHW. As soon as we are able to perform (at least) fitness calculation in hardware, we are also able to reduce design time substantially and potentially to support the continuous adaptation.

The idea of virtual RC initially appeared in [8], and was developed in the paper [11]. Some other authors have applied similar methods: Slorach and Sharman designed a simple fine-grained programmable logic array on top of an FPGA [12] and Torresen's group approached a problem of context switching in this way too [13]. Macias has also implemented an 8 × 8 Cell Matrix using Xilinx Spartan-2 XC2S-200 FPGA. Other types of specialized platforms have been developed [14,15]. Note that none of these approaches deal with real-world applications of EHW operating in time-varying environments.

3.1 Description of the Virtual RC

The CGP instance (introduced in Section 2) is in fact implemented in an FPGA. As Fig. 2 shows the RC under design operates with nine 8 bit inputs (I0 – I8) and a single 8 bit output. It consists of 29 CFBs allocated in a grid of 8 columns and 4 rows. The last column contains only a single CFB, which is always connected to circuit output (a connection of the circuit output is not evolved). Every CFB has two inputs and can be configured to perform one of 16 functions listed in Fig. 3. Examples of a hardware implementation of some typical functions are obvious from the same figure.

Every CFB is also equipped with a register to support pipeline processing. A column of CFBs is considered as a single stage of the pipeline. In order to

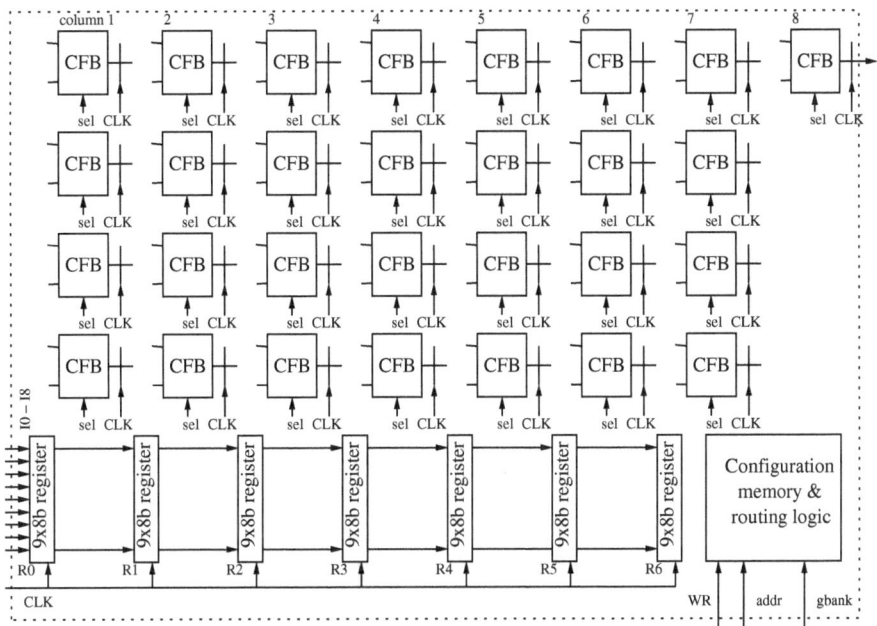

Fig. 2. A reconfigurable device for the evolutionary design of image operators.

synchronize CFBs with the processed inputs, we have to insert 72 bit (9 × 8 bits) registers R0 – R6 between neighboring columns. The CFBs' inputs are not connected to circuit inputs directly, but they are connected to the appropriate registers. It is more economical in hardware to allow a CFB's input is connected to one of 8 data sources than to one of 9 data sources because then only 3 address bits are needed for the routing multiplexer. Hence the following restrictions of conventional CGP (which are taken in account simultaneously) were introduced for this application to reduce the number of configuration bits:

- An input of each CFB placed in an odd row may be only connected to circuit's inputs I0 – I7 (via registers R0 – R6).
- An input of each CFB placed in an even row may be only connected to circuit's inputs I1 – I8 (via registers R0 – R6).
- An input of each CFB placed in the first (the second) column may be only connected to circuit's inputs via register R0 (R1, respectively).
- An input of each CFB placed in column 3 (4, 5, 6 or 7) may be connected to circuit's inputs via registers R1 and R2 (R2 and R3, R3 and R4, R4 and R5, R5 and R6, respectively) or to the output of a CFB, which is placed in some of the previous two columns.
- An input of the last CFB may be only connected to the output of a CFB, which is placed in the previous two columns.

Now we can derive the theoretical length of the configuration bitstream. Four bits are needed to determine the function in each CFB. Four bits also encode

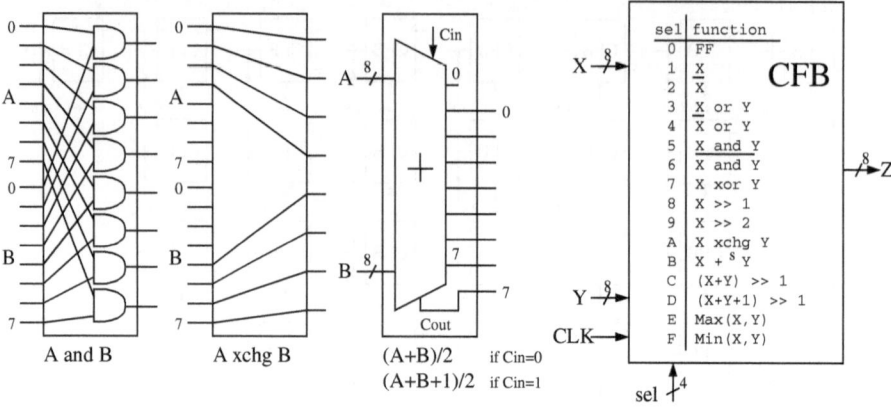

Fig. 3. An implementation of functions 5 (sel=0101), A (1010), C (1100), and D (1101) that are provided together with other functions in each CFB. Note that $>>$ is a shifter and $+^s$ is an adder with saturation.

connection of every CFB's input in columns $3 - 7$. We will utilize only three bits to do the same with inputs in columns 1, 2, and 8. In total, the length is: $9(3 + 3 + 4) + 20(4 + 4 + 4) = 330$ bits.

3.2   Routing Logic and Configuration Memory

The routing circuits are constructed using multiplexers that are controlled via bits of the configuration memory. Sixteen 16-input multiplexers are needed for selection of CFB's inputs. These inputs are usually taken either from the outputs of the eight CFBs in the previous two columns or from the previous two registers. Only 8-input multiplexers are required for CFBs of the first, the second, and the last columns since these CFBs utilize only eight possible inputs.

The configuration memory is implemented using flip-flops that are available in CLBs. Thus a two output CLB is employed as a two bit memory. Then all outputs of CLBs utilized for the configuration memory are connected to multiplexers that control the routing of CFBs. The configuration memory is depicted in Fig. 4. Four CFBs of a single column are configured simultaneously and their configuration is stored in a *cnfBank* configuration register. A column is selected via *addr* and a new configuration is uploaded using *gbank* configuration port and *WR* signal. Although the number of configuration bits needed for columns 1, 2 and 8 is lower than 48 bits, the configuration memory employs 48 bits per column in our prototype (in order to make the design easier). Hence the configuration memory consists of $48 \times 8 = 384$ bits.

The proposed configuration approach is a kind of partial reconfiguration that may be performed either form outside (if configuration signals are connected to circuit's pins) or from some internal circuits. Obviously, it is impossible to destroy the chip when a randomly generated configuration is uploaded.

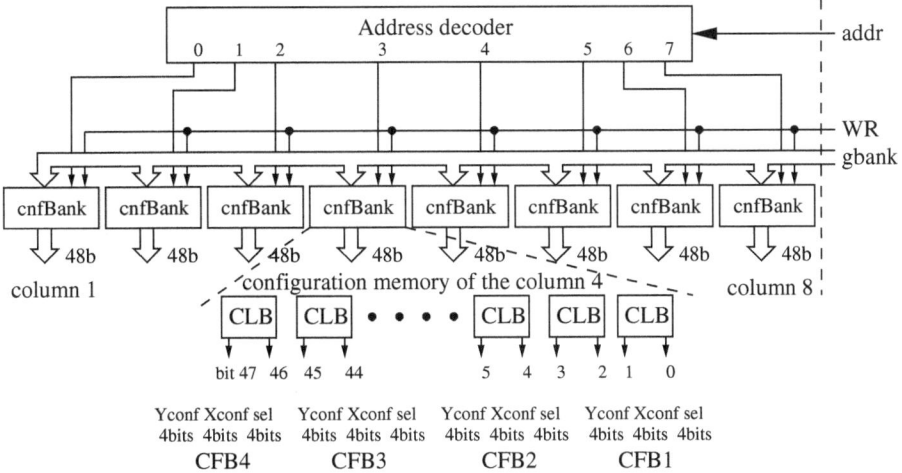

Fig. 4. The configuration memory is divided into eight configuration banks *cnfBank*. Each of them is responsible for configuration of a single column of CFBs.

3.3   Configuration Options

The virtual RC was described using VHDL at structural level but some elementary circuits (such as "Max" circuit) were defined behaviorally. Its functionality was tested in the *Active HDL v.3.6* environment.

Assume that the RC has been already configured and $L = 1$. Then the inputs (I0 – I8) are processed via the array of CFBs, and simultaneously, the inputs go through the register array R0 – R6. The first valid output is available in 9 clocks (a delay of R0 plus 8 CFBs), and the other ones each per a clock. Hence $K \times K$ pixels can be processed in $9 + (K \times K - 1)$ clocks. Maximal operational frequency is determined by a delay of a CFB and a delay of routing circuits (which is a single multiplexer).

It takes 8 clocks to configure the entire RC at the beginning of the computation. However, the reconfiguration process can be hidden totally. We can configure the first stage in the same time in which the first input values are stored into R0. Then every next stage can be configured while the previous stage processes data. Hence such a pipelined approach requires only 1 clock to reconfigure the entire device as seen in Fig. 5.

The RC in fact supports $L = 1$ as well as $L = 2$ setting. However if $L = 2$ is considered, every pixel must be sent to the RC two times to ensure synchronization of CFBs. Note that CFB's inputs may be connected to registers in two different stages (i.e. to values registered at two different moments) but a CFB must operate with the inputs registered at the same moment to produce a correct output. This leads to decreasing of the maximal frequency to a half if $L = 2$.

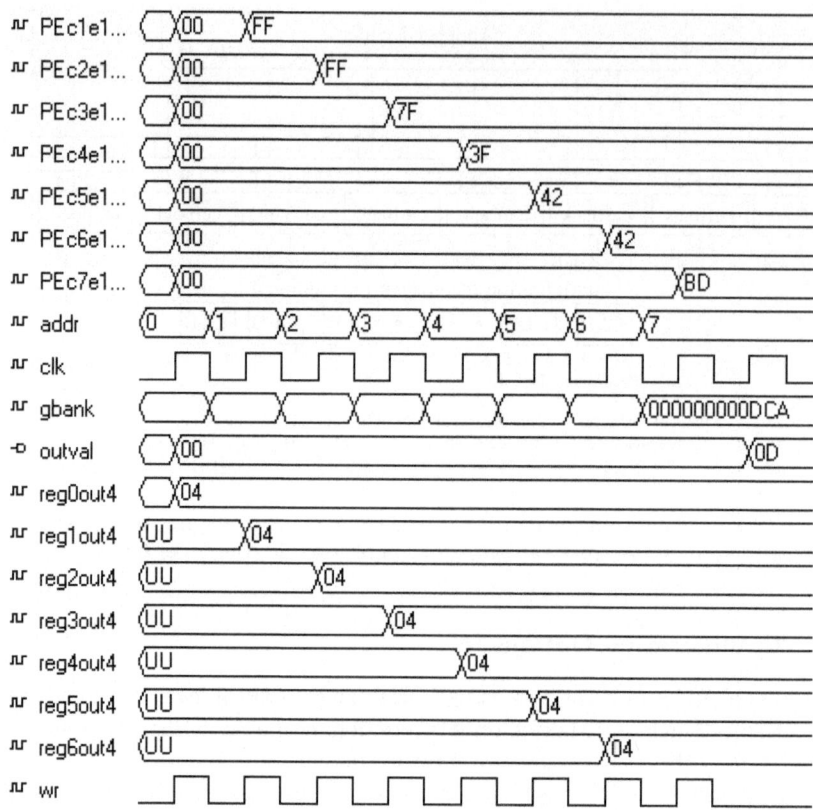

Fig. 5. An example of simultaneous pipelined reconfiguration and execution of the RC under design. We can observe the outputs of CFBs placed in the first row (*PEc*e1*) and the constant input I4 = 4, which goes via the register array (*reg*out4*). In the first clock, I4 is stored into *reg0out4*, and the first column of CFBs is configured using *gbank*, *addr* and *wr*. In the second clock, *PEc1e1* is calculated, I4 is stored into *reg1out4*, and the second column of CFBs is configured. The entire RC is configured in 8 clocks and the first output (*outval* = 0D) is available in the ninth clock.

3.4 Implementation Costs

In order to find out whether the design can be physically realized, we have tried to synthesize the RC into various FPGAs using *Xilinx Integrated Software Environment*. The results summarized in Table 1 qualify us to claim that it is possible to construct virtual reconfigurable devices for real-world applications. Furthermore it seems that genetic operators and fitness evaluation can be placed on the same FPGA (e.g. 77% XCV2000E's resources remain unused) and thus the entire application could be implemented in a single chip.

While we reached a sufficient operational frequency (at least for $L = 1$), higher implementation costs are main problem of the approach. If a fixed (either

Table 1. The results of the synthesis into various Xilinx Virtex FPGAs. Note that no timing constrains were defined for the synthesis program. The maximal operational frequency f_m is valid only for $L = 1$.

Family	Chip	Optimized for	Slices	Slices [%]	Equivalent gates	f_m [MHz]
Virtex E	XCV2000E	speed	4424	23	67937	96.4
Virtex 2	XC2V1000	speed	4879	95	74357	134.8
Virtex	XCV1000	speed	4489	36	68744	86.7
Virtex	XCV1000	area	4085	33	63773	81.3

conventional or evolved) operator is considered, about 2000 equivalent gates are needed for its implementation. The virtual RC (which can be configured to implement the same operator) requires 30 times more equivalent gates. A single CFB costs 1009 equivalent gates. So far our design consists of 29 CFBs, 55% gates needed to realize the entire virtual RC are devoted for implementation of routing multiplexers, configuration memory and the register array R0 – R6.

3.5 Fitness Calculation in Hardware

If $L = 1$ then the (virtual) RC can operate at $f_m = 134.8$ MHz. Thanks to pipelining, the RC is able to process each pixel at 134.8 MHz too. But furthermore, the *entire* evolvable system can operate at this frequency. Recall that a deterministic selection mechanism is applied in the evolutionary algorithm. Hence a new candidate circuit can always be generated simultaneously with fitness calculation of another circuit. As soon as the fitness calculation takes a half millisecond (e.g. it is 0.48 ms for 254×254 pixels), there is plenty of time to generate a new candidate circuit if the genetic unit takes place in the same FPGA. Because the reconfiguration time of the virtual RC is practically zero, the time of evolution is determined as

$$t_e = \frac{(K - 2)^2 . \mu . \rho}{f_m}$$

where $(K - 2)^2$ is the number of processed pixels, μ denotes population size, and ρ is the number of generations.

Fig. 6 shows that pipelining of the RC can be extended to pipelining of the entire evolvable system. Every output value of the RC is compared with the corresponding pixel value of the original image. Then an absolute value of the difference is added to the content of an accumulator (which corresponds to fitness) in next clock. The circuit, which selects neighboring values of each pixel, is taken from [16]. FIFO is a K-pixel data structure that may be easily implemented in an FPGA. The original image has to be synchronized in similar way as seen in Fig. 6.

We can estimate how much time is needed to evolve a single image operator. Assume the setting $K = 256, \mu = 16, \rho = 160000$, which requires 1 day on PentiumIII/800 MHz in average. If $f_m = 134.8$ MHz, then a filter can be designed

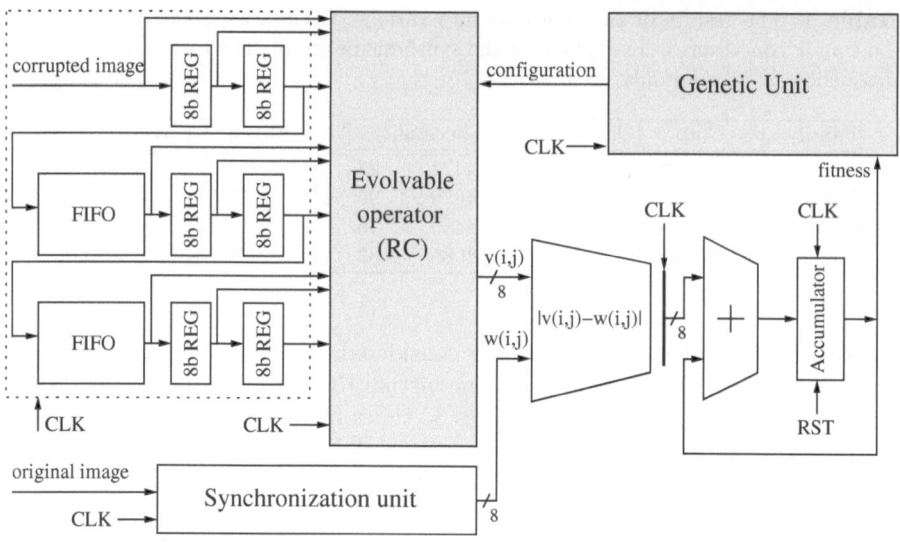

Fig. 6. Pipelining of the RC can be extended to the fitness calculation. FIFO ensures that each output pixel value is calculated using the right input values.

in $t_e = 20.4$ minutes in average using virtual RC. We obtained a speeding up more than 70. Some operators were evolved with $\rho = 20000$. If such a system is implemented in an FPGA, the adaptation time would be about $t_e = 2.6$ minutes. Assume that only $\frac{(K-2)^2}{2}$ pixels and $\rho = 10000$ are sufficient. Then we obtain the adaptation time $t_e = 38.3$ seconds which can be considered as a real-time adaptation in some applications. And finally, we can expect that it will be possible to implement in the future 16 virtual RCs in a single FPGA. Then all circuits of the population could be evaluated at the same time. Note that the fitness accumulator will have to be replicated 16 times, however only a single implementation of input FIFOs and registers is necessary. Such a system would be able to adapt to changing fitness function in 2.4 seconds.

4   Discussion

We did not implement the virtual RC in a physical FPGA yet. However, there is a company, which is interested to do it. Nevertheless the presented results show that an adaptive system for image pre-processing can be implemented in a common FPGA. If the approach is used only to speed up the evolutionary image operator design, then it should take several minutes to find a sufficient operator. If generality of the evolved operators is not required, it is possible to reduce the design time to tens of seconds. And that could be attractive for some real-time applications. Let us summarize the advantages of the proposed method for design of virtual reconfigurable devices:

– An architecture, which exactly fits to application requirements, can be developed for a given problem. It is mainly important for EHW, since hardware can reflect the representation used in the evolutionary algorithm.
– A style and granularity of reconfiguration may be defined according to the need of a given application.
– A virtual RC is described in a platform independent way. Hence it can be synthesized into various target devices and with various constraints.
– The process of a virtual RC design (i.e. its VHDL description) can be fully automatic. One can imagine a design environment, in which a designer is to press a button and a new virtual RC will be generated automatically according to the best parameters of RC and evolutionary algorithm found so far for a given EHW-based application.
– A description of a virtual RC can be offered as an IP macro.
– It is possible to implement very efficient context switching in Virtex FPGAs as it is evident in [13].

Higher implementation costs are the main disadvantages of the proposed method. Most resources available on an FPGA are wasted on configuration memory and routing logic implementation. It seems that the approach is suitable only for coarse-grained architectures since the scaling problem is not so painful. The scalability problem can be approached by reduction of configuration options (e.g. only local connection could be enabled only). The number of functions supported in a programmable element is not critical from a scalability viewpoint.

5   Conclusions

We have clarified that it is possible in principle to implement real-world applications of EHW in common FPGAs. We are going to realize the proposed RC in Virtex FPGA and to apply the method for design of other applications of EHW.

We have introduced a new level of abstraction to the hardware design. Virtual RCs are supposed to be beneficial not only for EHW, but also for the other research domains, e.g. reconfigurable computing.

Acknowledgment

The research was performed with the financial support of the Grant Agency of the Czech Republic under No. 102/01/1531 and 102/03/P004.

References

1. Yao, X., Higuchi, T.: Promises and Challenges of Evolvable Hardware. In: Proc. of the 1st International Conference on Evolvable Systems: From Biology to Hardware ICES'96, LNCS 1259, Springer-Verlag, Berlin (1997) 55–78
2. Higuchi, T. et al.: Real-World Applications of Analog and Digital Evolvable Hardware. IEEE Transactions on Evolutionary Computation, Vol. 3(3), (1999) 220–235

3. Vassilev, V., Miller, J.: Scalability Problems of Digital Circuit Evolution. In: Proc. of the 2nd NASA/DoD Workshop on Evolvable Hardware, IEEE Computer Society, Los Alamitos (2000) 55–64

4. Murakawa, M. et al.: Evolvable Hardware at Function Level. In: Proc. of the Parallel Problem Solving from Nature PPSN IV, LNCS 1141, Springer-Verlag Berlin (1996) 62–72

5. Hollingworth, G., Smith, S., Tyrrell, A.: The Intrinsic Evolution of Virtex Devices Through Internet Reconfigurable Logic. In: Proc. of the 3rd International Conference on Evolvable Systems: From Biology to Hardware ICES'00, LNCS 1801, Springer-Verlag, Berlin (2000) 72–79

6. Miller, J., Job, D., Vassilev, V.: Principles in the Evolutionary Design of Digital Circuits – Part I. Genetic Programming and Evolvable Machines, Vol. 1(1), Kluwer Academic Publisher (2000) 8–35

7. Sekanina, L.: Image Filter Design with Evolvable Hardware. In: Applications of Evolutionary Computing – Proc. of the 4th Workshop on Evolutionary Computation in Image Analysis and Signal Processing EvoIASP'02, LNCS 2279 Springer-Verlag, Berlin (2002) 255–266

8. Sekanina, L.: Modeling of Evolvable Hardware. MSc. Thesis, Brno University of Technology (1999) 62 pp.

9. Sekanina, L., Drábek, V.: Automatic Design of Image Operators Using Evolvable Hardware. In: Proc. of the 5th IEEE Design and Diagnostic of Electronic Circuits and Systems DDECS'02, Brno, Czech Republic (2002) 132–139

10. Sekanina, L.: Evolution of Digital Circuits Operating as Image Filters in Dynamically Changing Environment. In: Proc. of the 8th International Conference on Soft Computing Mendel'02, Brno, Czech Republic (2002) 33–38

11. Sekanina, L., Růžička, R.: Design of the Special Fast Reconfigurable Chip Using Common FPGA. In: Proc. of the 3rd IEEE Design and Diagnostic of Electronic Circuits and Systems DDECS'00, Polygrafia Bratislava, Slovakia (2000) 161–168

12. Sloarch, C., Sharman, K.: The Design and Implementation of Custom Architectures for Evolvable Hardware Using Off-the-Shelf Progarmmable Devices. In: Proc. of the 3rd International Conference on Evolvable Systems: From Biology to Hardware ICES'00, LNCS 1801, Springer-Verlag, Berlin (2000) 197–207

13. Torresen, J., Vinger, K. A.: High Performance Computing by Context Switching Reconfigurable Logic. In: Proc. of the 16th European Simulation Multiconference ESM 2002, Darmstadt, Germany (2002) 207–210

14. Haddow, P., Tufte, G.: Bridging the Genotype-Phenotype Mapping for Digital FPGAs. In: Proc of the 3rd NASA/DoD Workshop on Evolvable Hardware, IEEE Computer Society, Los Alamitos (2001) 109–115

15. Porter, R., McCabe, K., Bergmann, N.: An Application Approach to Evolvable Hardware. In: Proc. of the 1st NASA/DoD Workshop on Evolvable Hardware, IEEE Computer Society (1999)

16. Fučík, O.: Reconfigurable Embedded Systems. PhD Thesis, Brno University of Technology (1997) 65 pp.

Gene Finding Using Evolvable Reasoning Hardware

Moritoshi Yasunaga[1], Ikuo Yoshihara[2], and Jung H. Kim[3]

[1] Institute of Information Sciences and Electronics,
University of Tsukuba, Tsukuba Ibaraki, 305-8573,Japan
yasunaga@is.tsukuba.ac.jp
[2] Faculty of Engineering, Miyazaki University, Miyazaki, 889-2192 Japan
yoshiha@cs.miyazaki-u.ac.jp
[3] System Engineering Dept. University of Arkansas at Little Rock
Little Rock, Arkansas 72204-1099, USA
jhkim@ualr.edu

Abstract. We have proposed and developed an evolvable reasoning hardware (ERHW) already. In this paper, we apply the ERHW to the gene finding task that is one of the most important tasks in the genome informatics. We evaluate the ERHW using *human* and other living organisms' DNAs. The ERHW shows high reasoning accuracy of 95.8% and high reasoning speed of $280ns$.

Keywords: evolving hardware system, evolutionary hardware design methodologies, genome informatics.

1 Introduction

Everyday masses of DNA (deoxyribose nucleic acid) of various living organisms are sequenced using DNA sequencers. A sequenced DNA fragment, which is a long array of 4 bases (A, G, C, and T), however does not make any sense by itself, i.e. one cannot obtain any information from the sequenced result as it is [1][2]. A gene which is a sequence of DNA that represents a fundamental unit of heredity, is hidden in the sequenced fragments. Finding a gene is one of the most important tasks in the genome science. The gene finding however is biologically a burdensome task, because it has to be done through multiple biological experiments requiring long processing time (Human genes occupy less than 3% of the whole DNA containing 3 billion bases, but most of them are still hidden remaining to be decoded). Computational approach to high-precision and high-speed gene finding methods is thus in critical need in the genome science.

We have proposed an evolvable reasoning hardware (ERHW) and developed its prototype already [3]. In this paper, we apply the developed ERHW to the gene-finding task and evaluate its performance using DNAs of human and other living organisms.

A.M. Tyrrell, P.C. Haddow, and J. Torresen (Eds.): ICES 2003, LNCS 2606, pp. 198–207, 2003.

2   ERHW: Evolvable Reasoning Hardware

In this section, we describe an outline of the ERHW. Details of the ERHW and its performance (reasoning accuracy and speed) are well explained in [3]. Figure 1 shows the fundamental configuration of the ERHW. In the ERHW, kernels shown as ..GA******GTA.., or ..GGGGT*CG***.. in the figure play the most important role in the reasoning. Each kernel is generated from a record in the database which we used for the reasoning. For example, suppose three DNA sequences (records)

GAAAGTAGC
GTAAGAAGG
GCAAGTAGA

are recorded in the same category (e.g., in the *human-* related DNA fragments) in a database. We apply our genetic algorithm-based learning method (described later in this section) to these records in order to make kernels

GAAAG*AG*
G*AAGAAG*
G*AAG*AGA,

where meaningless letters in that category are changed to *don't care*(*) and the remaining letters show some rules.

Reasoning is carried out as follows. A query (an unknown sequenced DNA fragment) is compared with each kernel and the number of kernels that match the query is counted in every category ($C_0 \sim C_n$) in the counter shown in the figure. The category that has the largest number of matched kernels to the query is chosen (in the Maximum detector in the figure) and is inferred as the category that the query belongs to.

This simple reasoning algorithm is an extension of the Parzen window method [4] that is well known in the pattern recognition field, so that its validity is guaranteed theoretically and its reasoning accuracy increases as the number of kernels increases.

Furthermore, kernels can be simply implemented using fundamental combinatorial circuits in the same way as counters and the maximum detector (details are shown later).

The way to generate the kernel using the genetic algorithm (GA) is shown in Fig. 2. A population of N chromosomes is made for each record (DNA record in the database). Thus, if we use R_N records in the database, we make R_N populations and the following GA process is executed on each population. The chromosome is composed of l fields corresponding to the l letters in the record. Each field (locus) is filled with either operator (gene) '%' or '&', where '%' changes the letter to *don't care* while '&' remains the letter. And a potential kernel (kernel candidate) is generated by operating the chromosome onto the corresponding record and is evaluated under the following fitness using all R_N records. The kth chromosome in the population for jth record in the ith category evolves under the simple GA process with a fitness f of

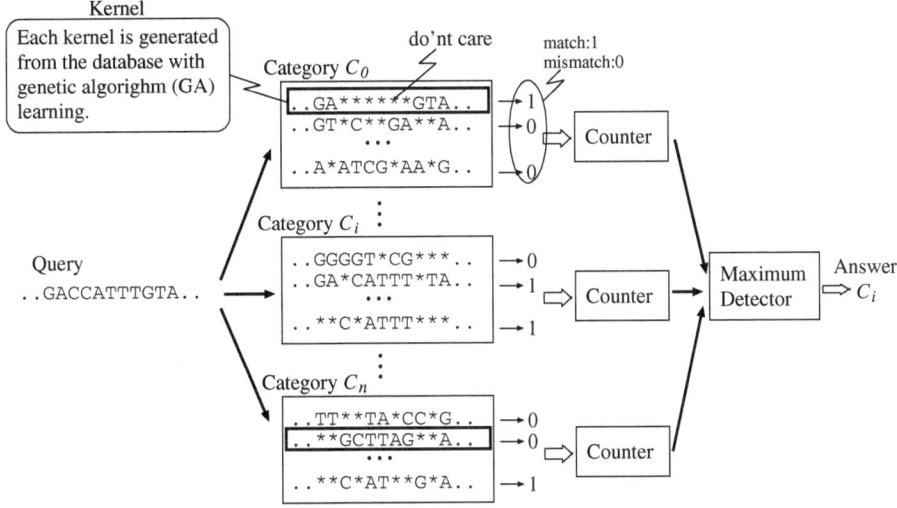

Fig. 1. ERHW configuration.

$$f\left(C_k^{ji}\right) = \frac{h}{1+g}, \tag{1}$$

where h increases as the number of records in the same category that match the kernel increases, while g increases as the number of records in the different category that match the kernel increases. Therefore, a chromosome of high fitness means that the kernel generated by the chromosome matches (includes) many records in the same category and mismatches (excludes) many records in the different category. This means that the kernel extracts some rule or a part of rule in that category. Finally in the GA process, the kernel is made up using the chromosome of the highest fitness by operating it onto the corresponding record. After the GA processes for all R_N records, R_N kernels are generated and assembled separately as shown in Fig. 1.

3   Splicing Site Prediction

Genes that carry hereditary information are hiding in the DNA, and their transcription regions consist of two kinds of segments (sub-sequences) called exon and intron (Fig. 3). The exons are translated via mRNAs (messenger rebo-nucleic acids) to organize specific proteins, while the introns are discarded as meaningless information. The exon-intron boundary, or splicing site is one of the most typical marks of gene. Therefore, finding splicing sites in the unknown DNA fragments is regarded as one of the best approaches to find gene.

It is already well known that the left (5' side) and the right (3' side) borders of the intron are necessarily GT and AG, respectively [5]. However, no further certain rules are currently known about exon-intron boundaries. The fact that 5'

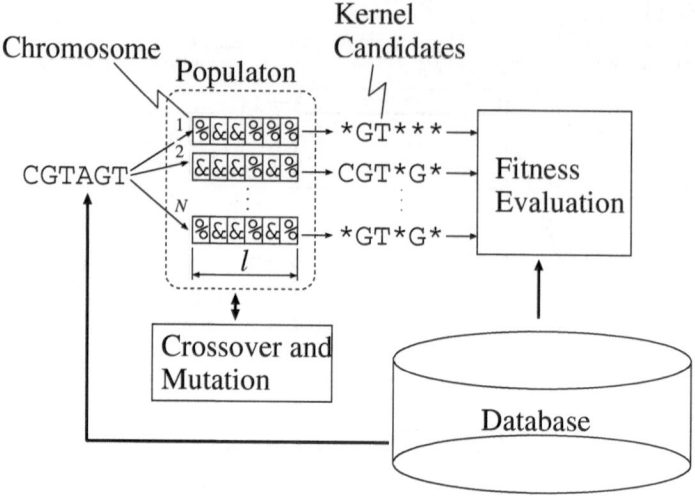

Fig. 2. GA process for kernel generation.

and 3' borders in introns are GT and AG respectively is the necessary condition (i.e., a boundary between an intron and an exon is necessarily GT or AG) but not the sufficient condition. A huge number of GTs and AGs are in the DNA sequences, and more than 99.99 % of them are not boundaries (Fig. 3). Thus, finding the true boundaries is one of the most difficult problems in the genome informatics.

Whether each GT/AG is boundary or not is expected to be subject to some particular arrangements of bases around the boundaries, i.e., some rules, or features are expected to be hiding in the sub-sequence of

XXXXXXXXGTXXXXXXXX or XXXXXXXXAGXXXXXXXX ,

where X represents A, G, C, or T (the number of Xs is discussed later).

4   Prototype Development

4.1   Genome Data and GA Operation

We used the GenBank database[1], which is one of the most well-known genome database, in the prototype development. A set of 11,000 records of GTs with their neighboring bases like XXXXGTXXXX (X represents a base of A, G, C, or T) was collected from more than 100 human-related DNA sequences (e.g., hemoglobin, interleukin, etc.), where 1,000 records out of 11,000 contained true boundaries (category 'true') and the remaining 10,000 records did not contain boundaries (category 'false'). We used the 10% of the total records (i.e., 100

---

[1] http://www.ncbi.nih.gov/ Genbank/GenbankOverview.html

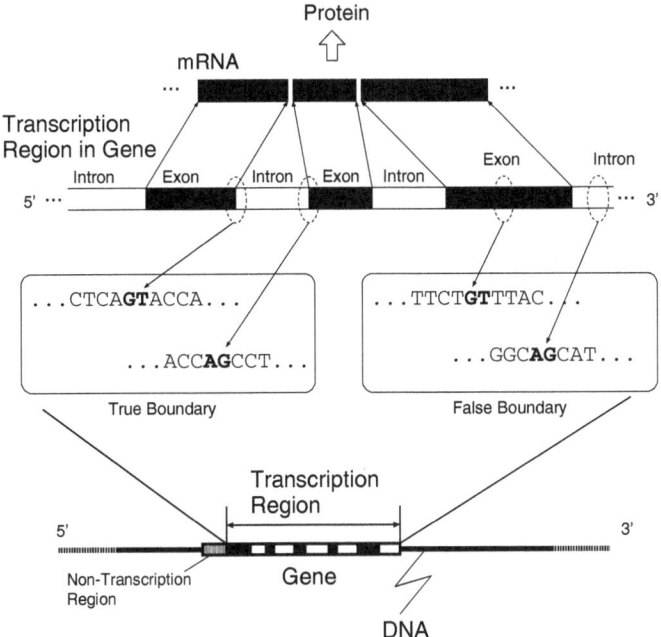

Fig. 3. DNA structure and splicing site.

Table 1. GA parameters.

Population	20
Crossover Rate	10%
Mutation Rate	0.1%
Number of Generation	100

'true' records and 1,000 'false' records) to construct the kernels in EHRW and the remaining 90% (i.e., 9,900 records) were used as queries (test data) to evaluate the reasoning accuracy.

In the GA processing, we used 20 ($=N$) individuals (chromosomes) in each population. In every generation, one pair of chromosomes p and q were chosen randomly for the crossover operation (Fig. 4). And two points in the chromosome were also chosen randomly. The fields between the chosen points were exchanged between the two chromosome p and q each other. In the mutation operation, 0.1% fields in all chromosomes were randomly selected, and operators (genes) in the fields were exchanged to the other one. In the selection operation, the elitist strategy was used. The GA operations above were iterated until the 100th generation. Then, the chromosome of the highest fitness was chosen in each population to make up a kernel. GA parameters are listed in Table 1.

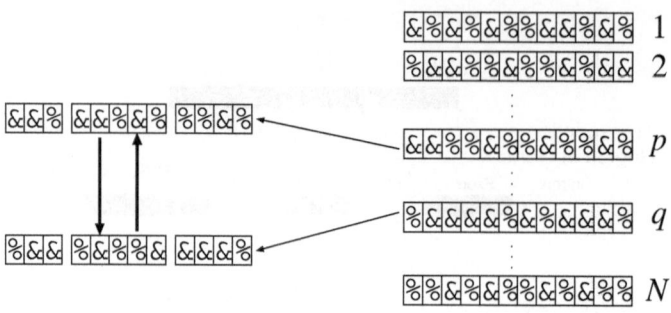

Fig. 4. Crossover operation. Two-point crossover between two chromosomes is used in the operation.

4.2   Hardware Configuration

Up to this point, we have used *don't care* for the letter (i.e., A, G, C, or T) in order to make our approach clearly (of course, this approach can be used practically). However, in the development we encoded the letters to the binary codes as A $\Rightarrow$ 1000, G $\Rightarrow$ 0100, C $\Rightarrow$ 0010, and T $\Rightarrow$ 0001 in order to increase the search space. Figure 5 shows a kernel example implemented using basic logic gates. One base is represented by the four-input AND gate in which only one input is positive and the others are negative (the reverse is also applicable). Therefore, if we use a record of 10 bases adjacent to the GT, its kernel circuit has 40 (10 × 4) inputs totally. In the prototype development and its evaluation, we could not know how many bases adjacent to GT affect the boundary, so that we carried out the GA process changing the number of the bases adjacent to the GT.

We developed a ERHW prototype (Fig. 6). The prototype contains seven FPGA chips (Xilinx XCV300-6GB432), data buffer memory chips, and some peripheral circuits in it. Four FPGAs out of seven are to implement kernels and counters, and one FPGA is to implement the maximum detector. Two remaining FPGAs (I/F chips) are for PC (host computer) interface circuits. The announced number of the system gates of XCV300 is 300k gates. Therefore, roughly 1,500k gates (300k × 5) can be used for the reasoning circuits, which are expected to be enough to implement many large practical reasoning applications.

The four FPGAs are connected with one of the I/F FPGA via the wide data-broadcast bus of 256-bit. Input and output data latches (flip-flops) are placed in each FPGA. Net reasoning time for one query thus needs three clock periods. And four clock periods, which are required in the I/F FPGAs to connect with the PC are added to the net reasoning time. Consequently, the reasoning time for one query in the prototype requires seven clock periods totally. A clock oscillator of 25 MHz is currently used in the prototype. Therefore, the reasoning time for one query is 280 ns (1/25MHz × 7 periods).

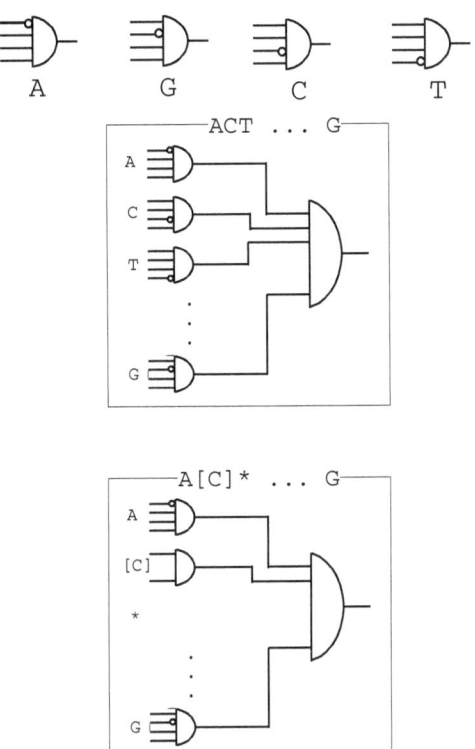

Fig. 5. Kernel Circuit. Four bases of A, G, C, and T are implemented using 4-input AND gates. [C] in the figure means a part of inputs of the AND gate for C are changed to *don't cares*.

5   Performance Evaluation

Figure 7 shows experimental results of reasoning accuracy as a function of the total number of bases adjacent to the GT. In the experiment, we used 9,900 test records that are not used to make kernels as mentioned above. The accuracy is saturated at 6 bases and the highest accuracy of 95.8 % is obtained at 6 bases. This saturation around 6 bases means that the important rules, or base-arrangement patterns to control whether the GT is true boundary or not are embedded in the 6 bases (3 bases on each side of the GT).

We measured the reasoning accuracy of the other living organisms using the *human* kernels. Results are listed in Table 2. High reasoning accuracy of 84.0 % was achieved in *gorilla*, while the reasoning accuracy considerably degreased in *yeast* and *dictyostelium discoideum* that is one of the most primitive organisms changing its form between animal and plant. Those experimental results mean that the kernels generated using *human* DNA preserve rules in the mammals, but the rules have been evolved and do not match the primitive organisms.

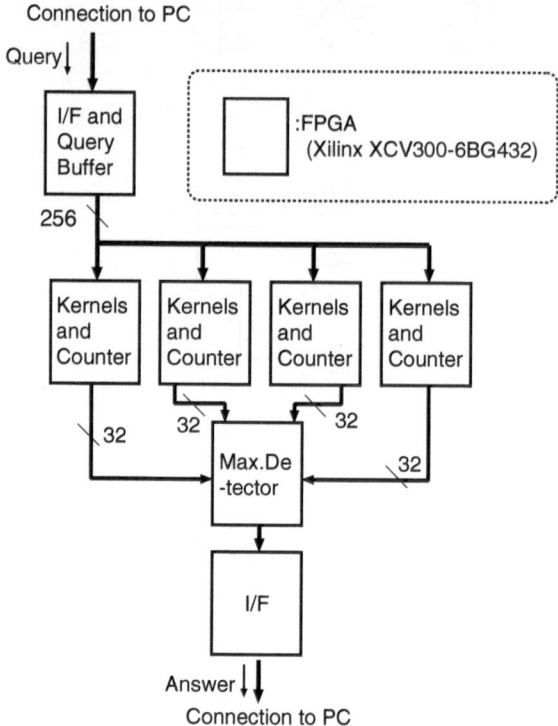

Fig. 6. Block diagram of the ERHW.

We also compared our result with those obtained using neural networks (Table 3) reported in [6]. In the table, "Neural Network" represents a single feed-forward multi-layered neural network and "Multi-modal Neural Network" represents multiple use of that network with a decision layer. We cannot compare the results each other directly because the data set used in our experiments was different from those used in [6]. However, we can say that ERHW may have higher reasoning accuracy than a single neural network, and have almost the same reasoning accuracy as the multi-modal neural network.

We used the Xilinx Foundation Logic Synthesizer[2] to synthesize the kernels, counters and the maximum detector. The ERHW designed using 1,100 DNA records requires 178,004 gates, and the prototype has enough capacity to implement those circuits.

Table 4 lists the measured reasoning time for one GT boundary. The reasoning speed of the prototype is by 1,000 times or more faster than that of the PC (Pentium III 1 GHz with 128 MB memory). The reason of this high speed processing in the ERHW in spite of low clock speed is that the circuits are designed

---

[2] http//:www.xilinx.co.jp/

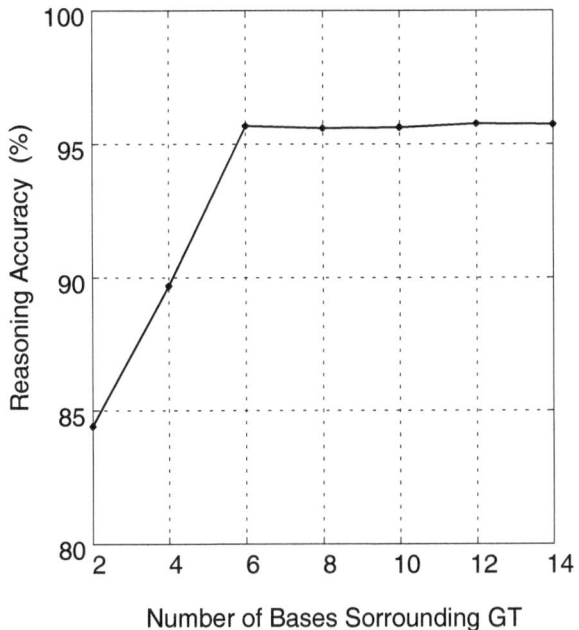

Fig. 7. Experimental results of reasoning accuracy as a function of the total number of bases adjacent to the GT.

Table 2. Reasoning accuracy of some organisms using the *human* kernels.

	human	*gorilla*	*dictyostelium discoideum*	*yeast*
Prediction Accuracy Using *Human* Kernels	95.8%	84.0%	48.0%	12.0%

directly from the DNA records, so that higher parallelism is achieved with less number of circuits.

6   Conclusions

The evolvable reasoning hardware (ERHW) was applied to the splicing site reasoning that was one of the best approaches to the gene-finding. Performance was evaluated using GenBank database, and the ERHW showed high reasoning accuracy of 95.8%. It also showed high reasoning speed of 280 ns that was faster than that of a personal computer (Pentium III 1GHz) by 1,000 times with a small circuit size of 178,004 gates. The kernels generated from the human's records also showed high reasoning accuracy to gorilla's records, but not to other primitive organisms' records. This result seems to indicate that some hereditary rules in the mammals are extracted in the generated human kernels.

Table 3. Comparison of reasoning accuracy with neural networks.

ERHW	Neural Network	Multi-modal Neural Network
95.8%	83.4%	95.0%

Table 4. Comparison of reasoning-time.

ERHW	PC (Pentium III 1 GHz)
280.0 ns	382 μ s

Acknowledgments

This research is supported in part by the 2002 MECSST grant in Japan (No. 14015206,14655145, and The City-Area Promotion Project) and the 2002 Japan Science and Technology Corporation grant(No. 13450163).

References

1. Moore, S. K.: Understanding the human genome. IEEE spectrum, **37** No. 11 (2001) 33–42.
2. Feitelson, D. G., and Treinin, M.: The Blueprint for Life? IEEE Computer, **35** No. 7 (2002) 34–40.
3. Yasunaga, M., Yoshihara, I., and Kim, J. H.: Evolvable Reasoning Hardware: Its Prototyping and Performance Evaluation. Journal of Genetic Programming and Evolvable Machines, Kluwer Academic Publishers, **2** No. 3 (2001) 211–230.
4. Fukunaga, K.: Introduction to Statistical Pattern Recognition. Academic Press (1990).
5. Brown, T.:Genomes. BIOS Scientific Publishers Ltd.(1999).
6. Yoshihara, I., Kamimai, Y., Yamamori, K., and Yasunaga, M.: A Multi-modal Neural Network for Identifying Exon-Intron Boundaries, Baba, N., Jain, L.C., and Howlett, R.J. (Eds), Knowledge-based Intelligent Information Engineering Systems and Allied Technologies, IOS Press, Holland, (2001) 998–1002.

Evolvable Fuzzy System for ATM Cell Scheduling

J.H. Li and M.H. Lim

School of EEE, Block S1
Nanyang Technological University
Singapore 639798
Tel: 65-67905408
emhlim@ntu.edu.sg

Abstract. In this paper, we propose an evolvable fuzzy system for ATM cell scheduling. When the scenarios of cell flows in an ATM network change dramatically, traditional scheduling algorithms, *first-in-first-out* (FIFO) and *static priority,* which employ static switching scheme may see their efficiency deteriorate. An alternative is to use *dynamically weighted priority*, which changes the priority based on the cell delay. Still, the algorithm *per se* is fixed and on the hardware level, nothing changes. Our goal in this paper is to achieve hardware adaptability and demonstrate its applicability in ATM switching. To do this, we proposed an evolvable fuzzy system [1]. With evolvable fuzzy system (EFS), the system undergoes transformation in response to changes in the cell patterns in order to maintain operational efficiency. We envisage that such a system is potentially useful for demanding real-time dynamic control applications.

1   Introduction

In recent years, ATM has become a universal network, because it can offer bandwidth support from telephony network to Internet. Due to the diversity of applications it can support, the bandwidth of ATM is critical. The efficiency of cell scheduling is therefore very important. In order to solve the problem of cell scheduling, many algorithms have been designed to cater for the desired QoS (quality of service) requirement. The common switching schemes are FIFO, static priority (SPR) and DWPS (dynamically weighted priority scheduling) [2]. The main attributes for determining QoS are cell loss and cell delay [3]. FIFO is a very simple method to do cell scheduling. It is easy to implement in hardware, but is not very good in terms of QoS performance. Static priority is also a simple method to do cell scheduling. However, it tends to be unfairly biased towards Class1 cell [1,2], which will always have a higher priority than other cell classes. DWPS is a significant improvement over the static SPR scheme. It adjusts the priority according to the cell flow scenarios [2]. But the adaptation scheme tends to be simple and may not be very efficient if the cell flows change dramatically. DWPS is also more difficult to implement in hardware.

Fuzzy logic is a powerful methodology in embedded control. It has been widely used in many real time applications since its inception. Cell flow scheduling is also a real time control problem, hence ideal for fuzzy control. But it can be tedious and complicated to manually formulate the fuzzy rule set for efficient cell scheduling in

A.M. Tyrrell, P.C. Haddow, and J. Torresen (Eds.): ICES 2003, LNCS 2606, pp. 208–217, 2003.

all scenarios. In this respect, an evolvable fuzzy system is very appropriate. With EFS, an optimal fuzzy rule set can be consistently maintained. In many fuzzy systems, rule set is manually derived based on intuition. This can be limited since the manually derived rule set may only be efficient for specific period of cell flows. Based on this premise, the principle behind EFS is therefore to evolve the rule set accordingly, which in this case is the changing patterns of cell flows. The adaptation of the rule set is achieved through symbiotic hybridization of an evolutionary algorithm embedded with fuzzy inference mechanisms [1,4].

If the pattern of cell flows changes, the fuzzy rule set should be changed accordingly to maintain consistent performance. To do this, online evolution to adapt to the cell flows can be implemented. Ultimately, it is desirable to implement the system as a form of intrinsic evolvable fuzzy hardware [1]. In principle, for such a scheme to be workable, it is just as important to address the architectural and hardware implementation issues pertaining to evolvable fuzzy system.

In this article, we will present some preliminary simulation results to demonstrate the potential usefulness of the proposed EFS for ATM scheduling. The QoS performance of the EFS is compared with the other three schemes, namely FIFO, SPR and DWPS.

This paper consists of 4 sections. In Section 2, we describe our proposed EFS in the context of ATM cell-switching problem. Section 3, outlines the simulation methodology. Results of simulation and comparisons with three other scheduling schemes will be presented. Finally in Section 4, we present some concluding remarks and outline some aspects of our future work.

2   Evolutionary Framework – ATM Cell-Scheduling

2.1   Problem Descriptions

The simplified block architecture of the ATM cell multiplexer is as shown in Fig.1. In the block diagram, BUF1 and BUF2 refer to buffers while MP refers to the multiplexing unit. The fuzzy switching control block is the part of the hardware that handles cell scheduling. The control scheme is derived through some evolutionary mechanisms, which forms the backbone of the EFS. For illustration, we classify the ATM services into two types, Class1 and Class2.

For ATM switching, Class1 can be a form of CBR (Constant Bit Rate) traffic, rt-VBR (real-time Variable Bit Rate) or both [5]. Class2 traffic type may refer to nrt-VBR (non-real-time Variable Bit Rate), UBR (Unspecified Bit Rate) or ABR (Available Bit Rate) [5]. Class1 type is delay sensitive while Class2 is considered to be not sensitive to delay. These two cell sources must be multiplexed on the OUTPUT by the MP unit through time division. We assume that the capacity of the OUTPUT to be the maximum of the input channels. It is also assumed that the capacities of the input buffers are finite and fixed.

To begin with, we define two symbols for the inputs, c_1 and c_2. The symbol c_1 refers to the status of Class1 cell, which is a function of V_1 and V_{max}. V_1 is the current cell rate of Class1 cell flow while V_{max} is the maximum cell rate of the line capacity. The symbol c_2 refers to buffer status of BUF2, which is a function of L_2 and L_{max}. L_2 is the number of empty units in BUF2 while L_{max} is the length of Class2 cell buffer. For

c_1 and c_2, the memberships are characterized by the linguistic term set {*very small*, *small*, *medium*, *large*, *very large*}. The output of the MP unit OUTPUT is characterized by the set {*true*, *false*}. It is clear that OUTPUT need not be fuzzy and hence characterized by two discrete values. Functionally, a *true* means that the MP unit allocates time packets to cater for the Class1 cell flow in BUF1. A *false* implies switching reverting to cells in BUF2.

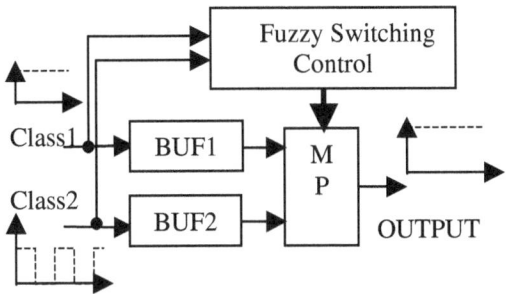

Fig. 1. Simplified multiplexer block diagram

Based on the above characterization of the switching network, it is possible to define the *n*-rule heuristics to control the switching behavior. With the fuzzy memberships defined, one can rely on intuitive logic to define the necessary input-output mappings as shown in Fig.2. The 25-rule system serves as the default ATM cell scheduling algorithm on system startup. We refer to these rules as the core rule set. It is noted that although our illustration shows an exhaustive rule set, actual systems may start off with a core rule set of *n* being less than 25.

2.2 Genetic Coding

As an illustration of the coding mechanism, consider the rule set in Table 1. The genetic code for the 25-rule system is represented as a genetic string structure "1222211221111121111211111". The allelic code 1 and 2 correspond to the labels *true* and *false* respectively. The position of the gene in the string corresponds to a specific rule in Table 1, when interpreted accordingly in a row wise manner.

2.3 Evolution of Rule Set

As rationalized above, we can simplify the maximum rule set system with one that utilizes fewer rules. For example, 0220001200011000011000011 refers to a 10-rule fuzzy system. In the string, 0 implies that there is no rule defined for the specific inputs scenario. Using GA, an optimal fuzzy rule set for the cell multiplexer can be derived through learning from past data on ATM traffic flow.

Table 1. A 25-rule fuzzy system for ATM cell scheduling.

c_2 \ c_1	very small	small	Medium	large	very large
very small	T	F	F	F	F
small	T	T	T	F	F
Medium	T	T	T	T	F
Large	T	T	T	T	F
very large	T	T	T	T	T

$T \equiv true, F \equiv false$

The most important component necessary for evolution is a mechanism to incorporate the fitness function in order to control the evolution. Based on the problem specifications, the goal is to use evolvable fuzzy system to schedule ATM cells such that the QoS requirements are satisfied. The QoS considered parameters are cell delay and cell loss ratio. The overall fitness function can be defined as follows:

$$F = \kappa - (\lambda \sum_{t=1}^{T} g(N(t), m(t)) + \theta \times |\gamma_1 - \gamma_2|) \tag{1}$$

Fitness is evaluated based on the scheduling performance for T number of cells. γ_1 and γ_2 are the cumulative number of cell loss for Class1 and Class2 respectively. λ is a weight parameter for the cell delay of Class1 cell flow. θ is a weight parameter to emphasize the significance of cell loss in the fitness function. Two additional parameters $N(t)$ and $m(t)$ are defined for the above fitness function. $N(t)$ refers to the maximum permitted delay time for Class1 cells, which can be determined. $m(t)$ is the waiting time of the cells in BUF1 before being sent. The function $g(N(t), m(t))$ which describes the measure of switching performance in terms of cell delay can be written as follows:

$$g(N(t), m(t)) = \begin{cases} N(t) - m(t) & N(t) \geq m(t) \\ (m(t) - N(t)) \times \omega & N(t) < m(t) \end{cases} \tag{2}$$

In the function $g(\cdot)$, ω is the penalty factor, a constant to account for cells that exceed the maximum permitted delay. Typically, ω should be greater than 1, depending on the level of tolerance. In general, the overall performance of the switching system corresponds to the fitness value. The parameter κ is a very large constant. For this fitness function, the parameters ω, θ and λ can be tuned accordingly based on the system's requirement.

2.4   Evolution Scheme

In order to satisfy the real time requirement and online evolution, we design the system architecture as in Fig.2.

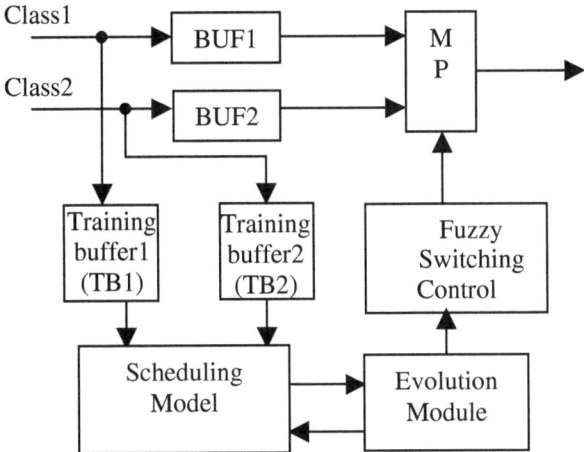

Fig. 2. Adaptation framework for EFS

In this system, the training buffers TB1 and TB2 are used to store Class1 and Class2 cells respectively. The size of TB1 and TB2, which corresponds to parameter T in Eq.1, is 2 or 3 times that of BUF1 and BUF2. When either TB1 or TB2 is full, the evolutionary process is triggered. Fitness evaluation is carried out by subjecting each chromosome to the scheduling model according to the cell flow stored in TB1 and TB2. The purpose of the scheduling model is to emulate the function of the ATM network as in Fig.1. If a chromosome that corresponds to a system rule set is better than the working chromosome, the working chromosome is replaced immediately. Functionally, the scheduling model emulates the ATM network to derive the cell delay and cell loss parameters. These parameters enable the fitness value to be calculated based on Eq.1. Basically, the evolution module serves to evolve the fuzzy rule set and it interacts with the scheduling model to evaluate the fitness of each evolved rule set. If evolution is triggered, it works in the background while the MP unit is in operation.

3   Simulation

The EFS for ATM cell scheduling proposed in this paper is simulated to study its QoS performance in terms of cell loss and cell delay. The EFS simulation is implemented in C++. Similarly, the FIFO, SPR and DWPS schemes are also simulated for performance comparison.

In our system simulator, cell flow traffic is simulated by reading from two input file streams. Each input stream corresponds to one cell class. The simulator reads from the files and checks if the arrival time of each cell is less than or equal to the simulation time. If it is, the cell is stored in BUF# (BUF1 or BUF2) and TB# (TB1 or TB2 respectively). The simulation time step T_{step} is set at 0.01 μs.

The data format of each cell in the input files is as shown in Fig.3. For Class1 cell flow, the Max_delay or maximum cell delay allowed can be specified based on system requirement. It is common to assume Max_delay to be 0. However, in order to eliminate the too good service problems [6], Max_delay can be given an appropriate value, which corresponds to the maximum delay that the receiver can tolerate. Bit_rate refers to the current bit rate. The Next_bit_rate is the bit rate of the next cell. The only difference between Class2 and Class1 cell is that Max_delay for Class2 cell is not used because in the fitness function, cell delay for Class2 is not considered.

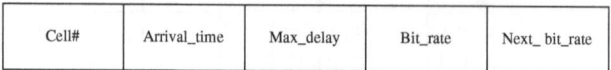

Cell#	Arrival_time	Max_delay	Bit_rate	Next_ bit_rate

Fig. 3. Cell format

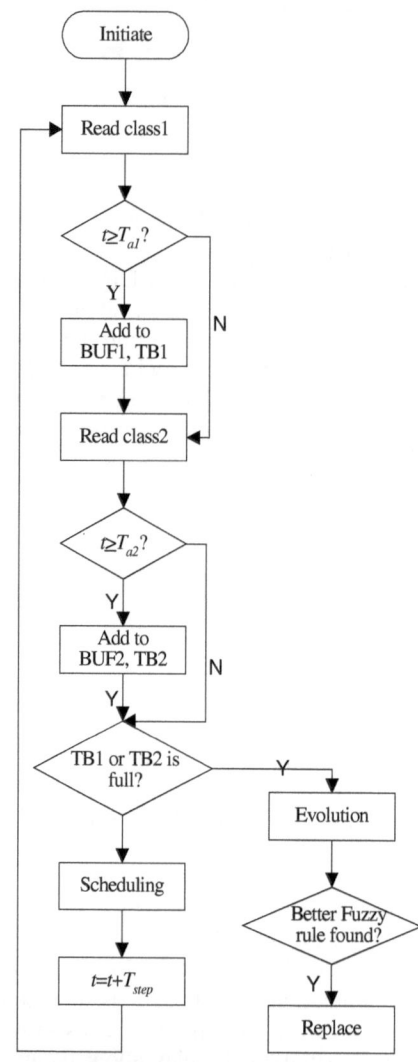

Fig. 4. Simulation flow diagram

The overall flow diagram of the system simulator is as shown in Fig.4. T_{a1} and T_{a2} refer to the cell arrival time for Class1 and Class2 respectively while t is the current running time of the simulation. For the EFS simulation, the parameters chosen are $\theta=2$ and $\lambda=4.5$. The size of BUF# is set at 100 units while TB# is 3 times that of BUF#.

A robust scheduling algorithm should perform well under various cell flow conditions [7]. Two kinds of cell flows representative of most cell flow scenarios are simulated. After the simulation, we compare the results of our algorithm with the other three scheduling methods that are popular in ATM cell scheduling. In both scenarios, the simulation time is 5 seconds.

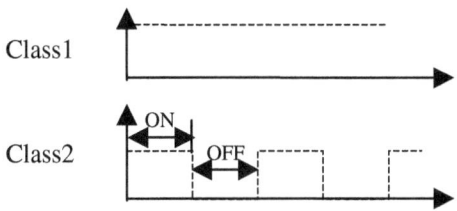

Fig. 5. Two classes of cell flows

3.1   Simulation Scenario 1

In this part of the simulation, we generate the traffic cell flows as in Fig.5.

Class1 is the CBR cell flow with cell rate of 155.52MHz. Class2 is VBR cell flow, also with a cell rate of 155.52MHz. The VBR specified has a 2ms ON period and a 2ms OFF period.

We compare the results of EFS scheduling with FIFO and dynamic priority cell scheduling. In this scenario, static priority scheduling can only send Class1 cells, so it need not be considered. The following are our simulation results. The comparison between the three schemes in terms of cell loss and cell delay is presented in Fig.6 and Fig.7.

In Fig.6, the difference in cell loss for Class1 and Class2 for the three scheduling algorithms is not significant. The total cell losses for the three scheduling algorithms are about the same.

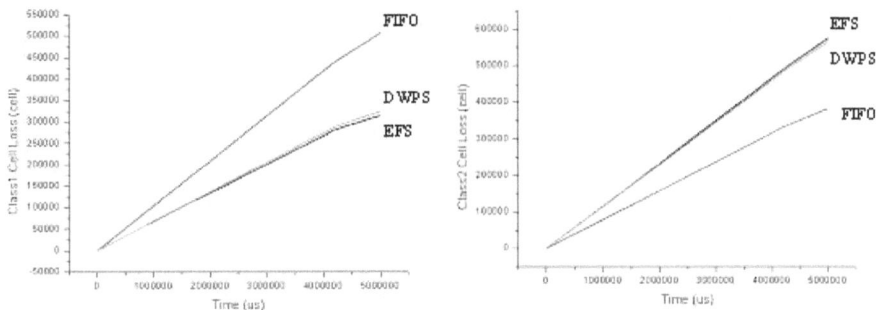

Fig. 6. Class1 cell loss and Class2 cell loss

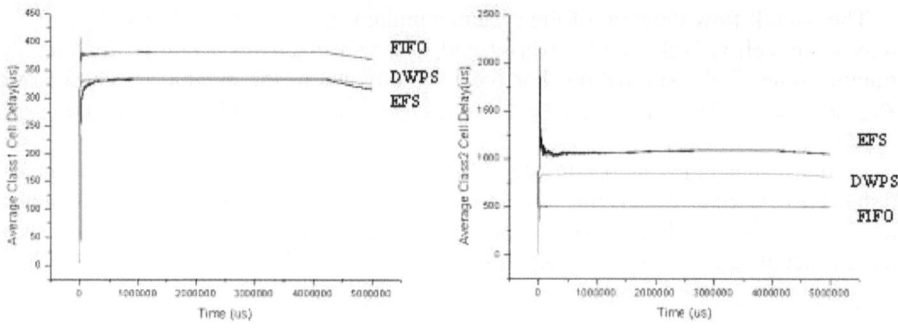

Fig. 7. Class1 and Class2 cell delay

From Fig.7, it can be seen that cell delay for Class1 cell using evolvable fuzzy system is slightly smaller than the cell delay of the other two algorithms. The cell delay of Class2 cell flow using evolvable fuzzy system is greater than that of the other two algorithms. But in terms of cell loss and cell delay in ATM network, EFS is able to achieve less Class1 cell delay and at the same time, achieve a balance of cell loss between Class1 and Class2 cell flow.

3.2 Simulation Scenario 2

In this part, we generate cell flows as in Fig.8.

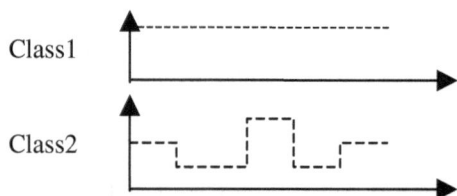

Fig. 8. Two classes of cell flows

Class1 refers to CBR cell flow with a cell rate of 100MHz. Class2 is VBR cell flow with unknown random cell rate. The minimum cell rate for VBR is 55.52MHz while the maximum is 155.52MHz. In this scenario, since the sum of the CBR and VBR cell rate is larger than the out channel's capacity, cell loss is unavoidable. Hence it is very useful to choose the appropriate scheduling method according to the cell loss and cell delay QoS parameters. From a practical point of view, this scenario is more likely to occur compared to the first scenario. The figures (Fig.9, Fig.10) that follow show the simulation results of EFS compared to the other three scheduling schemes.

In Fig.9, Class1 cell loss for the EFS is not zero, unlike that of DWPS and SPR. It is smaller than Class1 cell loss of FIFO. But in a way, EFS can maintain a desirable balance between Class1 and Class2 cell flow. The sums of cell losses of the two cell classes for the four algorithms are the same. Also, it is noticeable that Class1 cell delay for EFS is much smaller than that of FIFO and DWPS. Although Class1 cell delay of EFS is larger than SPR, but in terms of keeping a balance between Class1 cell and Class2 cell, EFS is much better than SPR. Considering the cell loss and cell delay on the whole, EFS performs better than the other three.

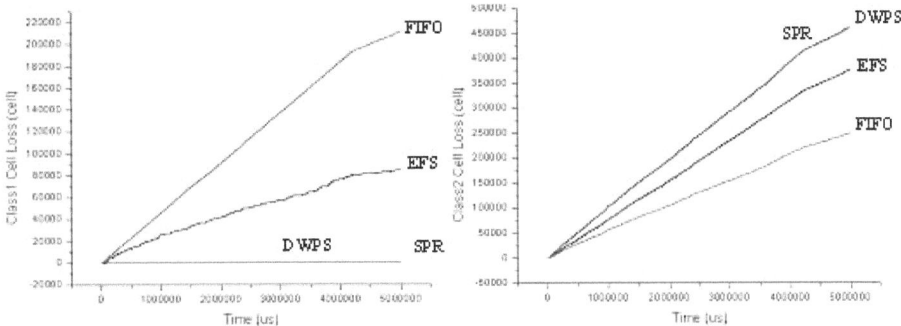

Fig. 9. Class1 and Class2 cell loss

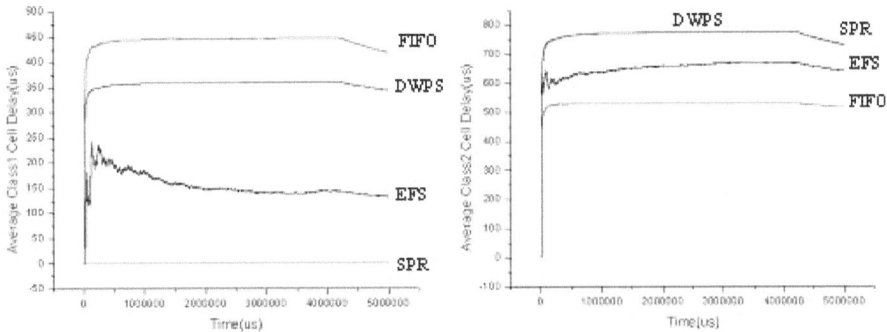

Fig. 10. Class1 and Class2 cell delay

It should also be noted that for EFS, it is convenient and easy to specify the relative importance of cell loss and cell delay. This can be achieved by changing the values of θ and λ. The values of these parameters can also be modified online, which is a special advantage that our proposed EFS framework enjoys.

4   Conclusion

From the simulation results presented in this paper, the overall performance of the proposed EFS is very encouraging. Our goal is to realize eventually the complete ATM cell scheduler on hardware, including the evolutionary mechanisms [1]. The simulation of the proposed framework has demonstrated the effectiveness and efficiency of the EFS scheduling scheme. There are essentially two types of evolvable hardware, extrinsic evolvable hardware and intrinsic evolvable hardware [8,13,14, 15]. From an implementation point of view, it is desirable to realize the evolutionary process within the hardware framework of the scheduler. This notion suggests the implementation of a form of intrinsic evolvable fuzzy hardware for dynamic real-time applications.

The results presented in this paper serve to demonstrate the viability of the proposed scheme. Further work on the evolutionary training algorithm needs to be carried out in order to improve the overall QoS performance. Detailed study on the architectural framework is required to derive a practical and efficient mode of intrinsically evolvable fuzzy hardware system. Our work discussed in this paper forms the basic framework of embedded fuzzy control and embedded evolutionary learning, which ultimately is realizable with the current system-on-chip (SoC) technology. Further ongoing work in this area will consider the direct implementation of the evolvable fuzzy hardware.

References

1. J.H. Li and M.H. Lim, "A Framework for Evolvable Fuzzy Hardware," Special session on Evolutionary Computing for Control and System Applications, ICONIP/SEAL/FSKD Joint Conference, 18-22 Nov 2002, Singapore.
2. T. Lizambri, F. Duran and S. Wakid, "Priority Scheduling and Buffer Management for ATM Traffic Shaping", The Seventh IEEE Workshop on Future Trends of Distributed Computing Systems 20, Dec. 1999.
3. M. P. Clark, *ATM networks: principles and use*, Chichester; New York, John Wiley, 1996
4. M.H. Lim, S. Rahardja and B.H. Gwee, "A GA paradigm for learning fuzzy rules", *Fuzzy Sets and Systems* 82(1996) pp. 177-186.
5. ATM Forum, "ATM Traffic Management Specification 4.0" April 1996, ftp://ftp.atmforum.com/pub/approved-specs/af-tm-056.000.pdf.
6. W. X. Liu, M. Murakawa and T. Higuchi, "ATM Cell Scheduling by Function Level Evolvable Hardware", *ICES 1996* : pp. 180-192.
7. D. B. Schwartz, "ATM Scheduling with Queuing Delay Predictions", http://ipoint.vlsi.uiuc.edu/wireless/papers-p/p205-schwartz.pdf.
8. Timothy G. W. Gordon and Peter J. Bentley, "On Evolvable Hardware", In Ovaska, S. and Sztandera, L. (Ed.) *Soft Computing in Industrial Electronics*. Physica-Verlag, Heidelberg, Germany, pp. 279-323.
9. H.D. Garis, "Evolvable Hardware: Principles and Practice", http://www.cs.usu.edu/~degaris/papers/CACM-E-Hard.html.
10. T. Higuchi, M. Iwata, I. Kajitani, M. Murakawa, S. Yoshizawa, and T. Furuya, "Hardware Evolution at Gate and Function Levels," Proc. Biologically Inspired Autonomous Systems: Computation, Cognition and Action, Durham, North Carolina, March, 1996.
11. D. Keymeulen, K. Konada, M. Iwata, Y. Kuniyoshi and T. Higuchi, "Robot Learning using Gate-Level Evolvable Hardware", In A. Birk and J. Demiris, (ed.), Proc of the Sixth European Workshop on Learning Robots, Lecture Notes in Artificial Intelligence, Springer-Verlag, 1998.
12. C. Z. Li, R. Bettati and W. Zhao, "Static Priority Scheduling for ATM Networks", Proc of the Real-Time Systems Symposium (RTSS'97), San Francisco. CA, Dec. 1997.
13. X. Yao and T. Higuchi. "Promises and Challenges of Evolvable Hardware", IEEE Transactions on Systems, Man and Cybernetics, Part C, Applications and Reviews, Vol.29, NO. 1, Feb. 1999, pp. 87 –97.
14. T.C. Fogarty, J.F. Miller and P. Thompson, "Evolving Digital Logic Circuits on Xilinx 6000 Family FPGA's", In P.K. Chawdry, R. Roy, R.K. Pant. Eds., Soft Computing in Engineering Design and Manufacturing, pp. 299-305, Springer-Verlag, London, 1998.
15. Y.S. Shi, R.Eberhart and Y.C. Chen, "Implementation of Evolutionary Fuzzy Systems", IEEE Transactions on Fuzzy Systems, Vol.7, No.2, Apr. 1999, pp. 109-119.

Synthesis of Boolean Functions
Using Information Theory

Arturo Hernández Aguirre[1],
Edgar C. González Equihua[1], and Carlos A. Coello Coello[2]

[1] CIMAT, Area de Computación, Callejón Jalisco s/n
Mineral de Valenciana, Guanajuato, Guanajuato 36240, Mexico
{artha,equihua}@cimat.mx
[2] CINVESTAV-IPN, Evolutionary Computation Group
Depto. Ing. Eléctrica, Sección de Computación
Av. Instituto Politécnico Nacional No. 2508
Col. San Pedro Zacatenco, México, D.F. 07300, Mexico
ccoello@cs.cinvestav.mx

Abstract. In this paper, we propose the use of Information Theory as the basis of the fitness function for Boolean circuit design. Boolean functions are implemented by means of multiplexers and genetic programming. Entropy based measures such as Mutual Information and Conditional Entropy are investigated as tools for similarity measures between circuits. A comparison of synthesized (through evolution) and minimized circuits through other methods denotes the advantages of the Information-Theoretical approach.

1 Introduction

Entropy is a measure of disorder and the basis of Information Theory (IT) [15]. Shannon [13] suggested the use of information entropy as a measure of the amount of information contained within a message. Thus, entropy tells us that there is a limit in the amount of information that can be removed from a random process without having any information loss. For instance, in theory, music can be compressed (in a lossless form) and reduced up to its entropy limit. Further reduction is only possible at the expense of information lost.

The ID3 algorithm for the construction of classifiers (based on decision trees) is probably the best-known computer science representative that relies on entropy measures [12]. For ID3, an attribute is more important for concept classification if it provides greater "information gain" than the others.

IT was first used by Hartamann et al. [6] to transform decision tables into decision trees. Boolean function minimization through IT techniques has been approached by several authors [7,8]. These methods are top-down, thus, the design strategy follows after a set of axioms in the knowledge domain. Luba et al. [10] address the synthesis of logic functions using a genetic algorithm and a fitness function based on conditional entropy. Their system needs heavy preprocessing of the search space (Shannon's expansion is applied to the target

A.M. Tyrrell, P.C. Haddow, and J. Torresen (Eds.): ICES 2003, LNCS 2606, pp. 218–227, 2003.

Boolean function as to find subexpressions whose purpose is to guide the genetic search. Only after that, the genetic algorithm is started).

In this paper we use multiplexers and genetic programming (GP) for the synthesis of Boolean functions. We propose a fitness function driven by the Normalized Mutual Information between the target function and the evolved function. Our system works exclusively in a bottom-up fashion, thus no preprocessing of the search space is needed. The paper is organized as follows. Section 2 describes the problem statement, Section 3 introduces basic concepts of information theory used throughout the article. In Section 4 we show how entropy based methods will prevent convergence of any evolutionary method if not used correctly. In Section 5 we propose three fitness function based on normalized mutual information and conditional entropy. Section 6 is devoted to experiments, and we finish with conclusions and final remarks in Section 7.

2 Problem Statement

The design problem is the following: find the smallest circuit that implements a Boolean function specified by its truth table [2,1,4]. The design metric adopted in this case is the number of components in a 100% functional circuit. The process works at "gate-level" and the only component replicated is the binary multiplexer. A binary multiplexers implements the Boolean function $f = ax + a'y$, where a is the control and $\{x,y\}$ the input signals. The use of multiplexers is a sound approach because: 1) they are universal generators of Boolean functions, and 2) any circuit in the population is the Shannon expansion of a Boolean function. The expansion takes the form of and-or sum of products (SOP) which are easily represented as decision trees. Therefore, circuits are encoded as trees and the approach follows the representation adopted by Genetic Programming. Leaves of the tree are only 1s and 0s (as in a decision tree), and the nodes are the variables of the Boolean function. Every variable of a node takes the place of the "pivot" variable used in the expansion.

Definition 1. Boolean Residue The residue of a Boolean function $f(x_1, x_2, \ldots, x_n)$ with respect to a variable x_j is the value of the function for a specific value of x_j. It is denoted by f_{x_j}, for $x_j = 1$ and by $f_{\bar{x}_j}$ for $x_j = 0$.

$$f = \bar{x}_j f|_{\bar{x}_j} + x_j f|_{x_j} \tag{1}$$

The pivot variable is x_j. For instance, for the function $f(a, b, c) = a'b'c + a'bc' + ab'c'$, the residue of the expansion over variable a is:

$$f(a, b, c) = a'F(a = 0) + aF(a = 1) = a'(b'c + bc') + a(b'c')$$

Therefore, pivot variable a takes the control of the multiplexer and the two residues form the inputs, as shown in Figure 1.

Further expansion of the functions at the mux inputs yields the complete tree of muxes implementing the target function.

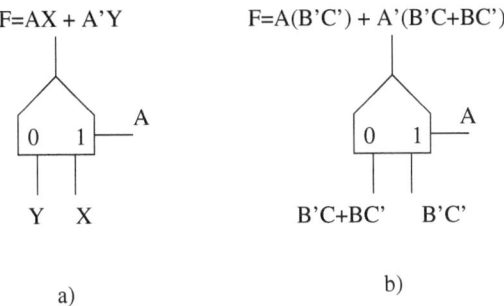

Fig. 1. a) the binary multiplexer, b) the Shannon expansion using a multiplexer

3   Basic Concepts of IT

Uncertainty and its measure provide the basis for developing ideas about Information Theory [5]. The most commonly used measure of information is Shannon's entropy.

Definition 2. Entropy The average information supplied by a set of k symbols whose probabilities are given by $\{p_1, p_2, \ldots, p_k\}$, can be expressed as,

$$H(p_1, p_2, \ldots, p_k) = -\sum_{s=1}^{k} p_k log_2 p_k \tag{2}$$

The information shared between the transmitter and the receiver at either end of the communication channel is estimated by its Mutual Information,

$$MI(T; R) = H(T) + H(R) - H(T, R) = H(T) - H(T|R) \tag{3}$$

The conditional entropy $H(T|R)$ can be calculated through the joint probability, as follows:

$$H(T|R) = -\sum_{i=1}^{n} \sum_{j=1}^{n} p(t_i r_j) log_2 \frac{p(t_i r_j)}{p(r_j)} \tag{4}$$

An alternative expression of mutual information is

$$MI(T; R) = \sum_{t \in T} \sum_{r \in R} p(t, r) log_2 \frac{p(t, r)}{p(t)p(r)} \tag{5}$$

Mutual information, Equation 3, is the difference between the marginal entropies $H(T) + H(R)$, and the joint entropy $H(T, R)$. We can explain it as a measure of the amount of information one random variable contains about another random variable, thus it is the reduction in the uncertainty of one random variable due to the knowledge of the other [5].

Table 1. Function $F = AB + BC$ used to compute MI(F;C)

A	B	C	F=AB+BC
0	0	0	0
0	0	1	0
0	1	0	0
0	1	1	1
1	0	0	0
1	0	1	0
1	1	0	1
1	1	1	1

Conditional entropy is used in top-down circuit minimization methods [3], and also in evolutionary approaches [10,9].

Mutual information is not an invariant measure between random variables because it contains the marginal entropies. Normalized Mutual Information is a better measure of the "prediction" that one variable can do about the other [14].

$$NMI(T; R) = \frac{H(T) + H(R)}{H(T, R)} \tag{6}$$

Normalized Mutual Information has been used in image registration with great success [11].

Example: We illustrate these concepts by computing the Mutual Information between two Boolean vectors F and C, shown in Table 1. Variable C is an argument of the Boolean function $F(A, B, C) = AB + BC$. We aim to estimate the description the variable C can do about variable F, that is, $MI(F; C)$.

We use Equations 3 and 4 to calculate $MI(F; C)$. Thus, we need the entropy $H(F)$ and the conditional entropy $H(F|C)$.

Entropy requires the discrete probabilities $p(F = 0)$ and $p(F = 1)$ which we find by counting their occurrences

$$H(F) = -\left(\frac{5}{8}log_2\frac{5}{8} + \frac{3}{8}log_2\frac{3}{8}\right) = 0.9544$$

The conditional entropy, Equation 4, uses the joint probability $p(f_i, c_j)$, which can be estimated through conditional probability, as follows: $p(f, c) = p(f)p(c|f)$. Since either vector F and C has two possible values, the discrete joint distribution has four entries, as follows:

$$p(F = 0, C = 0) = p(f = 0)p(c = 0|f = 0) = \frac{5}{8} \times \frac{3}{5} = 0.375$$

$$p(F = 0, C = 1) = p(f = 0)p(c = 1|f = 0) = \frac{5}{8} \times \frac{2}{5} = 0.25$$

$$p(F = 1, C = 0) = p(f = 1)p(c = 0|f = 1) = \frac{3}{8} \times \frac{1}{3} = 0.125$$

$$p(F = 1, C = 1) = p(f = 1)p(c = 1|f = 1) = \frac{3}{8} \times \frac{2}{3} = 0.25$$

Now we can compute the conditional entropy by using Equation 4. The double summation produces four terms (since $n = 2$):

$$H(F|C) = -\left(\frac{3}{8}log_2\frac{3}{4} + \frac{1}{4}log_2\frac{1}{2} + \frac{1}{8}log_2\frac{1}{4} + \frac{1}{4}log_2\frac{1}{2}\right)$$

$$H(F|C) = 0.9056$$

Therefore, $MI(F;C) = H(F) - H(F|C) = 0.9544 - 0.9056 = 0.0488$.

4   Entropy and Circuits

Entropy has to be carefuly applied to the synthesis of Boolean functions. Let's assume any two Boolean functions, $F1$ and $F2$, and a third $F3$ which is the one's complement of F2, then $F3 \neq F2$.

$$H(F2) = H(F3)$$

Also, Mutual Information shows a similar behavior.

$$MI(F1, F2) == MI(F1, F3)$$

The implications for Evolutionary Computation are important since careless use of mutual information can nullify the system's convergence. Assume the target Boolean function is T, then $MI(T, F2) = MI(T, F3)$, but only one of the circuits implementing $F2$ and $F3$ is close to the solution since their Boolean functions are complementary. A fitness function based on mutual information will reward both circuits with the same value, but one is better than the other. Things could worsen as evolution progresses because mutual information increases when the circuits get closer to the solution, but in fact, two complementary circuits are then given larger rewards. The scenario is one in which the population is driven by two equally strong attractors, hence convergence is never reached.

The fitness function of that scenario is as follows. Assume T is the target Boolean function (must be seen as a truth table), and C is output of any circuit in the population. Fitness function is either the maximization of mutual information or minimization of the conditional entropy term. This is,

$$badfitnessfunction\#1 = MI(T, C) = H(T) - H(T|C)$$

The entropy term $H(T)$ is constant since this is the expected target vector. Therefore, instead of maximizing mutual information the fitness function can minimize the conditional entropy,

$$badfitnessfunction\#2 = H(T) - H(T|C)$$

5   Fitness Function
Based on Normalized Mutual Information

So far, we have described the scenario where the population is driven by a fitness function based on the sole mutual information. We now propose two new fitness functions based on entropy. Let's assume a target Boolean function of m atributes $T(A_1, A_2, \ldots, A_m)$, and the circuit Boolean function C of the same size. We propose and report experiments using the two following fitness functions (higher fitness means that a better solution has been found).

$$fitness = (Length(T) - Hamming(T, C)) \times NMI(T, C) \qquad (7)$$

$$fitness1 = \sum_{i=1}^{m} \frac{fitness}{NMI(A_i, C)} \qquad (8)$$

$$fitness2 = \sum_{i=1}^{m} fitness \times NMI(A_i, C) \qquad (9)$$

$$fitness3 = (Length(T) - Hamming(T, C)) \times (10 - H(T|C)) \qquad (10)$$

Fitness1, Equation 7, is driven by $NMI(T, C)$ and adjusted by the factor $Length(T) - Hamming(T, C)$. This factor tends to zero when T and C are far in Hamming distance, and tends to $Length(T)$ when T and C are close in Hamming distance. The effect of the term is to give the correct rewarding of the NMI to a circuit C close to T. Equation 7 is designed to remove the convergence problems described in the previous section. Fitness1 and Fitness2, Equations 8 and 9, combine the NMI of T and C with NMI of C and the attributes A_k of the target function. Thus, fitness1 and fitness2 pretend to use more information available in the truth table in order to guide the search. Fitness3 is based on conditional entropy and it uses the mentioned factor to supress the reproduction of undesirable trees. Since conditional entropy has to be minimized we use the factor $10 - H(T|C)$ in order to miximize fitness.

6   Experiments

In the following experiments we find and contrast the convergence of our GP system for the three fitness functions defined above.

6.1   Experiment 1

Here we design the following (simple) Boolean function:

$$F(a, b, c, d) = \sum (0, 1, 2, 3, 4, 6, 8, 9, 12) = 1$$

We use a population size of 300 individuals, $p_c = 0.35$, $p_m = 0.65$, and we run our algorithm for 100 generations. The optimal solution has 6 nodes, thus we find the generation in which the first 100% functional solution appears, and the generation number where the optimal is found. The problem was solved 20 times for each fitness function.

Table 2 shows the results of these experiments.

Table 2. Generation number where the first 100% functional circuit is found, and the generation where the optimum is found, for three fitness functions

Event	Gen. at fitness1	Gen. at fitness2	Gen. at fitness3
100% Functional	13 ± 5	14 ± 7	18 ± 6
Optimum Solution	30 ± 7	30 ± 10	40 ± 20

6.2   Experiment 2

The next test function is:

$$F(a, b, c, d, e, f) = ab + cd + ef$$

In this case, we use a population size of 600 individuals, $p_c = 0.35$, $p_m = 0.65$, and we stop after 200 generations. The optimal solutions has 14 nodes. Each problem was solved 20 times for each fitnesss function.

Table 3 shows the results of these experiments.

Table 3. Generation number where the first 100% functional circuit is found, and the generation where the optimum is found, for three fitness functions

Event	Gen. at fitness1	Gen. at fitness2	Gen. at fitness3
100% Functional	39 ± 12	40 ± 11	50 ± 12
Optimum Solution	160 ± 15	167 ± 15	170 ± 20

6.3   Experiment 3

The last problem is related to partially specified Boolean functions [1]. With this experiment we address the ability of the system to design Boolean functions with "large" number of arguments and specific topology. For this, we have designed a synthetic problem were the topology is preserved when the number of variables increases.

Boolean functions with $2k$ variables are implemented with $(2 * 2k) - 1$ binary muxes *if* the truth table is specified as shown in Table 4.

We ran experiments for $k = 2, 3, 4$, thus 4,8, and 16 variables and we have contrasted these results with the best known solutions for this problem (reported in [1]). For completeness, all previous results are reported together with the results of the new experiments in Table 5, where we use the three fitness functions (Equations 8,9,10).

Table 4. Partially specified Boolean function of Example 3 needs $(2 * 2k) - 1$

ABCD	F(ABCD)
0 0 0 0	0
0 0 0 1	1
0 0 1 0	1
0 1 0 0	1
1 0 0 0	1
0 1 1 1	1
1 0 1 1	1
1 1 0 1	1
1 1 1 0	1
1 1 1 1	0

Table 5. Generation number where the first 100% functional circuit is found, and the generation where the optimum is found, for three fitness functions

k	variables	size	Avg(previous)	Avg(fitness1)	Avg(fitness2)	Avg(fitness3)
2	4	7	60	60	60	60
3	8	15	200	190	195	194
4	16	31	700	740	731	748
5	32	63	2000	2150	2138	2150

All parameters are kept with no change for similar experiments, average is computed for 20 runs. The previous experiments use a fitness function based on Hamming distance between the current solution of an individual and the target solution of the truth table. One important difference is the percentage of correct solution found. Previously we reported that in 90% of the runs we found the solution (for the case of fitness based on Hamming distance). For the three fitness functions based on entropy we found the solution in 99% of the runs.

7 Final Remarks and Conclusions

A fitness function using only conditional entropy was tested with no success at all. We believe this is a clear indication of a fitness function that does not take into account the properties of entropy. In general, the three fitness functions work quite well, all of them found the optimum in most cases, thus comparable to other fitness functions based on Hamming distances. Entropy based measures seem hard to adapt to Evolutionary Computation since the entropy of evolutionary systems is not well understood (after "creationists" would say evolution is imposible because entropy would not allow the development of a system). The final remark is that the convergence time and the quality of results produced is comparable with the many experiments we have done before in this area. Based on the results shown in Tables 2 and 3 we would give some advantage to normalized mutual information over simple mutual information because it is less biased. Results from Table 5 could imply that mutual information is able to capture "that" relationship between the data that the sole Hamming distance can not convey to the population.

Acknowledgements

The first author acknowledges partial support from CONACyT project No. I-39324-A. The second author acknowledge support from CONACyT through a scholarship to complete the Master in Science program at CIMAT The third author acknowledges support from CONACyT project No. NSF-CONACyT 32999-A.

References

1. Arturo Hernández Aguirre, Bill P. Buckles, and Carlos Coello Coello. Evolutionary synthesis of logic functions using multiplexers. In C. Dagli, A.L. Buczak, and et al., editors, *Proceedings of the 10th Conference Smart Engineering System Design*, pages 311–315, New York, 2000. ASME Press.
2. Arturo Hernández Aguirre, Carlos Coello Coello, and Bill P. Buckles. A genetic programming approach to logic function synthesis by means of multiplexers. In Adrian Stoica, Didier Keymeulen, and Jason Lohn, editors, *Proceedings of the First NASA/DoD Workshop on Evolvable Hardware*, pages 46–53, Los Alamitos, California, 1991. IEEE Computer Society.
3. V. Cheushev, S. Yanushkevith, and et al. Information theory method for flexible network synthesis. In *Proceedings of the IEEE 31st. International Symposium on Multiple-Valued Logic*, pages 201–206. IEEE Press, 2001.
4. Carlos Coello Coello and Arturo Hernández Aguirre. Design of combinational logic circuits through an evolutionary multiobjective optimization approach. *Artificial Intelligence for Engineering, Design, Analysis and Manufacture*, 16(1):39–53, January 2002.
5. T.M. Cover and J.A. Thomas. *Elements of Information Theory*. John Wiley & Sons, New York, 1991.
6. C.R.P. Hartmann, P.K. Varshney, K.G. Mehrotra, and C.L. Gerberich. Application of information theory to the construction of efficient decision trees. *IEEE Transactions on Information Theory*, 28(5):565–577, 1982.
7. A.M. Kabakcioglu, P.K. Varshney, and C.R.P. Hartmann. Application of information theory to switching function minimization. *IEE Proceedings, Part E*, 137:387–393, 1990.
8. A. Lloris, J.F. Gomez-Lopera, and R. Roman-Roldan. Using decision trees for the minimization of multiple-valued functions. *International Journal of Electronics*, 75(6):1035–1041, 1993.
9. T. Luba, C. Moraga, S. Yanushkevith, and et al. Application of design style in evolutionary multi-level network synthesis. In *Proceedings of the 26th EUROMICRO Conference Informatics:Inventing the Future*, pages 156–163. IEEE Press, 2000.
10. T. Luba, C. Moraga, S. Yanushkevith, and et al. Evolutionary multi-level network synthesis in given design style. In *Proceedings of the 30th IEEE International Symposium on Multiple valued Logic*, pages 253–258. IEEE Press, 2000.
11. Frederik Maes, André Collignon, Dirk Vandermeulen, Guy Marchal, and Paul Suetens. Multimodality image registration by maximization of mutual information. *IEEE Transactions on Medical Imaging*, 16(2):187–198, April 1997.

12. J.R. Quinlan. Learning efficient classification procedures and their application to chess games. In R. S. Michalski, J. G. Carbonell, and T. M. Mitchell, editors, *Machine Learning: An Artificial Intelligence Approach*, pages 463–482. Springer, Berlin, Heidelberg, 1983.
13. Claude E. Shannon. A Mathematical Theory of Information. *Bell System Technical Journal*, 27:379–423, July 1948.
14. C. Studholme, D.L.G. Hill, and D.J. Hawkes. An overlap invariant entropy measure of 3D medical image alignment. *Pattern Recognition*, 32:71–86, 1999.
15. W. Weaver and C. E. Shannon. *The Mathematical Theory of Communication*. University of Illinois Press, Urbana, Illinois, 1949.

Evolving Multiplier Circuits by Training Set and Training Vector Partitioning

Jim Torresen

Department of Informatics, University of Oslo
P.O. Box 1080 Blindern, N-0316 Oslo, Norway
jimtoer@ifi.uio.no
http://www.ifi.uio.no/~jimtoer

Abstract. Evolvable Hardware (EHW) has been proposed as a new method for evolving circuits automatically. One of the problems appearing is that only circuits of limited size are evolvable. In this paper it is shown that by applying an approach where the training set as well as each training vector is partitioned, large combinational circuits can be evolved. By applying the proposed scheme, it is shown that it is possible to evolve multiplier circuits larger then those evolved earlier.

1 Introduction

For the recent years, evolvable hardware (EHW) has become an important scheme for automatic circuit design. However, still there lack schemes to overcome the limitation in the chromosome string length [1,2]. A long string is required for representing a complex system. However, a larger number of generations are required by genetic algorithms (GA) as the string length increases. This often makes the search space *too* large and explains why only small circuits have been evolvable so far. Thus, work has been undertaken trying to diminish this limitation. Various experiments on *speeding up* the GA computation have been undertaken [3]. The schemes involve fitness computation in parallel or a partitioned population evolved in parallel. Experiments are focussed on speeding up the GA computation, rather than dividing the application into subtasks. This approach assumes that GA finds a solution if it is allowed to compute enough generations. When small applications require weeks of evolution time, there would probably be strict limitations on the systems evolvable even by parallel GA.

Other approaches to the problem have used variable length chromosomes [4]. Another option, called function level evolution, is to apply building blocks more complex than digital gates [5]. Most work is based on fixed functions. However, there has been work in Genetic Programming for *evolving* the functions – called Automatically Defined Functions (ADF) [6].

An improvement of artificial evolution – called co-evolution, has been proposed [7]. In co-evolution, a part of the data which defines the problem co-evolves simultaneously with a population of individuals solving the problem. This could lead to a solution with a better generalization than a solution based only on

A.M. Tyrrell, P.C. Haddow, and J. Torresen (Eds.): ICES 2003, LNCS 2606, pp. 228–237, 2003.

the initial data. A variant of co-evolution – called cooperative co-evolutionary algorithms, has been proposed by De Jong and Potter [8,9]. It consists of parallel evolution of sub-structures which interact to perform more complex higher level structures. Complete solutions are obtained by assembling representatives from each group of sub-structures together. In that way, the fitness measure can be computed for the top level system. However, by testing a number of individuals from each sub-structure population, the fitness of individuals in a sub-population can be sorted according to their performance in the top-level system. Thus, no explicit local fitness measure for the sub-populations are applied in this approach. However, a mechanism is provided for initially seeding each GA population with user-supplied rules. Darwen and Yao have proposed a co-evolution scheme where the subpopulations are divided without human intervention [10].

Incremental evolution for EHW was first introduced in [11] for a character recognition system. The approach is a divide-and-conquer on the evolution of the EHW system, and thus, named *increased complexity evolution*. It consists of a division of the *problem* domain together with incremental evolution of the hardware system. Evolution is first undertaken individually on a set of basic units. The evolved units are the building blocks used in further evolution of a larger and more complex system. The benefits of applying this scheme is both a *simpler* and *smaller* search space compared to conducting evolution in one single run. The goal is to develop a scheme that could evolve systems for complex real-world applications. Earlier work has shown that the approach is successful for character classification and prosthetic hand control [12,13]. There has been undertaken work on decomposition of logic functions by using evolution [14]. Results on evolving a 7-input and 10-output logic function show that such an approach is beneficial.

The largest correctly working multiplier circuit evolved – the author is aware of, is evolved by Vassilev et al [15]. It is a 4×4-bit multiplier evolved in a single run (8 outputs) by gates as building blocks. In this paper, the multiplier will not be evolved in a single run but rather be evolved as separate subsystems that together perform a correct multiplication. Normally, evolution is used to evolve a circuit with a smallest possible number of gates. In this work, we rather concentrate on reducing the evolution time at the same time as being able to solve problems not evolvable in a single run. This is motivated by the fact that the number of transistors becoming available in circuits increases according to Moores Law. Thus, this approach is based on "wasting" transistors to faster evolve a circuit or evolve circuits than are *not* evolvable in a single run.

The experiments are undertaken for evolving a 5×5-bit multiplier circuit. There is a substantial increase in complexity between a four-by-four bit multiplier (256 lines in the truth table) compared to a five-by-five bit multiplier (1024 lines in the truth table). Even though the experiments are for evolving multiplier circuits, the principles are valid for any combinational circuit defined by a complete truth table.

The next section introduces the concepts of the multiplier evolution. Results of experiments are reported in Section 3 with conclusions in Section 4.

2 Multiplier Evolution by Data Partitioning

In this section, the *increased complexity evolution* is applied to evolve a large multiplier circuit.

2.1 Approaches to Increased Complexity Evolution

Several alternative schemes on how to apply this method are present:

- **Partitioned training vector.** A first approach to incremental evolution is by partitioning each training vector. For evolving a truth table - i.e. like those used in digital design, each output could be evolved separately. In this method, the fitness function is given explicitly as a subset of the complete fitness function.
- **Partitioned training set.** A second approach is to divide the training set into several subsets. For evolving a truth table, this would correspond to distributing the rows into subsets, which are evolved separately [16]. This would be similar to the way humans learns: Learning to walk and learning to talk are two different learning tasks. The fitness function would have to be designed for each task individually and used together with a global fitness function. Later in this paper it is shown how this can be done in digital design.

In this paper, these two approaches are combined.

Table 1. The truth table for multiplying two by two bits.

X * Y = Z
00 * 00 = 0000
00 * 01 = 0000
00 * 10 = 0000
00 * 11 = 0000
01 * 00 = 0000
01 * 01 = 0001
01 * 10 = 0010
01 * 11 = 0011
10 * 00 = 0000
10 * 01 = 0010
10 * 10 = 0100
10 * 11 = 0110
11 * 00 = 0000
11 * 01 = 0011
11 * 10 = 0110
11 * 11 = 1001

Table 1 shows the truth table for a two by two bit multiplier circuit. In the context of evolving multiplier circuits, the training set consists of all possible permutations of the inputs. Evolving this circuit in a single run requires a circuit with four inputs and four outputs. If we instead would like to apply the *increased complexity evolution* we could have a partitioning like the one given in Table 2. This is for illustration only since there would be no problem in evolving such a small truth table in a single run. In this case, we would evolve a separate circuit for each of the four outputs – partitioned training vector (PTV). Further, partitioned training set (PTS) is also applied and divides the training set into two parts. This means that we end up with evolving eight subsystems in total (PTV equal to four multiplied by PTS equal to two). The most significant input bit (x_1) is used to distinguish between the two training set partitions. Thus, the final system can be assembled together – as seen in Fig. 1, to perform a correct multiplier function. This requires during evolution that the truth table vectors are assigned into two groups each applied to a evolve its separate groups of subsystems (A and B) as the horisontal line (and x_1) in Table 2 indicates. Within each such group there is no demand for sorting the vectors.

Table 2. The truth table for multiplying two by two bits.

x_1x_0	*	y_1y_0	z_3	z_2	z_1	z_0
00	*	00	0	0	0	0
00	*	01	0	0	0	0
00	*	10	0	0	0	0
00	*	11	0	0	0	0
01	*	00	0	0	0	0
01	*	01	0	0	0	1
01	*	10	0	0	1	0
01	*	11	0	0	1	1
10	*	00	0	0	0	0
10	*	01	0	0	1	0
10	*	10	0	1	0	0
10	*	11	0	1	1	0
11	*	00	0	0	0	0
11	*	01	0	0	1	1
11	*	10	0	1	1	0
11	*	11	1	0	0	1

The amount of hardware needed would be larger than that needed for evolving one system in a single run. However, the important issue – as will be shown in this paper, is the possibility of evolving circuits *larger* than those evolved by a single run evolution.

The number of generations required to evolve a given subsystem would be highly dependent on the training vectors for the given training set partition. Thus, to minimize both the evolution time and the amount of logic gates, an algorithm should be applied to find both the partitioning size of the training set as well as the partitioning of each training vector. The best way this can be undertaken is probably by starting with a fine grain partitioning to be able to find a working system – even though the amount of logic would be large. Then, the training vectors for consecutive subsystems that was easily evolved

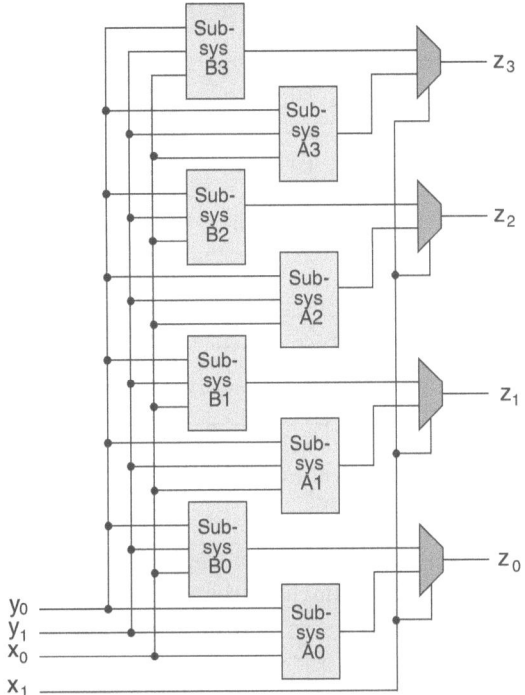

Fig. 1. An assembly of subsystems.

can be *merged* and applied to evolve *one* new subsystem that can substitute the group of subsystems earlier evolved. This, will reduce the hardware size. Another optimizing feature to be included is to look into the training set from the beginning to make dedicated partitions. That is, if a set of consecutive input vectors result in the same output value, there is no need to actually evolve a subsystem for this. In the following experiments we apply fixed partitioning sizes, however results will probably indicate that it will be helpful with an adaptive partitioning.

2.2   Target Hardware

Each subsystem evolved consists of a fixed-size array of logic gates. The array consists of n gate layers from input to output as seen in Fig. 2. Each logic gate (LG) is either an *AND* or an *XOR* gate. Each gate's two inputs in layer l is connected to the outputs of two gates in layer $l-1$. In the following experiments, the array consists of 8 layers each consisting of 16 gates. A 5×5-bit multiplier consists of 10 outputs which we are going to evolve by partitioning as described in the example above. Thus, only one gate is used in the layer 8 of the gate array. This results in a maximum number of gates for each subsystem: $N_{ss} = 4 \cdot 16 + 8 + 4 + 2 + 1 = 79$ gates.

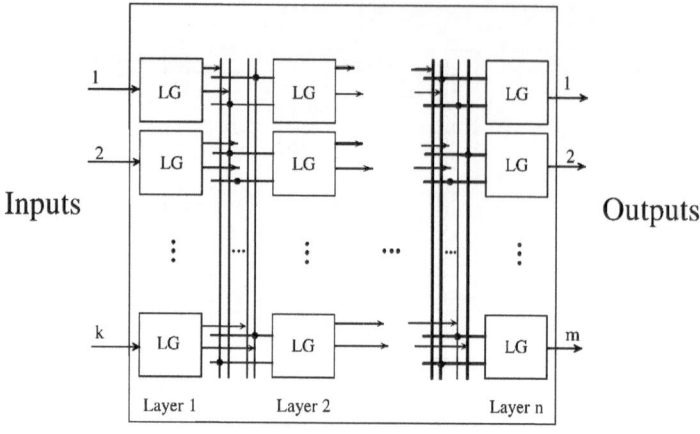

Fig. 2. The architecture of a gate array subsystem.

The *function* of each gate and its *two inputs* are determined by evolution. The encoding of each logic gate in the chromosome string is as follows:

Input 1 (4 bit)	Input 2 (4 bit)	Function (1 bit)

For the given array the chromosome string length becomes 1017 bit long which earlier experiments have shown to be appropriate.

2.3   GA Parameters and Fitness Function

Various experiments were undertaken to find appropriate GA parameters. The ones that seemed to give the best results were selected and fixed for all the experiments. This was necessary due to the large number of experiments that would have been required if GA parameters should be able vary through all the experiments. The preliminary experiments indicated that the parameter setting was not a major critical issue.

The simple GA style – given by Goldberg [17], was applied for the evolution with a population size of 50. For each new generation an entirely new population of individuals is generated. Elitism is used, thus, the best individuals from each generation are carried over to the next generation. The (single point) crossover rate is 0.5, thus the cloning rate is 0.5. Roulette wheel selection scheme is applied. The mutation rate – the probability of bit inversion for each bit in the binary chromosome string, is 0.005.

The fitness function is computed in the following way:

$$F = \sum_{\text{vec}} \sum_{\text{outp}} x \qquad \text{where } x = \begin{cases} 0 \text{ if } y \neq d \\ 1 \text{ if } y = d = 0 \\ 2 \text{ if } y = d = 1 \end{cases} \qquad (1)$$

For each output the computed output y is compared to the target d. If these equal and the value is equal to zero then 1 is added to the fitness function. On the other hand, if they equal and the value is equal to one, 2 is added. In this way, an emphasize is given to the outputs being one. This has shown to be important for getting faster evolution of well performing circuits. The function sum these values for the assigned outputs (outp) for the assigned truth table vectors (vec).

The proposed architecture fits into most FPGAs (Field Programmable Gate Arrays). The evolution is undertaken off-line using software simulation. However, since no feed-back connections are used and the number of gates between the input and output is limited, the real performance should equal the simulation. Any spikes could be removed using registers in the circuit.

Due to a large number of experiment (some requiring many generations to evolve) there has so far only been time to evolve one run of each reported experiment.

3   Results

This section reports the experiments undertaken to evolve multiplier circuits in an efficient way.

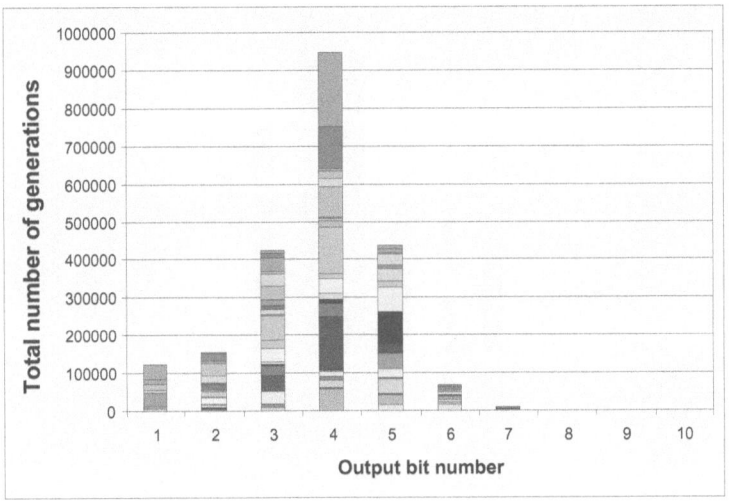

Fig. 3. Results of evolving one output at a time with 32 training set partitions (each column sums the number of generations required to evolve each output bit).

The experiments are based on using fixed partitioning sizes of the training set as well as evolving for one output bit at a time. Fig. 3 shows the results of evolving with 32 training set partitions. That correspond to 32 vectors from the truth table being used for evolving each subsystem. Each stacked column is the sum of the number of generations used for evolving subsystems for each of the 32 partitions for one output bit. Output bit number 1 is the most significant bit while bit number 10 is the least significant bit. For every partition it was possible to find a correctly working subsystem. This was by a single[1] run of each subsystem evolution, rather than a selection among multiple runs. Thus, the experiment has shown that by the given scheme it is possible to evolve multipliers larger that those reported earlier. There should be no limit in applying this approach for evolving even larger multiplier circuits. However, the amount of hardware will also probably then have to be substantially increased.

It is most difficult to evolve the five most significant bits with bit 4 requiring the largest number of generations (close to 1,000,000 generations). Bit 10 is the easiest to evolve and requires only 102 generations in total. This latter case would be a waste of gate resources since it (in another experiment) was evolved in a single run in 71 generations. That is, this system was evolved without training set partitions at all. The results indicate the need for an adaptive size in training set partitioning.

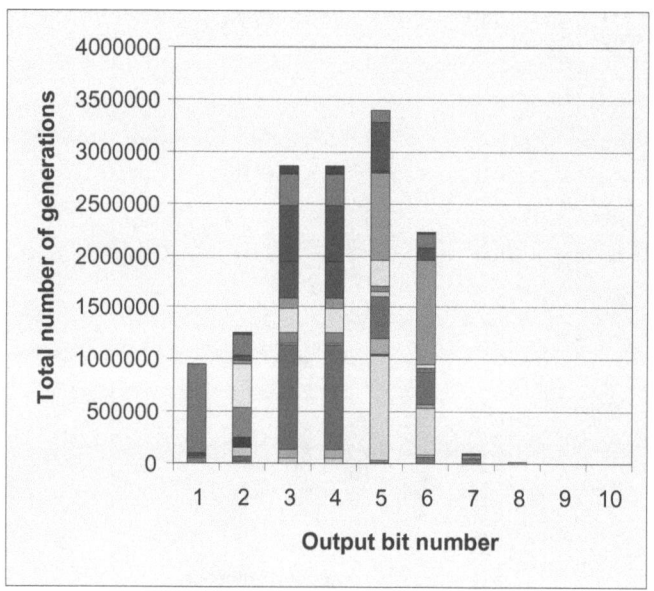

Fig. 4. Results of evolving one output at a time with 16 training set partitions (each column sums the number of generations required to evolve each output bit).

To show how the "difficult" parts of the truth table become harder to evolve with a smaller number of training set partitions, an experiment with 16 training set partitions was conducted. As seen in Fig. 4, output bit 3, 4 and 5 now require close to 3 million generations to evolve. Moreover, there was four instances (one

---

[1] With one exception – that reached the maximum 200,000 generations. It was easily evolved in a second run.

for each output bit 3, 4, 5 and 6) where no correctly working circuit was found in 1 million generations. Thus, here we see the need for small partitioning sizes to be able to find a correctly working circuit.

Computing the total number of generations results in 2.2 million for 32 partitions and 13.7 million for 16 partitions, respectively. Thus, about six times less number of generations are required with the cost of twice the circuits size. However, the important issue is that, applying the larger number of partitions is needed to find a correctly working circuit. The results correspond well with the earlier experiments for other applications mentioned in the introduction.

Even though it has been shown successful to evolve a large multiplier circuit, a 5×5 multiplier circuit is still much smaller than what can be designed in the traditional way. Thus, future work should concentrate on researching extension of this and other techniques for achieving automatic design of even larger logic circuits.

4   Conclusions

This paper has presented how incremental evolution can be applied for evolving multiplier circuits. The scheme is focused on evolution time and evolvability rather than minimizing hardware. Experiments verify that this is beneficial for solving more complex problems. A 5×5 multiplier circuit was evolved which is larger than any other reported evolved multiplier circuit.

References

1. W-P. Lee, J. Hallam, and H.H. Lund. Learning complex robot behaviours by evolutionary computing with task decomposition. In A. Birk and J. Demiris, editors, *Learning Robots: Proc. of 6th European Workshop, EWLR-6 Brighton*, volume 1545 of *Lecture Notes in Artificial Intelligence*, pages 155–172. Springer-Verlag, 1997.
2. X. Yao and T. Higuchi. Promises and challenges of evolvable hardware. In T. Higuchi et al., editors, *Evolvable Systems: From Biology to Hardware. First International Conference, ICES 96*, volume 1259 of *Lecture Notes in Computer Science*, pages 55–78. Springer-Verlag, 1997.
3. E. Cantu-Paz. A survey of parallel genetic algorithms. *Calculateurs Paralleles, Reseaux et Systems Repartis*, 10(2):141–171, 1998.
4. M. Iwata, I. Kajitani, H. Yamada, H. Iba, and T. Higuchi. A pattern recognition system using evolvable hardware. In *Proc. of Parallel Problem Solving from Nature IV (PPSN IV)*, volume 1141 of *Lecture Notes in Computer Science*, pages 761–770. Springer-Verlag, September 1996.
5. M. Murakawa, S. Yoshizawa, I. Kajitani, T. Furuya, M. Iwata, and T. Higuchi. Hardware evolution at function level. In *Proc. of Parallel Problem Solving from Nature IV (PPSN IV)*, volume 1141 of *Lecture Notes in Computer Science*, pages 62–71. Springer-Verlag, September 1996.
6. J. R. Koza. *Genetic Programming II: Automatic Discovery of Reusable Programs*. The MIT Press, 1994.
7. W.D. Hillis. Co-evolving parasites improve simulated evolution as an optimization procedure. *Physica D*, 42(1-3):228–234, 1990.

8. K.A. De Jong and M.A. Potter. Evolving complex structures via co-operative coevolution. In *Proc. of Fourth Annual Conference on Evolutionary Programming*, pages 307–317. MIT Press, 1995.

9. M.A. Potter and K.A. De Jong. Evolving neural networks with collaborative species. In *Proc. of Summer Computer Simulation Conference*. The Society for Computer Simulation, 1995.

10. P. Darwen and X. Yao. Automatic modularization by speciation. In *Proc. of 1996 IEEE International Conference on Evolutionary Computation*, pages 88–93, 1996.

11. J. Torresen. A divide-and-conquer approach to evolvable hardware. In M. Sipper et al., editors, *Evolvable Systems: From Biology to Hardware. Second International Conference, ICES 98*, volume 1478 of *Lecture Notes in Computer Science*, pages 57–65. Springer-Verlag, 1998.

12. J. Torresen. Two-step incremental evolution of a digital logic gate based prosthetic hand controller. In *Evolvable Systems: From Biology to Hardware. Fourth International Conference, (ICES'01)*, volume 2210 of *Lecture Notes in Computer Science*, pages 1–13. Springer-Verlag, 2001.

13. Jim Torresen. A scalable approach to evolvable hardware. *Journal of Genetic Programming and Evolvable Machines*, 3(3):259–282, 2002.

14. T. Kalganova. Bidirectional incremental evolution in extrinsic evolvable hardware. In J. Lohn et al., editor, *Proc. of the 2nd NASA/DoD Workshop on Evolvable Hardware*, pages 65–74. IEEE Computer Society, Silicon Valley, USA, July 2000.

15. D. Job V. Vassilev and J. Miller. Towards the automatic design of more efficient digital circuits. In J. Lohn et al., editor, *Proc. of the 2nd NASA/DoD Workshop on Evolvable Hardware*, pages 151–160. IEEE Computer Society, Silicon Valley, USA, July 2000.

16. J. F. Miller and P. Thomson. Aspects of digital evolution: Geometry and learning. In M. Sipper et al., editors, *Evolvable Systems: From Biology to Hardware. Second International Conference, ICES 98*, volume 1478 of *Lecture Notes in Computer Science*, pages 25–35. Springer-Verlag, 1998.

17. D. Goldberg. *Genetic Algorithms in search, optimization, and machine learning*. Addison–Wesley, 1989.

Evolution of Self-diagnosing Hardware

Miguel Garvie and Adrian Thompson

Centre for Computational Neuroscience and Robotics, School of Cognitive and
Computing Sciences, University of Sussex, Brighton BN1 9QH, UK.
mmg20, adrianth @cogs.susx.ac.uk

Abstract. The evolution of digital circuits performing built-in self-test
behaviour is attempted in simulation for a one bit adder and a two bit
multiplier. Promising results show evolved designs can perform a better
diagnosis using less resources than hand-designed equivalents. Future
extensions of the approach could allow the self-diagnosis of analog circuits
under failure and abnormal operating conditions.

Motivation

Self-diagnosis is important especially in mission critical systems exposed to radiation. Built-in self-test (BIST) is widely used yet commonly requires more than
100% overhead as in voting based systems [1, 14, 17] or off-line testing [23, 32].

Evolutionary methods [6, 9, 13] applied to hardware have produced circuits
comparable to those designed by experts [12, 18, 20, 30] and also unconventional
circuits [15, 26] in which hardware resources are used extremely efficiently. Moreover, many evolved systems in nature exhibit self-diagnostics such as the lym-
phatic system [2].

All this leads to the possibility that evolutionary methods could explore areas
of design space which reuse hardware components so that they contribute both
to the circuit's main functionality and its BIST, leading to a low overhead on-line
solution.

1 Introduction

Traditional approaches to BIST use the Test Pattern Generation (TPG) - Design Under Test (DUT) - Test Response Evaluation (TRE) model, with the first
and last being implemented using Linear Feedback Shift Registers (LFSR) [5].
Variations exist such as hierarchic, test-per-scan, BILBO (built-in logic block
observer), PRPG-MISR (pseudo random pattern generator - multi input signature register), circular BIST, and Reconfigurable Matrix Based Built-In Test
Processor (RMBITP) [7, 23, 32]. Even though techniques such as RMBITP are
successful at providing BIST for designs as large as 5 million gates with only
around 11% overhead, they all suffer from two main disadvantages. The first is
that they require the circuit's operation to go off-line periodically to feed in the
test patterns. The second is a bootstrapping problem: if the testing logic fails
we will never know if the rest of the circuit is functioning properly.

A.M. Tyrrell, P.C. Haddow, and J. Torresen (Eds.): ICES 2003, LNCS 2606, pp. 238–248, 2003.

Voting systems solve both these problems because on one hand faults are detected immediately during on-line operation and on the other hand the testing logic is usually small – all at the expense of complete redundant copies of the main circuit. Redundancy with on-line checking can be at the level of cells in a VLSI array [14, 17] or at the level of circuit modules, which could even be diverse designs of the same module, failing in different ways [1]. These, and other techniques for on-line diagnosis [3, 4, 21], share the problem that the benefits of self-checking must outweigh the higher fault rate due to increased silicon area.

There are examples in the evolutionary electronics literature of designs that operate in surprising or intricate ways [8, 12, 16, 18, 19, 22, 24]. This paper makes a first attempt to apply this creativity to the design of circuits having BIST. To establish a proof of principle two simple digital design problems were chosen, which may illuminate the following issues:

1. Can evolutionary search reach solutions to the BIST problem?
2. Do these solutions reuse logic for the main task and BIST?
3. Are these solutions competitive with conventionally designed ones?
4. Are there any principles of operation we could extract from them, perhaps to add to our own conventional design toolset?
5. Does this method scale up for larger problems?

Section 2 will describe the GA, the simulator, the fault model, the tasks to be evolved and the fitness evaluation mechanism for BIST. Section 3 presents the results achieved while section 4 discusses what was learned and future avenues.

2   Method

2.1   The Genetic Algorithm

A generational Genetic Algorithm (GA) is used with a population size of 32 with 2 elites where 60% of the next generation is created through mutation and the rest by single-point crossover. Following from earlier work [26] we adopted the model of a small genetically semi-converged population evolving for many generations. Fitness ranges from 0 (worst) to 1 (best). An adaptive mutation rate is used, analogous to a *Simulated Annealing* strategy moving from "exploring" to "exploiting". For each individual exactly m mutations are made, where $m = \lfloor k_r \times \ln\left(1/\overline{f}\right)\rfloor + m_{\text{floor}}$, $k_r = \frac{m_{\text{roof}}}{\ln(1/f_{\text{min}})}$, $\overline{f}$ is the current average fitness, f_{min} is minimum possible fitness above 0, $m_{\text{floor}} = 1$ is the number of mutations applied as $\overline{f}$ reaches 1, and $m_{\text{roof}} = 10$ is the number of mutations that would be applied if $\overline{f} = f_{\text{min}}$. This assumes that $\overline{f} > 0$ which is safe under the settings used in this paper. Informal preliminary experiments indicated solutions were found in less generations than with a constant mutation rate. Linear rank selection is used, such that the elite has twice the selective advantage of the median of the population.

The genotype-phenotype mapping used is similar to [18] excepting that: the locations of circuit outputs are fixed, there are no limitations on connectivity allowing sequential circuits, and the genotype is encoded in binary. The genotype length depends on the task being evolved. To allow implementation across a network of processors, an island based model was used [25] with a low migration rate. This population structure may have aided the search, but the details are not thought to be crucial to the results.

2.2   The Simulator

The simulator used is a simple version of an event driven digital logic simulator in which each logic unit is in charge of its own behaviour when given discrete time-slices and the state of its inputs. Logic units are Look-Up Tables (LUT) of two inputs capable of representing any two input logic gate. Any unit can be connected to any other allowing sequential circuits, so care must be taken to update all units "simultaneously". This is achieved by sending the time-slices to the logic units in two waves: the first to read their inputs and the second to update their outputs. During each evaluation, circuit inputs were kept stable for 20 time-slices and the outputs were read in the second half of this cycle allowing them time to settle.

Gate delays are simulated in time-slice units and are randomized with a Gaussian distribution ($\lfloor \mu = 1.5, \sigma^2 = 0.5 \rfloor$). This amounts to a noisy fitness evaluation with the intention to facilitate transfer of sequential circuits to real hardware as in a "Minimal Simulation" [10]. At the start of each generation, e different sets of logic delays are generated, simulating e different variations to the circuit delays from manufacturing or environmental variation. Each individual's performance is then evaluated in each of the e conditions, and its fitness is the mean value. The number e is occasionally adjusted by hand during the run, such that more evaluations are used as the population leaves the "exploring" stage and enters the "exploiting" stage. It was set such that twice the standard error of the series of fitness trials is significantly smaller than the difference in fitness between adjacent individuals in the rank table with non-equal fitnesses.

The Single-Stuck-At (SSA) Fault model was chosen because it simulates the most common type of on-line failure – that produced by radiation hazards in the absence of mishandling or manufacturing defects [23, 33]. It is also an industry standard and it has been shown that tests generated for SSA faults are also good at detecting other types of faults. SSA faults can be introduced at any of the logic units of the simulator simply by setting its output always to 0 or 1.

2.3   The Tasks

Two simple problems were chosen: a one bit adder with carry and a two bit multiplier. The adder was chosen as a small combinational circuit complex enough to be a good starting point for attempting to evolve BIST. It has three inputs A, B, C_{in} and two outputs S, C_{out} such that $S = A \oplus B \oplus C_{in}$ and $C_{out} = A \cdot B + A \cdot C_{in} + B \cdot C_{in}$. The multiplier, chosen as a step up from the adder, has four inputs $A_1 A_0 B_1 B_0$ and four outputs $P_3 P_2 P_1 P_0$ where $P = A \times B$.

The evolving networks may be recurrent and could show an unwanted dependence on the order in which inputs are presented, and on the networks' internal state. To demand insensitivity to input ordering, the same approach was taken as for the randomization of logic delays (above): at the start of each generation, e (the same number e defining the number of evaluations with random gate delays above) different orderings of the full set of possible inputs for that task were generated, and the individuals of that generation evaluated on all of them. On each of the e evaluations the circuit state was reset, then the ordering of the full set of inputs was presented twice in sequence, to prevent dependence on initial conditions.

The task evaluation score was measured as follows. Let Q_r be the concatenation of the series of values at the rth output bit for the final 10 time-slices of the presentation of each input vector during an evaluation, and Q'_r the desired response. We take the modulus of the correlation of Q_r and Q'_r, averaged over all N outputs:

$$f_{\text{t}} = \frac{\sum_{r=0}^{N-1} |corr(Q_r, Q'_r)|}{N} \tag{1}$$

2.4   Evolving BIST

An extra output E was recorded from circuits with the aim that it would go high whenever a fault affected any other output. The performance of a circuit at its main task f_{t} and at BIST behaviour f_{b} were evaluated separately. BIST behaviour itself was evaluated with two fitness measures:

1. BIST per fault f_{b_F}: Let u_{f} be the number of faults affecting task performance for which none of the possible input vectors raises E. Then f_{b_F} encourages faults to be "detectable": $f_{\text{b}_\text{F}} = 1/(1 + u_{\text{f}} \times k_{\text{f}})$ where k_{f} was chosen to be 25, to give f_{b_F} good sensitivity when u_{f} is small.
2. BIST per instance f_{b_I}: Let u_{i} be the number of instances out of all possible combinations of SSA faults and input vectors, for which the task output is incorrect but E is low. Then f_{b_I} encourages immediate detection of faults: $f_{\text{b}_\text{I}} = 1/(1 + u_{\text{i}} \times k_{\text{i}})$ where k_{i} was chosen to be 200.

Notice that having a high f_{b_F} but low f_{b_I} is similar to off-line BIST solutions while having a high f_{b_I} is like an on-line BIST detecting faults at the first instance they affect circuit behaviour.

u_{f} is measured by evaluating task fitness f_{t} separately under all SSA faults to every unit able to affect the task outputs. The same set of e evaluation conditions chosen for the current generation is used. If f_{t} falls by at least 0.001 due to a fault, then it is considered to affect task performance. u_{i} is measured by comparing the output of the circuit under each fault and input vector to its output for the same input vector under no faults. The output is deemed unaffected if it is the same at steps 10, 15 and 20 of the 20 time steps for which inputs are stable. Hence most faults that induce oscillations will be detected. If during the simulated time E goes high for more than $eSize$ consecutive time slices then it is considered to have gone high. $eSize$ is set at 7: greater than average race conditions yet not excluding a wide range of behaviours. A circuit exhibiting a high E when no faults were in place was deemed to have $f_{b_F} = f_{b_I} = 0$.

It seems foolish to try to detect faults at every instance when some faults are not detected at any instance. Early runs (§3.1) incorporated the objectives into one overall fitness value, but with more emphasis on f_{b_F} than on f_{b_I} and more on f_t than on f_b: $f = w_t \times f_t^2 + w_b \times f_b$ where $f_b = w_{b_F} \times f_{b_F}^2 + w_{b_I} \times f_{b_I}$ and weights $w_t \approx w_{b_F} \approx 0.85$, $w_b = 1 - w_t$ and $w_{b_I} = 1 - w_{b_F}$. Later runs (§3.2) exploited our use of rank selection. The objectives were given the priority $f_t > f_{b_F} > f_{b_I}$, and when sorting the individuals the comparison operator only considered an objective if higher priority objectives were equal. An extra objective encouraging parsimony was also used, having the lowest priority of all.

3   Results

3.1   One Bit Adder with Carry

The maximum number of logic units was constrained to 13 and the genotype length was 156 bits. This circuit can be implemented with a minimum of five logic units. It is worth mentioning that the straightforward TPG-DUT-TRE BIST architecture (§1) would require 3 flip-flops and 2 Xor gates for the TPG LFSR and 2 flip-flops and 1 Xor for the TRE LFSR totalling 13 extra gates. Even if unusual to apply this method to circuits of this size, it is still valuable as a comparison of BIST quality and underlying operation principles because it is an industry standard. The minimal on-line BIST solution – a voter with two copies – would require 8 extra gates, 5 for the extra copy and 3 for the comparison logic.

A) Hybrid Minimalist. This run was seeded with the elite of generation 10100 from a previous run which was started from a population of random individuals and did not take f_{b_I} into account. This elite was bred because it provided 82% off-line fault coverage being only 3 gates larger than the minimum adder.

After 4000 generations the elite was the circuit shown in Fig. 1a that can be trivially pruned to 7 gates by skipping unit 5 as shown. This circuit detects all SSA faults affecting task performance save unit 1 SA0 (clearly since $E = u_1 \wedge u_8$) with either input test vector $(A, B, C_{in}) = (0, 0, 0)$ detecting all SSA-1 faults or $(1, 0, 0)$ detecting all SSA-0. Note these are the *only* input vectors detecting faults since E is low whenever B or C_{in} are high. Due to its short test pattern this could be considered as a hybrid of the on-line and off-line approaches using the best of both worlds.

The formula for E is $\overline{B \vee C_{in}} \wedge C_{out}$ and acts as a consistency check: "C_{out} cannot be high if both B and C_{in} are low". The use of Xor gates – which propagate any bit flip – in the calculation of S, which is used for C_{out}, ensures all faults raise C_{out} when B and C_{in} are low and A is high or low, thus checking S implicitly. This method seems like a cross between *arithmetic coding* "checksums" and *state monitoring*.

Table 1 compares this to conventional solutions.

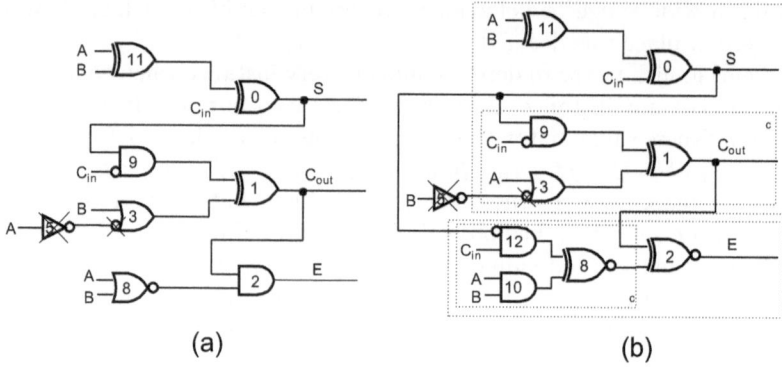

Fig. 1. Evolved BIST solutions for the adder: (a) nearly full off-line fault coverage with 2 extra gates and (b) full on-line solution with 3 extra gates. Unit 5 can be trivially removed in both – presumably the two necessary mutations would have eventually occurred. LUTs are displayed as their equivalent two input logic gate.

B) On-Line full. About 15000 generations later in the same run the full on-line BIST solution shown in Fig. 1b was the elite and can again be trivially pruned to 9 gates. The design can be decomposed into modules α (adder) and β (BIST). Sub-modules α_c and β_c calculate C_{out} and $\overline{C_{out}}$ respectively while unit 2 (whose output is E) acts as a voter. Errors in S are detected because they affect α_c and β_c differently. Here evolution has arrived at the mission critical system design concept of *design diversity* [1] and gone beyond by aiming for circuits having minimal re-design necessary for maximal fault-coverage. Even though it is not surprising that α_c and β_c are different, because there was no gene copy operator, the fact that their differences are exploited to evaluate the correctness of S and C_{out} simultaneously and under any fault and input vector is quite remarkable.

E also went high whenever the circuit failed due to pathological gate delays (*delay faults*). Hence this method could also be useful to catch out deviations of standard behaviour due to abnormal operating conditions (*parametric faults*) which could be useful in unconventional evolved designs such as [26]. Sequential faults are caught since the circuits are tested under more than ten different random orderings of the full input test sequence at each evaluation. Being an on-line solution transient faults will also be detected as they look like a temporary SSA.

Table 1 compares this to conventional solutions.

Adders Discussed. Evolutionary methods have found better solutions in terms of standard BIST design criteria than those of conventional methods. It is up to hardware designers to consider it worthwhile to have BIST on circuits of this size. Other runs minimized a hand-designed voting BIST solution from 8 overhead gates disappointingly just down to 7, arrived at a nearly full on-line solution with 8 overhead gates which included a low-pass filter ironing out race conditions

Table 1. Comparison of test quality attributes between conventional and evolved BIST solutions to the adder problem. Fault types: C=combinatorial, S=sequential, T=transient, D=delay. Test controller logic is not taken into account for off-line solutions.

Quality attribute	TPG-DUT-TRE	Evolved A	Voting	Evolved B
Fault Coverage	100%	90%	100%	100%
Test Scope	module	module	module	module
Fault types	C	CS	CST	CSTD
Area overhead	13 units	3 units	8 units	4 units
Pin overhead	2 (start/error)	2	1	1
Perf. penalty	0	1 gate delay	0	1 gate delay
Test time	8 test vectors	2 test vectors	0 (on-line)	0 (on-line)

because it was evolved with a small $eSize$ (meaning it could be clocked faster when deployed), evolved a full on-line solution with 6 overhead gates exhibiting a linear relation between the minimum $eSize$ it would work with and the standard deviation σ^2 of its delays, and evolved a circuit very similar to the one in Fig. 1b from a random initial population.

3.2 Two Bit Multiplier

The maximum size allowed for these circuits was 28 LUTs and the genotype length was 392 bits. The smallest implementations of this circuit use 7 two input logic gates [18], such as shown in Fig. 2a. Even though the difference in size between the adder and multiplier is not great, there is twice the number of outputs to check: a TPG-DUT-TRE BIST approach would require 11 gates for each of the LFSRs totalling 22 extra gates while a voter would require 7 gates for the comparators again totalling 14 extra gates.

Note that evolving this kind of circuit, even without any BIST functionality uses a considerable amount of CPU time as this seems to grow exponentially with the complexity of the specification [30].

Off-Line Full. This run was seeded with the conventionally designed multiplier shown in Fig. 2a. After around 150000 generations the circuit of Fig. 2b was the elite (after pruning as in Fig. 1). This is nearly equal to the seed except that one input of unit 0 comes from unit 2 instead of unit 5, which implements a similar strategy to that used in §3.1: one output is influenced by others through their reuse in a cascading fashion – here output P_2 is calculated using P_3 which is calculated using P_1 – so only this output need be checked. This is achieved by unit 14, which compares P_2 to a recalculation of its inverse by units 18 and 27. Only these gates (14, 18, 27) are used for BIST behaviour achieving full off-line fault coverage with the input vector pattern set $(A_1, A_0, B_1, B_0) = \{(0, 1, 1, 1), (1, 0, 1, 1), (1, 1, 1, 1)\}$, reducing the size of the LFSR to 5 gates while the signature analyzer LFSR is replaced by the error line. This run has shown that this method can also be used to add BIST behaviour

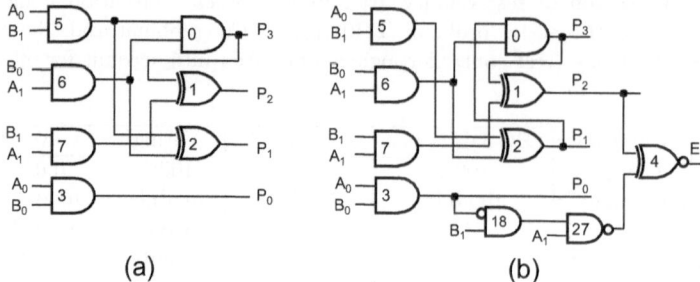

Fig. 2. (a) Hand designed two bit multiplier extracted from Fig. 15a of [18] is (b) post-evolved to incorporate BIST with 100% off-line coverage using only 3 gates overhead.

to existing hand designed circuits without substantial modifications to circuit structure thus leaving its properties unaltered.

4   Conclusion

A proof of principle has been established and the questions set out in §1 can be answered: circuits with BIST behaviour are within the reach of evolutionary search, can reuse components for the main task and BIST functionality, and they are competitive in overhead both to off-line and on-line solutions. Low overhead on-line and hybrid solutions have been found. These solutions would function correctly in real hardware because the slightly unconventional simulator does capture the processes and variations influencing combinatorial circuits. Evolution has explored design space containing established methods for BIST design and beyond. Some solutions use a "checksum" formula on the current circuit state to evaluate correctness (comparable to the "Immunotronics"[3] approach for sequential systems which evaluates correct circuit state transitions), others increase testability by cascading outputs so that errors are propagated to a single output and others exploit design diversity to minimize redundancy in a voting system. These methods could prove useful to be adopted by designers. For circuits larger than those considered here, it is still unclear whether the enhanced BIST strategies produced by evolution would be worth the computational effort needed to produce them since the evolution of large BIST circuits faces the same problems – and perhaps, solutions – as with other circuits [11, 28, 29, 31]. However, Vassilev et al. [30] have suggested that when large circuits can be evolved (perhaps from a hand-designed seed), their size leads to greater scope for evolutionary optimization.

Most cases arrived at modular designs (as in [18, 27]). Further understanding of this effect may aid future experiments. The modularity of the circuits can be useful in identifying design patterns or principles.

Any single circuit will have its own characteristics for which a particular BIST strategy is most suited. Evolution through blind variation and selection is capable of searching for this strategy without constraints. The evolved BIST circuits seem well suited to uncovering *delay faults* and may be useful for *parametric faults* such as those occurring in [26]. The evolution of self-diagnosing analog hardware is an interesting possibility where the E line could give an estimate of *how* wrong the outputs are. Future work could also include the adoption of a more comprehensive fault model where the memory of the LUTs is affected – as in FPGAs – and other common faults are simulated, and tackling more complex tasks such as sequential circuits – synchronous and asynchronous – perhaps using techniques set out in [11, 28, 29, 31].

Acknowledgments

Thanks to Matthew Quinn for helpful comments and to the COGS bursary that supports Michael Garvie's research.

References

1. A. Avizienis and John P. J. Kelly. Fault-tolerance by design diversity: Concepts and experiments. *Computer*, 17(8):67–80, August 1984.
2. A. Avizienis. Design diversity and the immune system paradigm: Cornerstones for information system survivability, 2000.
3. D.W. Bradley and A.M. Tyrrell. Immunotronics: Novel finite state machine architectures with built in self test using self-nonself differentiation. *IEEE Transactions on Evolutionary Computation*, 6(3):227–238, 2001.
4. R O Canham and A M Tyrrell. A multilayered immune system for hardware fault tolerance within an embryonic array. In J Timmis and P J Bentley, editors, *Proceedings of the 1st International Conference on Artificial Immune Systems (ICARIS)*, volume 1, pages 3–11, University of Kent at Canterbury, September 2002. University of Kent at Canterbury Printing Unit.
5. C. Dufaza. Theoretical properties of LFSRs for built-in self test. *INTEGRATION, the VLSI journal*, 25:17–35, 1998.
6. D. E. Goldberg. *Genetic Algorithms in Search, Optimization & Machine Learning*. Addison Wesley, 1989.
7. H. Golnabi and J. Provence. RMBITP: A reconfigurable matrix based built-in self-test processor. *Microelectronics Journal*, 28:115–127, 1997.
8. T. Higuchi, M. Iwata, and L. Weixin, editors. *Proc. 1st Int. Conf. on Evolvable Systems: From Biology to Hardware*, volume 1259 of *LNCS*. Springer-Verlag, 1997.
9. J. H. Holland. *Adaptation in Natural and Artificial Systems*. Ann Arbor: University of Michigan Press, 1975.
10. N. Jakobi. Half-baked, ad-hoc and noisy: Minimal simulations for evolutionary robotics. In Phil Husbands and Inman Harvey, editors, *Proc. 4th Eur. Conf. on Artificial Life (ECAL'97)*, pages 348–357. MIT Press, 1997.
11. T. Kalganova. Bidirectional incremental evolution in extrinsic evolvable hardware. In J. Lohn, A. Stoica, and D. Keymeulen, editors, *The Second NASA/DoD workshop on Evolvable Hardware*, pages 65–74, Palo Alto, California, 13-15 2000. IEEE Computer Society.

12. J. R. Koza, F. H. Bennett III, D. Andre, and M. A. Keane. Reuse, parameterized reuse, and hierarchical reuse of substructures in evolving electrical circuits using genetic programming. In T. Higuchi, M. Iwata, and L. Weixin, editors, *Proc. 1st Int. Conf. on Evolvable Systems: From biology to hardware (ICES-96)*, number 1259 in LNCS, pages 312–326. Springer-Verlag, 1996.

13. J. R. Koza. *Genetic Programming: On the programming of computers by means of natural selection.* MIT Press, Cambridge, Mass., 1992.

14. J. Lach, W. Mangione-Smith, and M. Potkonjak. Low overhead fault-tolerant FPGA systems, 1998.

15. P. Layzell. A new research tool for intrinsic hardware evolution. In M. Sipper, D. Mange, and A. Pérez-Uribe, editors, *Proc. 2nd Int. Conf. on Evolvable Systems (ICES'98)*, volume 1478 of *LNCS*, pages 47–56. Springer-Verlag, 1998.

16. J. Lohn, A. Stoica, D. Keymeulen, and S. Colombano, editors. *Proc. 2nd NASA/DoD workshop on Evolvable Hardware.* IEEE Computer Society, 2000.

17. D. Mange, A. Stauffer, and G. Tempesti. Embryonics: A microscopic view of the molecular architecture. In A. Perez-Uribe M. Sipper, D. Mange, editor, *Proc. 2nd Int. Conf. on Evolvable Systems (ICES1998: From biology to hardware*, volume 1478 of *LNCS*, pages 285–195. Springer-Verlag, 1998.

18. J. F. Miller, D. Job, and Vesselin K. Vassilev. Principles in the evolutionary design of digital circuits - part I. *Genetic Programming and Evolvable Machines*, 1(3), 2000.

19. J. Miller, A. Thompson, P. Thomson, and T. Fogarty, editors. *Proc. 3rd Int. Conf. on Evolvable Systems (ICES2000): From Biology to Hardware*, volume 1801 of *LNCS*. Springer-Verlag, 2000.

20. J. Miller. On the filtering properties of evolved gate arrays. In A. Stoica, J. Lohn, and D. Keymeulen, editors, *The First NASA/DoD Workshop on Evolvable Hardware*, pages 2–11, Pasadena, California, 1999. Jet Propulsion Laboratory, California Institute of Technology, IEEE Computer Society.

21. W. Mangione-Smith N. Shnidman and M. Potkonjak. On-line fault detection for bus-based field programmable gate arrays. *IEEE Transactions on VLSI systems*, 6(4):656–666, 1998.

22. M. Sipper, D. Mange, and A. Pérez-Uribe, editors. *Proc. 2nd Int. Conf. on Evolvable Systems (ICES98)*, volume 1478 of *LNCS*. Springer-Verlag, 1998.

23. A. Steininger. Testing and built-in self-test - a survey. *Journal of Systems Architecture*, 46:721–747, 200.

24. A. Stoica, D. Keymeulen, and J. Lohn, editors. *Proc. 1st NASA/DoD workshop on Evolvable Hardware.* IEEE Computer Society, 1999.

25. R. Tanese. Distributed genetic algorithms. In J.D. Schaffer, editor, *Proc. of the Third International Conference of Genetic Algorithms*, pages 434–439. Morgan Kauffmann, 1989.

26. A. Thompson, I. Harvey, and P. Husbands. Unconstrained evolution and hard consequences. In E. Sanchez and M. Tomassini, editors, *Towards Evolvable Hardware: The evolutionary engineering approach*, volume 1062 of *LNCS*, pages 136–165. Springer-Verlag, 1996.

27. A. Thompson and P. Layzell. Analysis of unconventional evolved electronics. *Communications of the ACM*, 42(4):71–79, April 1999.

28. P. Thomson. Circuit evolution and visualisation. In J. Miller, A. Thompson, P. Thomson, and T. Fogarty, editors, *Proc. 3rd Int. Conf. on Evolvable Systems (ICES2000): From biology to hardware*, volume 1801 of *LNCS*, pages 229–240. Springer-Verlag, 2000.

29. J. Torresen. A divide-and-conquer approach to evolvable hardware. *Lecture Notes in Computer Science*, 1478:57–??, 1998.
30. V. Vassilev, D. Job, and J. Miller. Towards the automatic design of more efficient digital circuits. In J. Lohn, A. Stoica, and D. Keymeulen, editors, *The Second NASA/DoD workshop on Evolvable Hardware*, pages 151–160, Palo Alto, California, 13-15 2000. IEEE Computer Society.
31. V. Vassilev and J. Miller. Scalability problems of digital circuit evolution: Evolvability and efficient designs. In J. Lohn, A. Stoica, and D. Keymeulen, editors, *The Second NASA/DoD workshop on Evolvable Hardware*, pages 55–64, Palo Alto, California, 13-15 2000. IEEE Computer Society.
32. H. Wunderlich. BIST for systems-on-a-chip. *INTEGRATION, the VLSI journal*, 26:55–78, 1998.
33. The NASA/GSFC Radiation Effects and Analysis Home Page. http://radhome.gsfc.nasa.gov/.

Routing of Embryonic Arrays
Using Genetic Algorithms

Cesar Ortega-Sanchez[1], Jose Torres-Jimenez[2], and Jorge Morales-Cruz[3]

[1]Institute of Electrical Research, Department of Control and Instrumentation
Av. Reforma 113, Palmira, Cuernavaca, Morelos, Mexico 62490
ortegas@iie.org.mx
[2]Computer Science Department, ITESM
Av. Reforma 182-A, Lomas de Cuernavaca, Temixco, Morelos, Mexico 62589
jtorres@campus.mor.itesm.mx
[3]Electronics Department, CENIDET
Interior Internado Palmira S/N, Palmira, Cuernavaca, Morelos, Mexico 62490
jmc_190777@hotmail.com

Abstract. This paper presents a genetic algorithm (GA) that solves the problem of routing a multiplexer network into a MUXTREE embryonic array. The procedure to translate the multiplexer network into a form suitable for the GA-based router is explained. The genetic algorithm works on a population of configuration registers (genome) that define the functionality and connectivity of the array. Fitness of each individual is evaluated and those closer to solving the required routing are selected for the next generation. A matrix-based method to evaluate the routing defined by each individual is also explained. The output of the genetic router is a VHDL program describing a look-up table that receives the cell co-ordinates as inputs and returns the value of the corresponding configuration register. The routing of a module-10 counter is presented as an example of the capabilities of the genetic router. The genetic algorithm approach provides not one, but multiple solutions to the routing problem, opening the road to a new level of redundancy where a new "genome" can be downloaded to the array when the conventional reconfiguration strategy runs out of spare cells.

1 Introduction

The Embryonics project was originally proposed by Mange et al [1] from the École Polytechnique Fédérale de Laussane in Switzerland, and soon after the University of York, UK, joined the efforts under the lead of Tyrrell [2]. During the past few years research on Embryonics has gained momentum and the first practical demonstrations of the technology have already been shown [3, 4]. Embryonics proposes the incorporation of biological concepts like growth, reproduction and healing into the realm of silicon processor-arrays [5]. One of the possible Embryonics implementations is the MUXTREE architecture [6, 7]. In MUXTREE implementations, a logic function is translated from truth tables into multiplexer networks via ordered binary decision diagrams (OBDDs). The multiplexer network has to be mapped into the structure of an embryonic array, which is a 2-D array of processing cells. Every cell performs the

A.M. Tyrrell, P.C. Haddow, and J. Torresen (Eds.): ICES 2003, LNCS 2606, pp. 249–261, 2003.

function of a two-input, one-output multiplexer. Connectivity between cells is limited to the nearest neighbours, and the number of data lines connecting cells to each other are scarce. Inputs to the multiplexer and routing of signals within each cell are set by a configuration register. In the MUXTREE architecture, each cell of the array contains a copy of the configuration registers of all the cells in the array [8].

The set of all configuration registers in an array is called "the genome" of the application. In each cell, a configuration register is selected by a unique pair of coordinates. When a cell fails, it is possible to reconfigure the array by changing the coordinates of cells so that the failing cell is logically eliminated and substituted by a healthy one. By these means embryonic arrays achieve fault-tolerance [9].

Connectivity limitations make mapping the multiplexer network into the embryonic array a task that grows in difficulty as the complexity (size) of the application grows. For small functions, it is possible, although time-consuming, to manually route the multiplexer network into the embryonic array. However, the time (and patience!) required to route medium-size applications is unacceptable for all practical purposes.

Given that Embryonics is a relatively new field, it lacks the CAD tools available for other, more mature bio-inspired technologies (artificial neural nets and genetic algorithms, for example). In its present state, MUXTREE applications can be synthesised and simulated using FPGA design software, like Xilinx's Foundation. However, the early stages of a design have to be developed by hand.

This paper presents a recently developed design tool that helps the designer in what is considered the most challenging and time-consuming task during the development of a MUXTREE application: the mapping of a multiplexer network into the fixed structure of an embryonic array. At the core of the router is a genetic algorithm (GA) that selects from a multitude of possible routings those that satisfy the connectivity of the multiplexer network given as input.

The paper is organised as follows: Section 2 introduces the MUXTREE architecture and the typical design flow of an application. Section 3 describes the genetic algorithm that solves the routing. In Section 4, a module-10 counter is presented as an example of the capabilities of the genetic algorithm employed as router. Results obtained with other examples are also presented. Section 5 resumes the conclusions and future work of this research.

2   Design Flow of a MUXTREE Application

Although the processing power of an individual MUXTREE cell is that of a multiplexer, an array of such cells can contain a multiplexer network representing any combinational or sequential logic function. The size of the network would be limited only by the physical dimensions of the array (allowing, of course, spare elements necessary for the embryonic reconfiguration).

Figure 1 shows a simplified diagram of the MUXTREE embryonic cell. A detailed description of the architecture can be found elsewhere [8].

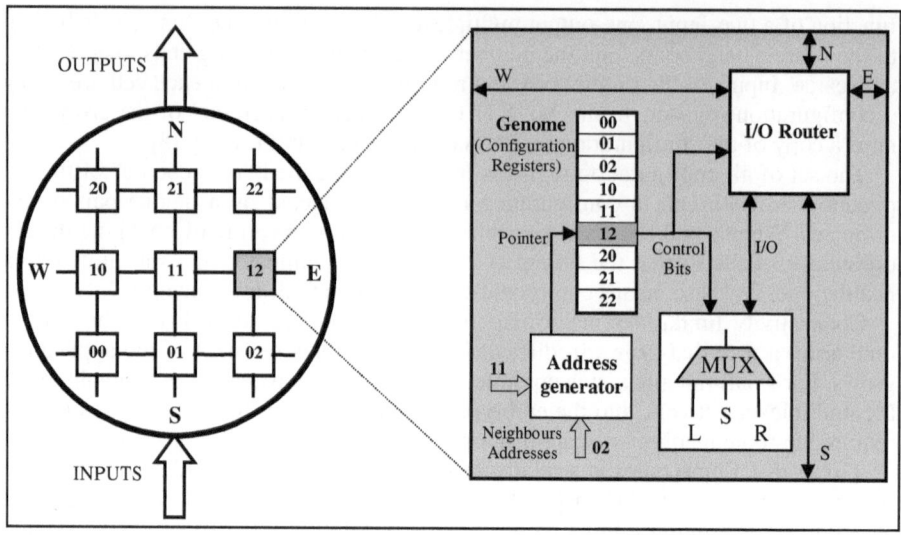

Fig. 1. Simplified diagram of a MUXTREE embryonic cell

In a typical application, the following steps are necessary to implement a combinational or sequential logic function using a MUXTREE embryonic array:

1. Describe the application as a set of logic functions or as a truth table.
2. From the description, construct an Ordered Binary Decision Diagram (OBDD).
3. Represent the OBDD as a multiplexer network.
4. Route the multiplexer network in an array of MUXTREE cells.
5. Obtain the configuration register of all the cells in the array.
6. Generate the "genome" of the application by grouping the configuration registers of all the cells in the array.
7. Use the genome as input for the software that is used to synthesise and simulate the design.

In a typical implementation, steps 1 to 6 are carried out by hand. Step 4 is the most difficult and time-consuming due to the limited interconnection between cells. A genetic algorithm-based, automatic router that implements steps 4 to 6 has been developed. The following section describes this tool in detail.

3   Routing Multiplexer Networks Using a Genetic Algorithm

During the past few years, genetic algorithms have been intensively used to solve problems whose solution either cannot be found analytically, or numeric methods take too long to find [10]. GAs are very good at finding solutions in vast search-spaces because they are inspired in the process of natural selection, where individuals who are fitter than others have more chances of passing their genes to the next generation [11].

The following elements are needed to solve a problem using GAs:

- A suitable representation of the problem so that an individual from a population can represent a possible solution.
- A randomly generated population of possible solutions.
- A fitness function that evaluates how close to solving the problem is every individual in the population.
- A selection criterion that discriminates individuals according to their fitness.
- Some genetic operators (typically mutation and crossover) to generate diversity within the population and to prevent the solutions from being stuck around local minima.
- A stop criterion that allows the system to decide when to finish the search. It can be either because a solution has been found, or because a specified number of generations have been tested.

Figure 2 shows a flow diagram representing the logical flow of a GA.

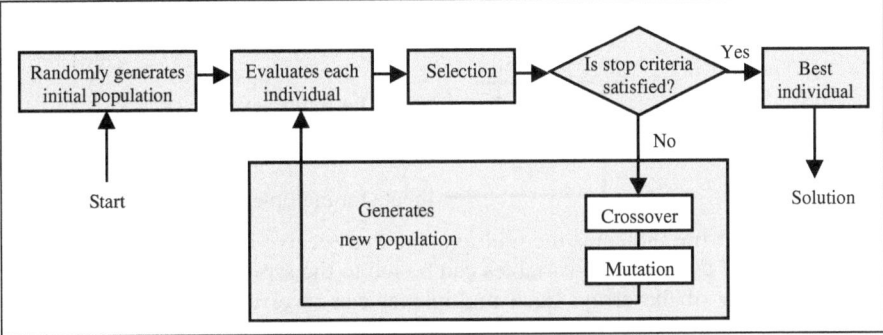

Fig. 2. Flow diagram of a typical genetic algorithm.

To solve the routing of multiplexer networks into MUXTREE arrays, the diagram in figure 2 was implemented in a C program. Three text input files indicate the topology of the target multiplexer network. Their content is presented next.

3.1   Input Files

For all practical purposes, a multiplexer network can be completely described by stating: the number of levels it contains, the number of multiplexers in each level and the inputs that arrive to each multiplexer (data and control). Additionally, in the MUXTREE architecture, it is necessary to distinguish the multiplexers that work in synchronous mode, i.e. with their outputs latched by a clock; from those that work asynchronously. The following example shows a simple multiplexer network and the input files associated to it.

In file number one, parameters pertaining to the GA are provided along with the size of the MUXTREE array that should contain the multiplexer network. Also the features that characterise the target network are given. Figure 3 shows the multiplexer network of a 3-input voter, along with the text of input file 1.

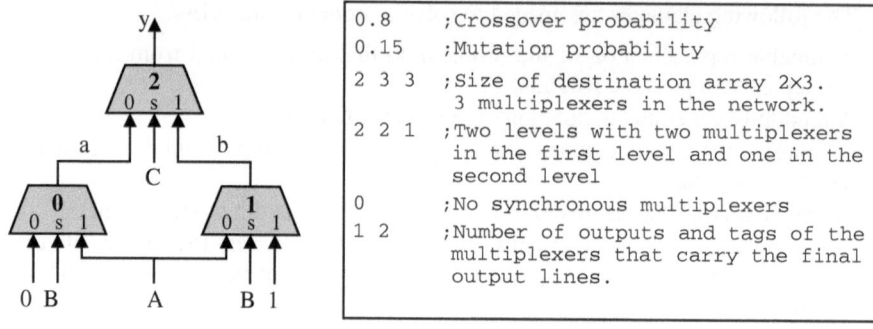

Fig. 3. 3-input voter multiplexer network and its input file 1 for the genetic router

The inputs for each multiplexer are indicated in a second input file. The file corresponding to the example in figure 3 is presented next. The order of inputs for each multiplexer is: Selection, Right input (**1**), Left input (**0**). In a MUXTREE array, multiplexers are numbered from left to right and from bottom to top.

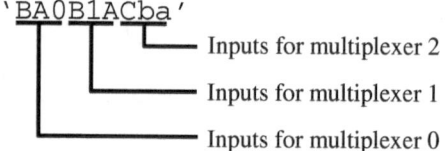

A third input file indicates the multiplexers that receive input variables, if any. In the MUXTREE architecture, variables can be fed to the array only through row 0, i.e. the bottom row of the array. Each multiplexer can receive up to two variables. In some difficult designs, the same variable can be fed in more than one input in order to simplify the routing. Following is the text of the third input file for the example in figure 3. For demonstration purposes, input variable A is fed through the two cells in the bottom row (the array size is 2×3).

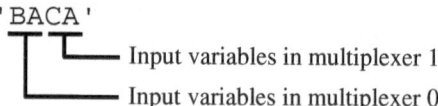

Generation of the three input files can be easily automated, but at the present stage they are still hand-generated.

3.2   Initial Population

The function of a MUXTREE cell is defined by a 17-bit configuration register [5]. Eight of those bits define the routing of signals within the cell. This byte is called the routing byte. Therefore, the flow of signals within a particular array will be defined by the routing bytes of all cells in the array (the routing genome). The size of this genome in bits is 8 times the number of cells in the array. For a 2×3 array, its routing would be completely defined by 2×3×8= 48 bits; therefore, any randomly generated 48-bit pattern will encode an arbitrary routing. In the genetic router proposed, every individual of the population is a routing genome. In the examples presented next, the initial population consists of 100 randomly generated individuals.

3.3   Fitness Function and Evaluation

The problem of evaluation is to find out how close an arbitrary routing is from the routing needed to implement the target multiplexer network. To achieve this, a carefully chosen fitness function must be used.

Definition 1: Two multiplexers are equivalent when they have the same inputs in the same order. Such inputs are defined as correct inputs.

Definition 2: Two multiplexers are partially equivalent when they have at least one input in the same position, i.e. at least one correct input.

By these definitions, a given routed MUXTREE array would be able to implement the same function as the target multiplexer network when it contains a set of multiplexers equivalent to those in the target network. In other words, the routing will be complete when it contains equivalent multiplexers for all multiplexers in the target network. Consequently, the fitness of a particular routing should be related to the number of equivalent and partially equivalent multiplexers. Hence, the fitness of a routing is defined as the number of correct inputs it contains. Since every multiplexer has three ordered inputs, the maximum fitness a routing can achieve is the number of multiplexers in the target network (n) multiplied by 3.

$$f_{max} = 3n .\tag{1}$$

To determine the fitness of a particular routing, it is necessary to locate all the correct inputs it contains. For this purpose, a connectivity matrix is used to "simulate" the propagation of signals within the network. Before propagation, the multiplexers in the embryonic array have no particular correspondent in the target network. After propagation, corresponding multiplexers are determined according to the number correct inputs every embryonic multiplexer has.

At the end of the evaluation process, every individual of a population will have a number associated to it. The bigger the number, the closer that routing is from solving the target network. For space reasons, a more detailed description of the connectivity matrix and the propagation process will be left for a future paper.

3.4   Selection

Selection is the process of choosing the individuals that will be parents of a new population. The genetic router presented in this paper implements selection by tournament, which consists in randomly selecting two genomes from the population and choosing the one with the highest fitness. For example,

If	$f_A > f_B$,	then:	$C_g = G_A$
If	$f_A < f_B$,	then:	$C_g = G_B$
If	$f_A = f_B$,	then:	$C_g = G_A$

Where:

f_A = Fitness of genome A; f_B = Fitness of genome B
G_A = Genome A; G_B = Genome B; C_g = Chosen genome

3.5   Crossover and Mutation

Crossover and mutation are used in the genetic router to prevent solutions from clustering around local minima.

Crossover consists of taking two 'parents' that have survived the selection process and mixing them to create two new individuals. Figure 4 shows two six-byte individuals being crossed over in two randomly selected points.

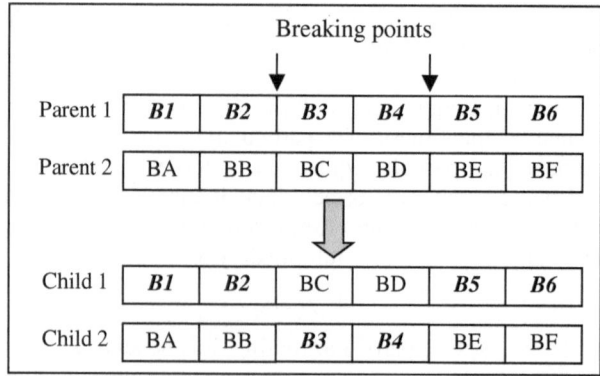

Fig. 4. Crossover between two individuals of a population

The crossover points are selected in such a way that bytes remain unaltered so that well-routed cells are preserved from one generation to the next.

Crossover is performed according to a crossover probability. Every time a crossover is going to take place, a random number is generated; if the number is smaller than crossover probability, then the crossover takes place, otherwise both parents pass to the next generation unchanged. In the examples presented in this paper, crossover probability is 0.8.

It is possible to apply mutation to children generated by crossover. By means of a mechanism similar to the one used in crossover, a mutation probability will define the possibility of mutation. Mutation probability in the examples of this paper is 0.15. When mutation is applied, the byte and bit to be mutated are selected randomly. Mutation inverts the logic value of the selected bit.

3.6   New Population

A new population is generated with the individuals selected for their fitness. It is possible that two new individuals end up being identical after they have passed through crossover and mutation. If that is the case, one of the repeat individuals is eliminated from the population and a brand new individual is randomly generated. All individuals in the new population are different to one another.

3.7   Output File

The genetic algorithm that solves the routing of multiplexer networks into MUXTREE arrays stops searching when one of the following conditions is met: A solution has been found, or the search has run a predetermined number of generations whether or not a solution has been found. The latter case can end up with none, one or multiple solutions to the routing.

If at least one solution is found, the genetic router generates a VHDL file containing the description of a look-up table that receives at its inputs the co-ordinates of a cell and returns the configuration register associated to that cell. This file has to be integrated to the design of the MUXTREE array that will implement the desired function. In section 4 there is a VHDL file generated by the genetic router.

If multiple solutions are found, a new level of fault-tolerance could be introduced to MUXTREE arrays. In its present implementation, MUXTREE arrays are disabled when spare cells have ran out and a new fault arises. However, with multiple routings capable of implementing the logic functions represented by the multiplexer network, it would be possible to download a new genome to the array every time spare cells run out. It is possible that one of the "spare genomes" can still implement the desired function.

4   Example: Routing of a Module-10 Counter

To demonstrate the functionality of the genetic router, the implementation of a module-10 counter is presented next. Table 1 shows the truth table of the counter and figure 5 the multiplexer network that implements it.

Table 1. Truth table of a module-10 counter

D	C	B	A	D^+	C^+	B^+	A^+
0	0	0	0	0	0	0	1
0	0	0	1	0	0	1	0
0	0	1	0	0	0	1	1
0	0	1	1	0	1	0	0
0	1	0	0	0	1	0	1
0	1	0	1	0	1	1	0
0	1	1	0	0	1	1	1
0	1	1	1	1	0	0	0
1	0	0	0	1	0	0	1
1	0	0	1	0	0	0	0
Other combinations				0	0	0	0

Table 2 presents the text of the three input files according to figure 5. The content of the third file is only a "0" because in a counter there are no external inputs, i.e. the application is purely sequential. This is indicated by the flip-flops at the outputs of the multiplexers that deliver the counting function in figure 5. These outputs are latched by a clock common to all cells.

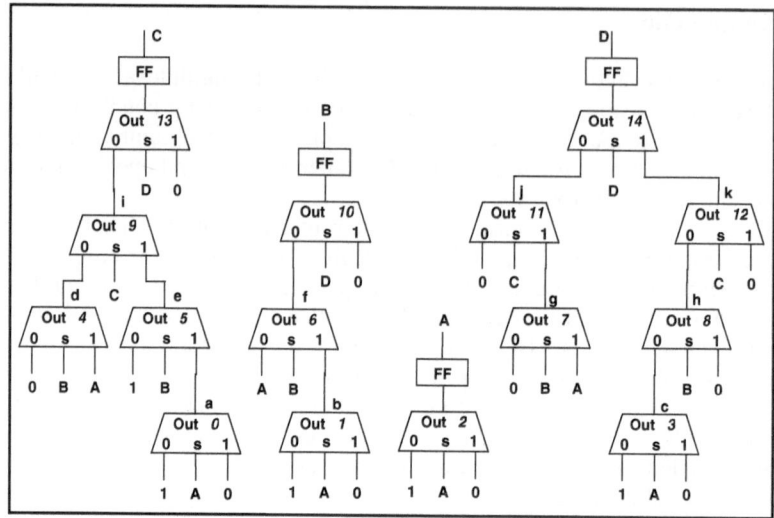

Fig. 5. Multiplexer network that implements a module-10 counter

Table 2. Input files for the genetic router

First input file	Second input file	Third input file
0.8 0.15 4 5 15 4 4 5 4 2 4 2 10 13 14 4 2 10 13 14	1A01A01A01A00BA1BaABb0B AcB0dCefD00CghC0iD0jDk	0

The following code is the VHDL file that the genetic router automatically generated when found the routing for the module-10 counter. The value of the configuration register for spare cells is specified by the case "when others".

```
library IEEE;
entity Mem_counter9 is
    port (
        okaux: in STD_LOGIC;
        xy: in STD_LOGIC_VECTOR (7 downto 0);
        conf: out STD_LOGIC_VECTOR (16 downto 0));
end Mem_counter9;
architecture Mem_counter9_arch of Mem_counter9 is
type REG is array (16 downto 0) of bit;
begin
  process(okaux,xy)
    begin
      if ( okaux = '0' ) then
        conf <= "01001000100000000";
```

```
    else
      case xy is
        when "00000000" => conf <= 01000000101000101";
        when "00000001" => conf <= 00000000111100100";
        when "00000010" => conf <= 00000100111010011";
        when "00000011" => conf <= 00000000111000001";
        when "00000100" => conf <= 00000000010110100";
        when "00000101" => conf <= 10000000100101101";
        when "00000110" => conf <= 11001011100100001";
        when "00010000" => conf <= 01110000001110010";
        when "00010001" => conf <= 00100000101010101";
        when "00010010" => conf <= 00100001000100010";
        when "00010011" => conf <= 00100000001110001";
        when "00010100" => conf <= 00000010000000000";
        when "00010101" => conf <= 00001001110001000";
        when "00010110" => conf <= 00000000000101011";
        when "00100000" => conf <= 00011001000100110";
        when "00100001" => conf <= 01000101101110010";
        when "00100010" => conf <= 00011000001110001";
        when "00100011" => conf <= 00000001110000010";
        when "00100100" => conf <= 00000000000111001";
        when "00100101" => conf <= 00000000000110001";
        when "00100110" => conf <= 00100000111111011";
        when "00110000" => conf <= 01000101011111010";
        when "00110001" => conf <= 00000000010011010";
        when "00110010" => conf <= 00011101011110010";
        when "00110011" => conf <= 00000000001100110";
        when "00110100" => conf <= 00000000000011000";
        when others      => conf <= 01001000100000000";
      end case;
    end if;
  end process;
end Mem_counter9_arch;
```

Figure 6 shows part of the simulation of the module-10 counter implemented in a MUXTREE array. Synthesis and simulation were carried out in Xilinx's Foundation.

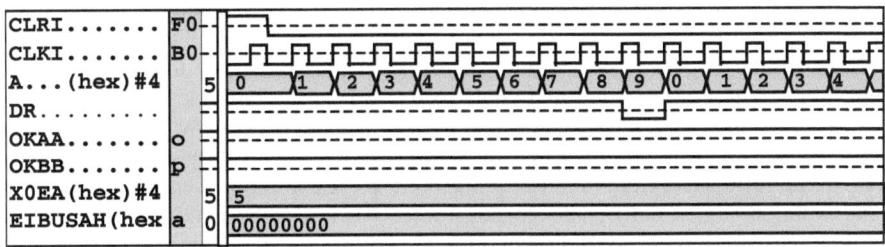

Fig. 6. Simulation of the module-10 counter implemented in a MUXTREE array

Figure 7 shows a graphical representation of one of the solutions found by the genetic router. The clock input is common to all cells. Solid lines represent fixed connections and broken lines are the connections programmed by the genome of the module-10 counter. Numbers in the multiplexers correspond to those in figure 5. A careful visual inspection of figure 7 will demonstrate that in fact, the multiplexer network in figure 5 is contained in the embryonic array presented.

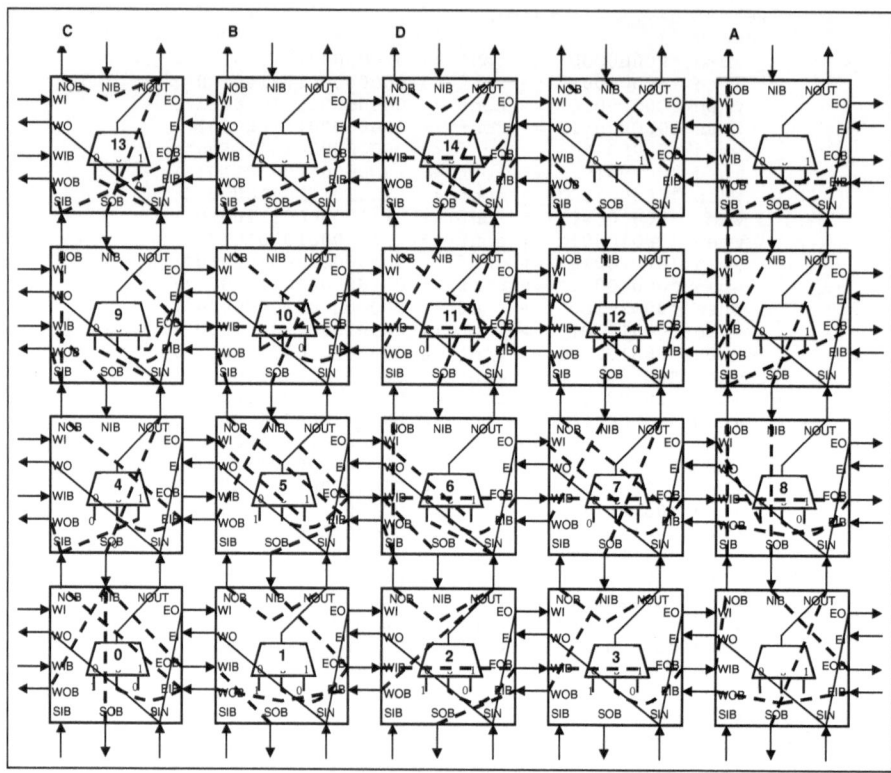

Fig. 7. Routing of a module-10 counter in a 4×5 MUXTREE array

4.1   Results with Other Examples

Table 3 resumes the results obtained when routing other applications. The column entitled *Type* indicates whether the application is combinational (C) or sequential (S). The *Array size* column indicates the size of the MUXTREE array that contained the application. The following column contains the size of the multiplexer network, followed by the maximum number of generations that the genetic algorithm searched for the solution. The following column indicates in which generation the first solution was found, followed by the time taken to find it. Next column presents the number of valid solutions delivered by the genetic router when ran the maximum number of generations. The last column presents the probability of successfully routing the corresponding application, e.g. the genetic router will find a solution to the 3-bit up/down counter, 7 out of 10 times that the program runs 10,000 generations.

5   Conclusions and Future Work

The genetic router presented in this paper is capable of mapping a given multiplexer network into a MUXTREE array. The router could also be applied to solve routing problems in networks presenting a topology similar to that of MUXTREE arrays.

Table 3. Results obtained when routing other applications.

Application	Type	Array size	Mux-net size	Num. of gen.	Gen. of 1st sol.	Time to 1st sol.	Num. of solutions	Routing prob.
3-bit voter	C	2×2	3	100	1	0.5 seg	82	1
3-bit U/D counter	S	3×3	5	10000	250	1.2 min	85	0.7
4-bit CRC generator	C	2×6	10	10000	80	45 seg	78	0.5
2-bit complete adder	C	4×8	15	30000	8500	60 min	75	0.4
4-bit parity generator	C	4×8	15	30000	2500	25 min	80	0.9
5-bit parity generator	C	5×6	11	20000	450	8 min	76	0.2
Module-10 counter	S	4×5	15	30000	6500	60 min	80	0.5

The multiple solutions delivered by the genetic router open the possibility of a new level of redundancy in applications requiring high levels of availability. Different routings could be tried in the same hardware until functionality of the array is restored.

There still are questions regarding the capabilities of the genetic router that require further research. How efficient in terms of resource-usage are the solutions found by the router? What is the maximum number of multiplexers that the router can solve? How sensitive the router's performance is to changes in the GA parameters? How well the genetic router compares against other routing techniques? [12-15] Embryonics is a very vast field where much research remains to be done.

Acknowledgements

We like to thank the Institute of Electrical Research and CONACyT for the facilities and financial support given to carry out this research. Thanks to Dr. Luis Schettino and the reviewers for their valuable comments.

References

[1] Mange D., Sanchez E., Stauffer A., Tempesti G., Durand S., Marchal P. and Piguet C., "Embryonics: A new methodology for designing FPGAs with self-repair and self-reproducing properties", Technical report 95/152, EPFL, Logic Systems Laboratory, 1995

[2] Ortega C. and Tyrrell A., "Fault-tolerant Systems: The way Biology does it!", Proceedings Euromicro 97 (Short Contributions), Budapest, IEEE CS Press, September, 1997, pp.146-151

[3] Tempesti G., Mange D., Stauffer A.and Teuscher C., "The BioWall: an Electronic Tissue for Prototyping Bio-Inspired Systems", in A. Stoica et al. (Eds.), Proceedings of the 2002 NASA/DoD Conference on Evolvable Hardware, IEEE Computer Society, Los Alamitos, Calif., 2002, pp.221-230

[4] Restrepo H.and Mange D., "An Embryonics Implementation of a Self-Replicating Universal Turing Machine", in Y. Liu, K. Tanaka, M. Iwata, T. Higuchi, M. Yasunaga (Eds.), Evolvable Systems: From Biology to Hardware, ICES 2001, volume 2210 of Lecture Notes in Computer Science, 2001, pp.74-87

[5]   Ortega C., Mange D., Smith S. and Tyrrell A., "Embryonics: A Bio-Inspired Cellular Architecture with Fault-Tolerant Properties", Genetic Programming and Evolvable Machines, Vol.1-3, July 2000, pp.187-215

[6]   Tempesti G., Mange D. and Stauffer A., "A Robust Multiplexer-based FPGA Inspired by Biological Systems", Special Issue of Journal of Systems Architecture on Dependable Parallel Computer Systems, February 1997, pp.719-733

[7]   Ortega C. and Tyrrell A., "MUXTREE revisited: Embryonics as a Reconfiguration Strategy in Fault-Tolerant Processor Arrays", Proceedings of ICES98, Lausanne, Switzerland, September, 1998, Lecture Notes in Computer Science 1478, Springer-Verlag, 1998, pp.206-217

[8]   Ortega C. and Tyrrell A., "Design of a Basic Cell to Construct Embryonic Arrays", IEE Transactions on Computers and Digital Techniques, Vol.145-3, May, 1998, pp.242-248

[9]   Ortega C. and Tyrrell A., "Reliability Analysis in Self-Repairing Embryonic Systems", in Stoica A., Keymeulen D. and Lohn J. (Eds.), Procs. of 1st NASA/DoD Workshop on Evolvable Hardware, Pasadena, CA, IEEE Computer Society, July 1999, pp.120-128

[10] Holland J., Adaptation in Natural and Artificial Systems, MIT Press, 1992

[11] Goldberg D., Genetic Algorithms in Search, Optimization and Machine Learning, Addison-Wesley, ISBN: 0201157675, 1989

[12] Drechsler, Evolutionary Algorithms in VLSI CAD, Kluwer, 1998

[13] Minato Shin-Ichi, Binary Decision Diagrams and Applications for Vlsi CAD,Kluwer International Series in Engineering and Computer Science, 342, 1996

[14] Bushnell Michael Lee, Design Automation: Automated Full-Custom Vlsi Layout Using the Ulysses Design Environment, Perspectives in Computing, Vol 21, 1988

[15] Cheng Chung-Kuan (Editor), Interconnect Analysis and Synthesis, Wiley-Interscience, October 1999

Exploiting Auto-adaptive μGP
for Highly Effective Test Programs Generation

F. Corno, F. Cumani, and G. Squillero

Politecnico di Torino
Dipartimento di Automatica e Informatica
Corso Duca degli Abruzzi 24 I-10129, Torino, Italy
http://www.cad.polito.it/

Abstract[*]. Integrated-circuit producers are shoved by competitive pressure; new devices require increasingly complex verifications to be performed at increasing pace. This paper presents a methodology to automatically induce a test program for a microprocessor that maximizes a given verification metric. The methodology is based on an auto-adaptive evolutionary algorithm and exploits a syntactical description of microprocessor assembly language and an RT-level functional model. Experimental results clearly show the effectiveness of the approach. Comparisons reveal how auto-adaptive mechanisms dramatically enhance both performances and quality of the results.

1 Introduction

Today competitive pressure compels all integrated-circuit producers cutting time to market. Engineers are required to design more and more complex devices, and all devices must be carefully checked before starting production. Thus, the current trend requires increasingly complex verifications to be performed at increasing pace. *Design verification* is a process similar to software debugging: device functionalities are tested, and characteristics examined for determining possible inconsistencies.

Concerning microprocessors, it has been maintained that about a third of the cost of developing a new one is devoted to this hardware debugging and testing [1]. Nevertheless, the inadequacy of existing verification methods is clearly illustrated by the Pentium's FDIV error, which cost its manufacturer an estimated $500 million.

The verification process of a microprocessor requires a deep knowledge of the instruction set and instruction format since only correct programs can internally perform meaningful operations. To verify microprocessor design, manufacturers usually rely on simulating the execution of test programs meticulously written by hand. However, hand-written test cases can be only exploited as a first line of defense against bugs, since they focus on basic functionalities and important but rarely-occurring corner cases. During the whole design process, exhaustive or nearly-exhaustive tests are often necessary, but the effort required to generate them manually may be practically unworkable.

---

[*] This work has been partially supported by *Intel Corporation* through the grant "GP Based Test Program Generation".

A.M. Tyrrell, P.C. Haddow, and J. Torresen (Eds.): ICES 2003, LNCS 2606, pp. 262–273, 2003.

Today, advances in simulation and emulation technology enabled the use of other sources of test stimuli such as existing application and system software [1]. Additionally, to increase test productivity sophisticated test-generation systems have been proposed [2] [3]. Nevertheless, although all these approaches can significantly increase design productivity, they are still biased towards corner cases, far from being fully automated and not broadly exploitable.

This paper presents an innovative methodology for generating a test case for validating a microprocessor core. An auto-adaptive evolutionary algorithm is exploited for generating an assembly program able to maximize a predefined verification metric. The algorithm relies on a description of the assembly language implemented by the processor and requires limited human intervention. At the end of the process, the test program can be simulated and designers are required to manually analyze only those parts of the description that the tool possibly failed to validate.

The proposed approach was tested on the i8051 microcontroller using the *statement coverage* as verification metric. The i8051, despite its relatively old age, is still one of the most popular microprocessors and can be considered a good example of a small microcontroller. The methodology is easily applicable to more complex designs, like pipelined microprocessors, and can easily exploit different design verification metrics, such as branch, conditional or path coverage.

Next Section introduces simulation-based design verification. Section 3 details the auto-adaptive evolutionary algorithm. Section 4 illustrates the case study and Section 5 reports experimental evaluation. Section 6 concludes the paper.

2 Simulation-Based Design Verification

Given an RT-level description of a microprocessor, simulation-based verification requires a test program and a tool able to simulate its execution. The goal is to uncover design errors, and the effectiveness of the test program is measured exploiting a given metric. The verification metric adopted in this paper is the *statement coverage*.

To avoid confusion, in the following the term "instruction" denotes an *instruction* in an assembly program, and the term "statement" refers to a *statement* in an RT-level description. The term "execute" is commonly used in both domains: instructions in a program are executed when the processor fetches them and operates accordingly; statements in a VHDL description are executed when the simulator evaluates them to infer design behavior. The verification metric exploited in this paper measures the percentage of executed (evaluated) RT-level statements over the total when the execution of a given test program is simulated.

Statement coverage can be considered as a required starting point for any design verification process. Such analysis ensures that no part of the design missed functional test during simulation, as well as reducing simulation effort from "over-verification" or redundant testing. Moreover, use of coverage analysis provides an easy and objective way of measuring simulation effectiveness to ensure that all bugs would be exposed with the minimum amount of effort. Indeed, most CAD vendors have recently added code-coverage features to their simulators.

Since no reliable automatic methodology exists, designers are used to write their own test programs to check the basic functionalities and critical corner cases. However, devising test cases is a difficult, time-consuming task: writing a test program requires a deep knowledge on the microprocessor architecture, but the designer is not the ideal verification engineer since he can be biased by his expectation, failing to check for some errors. Hand-written test cases are definitely required, but are not sufficient in the verification process.

The simplest method to obtain an assembly test program is probably to compile a high-level routine. This approach relies on a compiler or cross-compiler, but this requirement may be easily met. Despite their effortlessness, compiled problem-specific algorithms are not the best solution. They are severely inadequate to uncover design errors: due to the intrinsic nature of the algorithms or of compiler strategies, they are seldom able to execute all statements in a description testing all functionalities.

A better strategy is to generate random assembly programs. This approach is likely to cover more statements in the description, but is less straightforward. A random generated program may easily contain illegal operations, such as division by zero, or endless loops. However, the effort required to generate a syntactically correct assembly source is still moderate. The main drawback of this method is that a random program will hardly cover all corner cases, hence resulting in low statement coverage. In order to obtain a sufficient coverage a large number of long programs are needed, resulting in overlong simulation times.

This paper presents the *Auto-Adaptive µGP*, a new approach for devising a test program stemming from the genetic programming (GP) paradigm. The µGP algorithm cultivates Turing-complete programs; internally it represents individuals as directed acyclic graphs (DAGs), while maps them to assembly programs during evaluation. The theoretical framework was presented in [4], while its first application was shown in [5].

In this paper, the µGP is significantly enhanced by adding both recombination and auto-adaptation mechanisms, and fitness function has been changed. The induction of the test case requires a syntactical description of the assembly implemented by the microprocessor and, exploiting the new auto-adaptive mechanisms, necessitate less parameter tuning. The new features both speed-up the search process and yield better results compared to the previous approach.

3   Auto-adaptive µGP

The architecture is sketched in Figure 1. The microprocessor assembly language is described in an instruction library, and the µGP generates efficient test programs exploiting it. The execution of each assembly program is simulated with an external tool, and the corresponding statement coverage is used to drive the optimization process.

Genetic Programming (GP) was defined as a domain-independent problem-solving approach in which computer programs are evolved to solve, or approximately solve, problems [6]. GP addresses one of the more desired goals of computer science: creating, in an automated way, computer programs able to solve problems.

Traditional GP induced programs are mathematical functions that, after being evaluated, yield a specific result. The pioneering ideas of generating *real* (Turing complete) programs date back to [7]. More recently a genome compiler has been proposed in [8], which transforms standard GP trees into machine code before evaluation.

This paper exploits the versatile GP-like approach for inducing assembly programs presented in [4]. The methodology exploits a directed acyclic graph (DAG) for representing the *flow* of the program (Figure 2, left). The DAG is built with four kinds of nodes: prologue (followed by 1 child), epilogue (followed by no children), sequential instruction (followed by 1 child), and branch (followed by 2 children).

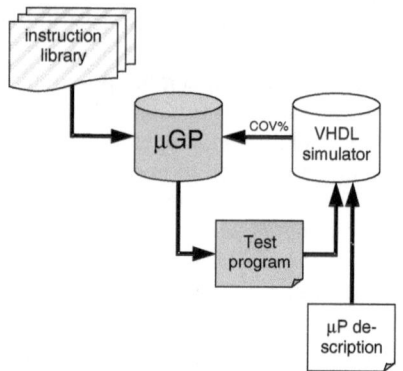

Fig. 1. System Architecture

- The **prologue** and **epilogue** nodes are always present and represent required operations, such as initializations. They depend both on the processor and on the operating environment, and they may be empty. The prologue has no parent node, while the epilogue has no children. These nodes may never be removed from the program, nor changed.
- **Sequential-instruction** nodes represent common operations, such as arithmetic or logic ones (e.g., node **B**). They are always followed by exactly one child. The number of parameters changes from instruction to instruction, following assembly specification. *Unconditional* branches are considered sequential, since execution flow does not split (e.g., node **D**).
- **Conditional-branch** nodes are translated to assembly-level conditional-branch instructions (e.g., node **A**). All common assembly languages implement some *jump-if-condition* mechanisms. All conditional branches implemented in the target assembly languages are included in the library.

Each node contains a pointer inside the instruction library and, when needed, its parameters (i.e., operand values or register specifications). For instance, (Figure 2. right) shows a sequential node that will be translated into an "*ORL A, R1*", i.e., a bitewise OR between accumulator and register R1. The instruction library may contain different entries corresponding to the same instruction. For instance, the entry referring to "*ORL int, A*", where the parameter is the data RAM address, is different from the entry "*ORL A, reg*" where the parameter may be @R0, @R1, R0, R1, R2, R3, R4, R5, R6 or R7.

The DAG is always translated to a syntactically correct, loop-less assembly program, although it is not possible to infer its semantic meaning. An induced program may perform operations on any register and any memory locations, and this exceptional freedom is essential to generate test programs.

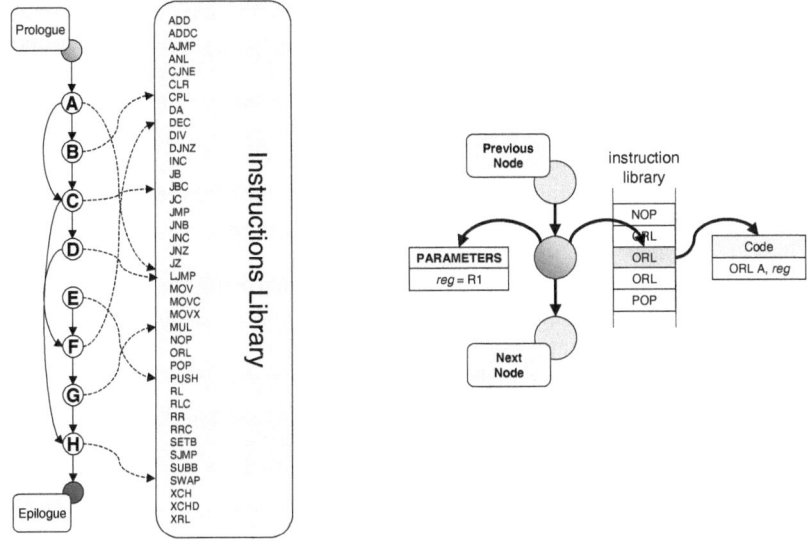

Fig. 2. DAG Representation

The library approach enables to exploit the genetic core and the DAG structure with different microcontrollers or microprocessors that not only implement different instruction sets, but also use different formalisms and conventions.

Adopted DAG representation prevents backward branches, either conditional or unconditional. This characteristic guarantees program termination, since no endless loop may be implemented, but introduces a small reduction in semantic power.

3.1 Genetic Operators

Test programs are induced by mutating the DAG topology and by mutating parameters inside DAG nodes. Both kinds of modifications are embedded in an evolutionary algorithm implementing a $(\mu+\lambda)$ strategy.

In more details, a population of μ individuals is cultivated, each individual representing a test program. In each step, λ new individuals are generated, parents are selected using tournament selection with tournament size τ. Each new individual is generated by applying one or more genetic operators. The cumulative probability of applying at least n consecutive operators is equal to $P(n) = p_c^n$, where parameter p_c is discussed later.

After creating new λ individuals, the best μ programs in the population of $(\mu+\lambda)$ are selected for surviving.

The initial population is generated creating μ empty programs (only prologue and epilogue) and then applying i_m consecutive random mutations to each.

The evolution process iterates until population reaches a *steady state* condition, i.e., no improvements are recorded for S_g generations.

Three mutation and one crossover operators are implemented and activated with probability p_{add}, p_{del}, p_{mod} and p_{xover} respectively.

- **Add node:** a new node is inserted into the DAG. The new node can be either a sequential instruction or a conditional branch. In both cases, the instruction referred by the node is randomly chosen. If the inserted node is a branch, either unconditional or conditional, one of the subsequent nodes is randomly chosen as the destination. Remarkably, when an unconditional branch is inserted, some nodes in the DAG may become unreachable (e.g., node **E** in Figure 2, left).
- **Remove node:** an existing internal node (except prologue or epilogue) is removed from the DAG. If the removed node was the target of one or more branch, parents' edges are updated.
- **Modify node:** all parameters of an existing internal node are randomly changed. Parameters include immediate values and register specifications.
- **Crossover:** two different programs are mated to generate a new one. First, parents are analyzed to detect potential cutting points, i.e., vertex in the DAG that if removed creates disjoint sub-graphs (e.g., node **C** in Figure 2, left). Then a standard 1-point crossover is exploited to generate the offspring.

3.2 Auto-adaptive Mechanisms

The auto-adaptive μGP internally tunes both the number of consecutive random mutations and the activation probabilities of all genetic operators. Modifying these parameters, the algorithm is able to shape the search process significantly improving its performances.

The number of consecutive random mutations is controlled by parameters p_c, that, intuitively, molds the mutation strength in the optimization process. The popular *1/5 rule*, attributed to Rechenberg, is a well-known example of modifying the *width* of each step to increase performance. Intuitively, in the beginning it is better to adopt a high value, allowing offspring to strongly differ from parents. On the other hand, in the end of the search process, it is preferable to reduce diversity around the local optimum, allowing only small mutations. The auto-adaptive μGP monitors improvements: let I_H be the number of newly created individuals attaining a fitness value higher than their parents over the last H generations. At the end of each generation, the new p_c value is calculated as $p_c^{new} = \alpha \cdot p_c + (1 - \alpha) \cdot \dfrac{I_H}{H \cdot \lambda}$. Then p_c is saturated to 0.9. And, initially, the maximum value is adopted ($p_c = 0.9$), considering all $H \cdot (\mu + \lambda)$ individuals as improvements. Finally, the coefficient α introduces an inertia to unexpected abrupt changes.

Regarding activation probabilities, initially they are set to the same value $p_{add} = p_{del} = p_{mod} = p_{xover} = 0.25$. During evolution, probability values are updated similarly to mutation strength: let O_1^{OP} be the number of successful invocation of genetic operator OP in the last generation, i.e., the number of invocations of OP where the resulting individual attained a fitness value higher than its parents; and let O_1 be the total number of operators invoked in the last generation. At the end of each generation, the new values are calculated as $p_{OP}^{new} = \alpha \cdot p_{OP} + (1-\alpha) \cdot \dfrac{O_1^{OP}}{O_1}$. Since it is possible that $p_c > 0$,

O_1 may be significantly larger than λ. Activation probabilities are normalized and forced to avoid values below .01 and over 0.9. If $O_1 = 0$, then all activation probabilities are pushed towards initial values.

3.3 Evaluation

Individuals are evaluated by simulation. The DAG is first translated into a syntactically correct assembly program and assembled to machine code. Then the execution of the test case is simulated on the RT-level description of microcontroller, gathering verification metric figures. The final value is considered as the *fitness* value of the individual, i.e., the extent to which it is able to produce offspring in the environment.

If two programs attain the same verification metric figures, the shortest is assumed to have a higher fitness.

Fitness values are used to select λ parents for generating new offspring through a tournament of size τ (i.e., τ individuals are randomly selected and the fittest one is picked). Moreover, fitness values are used to deterministically select the best μ individuals out of the $(\mu+\lambda)$ ones at the end of each evolution step.

Test program evaluation does not consider the internal structure of the microprocessor, nor does it include *hints* for increasing the coverage based on designers' knowledge.

4 Experimental Results

The proposed approach was tested on the i8051 microcontroller using the *statement coverage* as verification metric.

Despite its relatively old age, the i8051 is one of the most popular microcontrollers in use today, and many derivative microcontrollers are based on, and compatible with, it. The i8051 is an 8-bit microprocessor originally designed in the 80's by Intel that has gained great popularity since its introduction. Its standard form includes several on-chip peripherals, including timers, counters, and UART's, plus 4 Kbytes of on-chip program memory and 128 bytes of data memory, making single-chip implementations possible. Its hundreds of derivatives, manufactured by several different companies include even more on-chip peripherals, such as analog-digital converters, pulse-width modulators, I2C bus interfaces.

Table 1. i8051 RT-level description

NAME	MODULE	LINES	STMS
CTR	Processor core	5,206	2,121
ALU	Arithmetic Logic Unit	429	226
DEC	Decoder	270	220
XRM	External SRAM interface	77	11

The i8051 memory architecture includes 128 bytes of data memory that are accessible directly by its instructions. A 32-byte segment of this 128-byte memory block is bit addressable by a subset of the i8051 instructions, namely the bit-instructions. External data memory of up to 64 Kbytes is accessible by a special *"MOVX"* instruction. Up to 4 Kbytes of program instructions can be stored in the internal memory of the i8051, or the i8051 can be configured to use up to 64 Kbytes of external program memory. The majority of the i8051's instructions are executed within 12 clock cycles.

The i8051 instructions range from 0-operand ones, like *"DIV AB"* (divide accumulator A by B) where all operands are implicit, to 3-operand ones, like the *"CJNE Op1, Op2, RelAddr"* (compare Op1 with Op2 and jump if they are not equal). The i8051 allows 5 different addressing types: *immediate, direct, indirect, external direct* and *code indirect*. As in many CISC, registers are not orthogonal to the instructions and addressing modes.

A prototype of the proposed approach has been developed in ANSI C language in about 2,000 lines of code. The prototype exploits *Modelsim* v5.5a by Model Technology for simulating the design and getting coverage figures.

Table 2. μGP parameters

PAR	MEANINGS	VALUE
μ	Population size	5
λ	Offspring size	10
τ	Tournament size (selective pressure)	2
i_m	Initial mutations	100
H	History for auto-adaption	4
α	Auto-Adaption inertia	0.4
S_g	Steady state	500

The methodology was tested on a synthesizable RT-level implementation of the i8051 core consisting in about 7,500 VHDL lines (the corresponding gate-level netlist is about 12K gates). An external data RAM of 2 Kbytes was connected to the i8051, while no external program memory was used.

Four main blocks can be found in the RT-level description (Table 1): the processor core control logic (CTR), the arithmetic and logic unit (ALU), the instruction decoder (DEC), and the external SRAM interface (XRM). The CTR is described behaviorally as a sequential logic block; the ALU is described behaviorally as a combinational logic block; the DEC is described as a data-flow implementing a combinational logic block; the XRM models an external SRAM. For each block Table 1 reports the total number of VHDL lines [LINES] and the number of statements [STMS].

Inducing a test program with the proposed GP required about 1,200 generations, corresponding to the evaluation of about 12,000 programs. The experiments employed about 12 hours of CPU time on a SPARC ULTRA Workstation at 400MHz with 2GB of RAM. Table 2 shows the adopted parameters.

To assess the efficiency of the proposed method, the induced test program was compared with 5 programs devised with 3 different methodologies: compiled problem-specific algorithms; random test programs; and exhaustive functional test case. Table 3 compares the different programs in term of required program ROM bytes [SIZE] and instructions executed by the program [INST]. Statement coverage figures are reported in column [TOT]. Statistics are also detailed for the 4 blocks [ALU], [CTR], [DEC] and [XRM].

Table 3. Experimental Results

Test program	Program		Statement Coverage [%]				
	SIZE	INST	ALU	CTR	DEC	XRM	TOT
Fibonacci	324	1,176	49.60	30.20	62.70	81.80	34.70
int2bin	81	572	49.60	21.30	56.40	81.80	27.10
Random (size = GP)	648	334	86.70	87.60	93.20	100.00	88.20
Random (size = 4K)	4,096	2,373	96.50	94.60	97.70	100.00	95.00
TestAll (exhaustive)	2,834	52,953	95.10	99.40	100.00	100.00	99.10
ATS2002	469	228	99.60	99.70	100.00	100.00	99.70
Auto-adaptive μGP	292	183	100.00	99.80	100.00	100.00	99.84

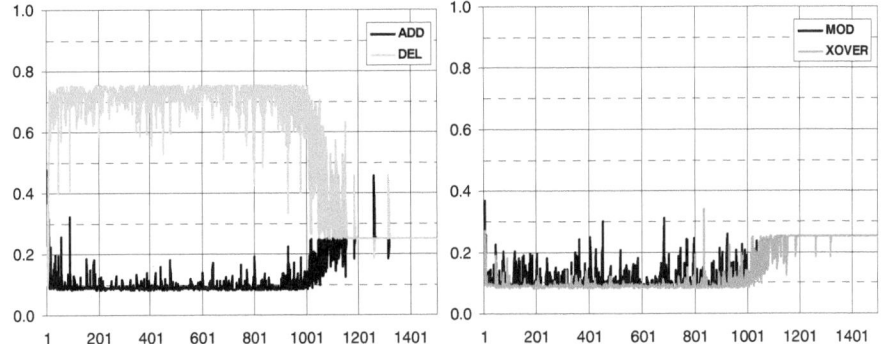

Fig. 3. Activation probabilities

The two problem-specific algorithms are *Fibonacci* and *int2bin*. The former calculates the Fibonacci series, while the latter converts an integer to a binary representation. As expected, the coverage figure is quite low. Looking at the decoder, it may be easily inferred that only a subset of the instruction set is used. Both programs execute loops (the number of executed instructions is higher than the number of stored ones). Neither program accesses the external data RAM.

Two different random test programs were considered. The former, *Random (size = GP)*, was devised using the same effort as the GP. 20,000 random programs of approximately 500 instructions were generated and the best one was chosen. The comparison allows evaluating the effectiveness of the evolutionary core in driving the search process. The latter, *Random (size = 4K)*, was devised generating 20,000 random programs filling all available ROM space and selecting the best one. It is included here to allow an estimation of the best result attainable with the random approach disregarding efficiency.

The exhaustive functional test program *TestAll* is also considered. The test case was devised by microprocessor designers; it is relatively long and includes several loops. It tests all possible instructions, although it is not able to check all possible corner cases in the implementation. For instance, "DIV AB" when A is less or equal than B was not taken into account.

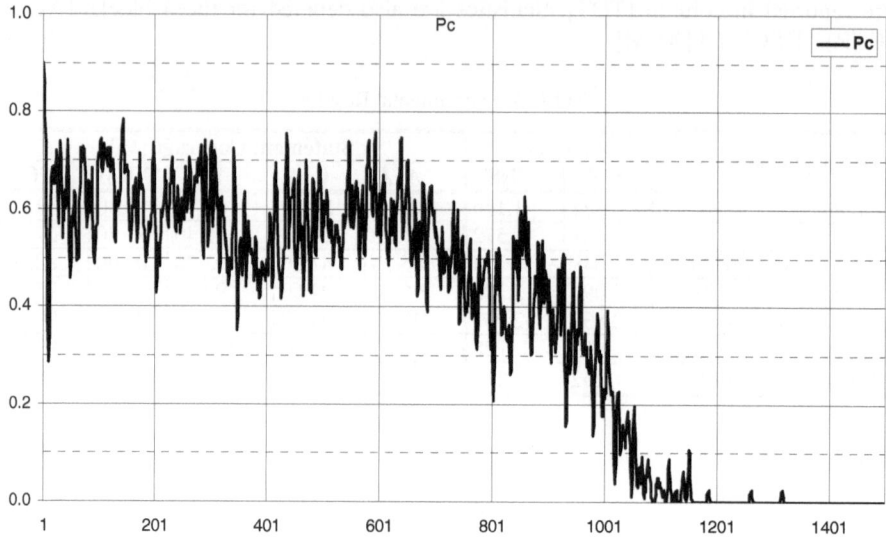

Fig. 4. Mutation strength

Finally, results attained by the basic μGP [5] are shown in *ATS2002*.

The induced test program got the highest statement coverage figure with the smallest size and the lowest run time. Execution of the test case is fast, since there are no loops by construction. Compared with original μGP, the auto-adaptive algorithm is able to produce better and shorter test programs, with even reduced human effort.

Figure 3 shows the evolution of activation probabilities during the experiment, the adapted mutation strength is shown in Figure 4. Parameter values are plotted against the generation number.

It may be surprising the high activation probability of the *Delete* operator. However, the initial population is generated applying i_m consecutive random mutations to empty programs and the average length of the initial programs is about 60 instructions. In this situation, it is useful for the μGP to delete instructions. Interestingly, experimental evidence suggests that when the μGP starts with a population of empty programs ($i_m = 0$) in the first hundreds of generations the p_{ADD} increases dramatically, but after this first phase the trend comes close to the one showed in Figure 3.

The mutation strength adaptation is more predictable: in the beginning is about $p_c = 0.6$ and then slowly decrease to 0. The last 300 generations are plotted even if the μGP already reached the steady-state condition to show activation probabilities reaching back their default values.

Performances of the Auto-Adaptive μGP are comparable with those attained by the standard μGP; however, results are significantly better. This is due to the fitness that takes into account program length and to the adaptive activation probabilities. It may be interesting to show how the simple mutation strength p_c influences performances.

Figure 5 compares performance of the standard μGP against the auto-adaptive one, where only parameter p_c is tuned. The statement coverage is plotted against the generation number, only the first 800 generations are considered. Graph shows clearly that the enhanced μGP is faster and able to reach a higher maximum.

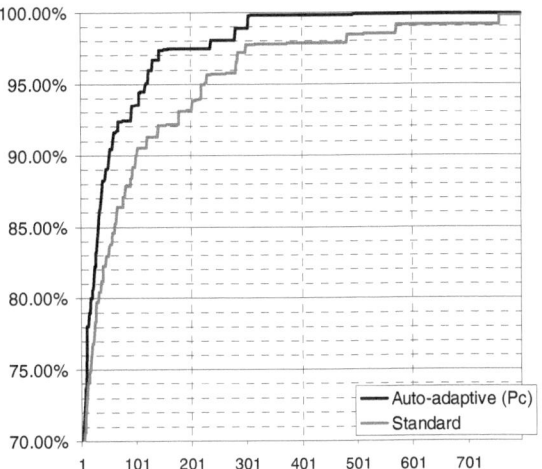

Fig. 5. Mutation strength effectiveness

5   Conclusions

This paper presents the *Auto-Adaptive μGP*, a methodology able to automatically induce an assembly test program for a microcontroller. The methodology is based on an auto-adaptive evolutionary algorithm and exploits the syntactical description of the language.

Unless commonly adopted methodology, it does not require a skilled programmer or additional analysis to be performed by verification engineers. The generated test case is able to efficiently maximize a given verification metric, and, at the end of the process, designers are required to manually analyze only those parts of the description that the tool failed to validate.

A prototype of the proposed approach has been developed in ANSI C language, and then it was tested on a synthesizable RT-level implementation of the i8051 using statement coverage as verification metric. Induced test programs outperformed test cases devised with alternative methodologies.

Experimental results show the efficiency of the methodology. Devising a test case able to reach the complete statement coverage is a difficult task, even on a small microcontroller like the i8051. Reported data show that random programs can hardly test all corner cases and also long, carefully designed hand-made test cases may not be exhaustive. The automatically induced test program, conversely, was able to cover almost all (99.84%) of the statements in the description.

The algorithm is able to tune internal parameters to optimize search process. New auto-adaptive mechanisms dramatically enhance both performances and quality of the results, as a result, the proposed approach betters the original μGP generating a test program that is both shorter and able to reach higher coverage.

Current work is targeted to apply the proposed approach to more complex processors.

References

1. J. Kumar, "Prototyping the M68060 for concurrent verification", *IEEE Design & Test of Computers*, 1997, pp. 34–41
2. A. K. Chandra et al., "AVPGEN - a test generator for architecture verification", *IEEE Transactions on Very Large Scale Integration (VLSI) Systems*, Vol. 3, 1995, pp. 188–200
3. A. Aharon et al. "Verification of the IBM RISC System/6000 by dynamic biased pseudo-random test program generator", *IBM Systems Journal*, 1991, pp. 527–538
4. F. Corno, G. Cumani, M. Sonza Reorda, G. Squillero, "Efficient Machine-Code Test-Program Induction", *Congress on Evolutionary Computation*, 2002, pp. 1486–1491
5. F. Corno, G. Cumani, M. Sonza Reorda, G. Squillero, "Evolutionary Test Program Induction for Microprocessor Design Verification", to appear in: *ATS02: The 11th Asian Test Symposium*, Guam (USA)
6. J. R. Koza, "Genetic programming", *Encyclopedia of Computer Science and Technology*, vol. 39, Marcel-Dekker, 1998, pp. 29–43
7. R. M. Friedberg, "A Learning Machine: Part I", *IBM Journal of Research and Development*, 1958, vol. 2, n. 1, pp 2–13
8. A. Fukunaga, A. Stechert, D. Mutz, "A genome compiler for high performance genetic programming", *Genetic Programming 1998: Proceedings of the 3rd Annual Conference*, 1998, pp. 86–94

Speeding up Hardware Evolution:
A Coprocessor for Evolutionary Algorithms

Tillmann Schmitz, Steffen Hohmann, Karlheinz Meier,
Johannes Schemmel, and Felix Schürmann

University of Heidelberg, Kirchhoff Institute for Physics,
INF 227, D-69120 Heidelberg
tschmitz@ix.urz.uni-heidelberg.de
http://www.kip.uni-heidelberg.de/vision

Abstract. This paper proposes a coprocessor architecture to speed up
hardware evolution. It is designed to be implemented in an FPGA with
an integrated microprocessor core. The coprocessor resides in the configurable logic, it can execute common genetic operators like crossover
and mutation with a targeted data throughput of 420 MByte/s. Together
with the microprocessor core, a complex evolutionary algorithm can be
developed in software, but is processed at the speed of dedicated hardware.

1 Introduction

An evolvable hardware platform usually consists of three main parts: the reconfigurable hardware (RH) and two processing units: the fitness calculation unit
(FCU) and the evolutionary algorithm unit (EAU). There are different possibilities to implement the processing units (FCU and EAU) like PC-based approaches
([1], [2]), DSPs [3], FPGAs [4] and ASICs.

These implementations can be classified whether they are software- (PC and
DSP) or hardware-oriented (FPGA and ASIC). While the first option provides
the advantage of variability, easier design and maintainability, it is relatively
slow. A hardware implementation allows a higher degree of parallelism and is
therefore faster, but usually needs more design effort.

Our group uses an analog neural network ASIC (ANN) [5] as the reconfigurable architecture (see Fig. 1). Its configuration stream consists of about 45 kByte
of data and it calculates one network layer every 20 ns. To exploit the network's
speed, substantial computing power and data-flow rate are needed to generate
new generations in time.

In this paper we propose a coprocessor architecture to bring up the evolutionary algorithm to the speed of dedicated hardware. It is customized to suit
the demands of our ANN, but the architecture is flexible enough to be transferred to other reconfigurable hardware systems. The coprocessor is designed for
a Virtex-II Pro FPGA manufactured by Xilinx [6]. This FPGA provides programmable logic cells, high speed serial links for off-chip communication and an

A.M. Tyrrell, P.C. Haddow, and J. Torresen (Eds.): ICES 2003, LNCS 2606, pp. 274–285, 2003.

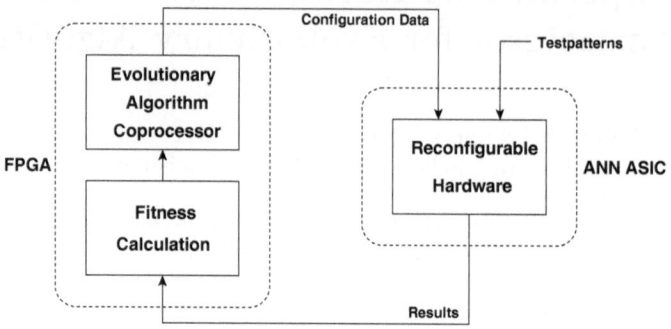

Fig. 1. Setup used for the evolvable hardware experiments

IBM PowerPC 405 core immersed in the FPGA fabric. With this device we are able to partition between soft- and hardware:

The evolutionary algorithm is written in software and executed in the microprocessor core. The coprocessor resides in the programmable logic cells and is coded in a hardware description language. The problem-specific fitness calculation is not a topic in this paper. It is done either by software in the microprocessor core or in specialized hardware in the FPGA.

The instruction set and the internal structure of the coprocessor are designed to allow a wide range of evolutionary operators. This makes it possible to use the same hardware configuration code for a large number of different experiments and training algorithms.

With this coprocessor system it is possible to design complex evolutionary algorithms in a software environment, but let the time-expensive parts be executed in hardware. This yields a data throughput of up to 420 MByte/s of genetic data.

2   Evolution System

Our hardware integrates a Xilinx FPGA Virtex-II Pro [6] together with 256 MByte of DDR-SDRAM [7] and an analog neural network ASIC on a single PCB[1], as shown in Fig. 2. It is interfaced to a host computer. This connection is used only to initialize the microprocessor core, the random generators, etc. and to monitor the evolution progress.

Analog Neural Network ASIC. The target of the evolution is a neural network. It is implemented in a mixed-signal ASIC developed by our group, partitioned in four network blocks containing 128 input neurons and 64 output neurons each. This sums up to 32768 synapses. Taking into account the resolution of 11 bit per synapse, our configuration bit stream, i.e. the genome, is 360448 bit (44 kByte) long (see [5]).

---

[1] Printed Circuit Board (PCB)

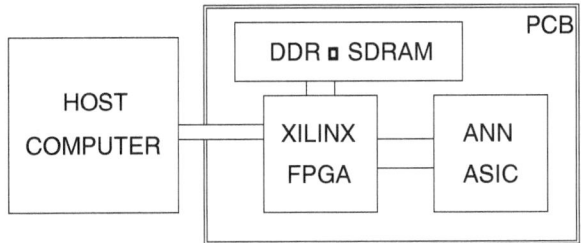

Fig. 2. Evolution system

This ASIC is connected to the FPGA by a parallel LVDS[2] link, capable of transferring up to 1.2 GByte/s of configuration or testpattern data. To evaluate an individual or genome, the configuration data has to be transferred to the ASIC first. After this is done, the testpatterns are applied successively and finally the results for each testpattern are sent back to the FPGA across the LVDS link.

Xilinx FPGA Virtex-II Pro. The coprocessor is designed for the Virtex-II Pro FPGA. Apart from the programmable logic, it contains 504 kBit internal dual-ported SRAM, hardware multiplier and high speed serial links for off-chip communication. Its outstanding feature is the PowerPC, a microprocessor core delivering 420 Dhrystone MIPS. As the PowerPC is immersed in the FPGA fabric, it can access part of the internal SRAM as fast as its own cache. Our coprocessor connects to the additional SRAM port, i.e. we use the dual ported internal SRAM to communicate between microprocessor core and coprocessor.

Peripherals and Scalability. The internal SRAM provided by the FPGA is neither sufficient to store the genome data nor the testpatterns. Therefore, we added a DDR-SDRAM module to the PCB for the FPGA to use up to 1 GByte of external memory at a transfer rate of up to 2.1 GByte/s.

The evolution system is designed in a way that allows to combine 16 systems on one backplane. These modules are connected to each other with the serial links described above. Each FPGA offers four links, we are targeting a 16 node, 2-dim torus, with each connection transferring up to 3.125 Gbit/s.

3   Evolutionary Algorithm Coprocessor (EAC)

3.1   Overview

The EAC pursues two aims: It must process a high data throughput and it has to be parameterized to allow a wide range of evolutionary algorithms without changing the hardware configuration. The first aim demands a pipelined structure. To satisfy the second this pipeline is controlled and managed by a set of

---

[2] Low-Voltage Differential Signalling (LVDS)

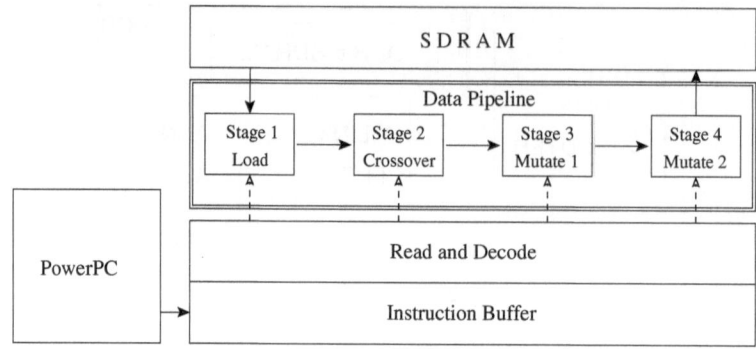

Fig. 3. Instruction buffer and command structure

instructions. This structure is shown in Fig. 3. The microprocessor core addresses the EAC by writing instructions to the *Instruction Buffer* (IB). They are read and decoded successively. Depending on the instruction they control different parts of the pipeline, indicated with dashed arrows in Fig. 3.

3.2   Data Flow Control

All genetic data is stored in the SDRAM. A population consists of a number of individuals, represented by their *genomes*. Each genome is divided into *chromosomes*. A chromosome is stored consecutively in the SDRAM and crossover takes place inside one chromosome. Each chromosome contains an arbitrary number of genes, the smallest unit of the genome. The mutation operator works on single genes. A typical ANN training setup uses 10-100 individuals and one chromosome per neuron. The coprocessor can handle any combination that fits into memory.

A more detailed view of the pipeline and the surrounding control structures is given in Fig. 4. The genes are read from the SDRAM as parents, modified in one crossover and two mutation stages and written back as children to a new address in the SDRAM.

In order to feed genetic data into the pipeline (or to write it back to the memory), the software must specify the SDRAM address and the number of genes to be transferred. The SDRAM works burst-oriented, i.e. consecutive addresses can be accessed very fast while random access is relatively slow. Therefore, the chromosomes are stored consecutively in the SDRAM, and FIFOs accumulate the data in order to address the SDRAM in bursts.

Parallelism. A gene is expressed with 11 bit resolution. As described below, we need two additional bits to control the evolutionary operators. To simplify the addressing, each gene occupies two bytes, so three bits are available for future use.

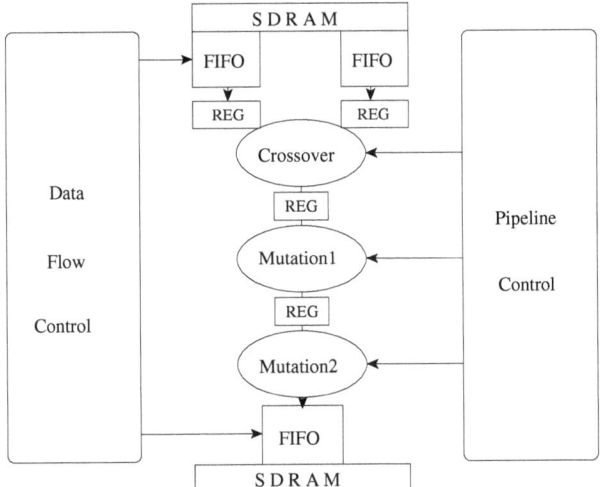

Fig. 4. The pipeline with its control structure

The design is targeted for 80 MHz. Four pipelines work in parallel to take advantage of the SDRAM transfer rate. Fig. 4 depicts one pipeline processing one gene per cycle. Provided, enough genetic data and instructions are supplied, the four pipelines can process four genes per cycle.

3.3   Combinatorial Logic

Fig. 5 shows a detailed description of the evolution pipeline. There are two different kinds of control options:

- *Multiplexer Control Bits*
- *Global Parameters*

The *Multiplexer Control Bits* (`SelectParent`, `SelectRandom`, `SelectScale`, `SelectConstant`, `SelectReplace` and `SelectMutate`) in Fig. 5 are numbered from one to six, they may change with every new gene. For example, `SelectMutate` decides whether the original gene is passed down unaltered or undergoes mutation. As this is done for each gene independently, `SelectMutate` changes almost with every new gene.

The *Global Parameters* (`RandomMaximum`, `Constant` and `Fraction`) are recognizable by dashed boxes. They are at least valid for a whole chromosome, in most cases for the entire evolution. `RandomMaximum` for example may be set to 1023 ($2^{10} - 1$) at the beginning to get mutated genes with every possible value. Later, as the fitness increases, `RandomMaximum` might be decreased to fine-tune the population.

This difference in the update frequency is important. The *Global Parameters* are updated infrequently, therefore, the software can set them with instructions.

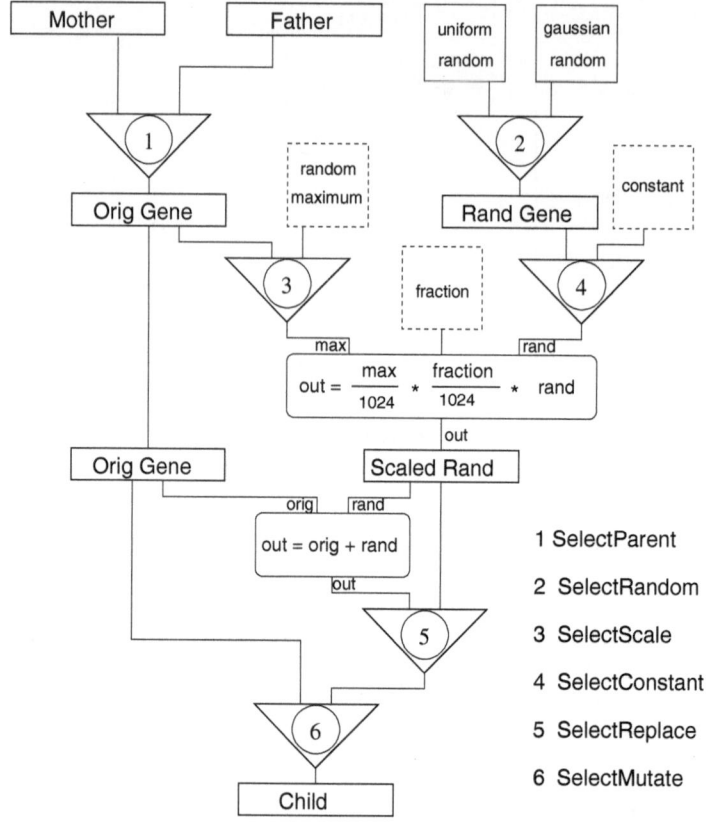

Fig. 5. The pipeline in detail

Since the *Multiplexer Control Bits* may be updated every cycle, they must be generated faster to achieve a substantial data throughput.

Multiplexer Control Bits (MCB). With the setting of the MCBs, one can

1. choose from which parent the child inherits the gene (`SelectParent`).
2. choose between a uniform or gaussian random distribution (`SelectRandom`).
3. scale the new gene value (`SelectScale`).
4. choose between a random or a constant new gene (`SelectConstant`).
5. choose whether the generated gene shall be added to the original or replace it (`SelectReplace`).
6. decide whether a mutation occurs or not (`SelectMutate`).

As stated above, the MCBs may change their values with every new gene, i.e. every new cycle. Thus they cannot be set individually by software instructions since this would be too slow. On the other hand, the MCBs actually define the

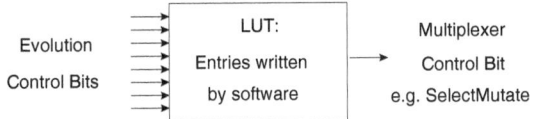

Fig. 6. A LUT sets the MCBs

underlying evolutionary algorithm. Still the description of the algorithm must be based in software to be variable.

This contradiction is solved by introducing an additional set of bits, the *Evolution Control Bits* (ECB). These ECBs generate the MCBs via a look-up table (LUT): The LUT address is given by the ECBs, the LUT entries are written by software, and the LUT outputs equals the MCBs (Fig. 6). There are three kinds of ECBs:

– Instruction dependent bits
  Two *Evolution Control Bits*, Mask1 and Mask2 are set by instructions. For example they can be used to set crossover points: To transfer the necessary data for a two-point crossover with crossings after the 27th and 78th bit for a chromosome with 128 genes, the run-length-coded mask would be (27,51,50).
– Random bits
  Two *Evolution Control Bits*, Rand1 and Rand2, are generated by a random generator each. They are used to decide whether a mutation occurs or not. Their output equals '1' with a probability given by an instruction.
– Gene dependent bits
  Four *Evolution Control Bits*, `Mother1`, `Mother2`, `Father1` and `Father2` are obtained from the parent genes. They can be used to steer the evolution, for example, to exclude a single gene from mutation if Mother1 and Father1 equals '0'. As stated above, our genes are 11 bit long plus two *Evolution Control Bits* for each gene.

Now each of the six *Multiplexer Control Bits* depends on the eight *Evolution Control Bits*. The exact dependency is controlled by the software. We are using a look-up table (LUT) based structure, where the LUT address is given by the *Evolution Control Bits*.

For each *Multiplexer Control Bit* one 8 bit LUT is needed. The ECBs form the input or address to that LUT, while the software specifies the LUT entries. Note that these entries are generally only written once, but can be changed if the necessity arises.

This is best understood with an example: A mutation algorithm shall use two different mutation rates, 2% and 5%. There are genes with the high mutation rate, genes subject to the low mutation rate and genes without any mutation at all. The Mother1 and Father1 bits are used to differentiate between the three cases. A gene shall be mutated with 5% probability if both father1 and mother1 bits are set to '1', but with a probability of 2% if either the father's bit or the mother's bit is set to '1'. No mutation shall occur if both bits are set to '0'.

The LUT implementing this behavior is shown in Tab. 1 (This example uses only a 4-bit, instead of an 8-bit LUT as described in the text. The remaining four *Evolution Control Bits* (Mother2, Father2, Mask1 and Mask2) are treated as don't cares for this setup).

Table 1. Example LUT with four LUT-Address-Bits. An 'X' in the table denotes 'don't care'. To obtain the full LUT, all rows with an 'X' must be duplicated and '0', resp. '1' has to be inserted instead of the 'X'

Father Bit1	Mother Bit1	Rand1 5%	Rand2 2%	Do Mutate	Explanation
0	0	X	X	0	Both gene bits '0': no mutation, independent of the state of the rand bits
1	0	X	0	0	Exactly one gene bit '1':
1	0	X	1	1	mutation occurs if Rand2 (2%)
0	1	X	0	0	equals '1', i.e. with a
0	1	X	1	1	probability of 2%
1	1	0	X	0	Both gene bits '1': mutation
1	1	1	X	1	occurs if Rand1 (5%) equals '1'

3.4   Memory Organization and Instruction Set

Memory Organization. Our system offers two kinds of memory, FPGA internal SRAM (504 kBit) and off-chip DDR-SDRAM (up to 1 GByte). The first is used to communicate between processor core and Coprocessor. It is dual ported and can be randomly accessed independently from both ports. The genetic data, i.e. the genomes, are stored in the off-chip SDRAM, they are far too big to fit into the FPGA.

The IB is 32-bit wide. It holds the instructions and additional data and addresses, which are used for indirect addressing, as described below.

Instruction Set. The IB does not only hold the instructions, but can also contain data and addresses. Some instructions do not transfer all their information directly. To transfer the LUT entries, the microprocessor core first writes them into the data section of the IB. After that, it issues the instruction SetLUTSelectParent(IB address). Now the coprocessor looks in the IB at the transmitted address for the data. This indirect addressing is done for two reasons:

First, some instructions must transfer more than 32 bit of data to the coprocessor. To keep a consistent instruction length of 32 bit, this data has to be transferred indirectly.

Second, using indirect addressing enables the microprocessor core to generate an address table in the IB. Depending on how many memory locations for

Table 2. Coprocessor instruction set

Instruction	Parameters	Representation
SetLUTSelectParent SetLUTSelectRandom SetLUTSelectScale SetLUTSelectConst SetLUTSelectReplace SetLUTSelectMutate	(IB address) (IB address) (IB address) (IB address) (IB address) (IB address)	Instruction identifier bit(0-7) Instruction Buffer Address: bit(8-18)
SetRandomMaximum SetFraction SetConstant SetOneProbability1/2	(RandomMaximum) (Fraction) (Constant) (probability1/2)	Instruction identifier bit(0-7) Value: bit(8-23)
MaskIndirectBin MaskIndirectRunLength	(IB adr, words to read) (IB adr, words to read)	Instruction identifier bit(0-7) Instruction Buffer Address: bit(8-18) words to read: bit (19-31)
MaskDirectRunLength0 MaskDirectRunLength1	(quantity of zeros) (quantity of ones)	Instruction identifier bit(0-7) quantity of ones/zeros: bit (8-23)
ReadParent1 ReadParent2 WriteChild	(IB adr) (IB adr) (IB adr)	Instruction identifier bit(0-7) Instruction Buffer Address bit(8-18)

chromosomes the algorithm needs, SDRAM addresses are written into the IB, followed by the length (i.e. the number of genes) per chromosome. To feed a sequence of genetic data into the pipeline during the evolution, the microprocessor core only transmits the corresponding IB addresses. The coprocessor then reads the SDRAM address and the number of genes to transfer from the IB. This simplifies the address management and reduces address transmitting overhead. The instruction set can be divided into two parts (see Tab. 2):

The first group of instructions must be given once to initialize the coprocessor with valid data. The LUT entries are written with an indirect addressing scheme. The *Global Parameters* must be specified directly. Both may change infrequently during the evolution.

The second group of instructions is given once for each chromosome. A mask must be given to define the crossover point(s). A mask can be defined directly or indirectly, using a run-length-code or by transferring all mask bits in binary.

Finally, the data flow control must be told where to get the parents, where to store the child and how many genes to transfer. This is done indirectly. As mentioned above, an address table in the IB is created. All data flow control instructions refer to that table.

This instruction set provides a wide range of possibilities for the software to implement various evolutionary algorithms. For example, the initial population can be generated just by setting the mutation rate to 100%, crossover operators using 3 or even 4 parent can be implemented with two (or three) 2-parent crossover operators in succession. It is also possible to just add a fraction or a constant value to each gene.

3.5   Performance Analysis

To estimate the performance of our evolution system we must consider not only the pipeline data throughput but the SDRAM transfer-rate and the microprocessor core performance as well.

- The pipeline is able to process four 11-bit-genes per cycle. With the targeted FPGA frequency of 80 MHz this sums up to 320 million genes or 420 MByte per second.
- Each gene must be read twice and written once, leading to an SDRAM transfer rate of approximately 1.26 GByte/s. This is 59% of the maximum transfer rate provided by the DDR-SDRAM.
- A chromosome on the average needs four instructions to be fed into the pipeline[3]. Considering the microprocessor core frequency of 300 MHz, a typical chromosome length of one chromosome per neuron (176 Byte for our ANN) and the pipeline data throughput of 420 MByte/s, the microprocessor core has around 125 cycles to compute and issue the four instructions mentioned above to process a chromosome and to keep pace with the pipeline.

This rough estimation shows that our system is able to run at the calculated speed of the pipeline.

4   Example: Crossover and Simple Mutation

This section presents an example of how to instruct the coprocessor to perform a crossover followed by a simple mutation. The software sets the crossover points directly and demands a mutation rate of 5%.

Tab. 3 shows the complete LUT, which is quite simple: All four parent bits, the MaskBit2 and the RandomBit2 are not used. The SelectParent equals MaskBit1 and the SelectMutate equals RandBit1. All other MCBs are constant, SelectRandom is '0' to choose a uniform distribution, SelectScale is '1' to scale with RandomMaximum, SelectConstant is '0' to use a randomly chosen new gene, SelectReplace is '1', i.e. in case of a mutation we replace the original gene, instead of adding something to it.

Global Parameters. Our mutated genes are completely random with mutation rate 5%, they cover all 11 bit, therefore we set the instructions:

```
SetRandomMaximum (1023);
SetOneProbability1 (51);
```

---

[3] These four instructions are: ReadParent1, ReadParent2, WriteChild and one mask instruction.

Table 3. Example LUT with four LUT-Address-Bits (X : don't care)

Mask Bit1	Mask Bit2	Rand Bit1	Rand Bit2	Mother Bit1/2	Father Bit1/2	Select Parent	Select Rand	Select Scale	Select Const	Select Replace	Select Mutate
0	X	0	X	X	X	0	1	0	1	0	0
1	X	0	X	X	X	1	1	0	1	0	0
0	X	1	X	X	X	0	1	0	1	0	1
1	X	1	X	X	X	1	1	0	1	0	1

Note that the remaining *Global Parameters* are not used in this setup and that the mutation probability is not expressed in per cent, but in per 1024 (51 / 1024 $\simeq$ 0.05).

Specifying the Crossover Mask. The SelectParent LUT is programmed to let the MaskBit1 decide from which parent the child inherits its gene. This MaskBit1 is set with:

```
for (ChromoCount=0; ChromoCount<MaxChromo; ChromoCount++){
    MyRand = Random(128); // between [1..128]
    MaskDirectRunLength0(MyRand);
    MaskDirectRunLength1(128-MyRand);}
```

Each loop cycle will take (MyRand) genes from the first parent and (128-MyRand) genes from the second.

Data Flow Commands. Prior to issuing any data flow commands, the software has to write an address table into the IB. With this table, all chromosomes are addressable by pointers. To cross, for example, the fittest with the second-fittest genome, we need to issue the instructions:

```
ReadParent1(IB_adr = Genome[Best]);
ReadParent2(IB_adr = Genome[Second]);
WriteChild (IB_adr = Genome[Unused1]);
```

Note that the pointers Best, Second, etc. in general change their value with every new generation.

A complete evolution starts with the creation of an initial population (by setting the mutation rate to 100%). This population is evaluated and ranked according to the fitness. Based on this ranking, the software decides which genomes are crossed and issues the appropriate data flow and mask instructions, as shown in the examples above. After all genomes have been crossed and mutated, the evaluation starts again. This circle is continued until a satisfactory fitness is reached.

5   Conclusion and Outlook

We have introduced a coprocessor architecture to speed up evolutionary algorithms designed for the FPGA Virtex-II Pro. The coprocessor is able to perform common genetic operators like crossover and mutation with a data throughput of up to 420 MByte/s due to pipelining and parallelism. It still has the flexibility of software. Therefore, wide range of evolutionary algorithms can be designed and maintained in software, but are processed at high speed in dedicated hardware.

In the future we plan to implement additional operators, e.g. the possibility to add two genes together during the crossover. Also, we are going to adapt the coprocessor to another reconfigurable hardware: The field programmable transistor array (FPTA) developed by our group (see [2]).

We also want to connect several of the proposed systems using the high speed serial links offered by the FPGA. This distributed, autonomous hardware evolution system will give us the resources to tackle large scale optimization problems.

References

1. F. Schürmann, S. Hohmann, J. Schemmel, and K. Meier. Towards an Artificial Neural Network Framework. In Adrian Stoica et.al., editor, *Proceedings of the 2002 NASA/DoD Conference an Evolvable Hardware*, pages 266–273, July 2002.
2. J. Langeheine, K. Meier, and J. Schemmel. Intrinsic Evolution of Quasi DC Solutions for Transistor Level Analog Electronic Circuits Using a CMOS FPTA chip. In *Proceedings of the 2002 NASA/DoD Conference an Evolvable Hardware*.
3. Adrian Stoica et. al. Evolving Circuits in Seconds: Experiments with a Stand-Alone Board-Level Evolvable System. In Adrian Stoica et.al., editor, *Proceedings of the 2002 NASA/DoD Conference an Evolvable Hardware*, pages 67–74, July 2002.
4. Barry Shackleford et. al. A high-performance, pipelined, FPGA-based genetic algorithm machine. *Genetic Programming and Evolvable Machines*, 1(2):33–60, March 2001.
5. J. Schemmel, F. Schürmann, S. Hohmann, and K. Meier. An integrated mixed-mode neural network architecture for megasynapse ANNs. In *Proceedings of the 2002 International Joint Conference on Neural Networks IJCNN'02*, page 2704. IEEE, May 2002.
6. Xilinx, Inc., www.xilinx.com. *Virtex-II Pro Platform FPGA Handbook*, 2002.
7. Micron Technology, Inc., www.micron.com. *Small-Outline DDR SDRAM Module*, Jan 2002.

Automatic Evolution of Signal Separators
Using Reconfigurable Hardware

Ricardo S. Zebulum, Adrian Stoica, Didier Keymeulen, M.I. Ferguson,
Vu Duong, Xin Guo*, and Vatche Vorperian

Jet Propulsion Laboratory
California Institute of Technology
Pasadena, CA 91109
ricardo@brain.jpl.nasa.gov
* Chromatech, Alameda CA 94501

Abstract. In this paper we describe the hardware evolution of analog circuits performing signal separation tasks using JPL's Stand-Alone Board-Level Evolvable System (SABLES). SABLES integrates a Field Programmable Transistor Array chip (FPTA-2) and a Digital Signal Processor (DSP) implementing the Evolutionary Platform (EP). The FPTA-2 is a second generation reconfigurable mixed signal array chip whose cells can be programmed at the transistor level. Its chip architecture consists of an 8x8 matrix of reconfigurable cells. The FPTA-2 is reconfigured by evolution to achieve circuits that can extract a target signal that is combined with an undesired component or to perform the separation of a combination of two signals. The paper considers also an adaptive filter where the fitness function depends on the input signal. The results demonstrate that SABLES is not only able to perform signal separation and extraction, but it is also flexible enough to adapt to different input signals without human intervention, such as in the case of self-tuning and adaptive filters.

1 Introduction

This work describes the evolution of circuits that perform signal separation using the most recent version of the Field Programmable Transistor Array chip, the FPTA-2. Source or signal separation is a classic signal processing problem [1]: given N physically distinct measurements which represent a priori unknown linear combinations of N independent signal sources, the objective is to auto-adaptively extract N equivalent statistically independent signals, e.g., arising from independent physical sources. One approach to solve this problem is the H-J adaptive network, which is based on the modified Hebbian learning rule [2]. Cohen and Andreou [1] proposed a hardware implementation for a Neural Network implementing the H-J adaptive algorithm and tested it for the case of separating two sine waves (782Hz and 1kHz) applied at the network input, after being mixed. The interest in the signal separation problem is motivated by applications in communication systems, such as adaptive noise reduction and adaptive echo cancellation.

Particularly, our focus is to demonstrate the synthesis of circuits that can perform signal extraction and separation using the SABLES platform. This class of applications encompasses experiments in which the reconfigurable hardware evolves to a

A.M. Tyrrell, P.C. Haddow, and J. Torresen (Eds.): ICES 2003, LNCS 2606, pp. 286–295, 2003.
© Springer-Verlag Berlin Heidelberg 2003

new configuration when the input signal characteristic changes, leading to potential applications such as adaptive noise elimination. Additionally, this paper also describes an experiment accomplishing the separation of two source signals from a mixture of these signals given at the circuit input.

SABLES was used in the evolutionary experiments. This system solution provides autonomous, fast (about 1,000 circuit evaluations per second), on-chip circuit reconfiguration. Its main components are a JPL Field Programmable Transistor Array chip as transistor-level reconfigurable hardware, and a TI DSP implementing the evolutionary algorithm as the controller for reconfiguration.

This article is organized as follows. Section 2 describes the SABLES and the FPTA-2 chip. Section 3 describes the experiments, including the evolution of tunable filters excited by an "unknown" combination of source signals; and the evolution of a circuits performing signal separation. Section 4 summarizes the conclusions and future applications.

2 SABLES

SABLES integrates an FPTA and a DSP implementing the Evolutionary Platform (EP) as shown in Figure 1. The system is stand-alone and is connected to the PC only for the purpose of receiving specifications and communicating back the results of evolution for analysis [3].

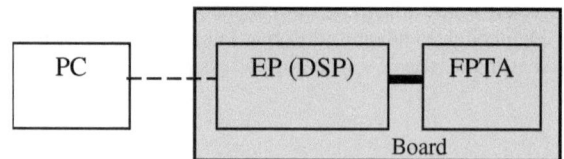

Fig. 1. Block diagram of a simple stand-alone evolvable system.

The FPTA is an implementation of an evolution-oriented reconfigurable architecture (EORA) [4]. The lack of evolution-oriented devices, in particular for analog, has been an important stumbling block for researchers attempting evolution in intrinsic mode (with evaluation directly in hardware). Extrinsic evolution (using simulated models) is slow and scales badly when performed accurately e.g. in SPICE, and less accurate models may lead to solutions that behave differently in hardware than in software simulations. The FPTA has transistor level reconfigurability, supports any arrangement of programming bits without danger of damage to the chip (as is the case with some commercial devices). Three generations of FPTA chips have been built and used in evolutionary experiments. The latest chip, FPTA-2, consists of an 8x8 array of reconfigurable cells. Each cell has a transistor array as well as a set of other programmable resources, including programmable resistors and static capacitors. Figure 2 provides a broad view of the chip architecture together with a detailed view of the reconfigurable transistor array cell. The reconfigurable circuitry consists of 14 transistors connected through 44 switches and is able to implement different building blocks for analog processing, such as two- and three-stage OpAmps, logarithmic photo detectors, or Gaussian computational circuits. It includes three capacitors, Cm1, Cm2 and Cc, of 100fF, 100fF and 5pF respectively. Details of the FPTA-2 can be found in [5]. Other implementations of reconfigurable analog devices can be found elsewhere [6-10].

The evolutionary algorithm was implemented in a DSP that directly controlled the FPTA-2, together forming a board-level evolvable system with fast internal communication ensured by a 32-bit bus operating at 7.5MHz. Details of the EP were presented in [11]. Over four orders of magnitude speed-up of evolution was obtained on the FPTA-2 chip compared to SPICE simulations on a Pentium processor (this performance figure was obtained for a circuit with approximately 100 transistors; the speed-up advantage increases with the size of the circuit). The evaluation time depends on the tests performed on the circuit. Many of the evaluation tests performed required less than two milliseconds per individual, which for example on a population of 100 individuals running for 200 generations required only 20 seconds. The bottleneck is now related to the complexity of the circuit and its intrinsic response time. SABLES fits in a box 8" x 8" x 3".

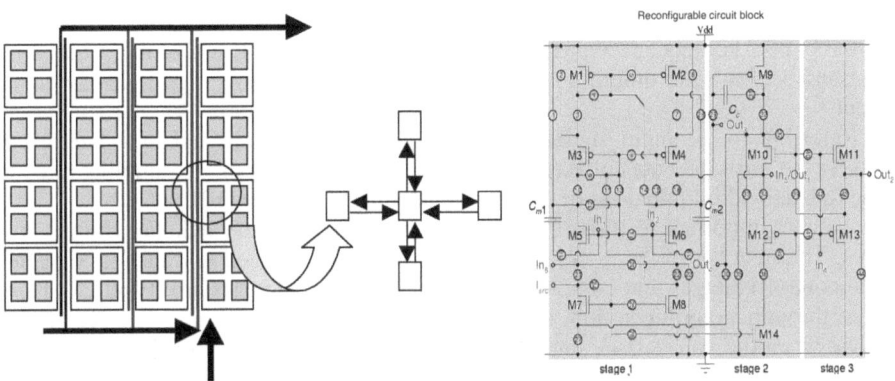

Fig. 2. FPTA 2 architecture in the left and schematic of cell transistor array (*re-configurable circuitry*) in the right. The cel contains additional capacitors and programmable resistors (not shown).

3   Experiments

The results for three different classes of experiments are presented in this section. The first two experiments demonstrate the evolution of circuits for signal extraction, while the last experiment shows the evolution of a circuit for signal separation. In the first case the evolving circuit is excited with combination of signal sources of known frequencies f_1 and f_2, e.g., the known input frequencies are used by the fitness evaluation function. The objective is to amplify the signal at frequency f_1 while attenuating the signal at frequency f_2. In the second case the circuit is excited with a combination of signal sources of {\it unknown frequencies, e.g., knowledge of the input sine waves frequencies is not used by the fitness evaluation function. The objective of the experiment is to evolve a circuit that amplify the strongest component of the input signal and attenuate the weakest one, therefore improving a hypothetical signal/noise ratio. It is shown that when the frequency of the strong and/or weak component in the input signal changes, the circuit is reconfigured by evolution to "adapt" to the new input profile. Finally, in the third experiment evolution performs signal separation: the evolving circuit is excited with two linear combinations of pure sine waves (source signals) of known frequencies, and the circuit outputs are the original source signals.

Based on prior experimentation, the GA parameters selected for these experiments were: 70% mutation rate; 20% crossover rate; replacement factor of 20%; population of 400; and 100 to 200 generations. A binary representation was used, where each bit determines the state (opened, closed) of a switch. Each execution took about 5 minutes in the SABLES system. More than 20 different GA executions were performed. In order to compute the fitness function, the FFT of the output signal(s) from the FPTA-2 was calculated. The fitness was a measure of the error of the FPTA-2 outputs to the target values in each experiment, as shown below.

$$Fitness = \sum_{i=1}^{N} | O_f(i) - T_f(i) | \qquad (1)$$

where N is the number of samples used in the FFT (usually 64), $O_f(i)$ is the magnitude of the ith FFT component of the FPTA-2 output, and $T_f(i)$ is the target magnitude of the ith FFT component. Other fitness measures such as the sum of the squared deviations between the output and the target were tried, without significant improvement.

Referring to the input signals frequencies, they were restricted to values below 50kHz, because the data converters at SABLES operate at 100kHz sampling rate and due to bandwidth limitations of reconfigurable analog devices [10].

Another important feature of these experiments refers to the switches control voltages. The switches of the re-configurable chip are implemented as transmission gates. The control voltages that completely open or close the switches are 0 and 2V. However, through experimentation, it has been observed that the results significantly improve when the values 0.4V and 1.6V are used to control the switches, meaning that they are now partly opened and closed (partly closed if the higher and lower control voltages are respectively applied to the NMOS and PMOS transistors of the switch, and partly opened in the other case).

Finally, another important issue is the search space size. If we allow a completely unconstrained evolution, we will end up with a very large search space size. One approach to reduce the search space size is to have the FPTA-2 cells constrained to a particular topology, so that only the interconnections among the cells and the feedback resistance values [5] are evolved. Through experimentation, it has been verified that the constrained approach delivered better results. In the following experiments the cell topologies were fixed to the one of inverting amplifiers.

3.1   Combination of Known Signal Sources

To evolve the circuit we introduce an input signal consisting of a linear combination of 10kHz and 25kHz sine waves. The objective is to eliminate the 25kHz tone and amplify the 10kHz tone. Four cells of the FPTA-2 were used in this experiment. Figure 3 provides a high level schematic of the input/output interconnections in this experiment. Figure 4 depicts the FFT of the input signal and of the evolved circuit output. Figure 5 shows the evolved circuit response in the time domain, when excited by 10kHz and 25kHz tones respectively.

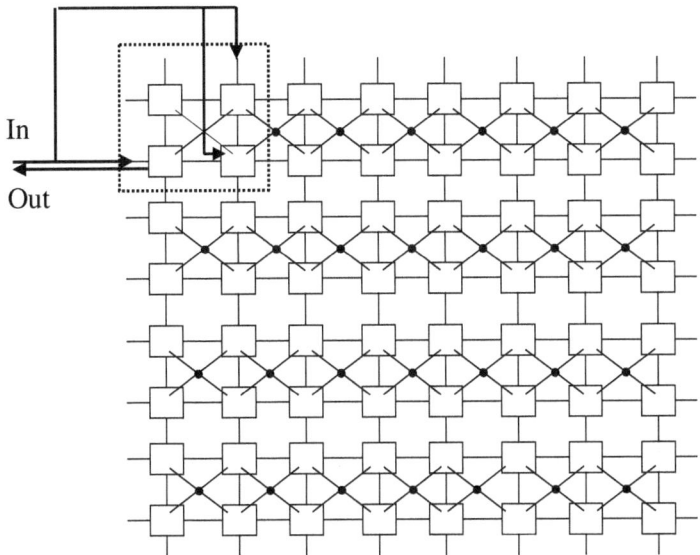

Fig. 3. Input (In)/Output(Out) connection in the signal extraction experiment. Only the four cells at the top-left of the array were used.

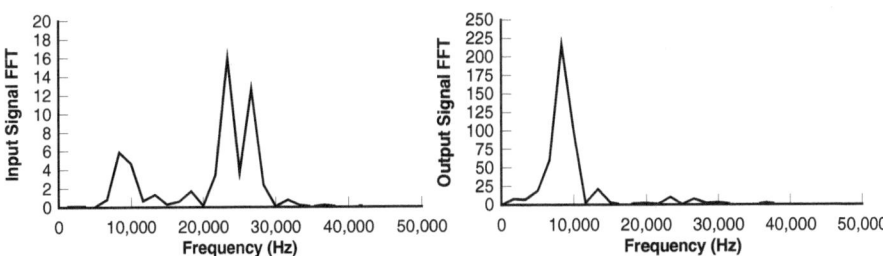

Fig. 4. Evolved circuit response in the frequency domain: FFT of the input (left); and output (right) signals.

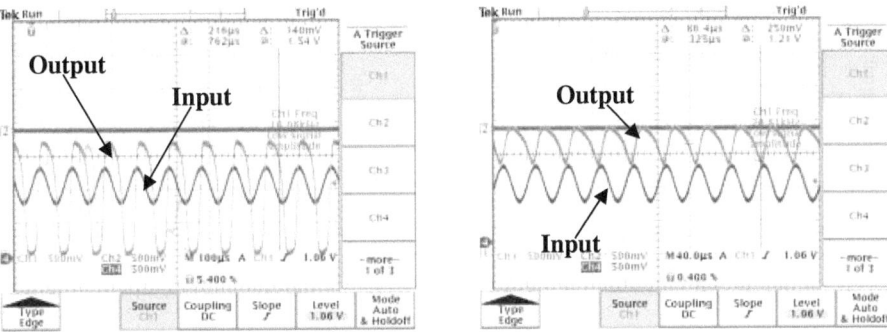

Fig. 5. Evolved circuit response in the time domain: when excited by a 10kHz input signal (left) and by a 25kHz input signal (right).

The evolved circuit displayed a 15dB gain at 10kHz and a 2dB attenuation at 25kHz.

Due to space limitation it is not possible to show in this paper the evolved circuits' schematics. The reader can refer to [12] to get the netlists of the evolved circuits.

3.2 Unknown Combination of Signal Sources

The linear combination of two source signals (10kHz and 20kHz) was applied to the FPTA-2 chip, as previously shown in Figure 3. Nevertheless, the source signals are unknown to the evolutionary system, in the sense that no information about the input signal is used by the fitness evaluation function. The fitness criteria employed here is to amplify the strongest tone at the input signal and attenuate the weakest one, which we *respectively* assume to be signal and noise. Four cells of the chip were used in the experiment. Figure 6 summarizes the performance of the evolved circuit. Initially, the 10kHz tone has about twice the amplitude of the 20kHz tone. The evolved circuit (C1) output increases the *"signal/noise"* ratio from 4.1dB to 16.9dB. In this case, the 10kHz tone is attenuated by 2.86dB, while the 20kHz tone is attenuated by 15.8dB. If we invert the source signal ratios, it is observed that a new circuit configuration, C2, is evolved: the *"signal/noise"* ratio increases from 10.8dB to 18.5dB. In this case, the 10kHz tone is attenuated by 12.5dB, while the 20kHz tone is attenuated by 4.8dB. It can be observed that the performance is better in the first case. Table 1 compares the results in terms of the achieved and target FFT values O_f and T_f.

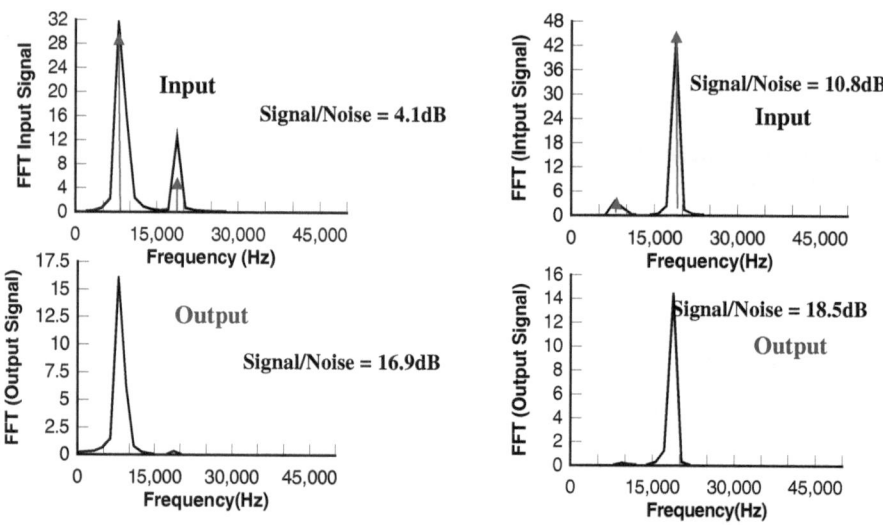

Fig. 6. In the left, FFT of the input (top) and output (bottom) of circuit C1 when the 10kHz tone has twice the amplitude of the 20kHz tone. In the right, the same information is shown for circuit C2, when the 20kHz tone has a higher amplitude.

Table 1. Comparison between the target and the attained FFT for the two circuits presented in this section: circuit 1 (passing 10kHz) and circuit2 (passing 20kHz).

	Wanted tone		Unwanted tone	
	Output(O_f)	Target (T_f)	Output (O_f)	Target (T_f)
Circuit 1	16.1	> 20	0.33	0
Circuit 2	14.5	> 20	0.21	0

3.3 Signal Separation Experiments

The goal of this experiment is to design a circuit able to separate two mixed signals, $E_1(t)$ and $E_2(t)$, obtained by the linear combinations of pure sine waves, $e_1(t)$ and $e_2(t)$, of known frequencies. The circuit outputs are the original pure sine waves. As in the previous case studies, we chose for the frequency of the pure sine wave $e_1(t)$, $f_1 =$ 10kHz and for the pure sine wave $e_2(t)$, $f_2 =$ 20kHz. These signals were linearly combined by a mixing matrix to produce the chip inputs $E_1(t)$ and $E_2(t)$:

$$\begin{bmatrix} E_1(t) \\ E_2(t) \end{bmatrix} = \begin{bmatrix} 0.25 & 0.5 \\ 0.5 & 0.25 \end{bmatrix} \begin{bmatrix} \sin(2\pi 10{,}000t) \\ \sin(2\pi 20{,}000t) \end{bmatrix}$$

The set of experiments was performed using 10 cells of the FPTA-2. Following the same procedure of the previous experiments, the cells were constrained to be inverting amplifiers reducing thereby the search space.

Figure 7 depicts the inputs and outputs of the best circuit achieved in this set of executions.

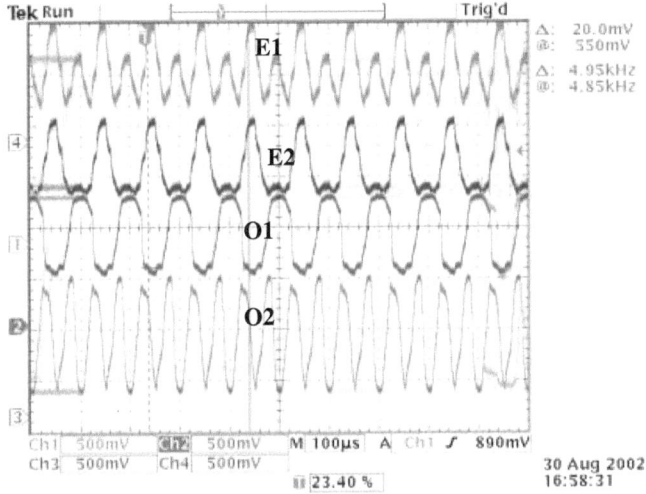

Fig. 7. Result of the signal separation experiment. At the top the inputs E1 and E2 are shown. At the bottom the outputs O1 (10kHz)) and O2 (20kHz) are shown.

Table 2 summarizes the evolved circuit performance in terms of the FFT and of the values measured in the frequency analyzer. The target FFT values used in the fitness function are also included in the table.

Table 2. Evolved circuits in the signal separation experiment. Amplitude of 10kHz and 20kHz tones as measured by the spectrum analyzer and calculated by an FFT algorithm used during evolution.

	10kHz Tone		20kHz Tone	
	FFT	Spectrum Analyzer	FFT	Spectrum Analyzer
Input E_1	33.6	-13dB	8.86	-20dB
Input E_2	7.6	-20dB	41.1	-15dB
Output O_1	36.9 *(Target:T_f>20)*	-13dB	1.07 *(Target:T_f= 0)*	-35.3dB
Output O_2	0.6 *(Target:T_f= 0)*	-30dB	84.5 *(Target:T_f>20)*	-13dB

From Table 2, it can be observed that the output O_1 attenuates the component e_2 (20kHz) by -15dB (from −20dB to −35.3dB), while keeping e_1 (10kHz) at the same level. On the other hand, the output O_2 attenuates the component e_1 by -10dB (from −20dB to −30dB), and amplifies e_2 by 2dB (from −15dB to −13dB). Finally Figure 8 plots the fitness of the best individual along the generations averaged over 3 GA runs. It can be observed from this graph that the fitness does not go to 0 for the best individuals. This is an expected systematic constant due to the way the fitness function was written – the minimum possible fitness in this experiment is 80.

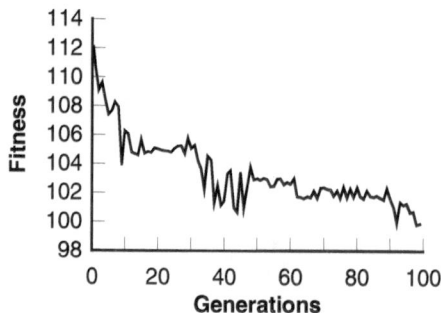

Fig. 8. Fitness against generations for the signal separation experiment.

This experiment is a first approach to tackle the independent component analyzer problem, which consists of recovering the original source signals from signals acquired from a set of sensors that pick up different linear combinations of source signals.

4 Discussion and Conclusions

Most of the experimental results could be conventionally implemented using low or high pass filters with zeros/poles in the range of 10-20kHz and a control or adaptation block. Supposing R=10K, a value that is in the range of the polysilicon on-chip resistances [4], we would then be talking about 0.1μ capacitors. In order to implement a capacitor of this size using the chip fabrication process (about 1fF/μm for metallic capacitors), a silicon area of 100mm^2 would be used, against only about 2mm^2 used by the evolved circuits. Several factors explain this discrepancy: the resistive effect of the switches contributes to increase the RC constant; to a small extent the exploration of parasitic capacitors of the MOS transistor; and the capacitances of the probing instruments (10p for the oscilloscopes). By cascading the filters evolved using a small number of cells, the FPTA-2 chip will be able to realize steeper frequency responses (possibly with roll-off superior to 60dB/dec).

It has also been demonstrated that the reconfigurable hardware can evolve to "adapt" to changes in the input signal. The task of separating a mixed sine wave input (with frequencies unknown a priori) has been approached using a 2x2 neuromorphic analog network implemented in VLSI using a total of 76 transistors [1]. The network training time was 80ms and they were able to separate the two sine waves by 15dB (worst) to 35dB (best). The evolved FPTA-2 hardware used 140 fixed transistors, taking a training time of the order of seconds and separating the sine waves by 17dB (worst) to 22dB (best). In the case of Evolvable Hardware, we point out that the reconfigurable chip was not specifically designed for the signal separation task, as it happens in the neuromorphic approach.

The results demonstrate that our evolutionary platform can perform signal separation and extraction for input signals consisting of a combination of sine waves, being also capable to adapt to different input signals without human intervention. Future applications will consist of the separation of more complex signals, such as mixed speech signals.

Among the lessons learned from the experiments reported in this paper, we emphasize the advantages of using partly opened/closed switches; and constraining the reconfigurable circuitry [4] to a particular topology. Future work will focus in extending these experiments to evolve filters with more stringent requirements; tackle more realistic case of adaptive noise cancellation; and tackle the problem of separating mixed speech signals.

Acknowledgements

The research described in this paper was performed at the Center for Integrated Space Microsystems, Jet Propulsion Laboratory, California Institute of Technology and was sponsored by the Defense Advanced Research Projects Agency (DARPA) and the National Aeronautics and Space Administration (NASA). We also thank the anonymous reviewers for their helpful comments.

References

1. Cohen, M. H., and Andreou, A. G., "Analog CMOS Integration and Experimentation with an Autoadaptive Independent Component Analyzer", IEEE trans. on Circuits and Systems, V. 42, N. 2, Feb., 1995.
2. J. Herault and C. Jutten, "Space or Time Adaptive Signal Processing by Neural Network Models", in Neural Networks for Computing, J. S. Denker, Ed. Snowbird, UT: AIP Conf. Proc., 1986.

3. Stoica, A., Zebulum, R.S., Ferguson, M.I., Keymeulen, D., Duong, V. "Evolving Circuits in Seconds: Experiments with a Stand-Alone Board Level Evolvable System". 2002 NASA/DoD Conference on Evolvable Hardware, Alexandria Virginia, USA, July 15-18, 2002, IEEE Computer Society.
4. A. Stoica, R. Zebulum, D. Keymeulen, R. Tawel, T. Daud, and A. Thakoor, Reconfigurable VLSI Architectures for Evolvable Hardware: from Experimental Field Programmable Transistor Arrays to Evolution-Oriented Chips. In IEEE Transactions on VLSI Systems, Special Issue on Reconfigurable and Adaptive VLSI Systems, vol. 9, No. 1,February 2001. (pp.227-232).
5. Stoica, R. Zebulum, and D. Keymeulen, "Progress and Challenges in Building Evolvable Devices", Third NASA/DoD Workshop on Evolvable Hardware, Long Beach, July, 12-14, 2001, (pp.33-35), IEEE Computer Society.
6. Langeheine, J., Meier, K., Schemmel, J. "Intrinsic Evolution of Quasi DC Solutions for Transistor Level Analog Electronic Circuit Using a CMOS FPTA Chip". In 2002 NASA/DoD Conference on Evolvable Hardware, IEEE Computer Society, pp. 75-84,2002.
7. Layzell, P.,"A New Research Tool for Intrinsic Hardware Evolution", In Proceedings of the Second International Conference on Evolvable Systems: From Biology to Hardware (ICES98), M.Sipper, D.Mange e A. Pérez-Uribe (editors), Springer-Verla vol. 1478: 47-56, 1998.
8. Murakawa, M., Yoshizawa, S., Adachi, T., Suzuki, S., Takasuka, K., Iwata, M., Higuchi, T., "Analogue EHW Chip for Intermediate Frequency Filters", In Proceedings of the Second International Conference on Evolvable Systems: From Biology to Hardware (ICES98), M.Sipper, D. Mange and A. Pérez-Uribe (editors), Springer-Verlag, vol. 1478: 134-143, 1998.
9. Santini, C., Zebulum, R. S., et. Al, "PAMA – Programmable Analog Multiplexer Array", In the Third NASA/DoD workshop on Evolvable Hardware, IEEE Computer Press, pp. 36-43, 2001.
10. V. Gaudet and G. Gulak, "Implementation Issues for High-Bandwidth Field-Programmable Analog Arrays," Journal of Circuits, Systems, and Computers Special Issue on Analog and Digital Arrays, World Scientific Publishing, Vol. 8, No. 5-6, pages 541-558, 1998.
11. Ferguson, M.I., Stoica A., Zebulum R., Keymeulen D. and Duong, V. "An Evolvable Hardware Platform based on DSP and FPTA", Proceedings of the Genetic and Evolutionary Computation Conference, July 9-13, 2002, pp145-152, New York, New York.
12. http://cism.jpl.nasa.gov/ehw/public/ices2003.

Distributed Control
in Self-reconfigurable Robots

Henrik Hautop Lund, Rasmus Lock Larsen, and Esben Hallundbæk Østergaard

The Maersk McKinney Moller Institute for Production Technology
University of Southern Denmark
Campusvej 55, 5230 Odense M., Denmark
{hhl,rasmus,esben}@mip.sdu.dk

Abstract. This paper explores self-assembling artefacts using distributed control. General design methods for distributed self-assembling artefacts are described and a simulator is developed to simulate an existing robotic unit (M-TRAN) in a self-assembling context. Two multi-agent behaviors are implemented in the simulation to test the performance of the distributed system. The behaviors are compared favorable against existing motion planning methods (e.g. cluster flow) for these self-assembling modular robots.

1 Introduction

Self-assembling artefacts are small robotic units that are able to cooperate, perhaps join together to form a more complex structure and change form to reach a complex task none of the individual units are able to reach without help from others. Such systems may be able to self-assemble and self-repair. Here, we implement a self-assembling system based on the physical realization of M-TRAN modules [MYT00]. However, contrary to the work by Yoshida et al. [YMK+01], our system will be totally distributed and the units will be able to cooperate to reach common goals. The system will be able to change it's form.

1.1 M-TRAN

Many researchers have tried to develop self-reconfigurable robots (CEBOT, Robotic Molecules, FRACTA, CONRO, etc.[1]) but here we will concentrate on the Modular Transformer (M-TRAN) module that was developed at AIST in collaboration with Prof. S. Murata. The M-TRAN module consists of two semi-cylindrical parts connected by a link. Two RC servo motors embedded in the link enable each part to rotate 180°. The rotational axis of the two parts are parallel giving the module 2 DOF. The module has 6 connecting surfaces, 3 on each part. The inactive part uses permanent magnets, the active part has the magnets fixed at a distance from the face using non-linear springs. The active

---

[1] Links to the research done on these modules are supplied at the Hydra web-page (http://hydra.mip.sdu.dk)

A.M. Tyrrell, P.C. Haddow, and J. Torresen (Eds.): ICES 2003, LNCS 2606, pp. 296–307, 2003.

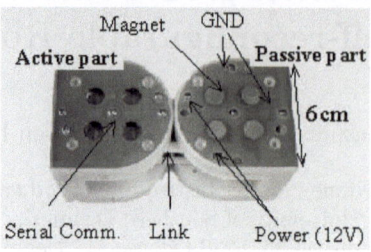

Fig. 1. The M-TRAN module.

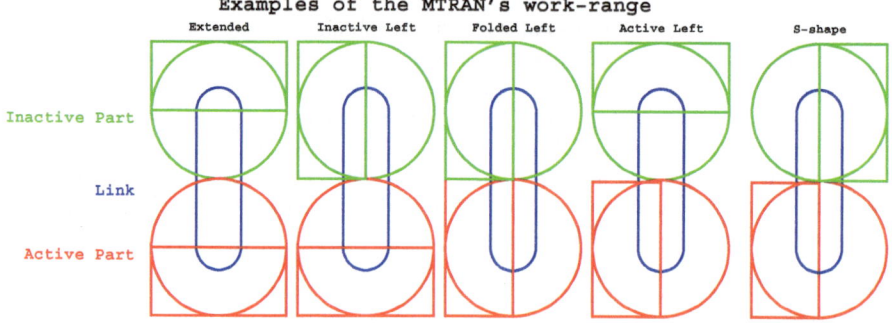

Fig. 2. Some examples of the work-range of the M-TRAN module (top view).

part can disconnect any part connected to it by heating a shape memory alloy spring, canceling the magnetic attraction between the permanent magnets [YMK$^+$01].

The M-TRAN modules are controlled from a central computer sending specific commands to each module. Although the modules do have on-board processing power, they are entirely controlled from the external computer. The second prototype of the M-TRAN module will have the inter-module-communication-capabilities needed for Multi-Agent behavior.

The M-TRAN modules have a very nice property when used in a lattice-type of structure. The way they connect ensures the lattice structure to follow a *checkerboard* pattern, and they are well suited for thread-like structures, although the mobility of the module is very limited.

1.2 The Work-Range

The module has two parts that are connected by a link. In the link, two actuators are embedded allowing each part to be moved in a 180° range with respect to the link (see fig. 2).

The M-TRAN module has two different operating modes with respect to the *base* of the module (the *base* is simply the part of the module that is kept fixed during a movement). 1) *Roll:* In this mode the actuators move the module in the vertical plane, the *base* is one of the *end-faces*. 2) *Pivot:* In this mode the actuators move the module in the horizontal plane, the *base* is one of the *side-faces*. With these two modes, the modules are able to move around on the ground (in one plane).

To summarize the properties of the M-TRAN Module:

- *Actuators*: 2, ±90°. Parallel rotational axis.
- *Connectors*: 6, 3 × permanent magnets, 3 × permanent magnets mounted on a plate with springs and one SMA spring.
- *Communication:* 6 bidirectional serial ports.

The M-TRAN modules use wired communication in both its prototypes. The first prototype only has one communication channel allowing the master computer to broadcast the commands to the modules using hard-coded IDs. The second prototype will have bidirectional serial communication allowing the modules to communicate with each other.

2   Simulation

For our simulations, we used Vortex, a commercial real-time physics engine for creating rigid bodies simulation and games, developed by CMLabs. It handles collision detection between the models as well as dynamic control (gravity, force-feedback, collisions etc.).

The physical model of the M-TRAN module is defined in the XML-language supported by Vortex, this model is then imported into the C/C++ program by calling the generic methods in the API to Vortex.

Modeling the Actuators. The actuators in the M-TRAN module are embedded in the link between the active and inactive parts. They are able to rotate each part 180°. In Vortex the actuators are implemented as *hinge*-joints with limits at ±90°, and positional encoders, enabling the programmer to ask the actuators at which angle they are positioned.

Modeling the Attachment Mechanism. The docking mechanism used in the M-TRAN modules is fairly simple. The central component in the attachment is the magnets. The magnets have enough attracting power to keep the connection between two modules while lifting one of the modules by the actuators on the base module. When the connection is to be broken, a small Shape-Memory-Alloy (SMA) spring is heated, the SMA's force combined with two non-linear springs is enough to cancel the magnetic force attracting the permanent magnets to each other.

First of all, for the attachment mechanism, a callback-function uses the collision library to detect collision between the active part and the inactive part. When this occurs the distance and angle between the two parts are compared, and if everything is ok, the two modules are locked together. However, this method must be extended, because before the collision library will try to establish a connection, the two faces must collide. A magnetic connection mechanism

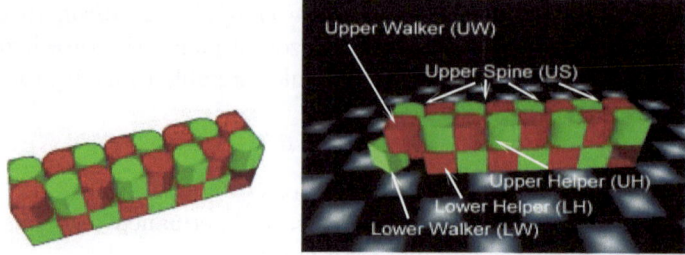

Fig. 3. Left: A 16-module structure of M-TRAN Modules. **Right:** Signatures and roles of the modules.

is able to attract the surfaces to each other even if they are not touching each other. Therefore, we introduce a periodical attempt at connection (the parts doesn't have to touch to be connected) and the possibility of using *tempera-ture* in the simulation, to make all bodies vibrate in random directions using *gaussian*-noise. Finally, we must model the magnetic force around the magnetic faces. The force follows eq. 1.

$$|f| = \text{maxforce} \cdot \frac{D^2}{D^2 + 9 \cdot r^2} \tag{1}$$

where f is the force acting on both surfaces, D is the range of the magnetic force and r is the distance between the magnets. This formula ensures that the magnetic force varies from 100% to 10% of the maxforce within the range of the magnets. Only one force is modelled for each connecting face (3 active faces per module).

Designing the Behaviors. Behaviors that are to be implemented in the Simulator are designed using Petri Nets [Pet62] and Class-diagrams from *Unified Modelling Language* (UML).

The structure used in this work will be constructed by using composite modules consisting of 4 M-TRAN modules connected 2 and 2 to form a firm structure with the ability to connect in all 6 directions (see left side of fig. 3). The composite unit contains two parallel "lower" modules that connects two parallel "upper" modules (perpendicular to the lower modules) to each other.

3   Walking Behavior

The walking behavior is the behavior that enables the cluster of modules to move in one direction. The strategy is to move modules from the rear to the front of the structure. The goal of the behavior is to start a migration of modules from the rear of the structure to the front. The moving modules should ask each module it visits whether it has reached its destination or not.

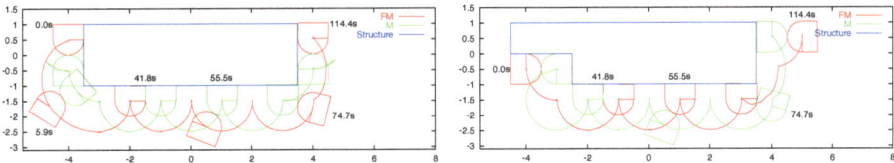

Fig. 4. Trace of the movement path for the walkers. **Left:** Lower Walker. **Right:** Upper Walker.

Related Work. One way of making the walking behavior is the "Cluster Flow" described in [YMK+01]. This behavior makes the modules move to the front one at a time. The upper modules use the other upper-modules to help them move along the side of the structure and the lower modules simply walks along the lower side of the structure. We also implemented this in our Vortex simulation, and a video of this behavior (`Walking.Alpha.mpg`) can be seen on the Hydra web-page (http://hydra.mip.sdu.dk). However, cluster flow is extremely ineffi-cient. A substantial amount of time is used to get the upper-modules to help the upper-walker because of the way the structure is organized.

4-unit Walker. Another possibility, that we explore here, is to let the upper and lower modules work together to reach the front of the structure. The idea here is that since the lower-module simply walks along the side of the structure – and the functional unit is a 4-module unit – the lower module should connect with the upper-module and together they should form a "micro"-walker and then walk to the front together.

3.1   Petri Net

The Walking Behavior is designed using Petri nets. The modules are assigned *roles* as the behavior is executed. The roles are illustrated in fig. 3. A module will change its role during the execution of a behavior. A LW will end up as a LH when the walking is complete. An Upper Helper UH will become an UW and finally an US (if it ends up in the "spine" row).

The path the LW and UW follows are shown in fig. 4.

As an example, the Petrin net for starting a walking movement is shown in fig. 5.

4   Morphing Behavior

The morphing behavior uses the hormone propagation to determine the center of the structure. This center is then used as the "stopping-point" for the walking behavior.

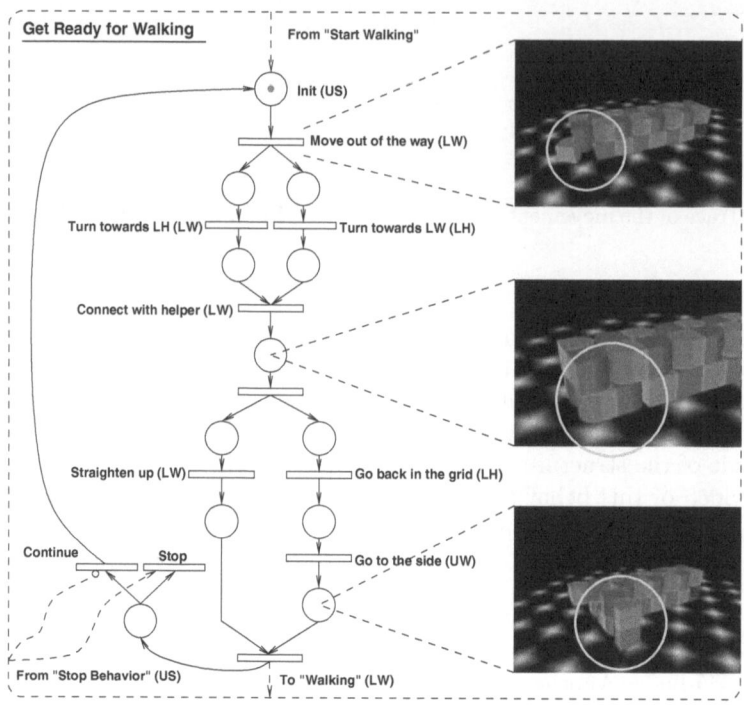

Fig. 5. An example petri net: The *Get Ready for Walking* behavior. The roles and signatures of the modules are shown in fig. 3, right.

4.1   Hormone Propagation

A program is propagated to the end of the spine. When a module determines that it is an end-module (no connections on the female end) a hormone is sent through the spine from the *active* to the *inactive* part on the next module, each module stores the value of the hormone in the memory, increments the value and sends the hormone to the next module (on the inactive connector). This procedure is executed until the other end of the spine is reached. Here another hormone is send back doing exactly the same, but storing the value of the hormone in another memory location (see fig. 6).

The end is determined to be the right-most *spine-module*. This module initializes the A hormone (see fig. 6) to 0, increments the value, and sends it to the module to the left. This procedure is repeated until the left-most module receives the hormone. This module has no other modules to send the incremented value to, it is the *end-module*. The *end-module* then sends a different hormone, B, back through the structure. When the *front-spine* receives the B hormone it knows that all modules (in the spine) have valid values for the A and B hormone, constituting two gradients through the spine. The mid modules can be determined by $|A - B| \leq 1$.

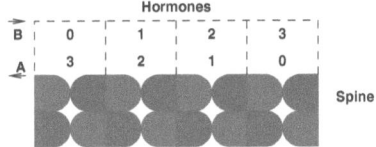

Fig. 6. Using hormones to determine the mid-point of the structure.

When this cycle is ended, all modules have two gradients indicating how far they are positioned from the structure's head and tale. If the spine-module determines it is in the center of the structure, it signals the lower modules that they are the "end-of-walking"-modules.

After the hormone propagation is ended, the structure starts the walking behavior again, here the *Stop-Walking* will indicate when the module has reached the new position. The "Return to Lattice" behavior is altered to reorganize at the stopping point.

5   Deadlocks and Collision Detection

The use of petri nets makes it possible to determine where dead-lock can occur. If a module fails while having a token that is supposed to be returned to another module (if there exists a transition in the net with arcs to other modules) the other module will never gain control of the behavior and the execution of the behavior will block. The *dynamic* design issue indicates that no module is allowed to go into a waiting loop for a substantial amount of time[2]. This constraint is applied to the system to be able to detect failing modules. If a module awaits a token from another module, a time-out mechanism can be build-in to determine when a module fails to accomplish a parallel task.

Researchers often run into the problem in a simulation with distributed control, the collision detection may be very difficult (or viewed as impossible). The simulator described in [YMK+01] uses a global scheme planner to traverse all the graphs representing possible movements of the individual modules. A solution is discarded if it makes two modules collide. This will ensure, that the program transferred from the central computer to the modules only do "valid" actions.

When using distributed control, collision detection becomes much harder, since no single entity knows the configuration of the whole structure. In the simulator used in this work, a module only knows if it is connected to a module or not. This makes collision detection extremely difficult, since the modules don't know anything about modules that are not their directly connected neighbours.

5.1   Locking as Collision Avoidance

To make sure a module doesn't execute an action that will make it collide with another module, the programming of the modules has to take the structure into

---

[2] *substantial amount of time* is context specific, no general value can be used

Fig. 7. The critical section, the *inquiring module* has to get a lock on the *controlling module* to avoid collisions with other modules wanting to occupy the critical section.

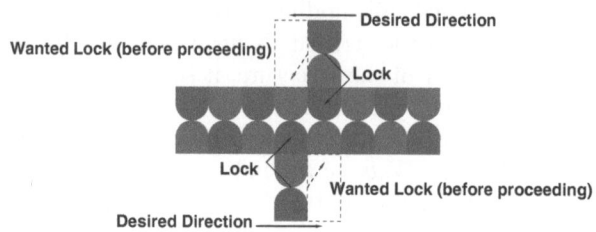

Fig. 8. Dead-lock condition with two moving modules, each modules blocks until they obtain the lock on the base to their respective critical sections, which will never happen.

account. Since a "critical" section is an empty space the critical section can not be locked directly. That is why the organization of the structure must be taken into account. Instead of locking the critical section a module that wants to use the critical section has to get a lock on the module that "controls" the critical section (see fig. 7).

Semaphores. In our case, a *critical section* in the simulator is not a piece of code, but a space in the structure, so the critical section can not be locked as easily as in the case of concurrent programming. Instead, we define each module as both a processing unit and a resource, this way a module can access another module exclusively, e.g. when using it as a base for a movement.

However, semaphores introduce the possibility of dead-locks. In fig. 8 a possible dead-lock situation is depicted where two module move in opposite directions in the structure, each module has a lock on its "base" and want to obtain a lock on the next module in line (unfortunately this module constitutes the base of the other moving-module).

5.2 The "Walking" Behavior and Collision

The walking behavior described earlier uses the locking to ensure an area is free before moving into it. This method uses extensive knowledge of the structure. It is divided into the following steps.

Base Lock. Whenever a module uses another module as a base, the active module will lock the base-module.

Self Lock. Whenever a module wants to move into an area with the intend of connecting to a base-module, the module is queried (through the structure) and asked to obtain a *self-lock*, to ensure no other modules uses it as a base or intend to move to the module as a base.

This schema works as long as the programming of the modules are carefully planned. The locking mechanism locks a whole module, not only the *active* or *inactive* part. Though this method can be used to avoid collision between modules, it will introduce unnecessary stalling of the system, if a module is locked without being critical (e.g. to be used as a mediator of messages).

6   Results

The walking behavior will move at least 4 units from the rear of the structure to the front. The 4 units will move in pairs forming *micro-walkers*.

The walking behavior is made of three parts

1. *Get Ready for Walking* – This is the first stage of the walking behavior, here the *micro-walkers* are connected on the side of the structure. The assembling of the *micro-walker* is different for the two *micro-walkers* constituting the 4-unit composite.
2. *Walking* – This is the universal part of the behavior, it applies to all the *micro-walkers*. This sub-behavior will obtain a lock on the next lower-module (to make sure no other walking modules are in the position the *micro-walker* desires to go to), then the step is completed and the lower-helper module is asked whether it is an *end-module*, if no, the walking continues.
3. *Return to Lattice* – This is the last stage of the walking, here the *micro-walkers* are returned to the lattice. This sub-behavior has two implementations; one for the *Walking Behavior* and one for the *Morphing Behavior*. Each *micro-walker* is treated differently (depending on the space left for the walker).

The behavior is monitored by tracing the single module's two parts. A trace of the first *micro-walker* step in a 16-module configuration is shown in fig. 4.

The morphing behavior transforms one row into two rows. It utilizes the *Walking Behavior* to move the modules to the correct position. In the design phase the design of the morphing behavior was the following:

1. *Detect Mid-point in structure* – Use *hormone-propagation* to create two gradients in the upper *spine* modules, and use these gradients to detect the mid-point of the structure, then make sure the *lower-helper* modules that are in the middle of the structure knows they are *end-point* modules (when queried from the *lower-walker*).
2. *Initiate walking from both sides* when the walking is initiated at both ends of the structure, the morphing behavior is fast compared to the walking (two walking behaviors are executed in parallel).

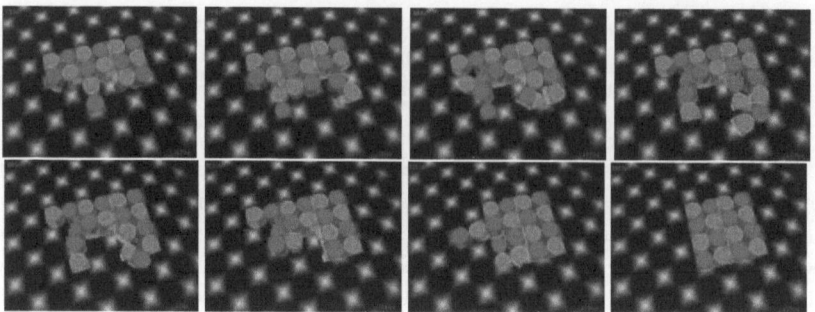

Fig. 9. Stills of the *Morphing Behavior* with 16 modules.

This was the initial design of the *Morphing Behavior*. When the behavior finally was implemented the detection of the mid-point was done as described, and posed no problems. But the walking from both sides was discarded because it would require a "walking from front to rear" behavior that would be trivial to implement, but very tiresome since it would be a mirror of the "walking from rear to front" behavior. Had it been implemented, the morphing behavior would be prettier to look at, but the resulting state of the structure would still be the same. A morphing behavior can be seen in fig. 9.

6.1   Communication

One of the topics of concern when discussing distributed computing, is the communication load of the system. When executing a behavior, 3 different types of communication is used: 1) *Requests* – When a module wants another module to perform some action or behavior, the module will send a request to the helping module. 2) *Synchronization* – When two modules interact, and their actions need to be synchronized, they communicate using simple synchronization commands. 3) *Propagation* – When a module has no knowledge of the program it was requested to execute, the module will ask the initiator to send that program, propagating the knowledge to the module that needs it.

Requests and *Synchronization* commands are fairly simple and the data transferred is small compared to the *Propagation* of knowledge.

Walking. This section describes the results obtained from the simulation using a fully distributed setup. That is, all modules are preprogrammed to contain the programs they need to complete the "Walking Behavior". In the experiment the lower front module is asked to start the walking. The Walking Behavior is continued until 12 modules has been moved from the rear to the front. The experiment is run for 12, 16, 20, ..., 36 modules.

As the graphs in fig. 10 shows, the communication doesn't explode as the number of modules increases. The correlation between the communication load and the number of modules is almost linear ($R^2_{time} = 0.990$ and $R^2_{comm} = 0.998$). The average communication per module is approximately constant.

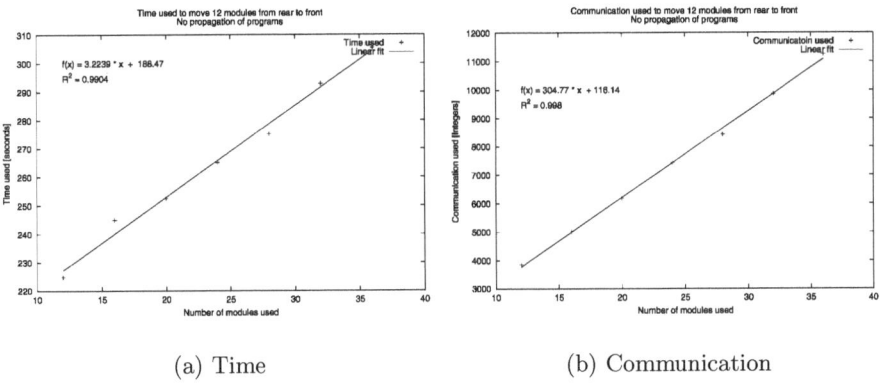

(a) Time   (b) Communication

Fig. 10. The time and communication used to move 12 modules with no propagation of programs.

Table 1. Data from morphing experiment with 16, 24 and 32 modules.

No. Modules	Time used	Communication	Comm. per module
16	146.425	1901	119
24	214.600	3778	157
32	282.775	6196	194

Table 2. The linear fits to the data collected in table 1.

type	slope	offset	R^2
Time used	4.68	44.4	0.9997
Communication	268.44	2484.2	0.9947
Comm. per module	8.52	10.1	1.0000

Morphing. The morphing behavior is executed from the lower front module with 16, 24 and 32 modules. The measured data is the *time* for the morphing behavior to be completed, that is for the last module in the second row to connect fully with the structure and the *amount of data transferred* to accomplish the goal, these values are listed in table 1 with the addition of a column containing the average communication per module (to see how well the behavior scales).

The three value are still approximately linear (all thought the test-set still is too small to say for sure). The linear fits from the three series are shown in table 2.

The slope of the communication per module is positive, meaning that the communication per module increases as the number of modules increase. Since the modules can only by increased by multiples of 8 (2 composite units ~ 4 micro-walkers ~ 8 modules) the number of modules that need to be moved increases as well as the distance they need to travel.

In a 16-module structure 8 modules need to be relocated from the rear to the side of the structure, meaning 4 micro-walkers need to travel across 7,5,3 and

1 module (the distance is shortened in both ends). In a 24-module structure 12 modules need to be relocated, i.e. 6 micro-walkers will travel across 11,9,7,5,3 and 1 module.

7 Conclusion

The "Walking Behavior" can be compared to the "Cluster Flow" described in [YMK+01]. The "Cluster Flow" sends one module to the front before sending the next module. The modules move in the layer they belong to (the upper modules use the upper modules to move, the lower modules only use the lower modules to move).

Compared to this "Cluster Flow", the "Walking Behavior" in this work is much more efficient.

1. The modules are sent on their way continuously from the rear using local communication.
2. The modules from the two layers cooperate to make a "micro"-walker that are able to walk on the side of the structure. This enables the modules to use the lower-modules as connecting surface for the entire walk.

On the other hand, the behavior implemented in the simulation has a delay from the stop-condition is set to the behavior actually stops (because the last two *micro-walkers* have to get to the rear, and there may be several "walkers" on the side of the structure).

Regarding communication load, the presented results for the *Walking Behavior* and the *Morphing Behavior* show that the system scales very well in number of units. The communication load increases linearly with the number of units, and the average communication per module goes toward a constant value as the number of modules increase. This is due to the localized communication.

Acknowledgement

The work was performed as part of the HYDRA project funded by the EU Future and Emergent Technology.

References

MYT00. Satoshi Murata, Eiichi Yoshida, and Kohji Tomita. Hardware design of modular robotic system. In *2000 IEEE/RSJ International Conference on Intelligent Robots and Systems (IROS 2000)*, 2000.

Pet62. Carl Adam Petri. *Kommunikation mit Automaten*. PhD thesis, Universität Bonn, 1962.

YMK+ 01. Eiichi Yoshida, Satoshi Murata, Akiya Kamimura, Kohji Tomita, Haruhisa Kurokawa, and Shigeru Kokaji. A Motion Planning Method for a Self-Reconfigurable Modular Robot. *IROS2001*, 2001.

Co-evolving Complex Robot Behavior

Esben Hallundbæk Østergaard and Henrik Hautop Lund

The Maersk McKinney Moller Institute for Production Technology
University of Southern Denmark
Campusvej 55, 5230 Odense M., Denmark
{esben,hhl}@mip.sdu.dk

Abstract. Reports on evolutionary robotics systems have so far been on evolving controllers that make simple robots do simple tasks in simple environments. In this paper we try to stress the evolutionary robotics approach by evolving a controller for a more complex task, namely Khepera robot soccer, and evaluate evolved controller performance against hand-coded controllers. We present a system that uses competitive co-evolution to develop robot controllers for the task. The system is described, and performance of the system is documented. Co-evolution is tested against single-population evolution, and it is concluded that co-evolution has the ability to produce more robust individuals with respect to opponent strategies.

1 Introduction

Many of the evolutionary robotics systems that have been reported so far, has documented successful generation of robot controllers for simple robots performing simple tasks in simple environments. In this paper we will try to go one step further and apply the method to a more complex and competitive task, namely the Khepera Robot Soccer task. Using this task as a testbed has two advantages: 1) No known optimal strategy exists. 2) Real-world competitions exist, so that evolved controllers can be tested against hand-coded controllers in a neutral competitive scenario. Given the competitive nature of the task, the task lends itself to artificial co-evolution. The emphasis of this paper is on co-evolution applied to this task, and on the viability of the evolutionary robotics approach in general.

1.1 The Khepera Robot Soccer Task

In a Khepera robot soccer match two Khepera robots are pitted against each other in an arena. The arena is 105cm long and 68cm wide with cut off corners, as shown in figure 1. The Khepera robot is a widely used miniature battery-driven robot with on-board processing capabilities from K-Team. The Khepera used for soccer playing is equipped with 8 IR-sensors, wheel encoders and a camera.

A match consists of five rounds. A round lasts until a goal is scored, until the ball has not moved for thirty seconds, or until four minutes have elapsed. At

A.M. Tyrrell, P.C. Haddow, and J. Torresen (Eds.): ICES 2003, LNCS 2606, pp. 308–319, 2003.
© Springer-Verlag Berlin Heidelberg 2003

Fig. 1. A picture of the football arena.

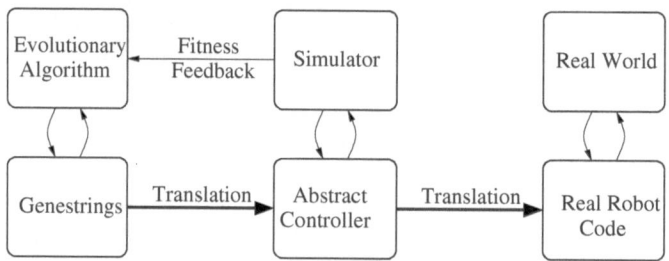

Fig. 2. The implemented system is an evolutionary robotics system using a simulator to speed up evolution.

the beginning of each round, the players are positioned on random symmetric starting points. The ball starts at the center of the pitch. Each player starts facing the opponent's goal line at random, symmetric starting positions.

2   Approach

We decided to use an evolutionary robotics approach, and to construct a simulator to speed up evolution. The implemented system is shown in figure 2. After a pre-specified number of generations in the simulator, the best performing controller is translated into a binary file that can be executed directly on a Khepera robot. The following sections will go into more detail about each of the components of the system.

3   Co-evolution

Since D. Hills, [Hil92], showed that better sorting networks could be produced when using co-evolution, co-evolutionary methods have spread to other fields of research where Evolutionary Algorithms are being used, [AP93], [CM96] and [BLP96]. The task of evolving a robot football player intuitively lends itself to co-evolutionary methods because of the competitive nature of the problem. Given

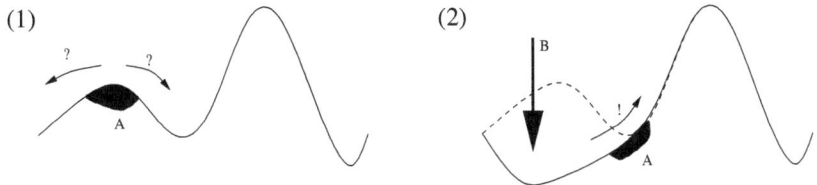

Fig. 3. Illustration of the idea behind co-evolution.

a specific opponent, it might be possible to find the best strategy, but there is no single strategy that is best against all opponents. The task is thus to find a player that performs well against a large number of opponent strategies.

An attempt to illustrate competitive co-evolution for two populations can be seen in figure 3. When one population (A) finds a good locally optimal strategy, the competing population (B) is encouraged to find a solution that performs especially well against the other populations strategy. This changes the fitness landscape for population (A), forcing evolution to find another solution to the problem.

The effect of one population changing the fitness landscape of the other population is called the *Red Queen Effect*, described in [CM95]. The Red Queen Effect has been examined by co-evolving pursuit and evasion behavior, both in simulation [CM96] and on real robots [FNM01]. In [FNM98] and [FN98], D. Floreano et al. investigate the dynamics of competitive co-evolutionary systems, and show that the Red Queen Effect can give rise to "cycles" between alternative classes of strategies and can make it difficult to monitor evolutionary progress. They also state that co-evolution can be used with benefit when the task is such, that the best strategy of one population is dependent on the strategy of the other population.

3.1   The Evolutionary Algorithm

The co-evolutionary algorithm was implemented as two separate GA's. There is no exchange of gene material between the two populations. After much trial and error, good evolution parameters were found to be a population size of 50, an elite of 10, and these 10 were also used for reproduction. The probability of mutation was 5% per gene (byte), and there was no cross-over.

For evaluation, individuals played 25 simulated rounds against individuals from the opponent population (every second player, alternating). The following fitness fuction was used;

- 10.000 per goal, -10.000 for own goal[1].
- 2.5 per second remaining of the round when scored a goal.
- -0.01 per cm to the ball when the round ends.

---

[1] The player getting (mis)credit for the goal, is the player that most recently touched the ball.

The fitness for one round was the sum of the three components. The first component dominates the fitness. The second and third components are only relevant in cases where an equal number of goals have been scored. The second component favors quick goals, and the third component favors robots that stay close to the ball. The third component is meant for bootstrapping the evolutionary process.

Design of the fitness function was a guess based on intuition. It turned out to serve it's purpose, and no further experimentations were performed.

4   The Simulator

A central issue in any evolutionary robotics systems is the evaluation of candidate controllers during evolution. It was decided to perform the evaluations in a simulation modelling the Khperea soccer world for a number of practical reasons. Using a table based model [MNL95] was rejected due to the large statespace of the world. Instead a hybrid between the minimal simulation approach, suggested by Nick Jacobi [Jac98a], and a geometric model is used. Basically everything that easily could be modeled geometrical is modeled geometrically, and then the rest is made *unreliable* by adding noise[2].

The simulator was constructed using iterative refinements, where each version of the simulator was tested by observing differences in the simulated robot behavior and the real world robot behavior.

4.1   The Use of Minimal Simulation

The interaction between the different world objects in the soccer arena is very difficult to model accurately [Smi98]. A minimal simulation approach could be to make the contact between world objects unreliable for the robot controllers, as described in [Jac98b]. By making the simulated environment more unreliable than the real environment, evolved controllers would be forced to be robust, thus increasing the chance that the controller will perform well when transferred to a real robot. In the simulation, interaction between world objects is split up into cases, and specialized code deals with each specific case. For each cases, interaction was modeled according to observation as closely as possible and then noise was applied both to robot and ball movement.

As stated in [HCH92] and [MC96], producing simulations of visual sensing is a very time-consuming task, both for the programmer when building the simulator, and for the computer during simulation. Instead of simulating the Khepera camera pixel by pixel, the output from the pre-processors were simulated.

5   Behavior Representation

The robot controller is a fixed-structure tree of arbitrators, as shown in figure 4. Modules 0-6 are arbitrator modules, and modules 7-42 are primitive actions. At

---

[2] An applet can be seen at http://www.mip.sdu.dk/~esben/EvoRobSoc/applet/

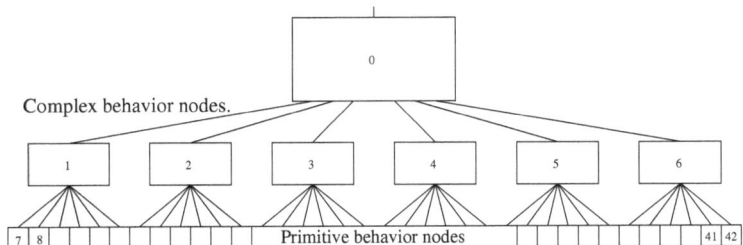

Fig. 4. Architecture of the controller. A tree of behavior modules.

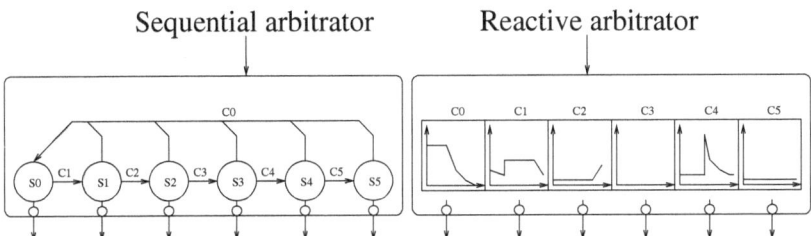

Fig. 5. The two types of arbitrators. Control is propagated down to a sub module. In the sequential arbitrator, control is transferred to the sub module corresponding to the state of a finite state automaton. In the "reactive" arbitrator, control is transferred to the sub module corresponding to condition with the highest activation. The activation goes high when the condition is true, and then decays exponentially when the condition is false. The height and slope is determined by genes.

each 100 ms time step, control is propagated down through the tree to one of the 36 primitives, which then has full control of the robot for the duration of the time step. The architecture is based on the task decomposition approach described by W. Lee, J. Hallam and H. H. Lund [LHL97].

Two types of arbitrators were used, implementing sequential and reactive arbitration, both illustrated in figure 5. Both arbitrators uses a set of conditions to arbitrate between six sub modules, and transfer control to only one of them. The conditions are implemented as fixed size boolean expressions. Figure 7 shows the structure of the condition trees (a), and an example (b). The bottom four nodes of the condition trees can be constant values or current sensor reading of the robot.

The contents of the nodes in the behavior and condition trees was coded by the gene strings. A genotype is a string of 638 bytes, corresponding to a search space of about 10^{1500}. 36 bytes code for the primitve actions, whereas the rest code for the conditions.

5.1 The Behavior Primitives

Generalizations of the behavior primitives used in a successful hand-coded robot football player are used as building blocks for the evolved controllers. The mo-

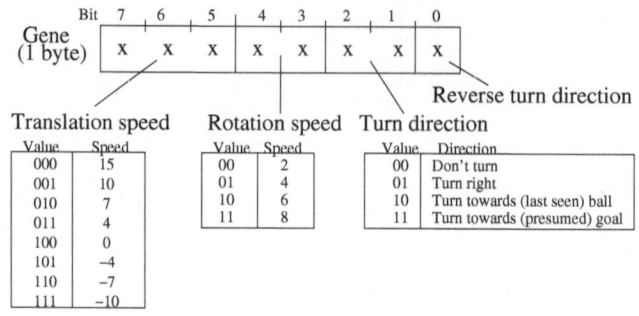

Fig. 6. A gene coding for a primitive action. All speeds are in Khepera speed units. Rotation speed is the difference in speeds for the two motors.

tor primitive are combinations of three parameters; translation speed, rotation speed, and rotation direction, as shown in figure 6. Note that for example action number $001 - 01 - 10 - 0$ will make the robot move forward while turning toward the ball.

The sensor primitive are output from preprocessors, not the raw Khepera sensor data.

- The proximity sensors, filtered so the returned value is the minimum of the last three measurements, and then paired two and two, as in [VPB99]
- The output of the image processing algorithm. The algorithm returns the centroid and width of the ball-blob, if the ball is in the image.
- A "stuck sensor" that uses the position counters to determine whether the robot is stuck or not.
- A sense of direction, that uses the incremental encoders to give an approximated heading of the robot.

These motor and sensor primitives are quite high level compared to what is described in other systems in the studied literature. M. Matarić and D. Cliff wrote: "This abstraction of representation allows for significantly reducing the search space and can greatly accelerate the evolutionary process" [MC96]. A counter argument is, that using higher level primitives reduces the search space so fewer potential solutions are available to evolution. Also, higher level primitives require more work for the programmer.

6   Tests and Results

The system was tested both on performance in the simulation, an on the performance in the real world. Also, a number of tests were performed to evaluate the use of co-evolution.

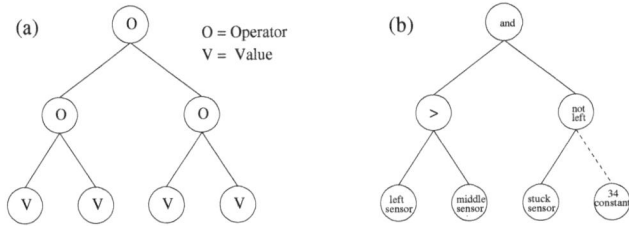

Fig. 7. The structure of a condition. (a) the general structure, (b) an example.

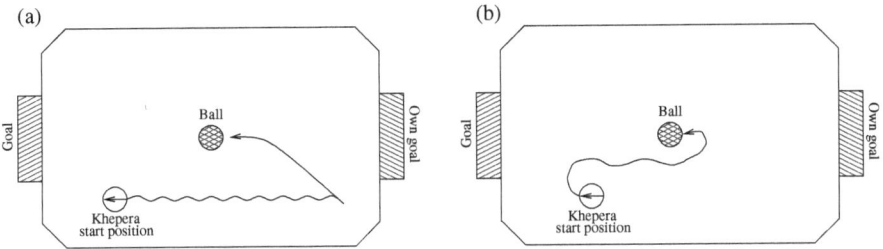

Fig. 8. Different evolved football strategies.

6.1   Behavior in Simulator

The first test was to test weather the given architecture and primitives were sufficient to make a simulated robots play soccer. Initial runs revealed that after about 250 generations the best individuals of both populations were able to score about 80% of the time in the simulator, when alone in the arena. This certainly shows that the set of primitives used is sufficiently expressive, but it also gives rise to some concern with respect to the realism of the simulator, since 80% scoring rate is very good, even for a hand coded robot.

Evolved Strategy. Two distinct strategies for approaching the ball were observed in the evolved robots. The two types are shown in figure 8 (a) and (b). The figure shows one of the difficult situations, where the robot has to go back and get behind the ball to avoid scoring an own goal. Neither of these strategies resemble the typical hand-coded strategies.

6.2   Testing Real-World Performance

A severe drop in fitness was observed when robots are transferred to reality. In the real world, the evolved robots would typically get a fitness of about 1000, which is quite far from the about 8000 typically obtained in simulation. This difference, known as the *reality gap*, is in this application probably mainly due to imperfect modelling of skidding and world object interactions.

The system was really put to the test, when an evolved robot controller participated in the Danish Robot Football Competition, Dec. 1999 under the

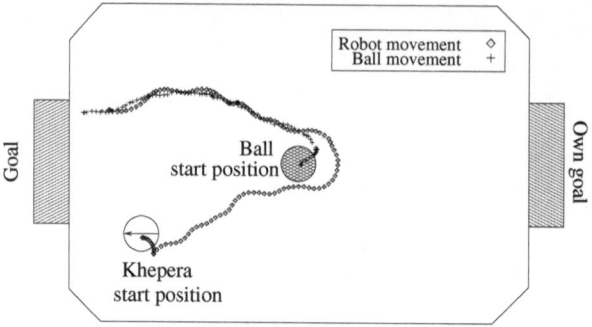

Fig. 9. The plotted position of the robot *Brute Force* and the ball in the simulator. The ball starts to move when *Brute Force* pushes the ball.

name *Brute Force*. Videos can be found on `http://www.mip.sdu.dk/~esben/EvoRobSoc`, and is the main source of documentation for the real robot performance. In the preliminary matches, *Brute Force* won two times and had one draw, which was enough to qualify for the semi finals. In the semi final *Brute Force* won 1-0 without much difficulties, but lost in the final against *KITT*, partly due to a dead battery in the second round (The evening before the final, *Brute Force* won 3-0 against *KITT* in a test match). An other robot soccer player evolved with the system participated in the FIRA2002 WorldCup KheperaSot tournament, where it won the second place.

The behavior of *Brute Force* is quite different from the behavior of the hand-coded competitors, being more in direct contact with the ball. Even though *Brute Force* was moving slower than its opponents, it still won many games by better maneuvering. *Brute Force*'s behavior in the simulator is shown in figure 9. By observation, this behavior corresponds to the behavior in the real world. An applet showing more *Brute Force* behavior can be seen on the papers home page[3].

6.3   Testing Co-evolution, Chasing the Red Queen

Two different approaches were tried to track the Red Queen. One approach is evaluating individuals against reference players. The other approach is the Master Tournament method, described in [FN98]. Four tests were set up to test the effect of co-evolution.

Is Co-evolution More Efficient than Single Population Evolution with Respect to Use of Computing Power? Two evolutionary setups were implemented. One with co-evolution of 2×50 individuals in the populations, and one single population evolution with 100 individuals in a population (players

---

[3] http://www.mip.sdu.dk/~esben/EvoRobSoc/

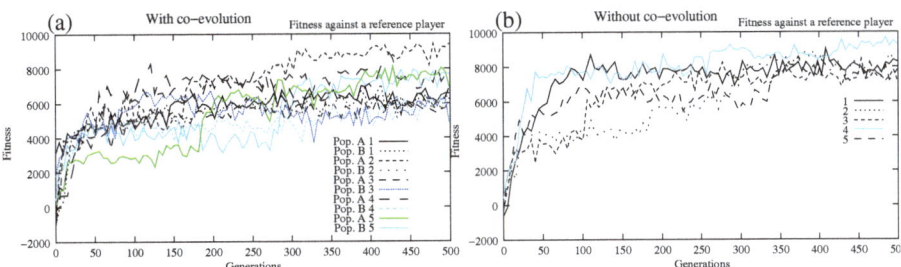

Fig. 10. Co-evolution vs. single population evolution, test 1. Performance against a reference player of the co-evolution with two populations of 50 robots (a), and a single population of 100 robots (b).

are siblings). In both cases, performance was measured a-posteriori against a reference player.

Five runs were started with each setup, results are shown in figure 10. The single population evolutionary algorithm seems to climb faster, and seems to reach higher values. To test whether there is a statistical significant difference between the performance when using evolution or co-evolution, the average fitness of the last 100 generations was calculated for each population. From these data, statistics show that co-evolution performs significantly worse than single population evolution.

Does Co-evolution Hurt Evolution? Another test was designed to test whether co-evolution made the evolutionary algorithm worse. In this test, the population size for the single population evolutionary algorithm was decreased to 50 and compared to 2×50 individuals in the co-evolutionary algorithm, so that the single population evolutionary algorithm runs with half as many fitness evaluations as the co-evolutionary algorithm.

Six runs, each 1000 generations, were made with the single population algorithm, and compared to five co-evolution runs. Results can be seen on the left half of figure 11.

The data shows that the difference between the two algorithms is not significant. The co-evolutionary algorithm shows no advantage over the single population algorithm in our data. Since the co-evolutionary algorithm uses twice as many fitness evaluations, the single population algorithm is to be preferred.

Does Co-evolution Win in the Long Run? Due to arms-race dynamics, co-evolution might have an advantage after some time, when the single-population evolution reaches a plateau. A test was made in which evolutions from the left graph of figure 11 were continued for another 2000 generations. The result of this run can be seen on the right graph in figure 11, and shows no significant difference in the two methods (at 5% confidence level). Over the last 100 generations, the single population evolutionary algorithm has a higher average value than the co-evolutionary, but the difference is not significant.

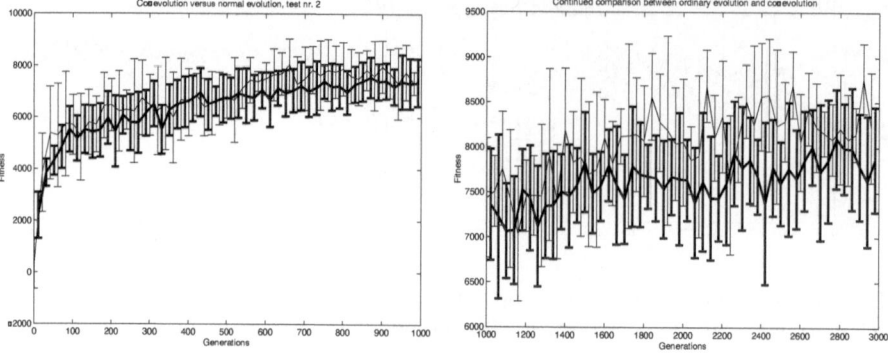

Fig. 11. Left: Co-evolution vs. single population evolution. The graph shows the average and the 95% confidence interval of the elite fitness-evaluations averaged over both populations. The thin line is the single population evolutionary algorithm (6 runs), the thicker line is the co-evolution one (5 runs). **Right:** A continuation of the graph shown on the left.

Does Co-evolution Produce More Robust Results? A fourth test was run to test whether the robots evolved with co-evolution were more robust with respect to handling different opponent strategies than the ones evolved with single population evolution, as argued by Stefano Nolfi and Dario Floreano in [NF98]. An experiment was set up, in which 10 co-evolved individuals played against 6 single population evolved individuals. The co-evolved individuals were chosen to be the best individuals from each population from the previous experiment, according to the master tournament method. For the 6 single population runs, the chosen individuals were the best individuals from the final generation of each run. Each of the co-evolved individuals was set to play 100 rounds against each of the individuals evolved with single population evolution. Each co-evolved individual thus played 600 rounds, while the other individuals played 1000 rounds. The average fitnesses obtained from these rounds are shown in the table in figure 12. The table shows a significantly higher average fitness for co-evolved individuals.

7   Discussion and Conclusion

A system has been implemented, that successfully uses evolution to generates controllers for the Khepera Robot Soccer task. Using this system, co-evolution is compared to single population evolution through four tests. The first three tests showed no benefit from using co-evolution. Single population evolution seems to converge faster to good solutions than co-evolutionary algorithms. However, the fourth test revealed, that the co-evolved individuals have developed more robust strategies than the single population evolved individuals. This supports D. Floreano and S. Nolfi's hypothesis, that "co-evolution can have a higher adaptive power than evolution" [FN98]. This series of experiments also shows, that the effect of co-evolution can be very difficult to measure. Under what

	Average fitness for individuals		
Player No.	Without co-evolution	With co-evolution	
		Pop. A	Pop. B
1	2306.11	3298.83	1481.15
2	2277.18	5349.24	4278.84
3	2772.82	3200.50	3432.08
4	1110.28	3011.15	6223.48
5	3964.32	4182.49	4476.53
6	245.54		
Average	**2112.71**	**3893.43**	
Std.Dev.	1297.68	1323.49	

Fig. 12. Single population evolved players playing against co-evolved players.

prerequisites the positive effect of co-evolution will take place is more or less unexplored territory.

The objective of the work was to explore the co-evolutionary robotics approach and to test whether it could be used to evolve behavior for the Khepera robot soccer task. This seems to be the case. However, existing theories still need further development in several areas to reduce the amount of intuition required to build such a system. The problems of constructing simulators of a sufficiently high fidelity seems to be the major obstacle for the viability of the approach taken in this paper. Other problems are the heuristics involved in deciding behavior architecture and sensor and motor primitives.

In 1992, R. Brooks wrote "To compete with hand coding techniques it will be necessary to automatically evolve programs that are one to two orders of magnitude more complex than those previously reported in any domain." [Bro92]. This goal seems to have been reached for the present task, since the controllers evolved with this system plays football at least as well as most of the hand coded controllers in the Khepera robot soccer competitions.

References

AP93.   Peter J. Angeline and Jordan B. Pollack. Competitive Environments Evolve Better Solutions for Complex Tasks. In Stephanie Forrest, editor, *Genetic Algorithm: Proceedings of the Fifth International Conference (GA93)*, 1993.

BLP96.   Alan D. Blair, Mark Land, and Jordan B. Pollack. Coevolution of a backgammon player. In *Proceedings of the Fifth Artificial Life Conference*, 1996.

Bro92.   R. Brooks. Artificial life and real robots. In *European Conference on Artificial Life*, pages 3–10, 1992.

CM95.   Dave Cliff and Geoffrey F. Miller. Tracking the red queen: Measurements of adaptive progress in co-evolutionary simulations. In F. Morán, A. Moreno, J. J. Merelo, and P. Chacón, editors, *Proceedings of the Third European Conference on Artificial Life : Advances in Artificial Life*, volume 929 of *LNAI*, pages 200–218, Berlin, June 1995. Springer Verlag.

CM96.   Dave Cliff and George Miller. Co-evolution of Pursuit and Evasion II: Simu-
         lation Methods and Results. In Maes, P., Matarič, M., Meyer, J.-A., Pollack,
         J., and Wilson, S. W., editors, *From Animals to Animats 4: Proceedings of the
         fourth Internation Conference on Simulation of Adaptive Behavior (SAB96)*.
         MIT Press Bradford Books, 1996.

FN98.   Dario Floreano and Stefano Nolfi. Co-evolving predator and prey robots: Do
         'arms races' arise in artificial evolution? *Artificial Life, 4 (4)*, pages 311–335,
         1998.

FNM98.   Dario Floreano, Stefano Nolfi, and Francesco Mondada. Competitive co-
         evolutionary robotics: From theory to practice. In R. Pfeifer, B. Blumberg,
         J-A. Meyer, and S.W. Wilson, editors, *Animals to Animats V*, pages 512–524.
         Cambridge, MA: MIT Press, 1998.

FNM01.   Dario Floreano, Stefano Nolfi, and Francesco Mondada. Co-evolution and
         ontogenetic change in competing robots. In M. Patel, V. Honavarand, and
         K. Balakrishnan, editors, *Advances in the Evolutionary Synthesis of Intelli-
         gent Agents, Cambridge (MA)*. MIT Press, 2001.

HCH92.   I. Harvey, D. Cliff, and P. Husbands. Issues in evolutioanry robotics. In
         Roitblat, H. Meyer, J.-A. and Wilson, S., editors, *Proceedings of SAB92*.
         MIT Press Bradford Books, Cambridge, MA, jul 1992.

Hil92.   Daniel W. Hillis. Co-evolving parasites improve simulated evolution as an
         optimization procedure. In Chris Langton et al., editor, *Artificial Life II*,
         pages 313–324, 1992.

Jac98a. Nick Jacobi. The minimal simulation approach to evolutionary robotics. In
         Takashi Gomi, editor, *Evolutionary Robotics, Volume II*, 1998.

Jac98b. Nick Jacobi. Running across the reality gap: Octopod locomotion evolved in
         a minimal simulation. *Lecture Notes in Computer Science*, 1468:39–??, 1998.

LHL97.   Wei-Po Lee, John Hallam, and Henrik Hautop Lund. Learning complex robot
         behaviours by evolutionary approaches. *6th European Workshop on Learning
         Robots, EWLR-6*, aug 1997.

MC96.   Maja Matarič and Dave Cliff. Challenges in evolving controllers for physical
         robots. In *Proceedings, AAAI-92*, 1996.

MNL95.   Orazio Miglino, Stefano Nolfi, and Henrik Hautop Lund. Evolving mobile
         robots in simulated and real environments. *Artificial Life 2*, pages 417–434,
         1995.

NF98.   Stefano Nolfi and Dario Floreano. How co-evolution can enhance the adap-
         tive power of artificial evolution: Implications for evolutionary robotics. In
         Philip Husbands and Jean-Arcady Meyer, editors, *Lecture Notes in Com-
         puter Science*, volume 1468, 1998.

Smi98.   T.M.C. Smith. Blurred vision: Simulation-reality transfer of a visually guided
         robot. In P. Husbands and J.-A. Meyer, editors, *Evolutionary Robotics, First
         European Workshop: EvoRobot98*, Lecture Notes in Computer Science 1468,
         pages 152–164. Springer-Verlag, 1998.

VPB99.   M.M.B.R. Vellasco, M.A.C. Pchesco, and I.L Brasil. Mobile robot control
         using fuzzy logic. In Mondada, F. Löffler, A. and Rückert, U., editors, *Ex-
         periments with the Mini-Robot Khepera*, 1999.

Evolving Reinforcement Learning-Like Abilities for Robots

Jesper Blynel

Autonomous Systems Lab
Institute of Systems Engineering
Swiss Federal Institute of Technology (EPFL)
CH-1015, Lausanne, Switzerland
Jesper.Blynel@epfl.ch

Abstract. In [8] Yamauchi and Beer explored the abilities of continuous time recurrent neural networks (CTRNNs) to display reinforcement-learning like abilities. The investigated tasks were generation and learning of short bit sequences. This "learning" came about without modifications of synaptic strengths, but simply from internal dynamics of the evolved networks. In this paper this approach will be extended to two embodied agent tasks, where simulated robots have acquire and retain "knowledge" while moving around different mazes. The evolved controllers are analyzed and the results are discussed.

1 Introduction

Yamauchi and Beer [8] showed in 1994 that continuous time recurrent neural networks (CTRNNs) can be evolved to display reinforcement learning-like abilities. The investigated tasks were generation and learning of short bit sequences. These tasks were accomplished without modifications of synaptic strengths, but simply by exploiting the rich internal dynamics of this class of networks. The claim was that the kind of "learning" showed by the networks is more biological plausible than more traditional approaches to learning such as reinforcement learning [4]. In order to justify their choice of bit sequence tasks, the authors made the analogy that a rat, when navigating a maze, is faced with a sequence binary decisions of which way to turn. Despite of this analogy the tasks studied were very artificial seen from an evolutionary robotics perspective.

This issue will be picked up upon in this paper by applying a similar approach in a more "realistic" setting of a simulated robot navigating a maze. In the first of several trials in an environment the robot has to a acquire and retain "knowledge" which can then later on be exploited. In order to show the feasibility of the approach the first task investigated is a simple setup where the movement of the robot is constrained to one line. In the second task the robot has to navigate in the more challenging environment of a T-maze. The learning-like properties of a successful controller is analyzed and it is shown how to modulate the robot behavior by changing the activity of one neuron of an evolved network.

A.M. Tyrrell, P.C. Haddow, and J. Torresen (Eds.): ICES 2003, LNCS 2606, pp. 320–331, 2003.

Tuci et. al. [6] have recently, in parallel to the work presented in this paper, applied a similar approach in order to solve an extended version of a landmark task also originally studied by Yamauchi and Beer [7]. A simulated robot had to sometimes approach a light source and sometime avoid it, based on the position a reward-zone in the environment. A very complex and task specific fitness function, however, had to be used in order to evolve successful controllers. The reason for this might be that task studied required evolved controllers to use the same sensory modality for the two very different behaviors of light approaching and light avoidance. One could suspect that this fact made the search problem for the genetic algorithm much harder. In the work presented in this paper the focus is on applying much simple fitness functions, but shaping the experimental settings in oder to make to search task for the genetic algorithm accomplishable.

We have previously compared the class of CTRNNs used in this article to Plastic Neural Networks (PNNs) with online modification of network weights, on a set of tasks requiring sensory-motor adaptivity and learning [1]. The results obtained were that the PNNs were more adaptive with respect to changes applied to the environment after the evolutionary process, but the CTRNNs, on the other hand were easier to evolve to display learning-like abilities. For this reason focus of this paper will be on CTRNNs.

The rest of this paper is organized as follows. Section 2 describes the neural network used, section 3 describes the first simple experiment, section 4 describes and analyses the more complex T-Maze experiment. Section 5 discusses the results obtained and concludes the paper.

2   Continuous-Time Recurrent Neural Networks

In the continuous-time recurrent neural networks used for the experiments in this paper the state of each neuron can be described by the following differential equation:

$$\frac{d\gamma_i}{dt} = \frac{1}{\tau_i}\left(-\gamma_i + \sum_{j=1}^{N} w_{ij}A_j + \sum_{k=1}^{S} w_{ik}I_k\right) \quad , \tag{1}$$

where N is the number of neurons, i ($= 1, 2, ..., N$) is the index, γ_i describes the neuron state (cell potential), τ_i is the time constant, w_{ij} is the strength of the synapse from the presynaptic neuron j to the postsynaptic neuron i, $A_j = \sigma(\gamma_j - \theta_j)$ is the activation of the presynaptic neuron where $\sigma(x) = 1/(1 + e^{-x})$ is the standard logistic function and θ_j is a bias term. Finally, S is the number of sensory receptors, w_{ik} is the strength of the synapse from the presynaptic sensory receptor k to the postsynaptic neuron i and I_k is the activation of the sensory receptor ($I_k \in [0, 1]$). As in [8] the *Forward Euler* numerical integration method was used. The iterative update rule for the state of each neuron becomes:

$$\gamma_i(n+1) = \gamma_i(n) + \frac{\Delta t}{\tau_i}\left(-\gamma_i(n) + \sum_{j=1}^{N} w_{ij}A_j(n) + \sum_{k=1}^{S} w_{ik}I_k\right) \quad , \tag{2}$$

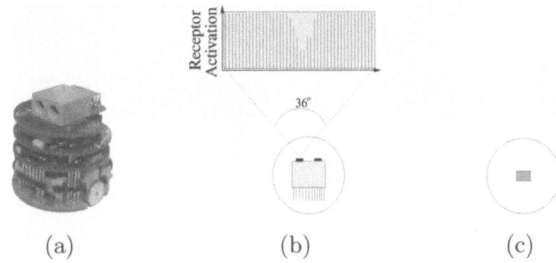

(a)   (b)   (c)

Fig. 1. The Khepera robot simulated in the experiments. The robot has (a) 8 infrared sensors distributed around the body used for object proximity and ambient light intensity measurements; (b) a linear vision module with 64 equally-spaced photoreceptors covering a visual field of 36°; and (c) a floor sensor measuring surface brightness.

where n is the iteration step number and Δt is the step size. Initially the state of each neuron was $\gamma_i(0) = 0 \ \forall i$, the step size set to $\Delta t = 1$. The range of the other parameters were the following:

$$\tau \in [1, 70], \ \theta \in [-1, 1] \ and \ w \in [-5, 5]$$

Notice from equation (2) how the time constant τ determines the dynamics of the neuron. A high time constant results in a slowly changing neuron state while decreasing it makes the neuron more responsive to current synaptic inputs approximating a reactive neuron with no internal state in the limit.

3   The Simple "Reinforcement Learning" Task

The first experiment is a basic setup which requires acquisition and storage of "knowledge". A simulated Khepera robot (figure 1) is placed in a rectangular arena with two potential reward-zone positions (figure 2), a light bulb to the left and a black vertical stripe on the right. The robot has 6 trials to find out where the reward-zone is, go there and stay there. At the beginning, the position of the reward-zone (grey half-circle in figure 2) is randomly chosen, either to the left (below the light bulb) or to the right (below the stripe), and remains the same for 3 consecutive trials. After 3 trials, the reward-zone is switched to the other end of the environment. At the beginning of each trial, the robot is randomly positioned within the center third of the dashed line shown in figure 2, always facing the black stripe. In order to make the task simpler, the robot can only move back and forth along this line, i.e. it cannot rotate. This is realized by having one motor output neuron controlling the speed of *both* wheels of the robot.

The reinforcement signal comes from a floor sensor[1] (figure 1(c)) which is *on* when the robot is inside a gray reward-zone and *off* otherwise. Notice that this

---

[1] This sensor can be easily be mounted on a standard Khepera robot by placing an additional IR sensor under the robot body and connecting it one of the A/D channels.

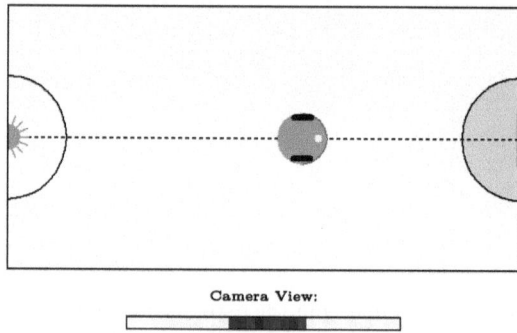

Fig. 2. The environment used in the simple reinforcement learning task. To the left there is a light bulb and to the right there is black stripe on the wall. A gray reward-zone is randomly placed in either end (here shown to the right). The robot is constrained to move along the dashed line always facing the black stripe on the wall. Below the environment the view from the linear camera is shown.

information is a sensory input just like any other, in contrast to conventional reinforcement learning systems where the reward signal plays a special role in the architecture and in the learning algorithm [4].

3.1 Network Architectures and Genetic Encoding

The neural network architecture is shown in figure 3. The network consists of 5 fully interconnected neurons (4 hidden + 1 motor output) and 6 sensory receptors. Every neuron has synaptic connections from all neurons and all sensory receptors. The sensory receptors are configured the following way (see also figure 4):

- *2 Light receptors*: The robots infrared sensors in passive mode are used to measure the ambient light. Only the two sensors on the back of the robot are used.
- *3 Visual receptors*: The linear vision module has 64 equally spaced photoreceptors spanning a visual field of 36° (figure f1(b)). The visual field is divided into 3 sectors and the average pixel-value in each sector is passed to the corresponding visual receptor.
- *1 Floor receptor*: An infrared sensor in the center of the robot body pointing downwards (figure 1(c)) measures surface brightness. If the robot is inside the reward-zone the corresponding receptor is *on* (=1) and otherwise it's *off* (=0).

The value from each receptor is scaled into the range $[0, 1]$. The activation of the motor output neuron in scaled linearly in the range $[-10, 10]$ and used to set the speed of *both* wheels of the simulated robot. The parameters of the network are encoded in a bitstring genotype (figure 3, top). Each neuron has 13 encoded parameters: a time constant (τ), a threshold (θ), and 11 synaptic

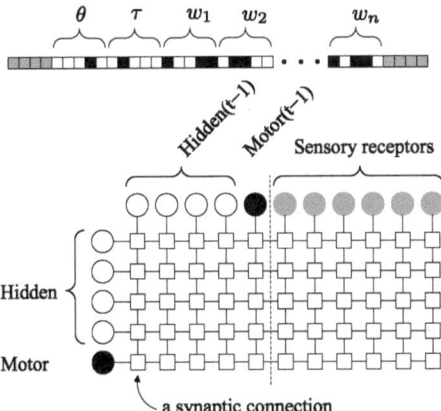

Fig. 3. Genetic encoding of the parameters for one neuron (top) and architecture of the neural network used in the simple reinforcement learning task. *Genetic Encoding*: Each neuron parameter is encoded using 5 bits. θ is the bias, τ is the time constant, and $w_1 \ldots w_n$ are the strengths of the incoming synapses to this neuron. *Neural Architecture*: The network consists of 5 neurons (4 hidden + 1 motor output) and 6 sensory receptors. Every neuron has synaptic connections from all neurons and all sensory receptors.

strengths (w_{ij}). Each parameter is encoded linearly within it's range using 5 bits, giving a total genotype length of 325 bits.

3.2   Experiments

The experiments were carried out in a realistic simulation of the Khepera robot based on samplings of infrared sensor values [3], computing geometric projections for linear camera inputs and adding 5% uniform noise to every value. A simple genetic algorithm with rank-based selection was used. A population of 100 neural controllers was evolved for 100 generations. At every generation the best 20 individuals made 5 copies each. One copy of the best individual remained unchanged (elitism). Single-point crossover with a 0.04 probability and bit-switch mutation with a 0.02 probability per bit was used. Every neural controller was

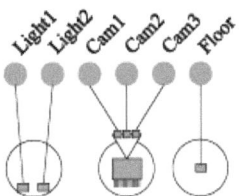

Fig. 4. Configuration of the 6 sensory receptors used in the simple reinforcement learning task.

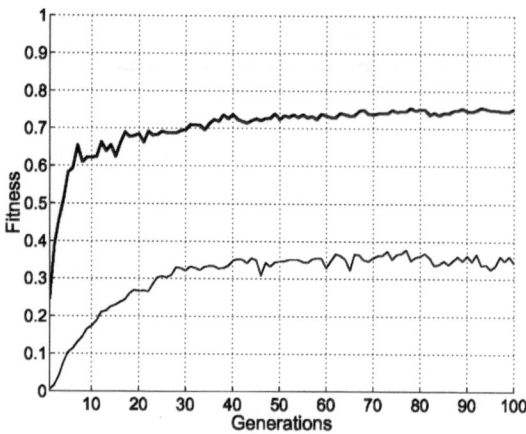

Fig. 5. *Simple Reinforcement Learning Task.* Thick line shows best fitness and thin line shows population mean (both are averages over 10 replications of the experiment).

tested for 3 epochs of 6 trials each. Each trial lasted 150 sensory-motor steps (corresponding to 15 seconds of simulated time). At the beginning of each *epoch*, the neural controller was re-initialized by setting the state of each neuron to 0. This means that the robot could potentially build up and store information in the dynamic neuron states between trials within the same epoch. The fitness of an individual was proportional to the amount of time spend on the reward-zones subtracted a penalty for spending time in only one of the two zones:

$$fitness = \frac{\sum_{i=1}^{epochs} f_1(i) + f_2(i) - |f_1(i) - f_2(i)|}{epochs \times trials\_per\_epoch \times steps\_per\_trial} ,$$

where f_1 is the number of steps spend in the reward-zone in trials with reward to the left and f_2 is the number of steps spent in the reward-zone in trials with reward to the right. Notice that if f_1 or f_2 is zero in an epoch the total fitness will also be zero in that epoch. Without taking the absolute difference between f_1 and f_2, evolved controllers found the sub-optimal solution of only accumulating fitness in one end of the environment.

The experiment was repeated 10 times with different initializations of the computer's pseudo-random number generator. The results of the experiments are shown in figure 5. The thick line shows best fitness and the thin line shows population mean, both are averages over 10 replications of the experiment. As can be seen from the graph only about 40 generations on average were required for the fitness values to reach a stable high level.

The typical behavior of an evolved controller is visualized in figure 6. The x-position of the robot is plotted over time for an entire epoch of 6 trials[2]. Before

---

[2] Each trial is shortened from 150 to 100 steps in figure 6 because the position of the robot always stabilizes within this time-frame for this individual.

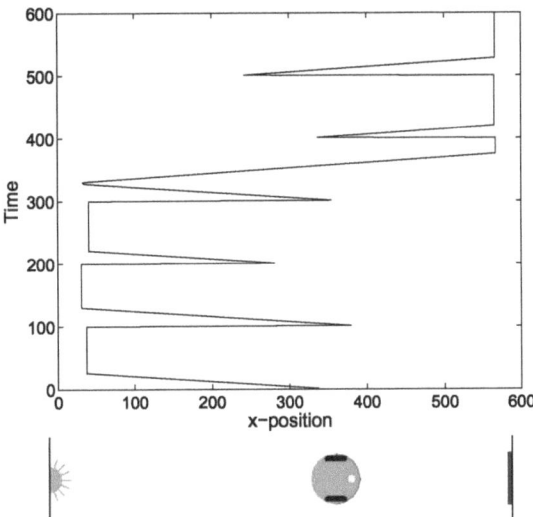

Fig. 6. Typical behavior of an evolved robot in the simple reinforcement learning task. *Top*: Robot x-position is plotted against time over 6 trials. *Bottom*: Environment layout. See text for description.

the first step of each trial the robot is randomly placed within the center third of the x-axis. This evolved robot begins to move to the left where the reward-zone is positioned for the first 3 trials. After 3 trials the reward position is switched. The robot still moves to the left by default, but when it discovers that the floor sensor remains off, it reverses direction and moves right. For the remaining trials, it "remembers" that the reward-zone is to the right and moves there directly.

4   The T-Maze Task

In the second experiment the robot had to navigate the T-maze shown in figure 7. The task for the robot was in principle the same as in the previous experiment, but because of the increased complexity of the environment the robot was now allowed to move both wheels independently. The reward-zone (black square in figure 7) could be positioned in either the left or the right arm of the maze. The position of the reward-zone stayed fixed during each epoch. The robot was tested for 4 epochs of 5 trials each - two epochs with the reward-zone in each arm of the maze As before the neural network is re-initialized between epochs but *not* between trials. The optimal behavior of the robot in this environment is to use the first trial of each epoch to locate and "remember" the position of the reward-zone, and thereafter move directly towards it for the remaining trials of the epoch. To put additional evolutionary pressure on this behavior, the number of available sensory-motor steps was 360 in the first trial of each epoch and only 180 in the remaining 4 trials. Given the size of the maze this

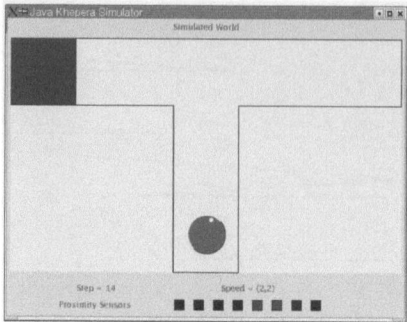

Fig. 7. The T-maze environment used in the second experiment. The reward-zone (black square) can be positioned in either the left arm (shown above) or in the right arm of the maze.

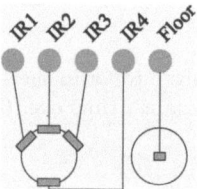

Fig. 8. Configuration of the 5 sensory neurons used in the T-maze task. The eight IR proximity sensors are paired two by and the fifth input is from the floor sensor.

means that the robot only had time the explore the whole maze during the first trial of each epoch. In addition a poison-zone (white square in figure 10(b)) was positioned opposite the reward-zone in the last 4 but *not* the first trial of each epoch. An individual was immediately killed if it stepped over the poison-zone. With these changes the *fitness function* could be simplified to the sum of trials an individual ends it's life inside the reward-zone. Since each individual was tested for a total of 20 trials the maximal possible fitness was 20. Notice that there is no direct pressure on evolving fast moving robots given this fitness function. This is however compensated by the fact the the number of steps in each trial is limited and has been adjusted to fit the size of the environment.

In this experiment, in addition to the floor sensor, only one sensory modality, the infra proximity sensors of the simulated robot, was used (see figure 8). The network architecture was similar to the one shown in figure 3. The network still contained 4 hidden neurons, but now had 2 motor output neurons and only 5 sensory receptors, giving a total of 66 connection weights and a total genotype size of 390 bits. Because of the increased complexity of the task the population size was increased to 200, where 40 parents make 5 copies each. The elite size was set to 5 and the range of the time-constant τ was changed to $[0, 50]$. The evolution was run for 200 generations and replicated 10 times.

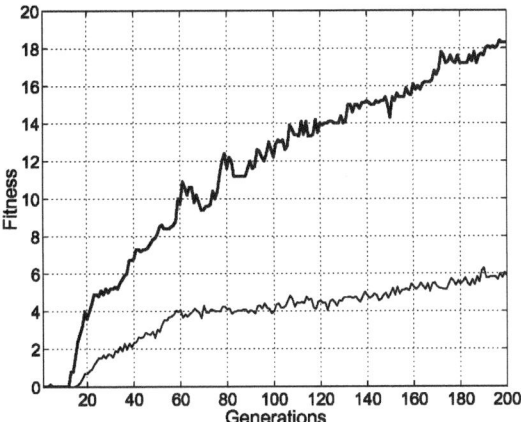

Fig. 9. *T-Maze Task.* Thick line shows best fitness and thin line shows population mean (both are averages over 10 replications of the experiment).

The fitness results of the evolutionary runs on this T-maze experiment are shown in figure 9. The evolutionary process found individuals able to collect the maximal fitness of 20 in 6 out of the 10 replications of the experiment. The maximal fitness in the 4 remaining runs was around 16. The behavior of an individual from the final generation of one of the successful runs is shown in figure 10 and 11. The robot starts out in trial 1 of the first epoch (figure 10(a)) by exploring the maze until it locates the reward-zone where it stays the remaining time of the trial. In the following trials (figure 10(b)), the robot is able to retain the "knowledge" gathered during trial 1 and always turns left at the T-junction in order to move towards and stay on the reward-zone. In the epochs with reward to the right the robot moves directly towards the reward-zone in trial 1 (figure 11(a)), since the default behavior of this individual is to turn right at the first T-junction after a re-initialization of the neural controller. For the remaining trials this successful behavior is repeated (figure 11(b)).

4.1   Analysis

In order to better understand how the evolved neural networks worked, some further analysis on an individual from one of the successful evolutions on the T-maze task was done. The neuron activities of the network was recorded over two epochs, one with the reward in each end. The plots of the first trial of each of these two epochs are shown in figure 12(a) and 12(b), both these trial lasting 360 sensory-motor cycles as during evolution. The top four graphs shows the activity of the hidden neurons (H0 to H3) and the next two shows the activity of the motor output neurons (M0 and M1). The output of the floor sensor (F) is added at the bottom for clarification. The activities vary between 0 and 1 as defined by the logistic activation function. The neuron activities of figure 12(a) correspond to the robot trace shown in figure 10(a) and figure 12(b) corresponds to trace

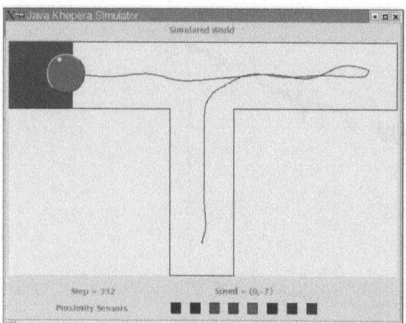

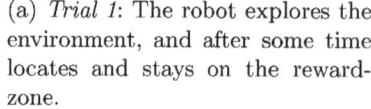

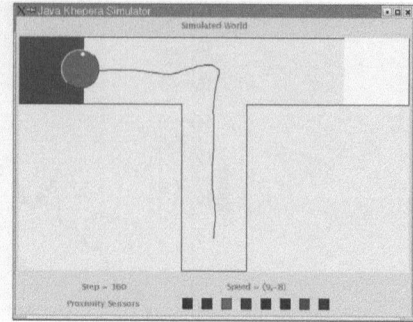

(a) *Trial 1*: The robot explores the environment, and after some time locates and stays on the reward-zone.

(b) *Trials 2-5*: For the remaining trials the robot exploits the "knowledge" gained in trial 1 and moves directly towards the reward-zone.

Fig. 10. Robot traces of an epoch with the reward-zone to the left.

shown in figure 11(a). In the first case (reward left) the robot searches the whole environment before if finds the reward-zone, whereas in the second case it finds it directly. This can be seen by comparing the plots of the floor sensor (F) which goes high when the robot enters the reward-zone. In order to better understand how the network retains "knowledge" between trials it is useful to compare the activity of the hidden neurons towards the end of each trial.

This can tell something about how the information which the network seems to retain is captured. It can be seen that the dynamics of neurons H0, H1 and H3 are very similar in the two plots whereas the activity of H2 differs significantly. In figure 12(a) the activity of H2 falls almost linearly from about 0.5 to about 0 across the trial. In figure 12(a), on the other hand, the graph of H2 is similar for the first 150 sensory-motor steps, where paths of the two robots are the same, but then starts to rise when the robot enters the reward-zone, and is about 1 at end of the trial. The state of H2 at the end of the trials was -1.27 in the first case and 2.11 in the second. It seems that this neuron play a main role in the memory capabilities of this controller. In order to further test this hypothesis, an additional set of tests was done. 51 epochs of the robot moving in the maze for 1 trial were performed. Every neuron state was set to 0 (as during evolution) in the beginning of each epoch except for neuron H2 which initial state value was increased by 0.1 for each epoch, starting at -2.5 and ending at +2.5. The behavior of the robot when encountering the T-junction was monitored during these epochs. The robot was able to successfully navigate the maze in every case, although a bit slower than before. It generally took a couple of seconds of adaptation by jiggering movements before the robot was at normal speed. The results showed that the robot turned left at then T-junction whenever the initial state value of H2 was set below -0.3, and it turned right every time when the value was above -0.1. This result shows as hypothesized above, that the overall

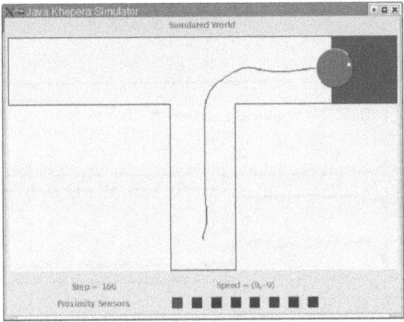

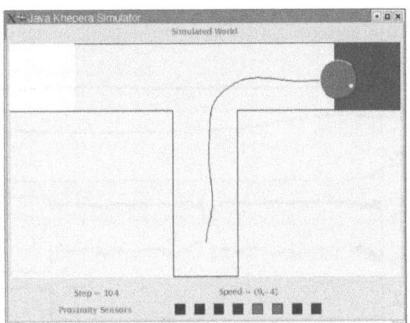

(a) *Trial 1*: The default behavior of this individual of turning right takes the robot directly to the reward-zone in trial 1.

(b) *Trials 2-5*: For the remaining trials the robot repeats this behavior.

Fig. 11. Robot traces from an epoch with the reward-zone to the right.

behavior of the robot can be modulated by changing the state of single neuron which plays a crucial role in this network's "memory capabilities".

5   Conclusion

The experiments in this paper have shown that the approach of evolving continuous time recurrent neural networks to display reinforcement learning-like abilities initiated by Yamauchi and Beer [8] can indeed be extended to robotics task where an embodied agent has to move around in an environment. The first experiment was on purpose kept simple in order to show the feasibility of the approach. The second experiment showed that similar results could be obtained in a more complex environment. Successful individuals were in both cases able acquire and retain "knowledge" by in trial 1 exploring the environment it happened to be situated in. This information was thereafter exploited in the following trials where direct traces towards the reward-zone was seen. This kind of "learning" resembles biological learning in animals much more than more traditional approaches to learning such as reinforcement learning [4], where an agent normally need a substantial number of learning trials.

The analysis of an individual in T-maze experiment showed that one of the hidden neurons in the network was responsible for storing crucial "knowledge" and thus generating the learning-like behavior. This shows that, at least in this kind of environments investigated in this paper, it is not necessary for an successful agent to build a cognitive map of the environment. If this is also the case in more complex environments needs to be investigated. Current research focuses on extensions to double T-maze environments similar to those recently investigated in [5] on a series of delayed response tasks. Future work also includes transfers of the evolved behaviors to a physical Khepera robot.

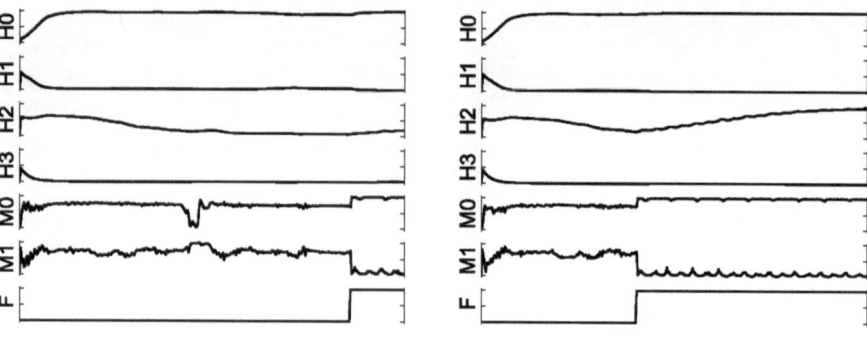

(a) Activity of each neuron plotted over trail 1 of an epoch with reward to the left.

(b) Activity of each neuron plotted over trail 1 of an epoch with reward to the right.

Fig. 12. Neuron activities.

Acknowledgements

This work was supported by the Swiss National Science Foundation, grant nr. 620-58049. Parts of this paper was written during a short stay at the Adaptronics Group, The Maersk Institute, University of Southern Denmark.

References

1. J. Blynel and D. Floreano. Levels of dynamics and adaptive behaviour in evolutionary neural controllers. In Hallam et al. [2], pages 272–281.
2. B. Hallam, D. Floreano, J. Hallam, G Hayes, and J-A Meyer, editors. *From Animals to Animats 7: Proceedings of the Seventh International Conference on Simulation of Adaptive Behavior*. MIT Press-Bradford Books, Cambridge, MA, 2002.
3. O. Miglino, H. H. Lund, and S. Nolfi. Evolving Mobile Robots in Simulated and Real Environments. *Artificial Life*, 2(4):417–434, 1995.
4. R. Sutton and A. Barto. *Introduction to Reinforcement Learning*. MIT Press, Cambridge, MA, 1998.
5. M. Thieme and T. Ziemke. The road sign problem revisited: Handling delayed response tasks with neural robot controllers. In Hallam et al. [2], pages 228–229.
6. E. Tuci, I. Harvey, and M. Quinn. Evolving integrated controllers for autonomous learning robots using dynamic neural networks. In Hallam et al. [2], pages 282–291.
7. B. Yamauchi and R. D. Beer. Integrating reactive, sequential, and learning behaviour using dynamical neural networks. In D. Cliff, P. Husbands, J. Meyer, and S. W. Wilson, editors, *From Animals to Animats III: Proceedings of the Third International Conference on Simulation of Adaptive Behavior*, pages 382–391. MIT Press-Bradford Books, Cambridge, MA, 1994.
8. B. Yamauchi and R. D. Beer. Sequential behavior and learning in evolved dynamical neural networks. *Adaptive Behavior*, 2(3):219–246, 1994.

Evolving Image Processing Operations
for an Evolvable Hardware Environment

Stephen L. Smith, David P. Crouch, and Andy M. Tyrrell

Department of Electronics, The University of York, Heslington, York YO10 5DD, UK
{Steve.Smith,Andy.Tyrrell}@bioinspired.com
http://www.bioinspired.com

Abstract. This paper describes the application of genetic algorithms to evolve new spatial masks for non-linear image processing operations, which are ultimately to be implemented on evolvable hardware. The development environment was custom-built to allow full control over the evolution process and enable the importance of the evolution strategy (including the representation scheme, parameters and fitness function) to be investigated and understood. Results of applying the evolved mask to threshold real-world images are provided and are shown to be an improvement on conventional image processing operations. The envisaged infrastructure for the evolvable hardware is also considered, and the implementation of the image processing operations discussed.

1 Introduction

The use of genetic algorithms in image processing is widespread and diverse, but not particularly well understood. The aim of this paper is to investigate the application of a genetic algorithm (GA) to a simple image processing operation so that the evolutionary strategy and the parameters that govern its operation may be studied in detail.

A spatial mask is evolved to perform the adaptive thresholding of gray scale images. A custom-built environment for the development of the GA is described which gives full control over the representation scheme, the GA, its parameters and fitness function. Results are provided illustrating the evolving of a new spatial mask and its application to real-world images.

The ultimate aim of this work is to implement the algorithms within an evolvable hardware system which will provide maximum flexibility and speed in the solution of demanding image processing problems.

1.1 Evolutionary Algorithms in Image Processing

A broad and disparate range of examples of the use of genetic algorithms and genetic programs in image processing may be found in the literature. Notable examples are the use of genetic programs in the segmentation of medical resonance imaging scans

A.M. Tyrrell, P.C. Haddow, and J. Torresen (Eds.): ICES 2003, LNCS 2606, pp. 332–343, 2003.

[1], a genetic program that performs edge detection on one-dimensional signals [2] and the evolution of genetic programs to detect edges in petrographic images [3].

The work described in this paper is an extension to that first undertaken by Hollingworth and the authors [4]. This was concerned with the evolution of spatial masks to detect edges within gray scale images, which subsequently, could be implemented within an evolvable hardware environment. The latter is described below.

1.2 Evolvable Hardware Architecture to Support Image Processing

The evolvable hardware architecture proposed by the authors for implementation of the evolutionary algorithms is shown in Figure 1. Each processing block in the network is preloaded with a single pixel from the image to be processed.

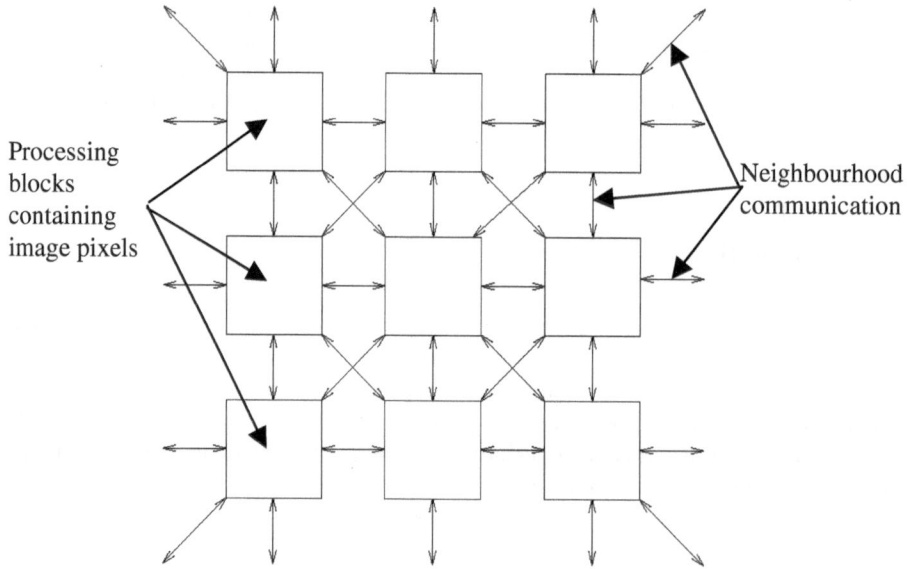

Fig. 1. Evolvable hardware architecture for image processing operations

The common processing block is shown in Figure 2. The main components of this processing block are:

- the particular image pixel under consideration;
- the values of the neighboring eight pixels in the image;
- the logic function to be applied to these (defined by the genotype string), and
- the resulting output pixel that will replace that under consideration.

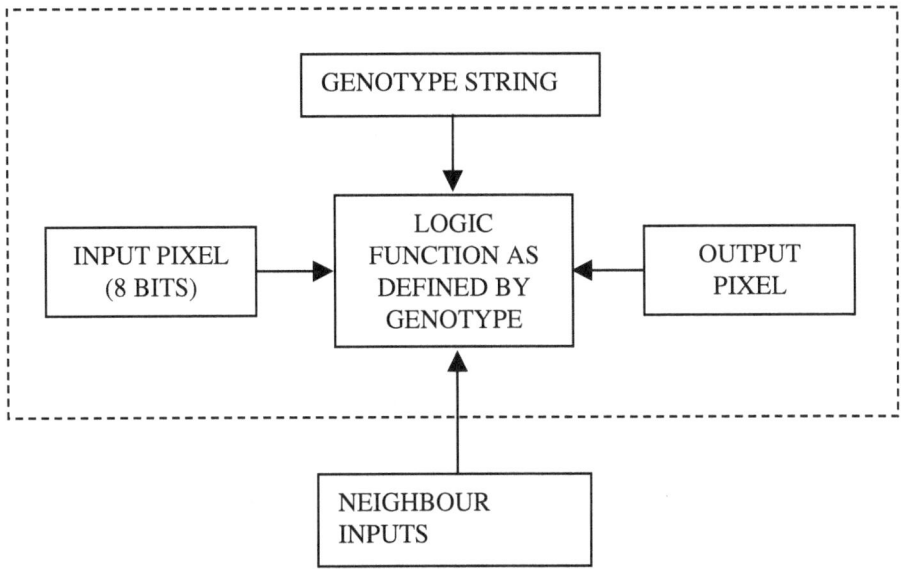

Fig. 2. Single pixel processing block

The genotype string is used to 'program' the logic function, which then calculates the value of the output pixel using the values of the input and neighbouring pixels.

1.3 Evolving New Image Processing Operations

The new image processing operations reported in this paper were evolved using genetic algorithms in a software environment, but were specified such that a future hardware implementation will be possible. The image processing operation chosen to investigate was *dynamic* or *adaptive* thresholding [7] as this is a simple yet clearly non-linear local neighbourhood operation that is not traditionally implemented with a single spatial mask. The conventional method is to apply a different threshold to each local area within the image, based upon some statistical function of the values of the pixels contained therein. The method chosen for comparison in this investigation is the *mean-C* procedure. This simply calculates a threshold value by subtracting a constant C from the mean pixel value for the local area of the image currently being processed. The aim of evolving a single spatial mask to perform adaptive thresholding may appear illogical but provides an interesting platform in which to investigate the evolution of such image processing operators.

2 Method

Although purpose written environments such as GALib [5] and Gpc++ [6] are available for the development of genetic algorithms and genetic programs respectively, a specially written software environment was developed to enable full flexibility and control over the design of the evolutionary algorithms (EA). For this work a standard approach to the implementation of a genetic algorithm (GA) was employed. This requires a population of possible solutions to the problem to be created, which evolves over a number of generations. The fitness of each member of the population is calculated and will determine whether it survives to the next generation or dies. To help generate diversity in the population operations to simulate meiosis, *crossover* (the sharing of genetic information between parents) and *mutation* are often implemented. The main components of the GA, such as the representation of the EA, the parameters governing its operation and the evaluation of a fitness function are now considered.

2.1 Representation

The basic structure of an image processing mask or window operator is simply a matrix of weights which are applied to a number of neigbouring pixels within the image on a repetitive basis until the entire image has been processed. A typical 3x3 pixel mask is shown in Figure 3.

W1	W2	W3
W4	W5	W6
W7	W8	W9

Fig.3. Typical image processing mask operator

Typically, the weights W1-W9 are applied to the respective pixel values and summed. The resulting value is then used to replace that of the pixel of interest, usually the centre pixel. Other manipulations of the pixel element are also common, as are larger masks such as 5x5 and 7x7 pixels.

The representation of the GA used in the evolution of the new image processing operation is simply the weights W1-W9 arranged in a single binary string as shown in Figure 4.

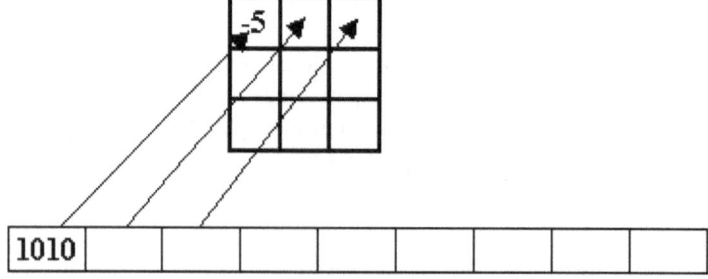

Fig.4. Encoding of a 3x3 spatial mask to a 36-bit genome string

This has the distinct advantages of simplicity in application of the evolutionary process and implementation on the proposed evolvable hardware platform.

2.2 Selection Method

Standard implementations of both rank and tournament selection have been implemented to select individuals for the next generation.

In tournament selection, a specified number of individuals from the population are chosen randomly and the fittest of these is used in the next generation.

Rank selection requires the members of the population to be arranged in an array in order of fitness. Element 0 is the fittest member and element P-1 (P is the population size) is the least fit. A random number *r* is generated between 0 and 1and used within the following formula to generate the position of a 'fit' individual in the population array:

$$i = (\text{int})P \times r^{\text{bias}}$$

bias is the selection strength that varies from 1 upwards such that **bias**=1 means there is an equal chance of selecting every member of the population. Increasing **bias** gives strong biasing towards the fittest individual. *i* is the position in the array of the chosen individual.

2.3 Mutation Method

Three methods of mutation have been implemented to modify individuals for the next generation: *unbalanced mutation, balanced mutation* and *create new population.*

Unbalanced mutation simply loops through each node of the genome and depending on the predetermined mutation rate, replaces the node with a new random value, within a set range. In this implementation, the genome is the binary string containing the spatial mask weights, and each node relates to an individual weight within this.

Balanced mutation is similar to that of unbalanced mutation but an algorithm is employed to ensure the sum of all the nodes within the genome is preserved.

Create new population simply replaces the entire genome (the binary string) with a new, randomly generated one.

In the custom software environment developed for this work, it is possible to employ all three mutation methods randomly in the generation of new members of the population.

2.4 Crossover Method

Crossover is more controlled than mutation for modifying individuals for the next generation as it replaces a node of the genome of one individual with that of another, and not purely a random value. Two methods of crossover have been implemented here: *in-place* and *shuffle positions.*

In-place ensures that the nodes swapped between the *parent* genomes are kept in the same relative positions.

Shuffle positions, however, allows nodes to be placed without regard to their relative positions in the parents' genomes.

As with the mutation facility, within the current evolutionary environment it is possible to employ both crossover methods on a random basis.

2.5 Fitness Function

The role of the fitness function is to evaluate the performance of each individual within the population. In the current context of evolving image processing mask operators, the fitness is determined by applying the evolved mask operator to one or more images and determining the success with which the desired operation has been achieved.

To facilitate an objective measure of fitness a number of good quality images are selected and then purposely degraded to provide test images to which the evolved mask operator can be applied. The resulting images can then be directly compared with the respective original images and a measure of fitness calculated based on the number of falsely identified white and black pixels as follows:

$$fitness = bias*falseBlackPixels + (1-bias)*falseWhitePixels$$

where **bias** is a value ranging from 0 to 1 and is used to weight the fitness according to the desire to deter the identification of false black pixels, false white pixels, or with a value of 0.5 give equal regard.

The final fitness value is simply the sum of fitness values for each test image employed.

3 Results

New spatial masks for adaptive thresholding were evolved using a set of eight test images shown in Figure 5. These 64x64 pixel grayscale images were generated by degrading the set of black and white images shown in Figure 6. The latter are subsequently used as *target images* in evaluating the fitness of each individual of the population.

The set of parameters for the GA employed to evolve the masks are listed in Table 1.

The range of permissible values for each node in the genotype (which equates to the weight in the resulting spatial mask) was of particular interest and a comparison of five different ranges was considered: +/-10, +/-50, +/-100, +/-1000 and +/-10000. The evolved spatial masks and the result of applying each mask to the sixth test image is shown in Figure 7. The relative effect in terms of the best fitness function is shown in Table 2.

Fig. 5. Test images used in evolving adaptive threshold spatial masks

T"lQ9iO	VPldwp	%vWJUy	6D[s(Kn
eR"rz$M	£2&uX.?	okt8,YB	ZjA'5Lc)
N;h7Sa	IFbmx4C	G@qgfE:	]3H-0
%wUT2	o.uzxM5	YSDmozt	Ot8vPJ"
kJD'Ed9	5bdmjK6	yU?mp:m	OQoJ'"2
.g9vx1Rh	OWD?K	[49VxY	?jWZ"W

Fig. 6. Target images used in evaluating the fitness of the evolved spatial masks

Table 1. Parameters employed for first evolution of adaptive threshold masks

Parameter	Value
Dimension of evolved mask	3x3
Population size	100
Generations evolved	1000
Selection method	Tournament
Mutation method	All
Mutation rate	0.2
Crossover method	All
Crossover rate	0.95

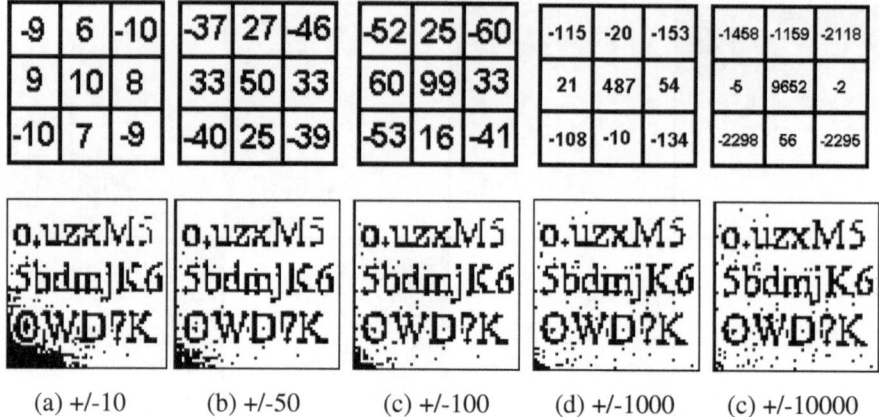

| (a) +/-10 | (b) +/-50 | (c) +/-100 | (d) +/-1000 | (c) +/-10000 |

Fig. 7. Effect of node range value on evolved spatial masks and subsequent application to test image 6

Table 2. Effect of node range on average and best fitness

Node range (+/-)	Best fitness (%)
10	65.16
50	72.50
100	74.48
1000	77.17
10000	77.60

The evolved masks were then applied to real world images which are considered demanding for successful adaptive thresholding using conventional approaches. The original images are shown in Figure 8 and the result of conventional global and adaptive thresholding (using the mean-C technique) are shown in Figures 9 and 10 respectively.

Fig. 8. Original 'real-world' images

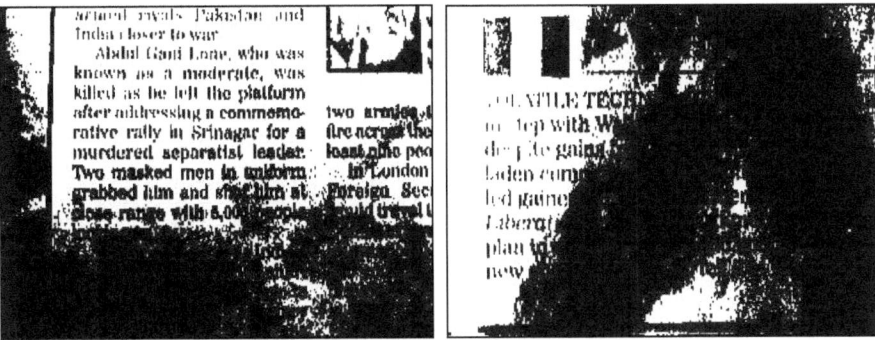

Fig. 9. Conventional global thresholding of images

Fig. 10. Conventional adaptive thresholding of images (3x3 mask)

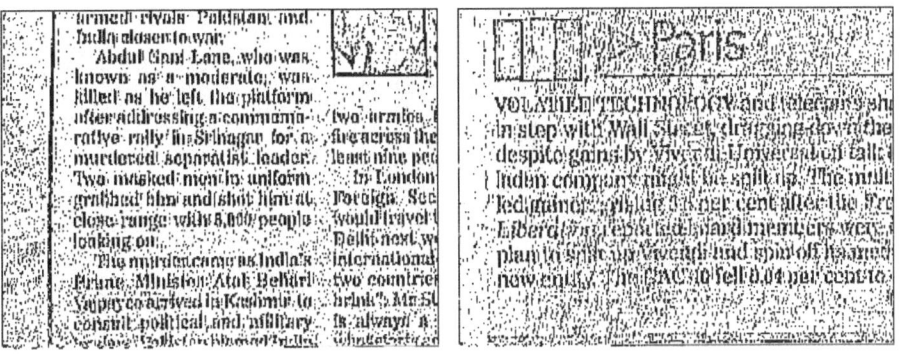

Fig. 11. Evolved 3x3 mask (50% bias) applied to images

The results of applying the evolved 3x3 spatial mask with node range of +/-10000 to the real world images is shown Figure 11. It is arguable whether the text is easier to read after applying the evolved mask or the conventional mean-C operator. Although there is more noise with the evolved mask, there are no 'faded out' patches.

In an attempt to reduce this noise another 3x3 mask was evolved, but this time with a node range of +/-100 (rather than +/-10000) and the fitness function bias set to

75% to generated less false black pixels. The results of this can be seen in Figure 12 and are arguably an improvement on the conventional adaptive threshold and the previously evolved mask.

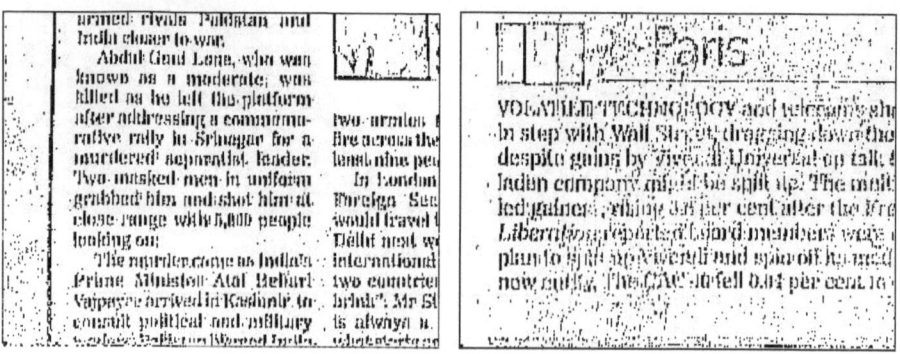

Fig. 12. Evolved 3x3 mask (75% bias) applied to images

Finally, evolution of a 5x5 pixel masks was investigated, again with a node range of +/-100 and a fitness function bias of 75% (other parameters remained the same as listed in Table 1). The evolved mask is shown in Figure 13 and the result of applying the evolved mask to the 'real-world' images is shown in Figure 15. For comparison, the result of applying a conventional 5x5 mean-C adaptive threshold mask is shown in Figure 14. It is again arguable that this evolved mask is an improvement on both the conventional adaptive thresholding and previously considered evolved masks.

For information, the evolution of the 5x5 mask in terms of best and average fitness per generation is shown in Figure 16.

-9	-1	-53	-9	21
-32	-19	78	-37	-66
-32	89	100	74	26
-20	-39	94	-15	-58
8	-13	-62	6	-8

Fig. 13. Evolved 5x5 adaptive threshold mask

4   Conclusions

The aim of this paper is to demonstrate that simple, yet effective, image processing spatial mask operators may be evolved, and that these may be easily implemented on an evolvable hardware platform. The adaptive thresholding mask evolved is comparable, if not an improvement, on a conventional mean-C thresholding mask and far superior to conventional global thresholding operation. The evolvable hardware platform considered for hosting these masks, which is currently under development, will enable further masks to be evolved in real-time to provide unique solutions to demanding image processing applications.

Fig. 14. Conventional adaptive thresholding of images (5x5 mask)

Fig. 15. Evolved 5x5 mask (75% bias) applied to images

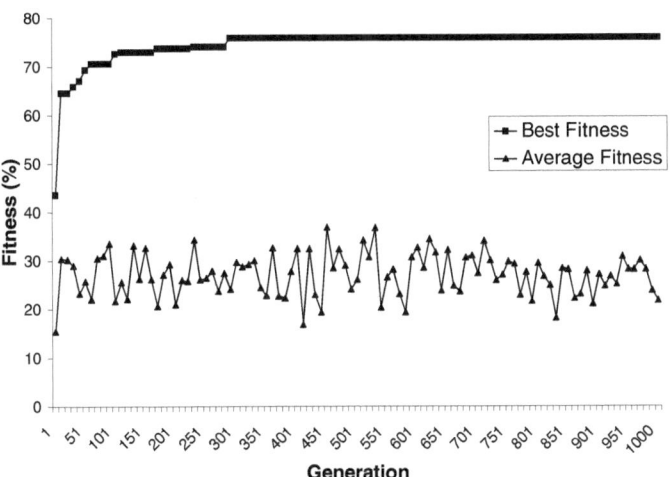

Fig. 16. Evolution of 5x5 spatial mask: best and average fitness per generation

References

1. R. Poli: Genetic Programming for Image Analysis. Genetic Programming: Proceedings of the First Annual Conference (1996) 363-368.
2. C. Harris and B. Buxton: Evolving edge detectors. Research Note RN/96/3. University College London, Department of Computer Science (1996).
3. B. Ross, F. Feuten and D. Yashkir: Edge Detection of Petrographic Images Using Genetic Programming. Brock Computer Science Technical Reports, Brock University, Ontario, Canada CS-00-01 (2000).
4. G. Hollingworth, A. Tyrrell and S. Smith: Simulation of Evolvable Hardware to Solve Low Level Image Processing Tasks. In: R.Poli, et al. (eds.): Evolutionary Image Analysis, Signal Processing and Telecommunications. Lecture Notes in Computer Science, Vol. 1596. Springer-Verlag, Berlin Heidelberg New York (1999) 46–58.
5. M. Wall: GALib. Massachusetts Institute of Technology (MIT), http://lancet.mit.edu/galib-2.4/ (2002).
6. A. Fraser: Genetic Programming in C++. Technical Report 040, Cybernetics Research Institute, University of Salford (1994).
7. K. Castleman: Digital Image Processing. Prentice Hall, New Jersey (1996).

Hardware Implementation of a Genetic Controller and Effects of Training on Evolution

M.A. Hannan Bin Azhar and K.R. Dimond

Department of Electronics, University of Kent, Canterbury, Kent, CT2 7NT, England
Tel: +44(0)1227823486
{maha2,k.r.dimond}@ukc.ac.uk

Abstract. This article describes an FPGA (Field Programmable Gate Array) based hardware implementation of a genetic controller to be applied for the evolution of an Artificial Neural Network (ANN) [3] for collision-free navigation task of mobile robots. The adaptive nature of ANN enables it to train itself while the robot interacts with the environment. In addition to online training, the genetic evolution in neuron bits will be examined in an experiment to understand the interaction between evolution and lifetime adaptation of the ANN. The concept of chromosome for navigation task, design techniques of various blocks inside the GA controller will be elaborately described here.

1 Introduction

Genetic Algorithms (GAs) introduced by John Holland [6] are adaptive search strategies based on a highly abstract model of biological evolution to find a possible solution in a given problem space. The genetic process can evolve the neural architecture of an ANN. It starts with a population of robots with random configuration bits known as chromosomes. Genetically-inspired operators like crossover and mutation are used to introduce new individuals into the population [6]. To realize the GA operation inside an FPGA, a random generator, a crossover block, a partner selector and a fitness block were modeled in VHDL (Very High Speed Hardware Description Language). GA based approach is the basis for most evolutionary robotics applications [6, 10]. Neural networks can easily exploit various form of learning during lifetime and this learning process may help and speed up the evolutionary process [1]. The neural network will undergo change as a result of two forces: learning during the "lifetime" of a network and evolutionary change over the course of several "generations" of networks. It will be investigated how learning at the individual level can have an influence on evolution at the population level. In a sensory-motor navigation task the inputs to an ANN [3] are object patterns from the sensors which tell the ANN about obstacle-scenarios around the robot. Depending on the pattern a decision is made by the ANN and appropriate commands are sent to the motors for proper manoeuvre. Each robot is evaluated according to an objective fitness function derived from its ability to avoid collisions. The larger the fitness the more chance the robot has to produce offspring in the next generation of population. Over a number of generations, the fitness of the population increases and successful architectures are created. The use of an FPGA can allow high speed parallel processing [2, 9, 13] of the neural network and the genetic controller. VHDL was used for modeling devices as it

A.M. Tyrrell, P.C. Haddow, and J. Torresen (Eds.): ICES 2003, LNCS 2606, pp. 344–354, 2003.

allows simulation over many different levels of abstraction and is widely accepted as a standard for modelling hardware [12, 11, 7]. The design package available for the project, MAX+Plus II version 10.1, converts the VHDL design file into a bitstream file which configures the FPGA. The ANN that will be evolved was implemented by the same authors and can be found in [3].

2 Chromosome

The heart of the control structure of the robot is its Chromosome. This consists of an array of bits of storage. These bits are grouped together to form neurons. The size of the neurons or the number of bits per neuron, neurons per class and number of classes in a network actually define the architecture of the network. Neuron contents or the bits of the neurons define the behaviour of the network. All these behavioural bits and network architecture are stored in the chromosome in binary values. In a typical design [3] for a network with 3 classes, 5 neurons per class and 8 bits per neurons the number of bits in a chromosome will be 120. The pattern of these 120 bits in chromosome will define the behaviour of the ANN. The next part of the chromosome is the configuration bits for sensor modules. For each sensor two bits are required, so for 8 sensors the total number of configuration bits are 16. Two bits of each sensor will determine the presence of that sensor in the next generation. The different combinations of bits could be "00" ,"01", "10" and "11". The control strategy was such that the sensor would be enabled only when the genetic pair is "00", for other three cases it is disabled. Thus the probability of the phenotype of having that sensor to the next generation is 25%. This approach is called the "dominance approach" [8] where all features are recessive [4].

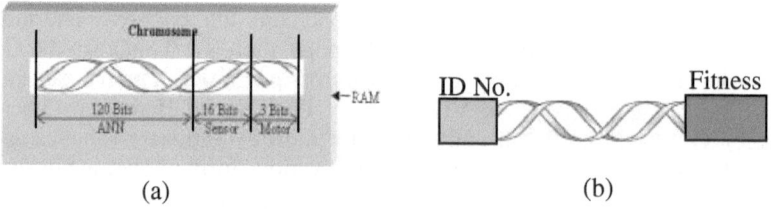

Fig. 1. (a) Chromosome in RAM (b) Identification number and Fitness value with the chromosome

The last three bits in the chromosome are reserved for the speed control of motors. 3 bits give 8 levels of speed. In a group of robots a number is associated with each robot to identify it from other robots. This identification number can be stored in the RAM at one side of the chromosome. The length of the identification number depends on the number of robots. For three or four robots the length is two bits, whereas for five to eight robots the length should be three bits. In general for $2^{n-1} + 1$ to 2^n number of robots the length of the identification number will be n bits. Fitness value of the robot can be stored in the RAM at other side of the chromosome as shown in the Figure 1. Thus the fitness which measures the robot's performance will be written to this particular location of the RAM. The length of the fitness value could be as long as 17 bits as used in the designed system.

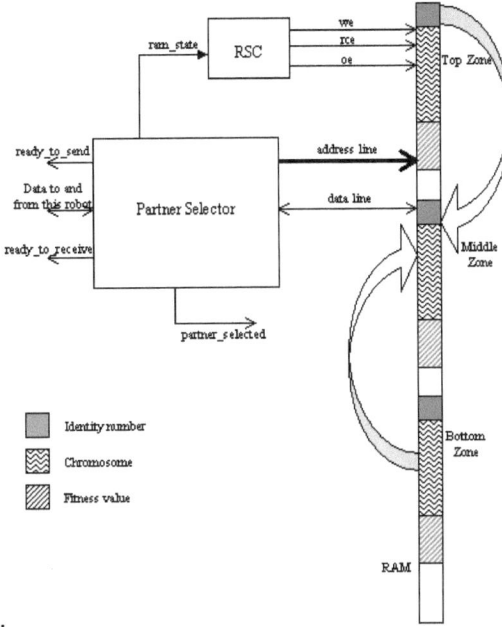

Fig. 2. Partner Selector Block

3   Partner Selector

Among a group of mobile robots partner selection criteria could be as simple as choosing the robot with highest fitness value to mate with all other robots. In other ways it could be more complex like the roulette-wheel technique [10]. Due to the restriction in number of cells in the FPGA and to avoid the complexity the simplest approach should be preferred. In a group of robots each robot could be identified by a certain number. For example in a small group of three robots it could be robot 1, robot 2 and robot 3. A similar block as shown in the Figure 2 could be modelled in VHDL to select the best robot to breed. A comparator in this block compares the fitness values and singles out the highest one. In the RAM three different zones can be used to store three different robot's information. Information consists of a robot's identification number, chromosome and fitness value. As shown in Figure 2 in the top zone the robot's own information is kept. In the middle and bottom zone information from the two other robots can be stored. The information with a greater fitness value between the middle and bottom zone is always swapped to the middle. When the robot receives the information data from another robot it stores the data in the bottom zone of the RAM. This means that previous data in the bottom zone can be overwritten by the new data. Upon receiving the next robot's data again the fitness value of the bottom zone is compared with the fitness value in the middle zone and if it is greater, then all the information stored in the bottom zone are swapped to the middle zone. After swapping the "ready_to_receive" signal goes to HIGH to indicate that this robot is now ready to receive data from another robot. If this is the turn of this robot to send its own information, then the 'ready_to_send' signal goes to HIGH instead of the 'ready_to_receive'. When all the 'ready_to_receive' signals from all other robots are HIGH then it sends its own information data from top zone of the RAM. When all the robots' information are received and compared and swapped to the middle zone, the middle zone contains the information data of the robot which has the highest fitness value. At the beginning of the partner selection process the robot's own information data is copied from the first zone to the middle, so if its own fitness is the highest then still it will be preserved in the middle zone as the comparator always keeps the data with higher fitness value in the middle. After the selection process the 'partner_selected' goes to HIGH which triggers the crossover. To read or write in the RAM the 'ram_state' signal is sent to the RAM State Controller (RSC). Depending on the 'ram_state' signal 'we', 'oe' and 'rce' signals are set by the RSC. Details of the RSC can be found in [3].

4 Crossover

The Partner selector block selects the best chromosome, that with the highest fitness value, and saves the chromosome in the middle zone of the RAM (Figure 2). The crossover block now copies a portion of the best chromosome and then pastes it on the robot's own chromosome stored in the top zone. So some characteristics of the best chromosome will be passed to the population in the next generation. A logic HIGH on the 'partner_selected' signal enables the crossover block and swapping of bits between best chromosome and the robot's own chromosome starts. First a point of crossover is selected randomly. The modelled crossover block is shown in the Figure 3. The block gets a random number input within the length of the chromosome to start the crossing over at a random point. If the chromosome has a length of 139 bits then the random number could be between 1 to 138. For example if the number is 136 (as shown in simulation in Figure 4) then the four bits from 136 to 139 of best chromosome will be swapped sequentially to the same locations of the robot's own chromosome. In the Figure 3 the part that has to be swapped is identified by the shaded area. For accessing the RAM proper RAM control signals are defined by the 'ram_state' signal to the Ram State Controller (RSC) . This signal is also shown in simulation as 0 (for idle), 1 (for writing) or 2 (for reading).

Once the crossover is completed the 'crossdone' signal is set to HIGH for one clock period which disables the crossover block until the 'partner_selected' signal again goes to HIGH. The simulation of this block shows proper operation (Figure 4). Random number was 136 in the simulation and while crossing over, it first reads one location of a chromosome and then writes it to the same location of the other chromosome and then it copies and pastes the next location and this way it sequentially progresses until the last bit of the chromosome is moved.

5   Fitness Block

A reward-punishment scheme was applied while evaluating fitness. A robot's fitness is evaluated while a generation is running. The fitness function measures the performance of the robot with respect to the desired behaviour and task. For the obstacle avoidance problem a simple rule can be applied: fitness will be increased (rewarding) each time the robot successfully avoids any obstacle. Its fitness will be decreased (punishing) each time it collides. For example a fitness function can be programmed like this:

1. Fitness is increased by one every second the robot moves avoiding collision,
2. Fitness is decreased by three in every operating cycle if the robot has a collision ,
3. If the robot starts turning for more than five seconds then the fitness is again decreased by two.

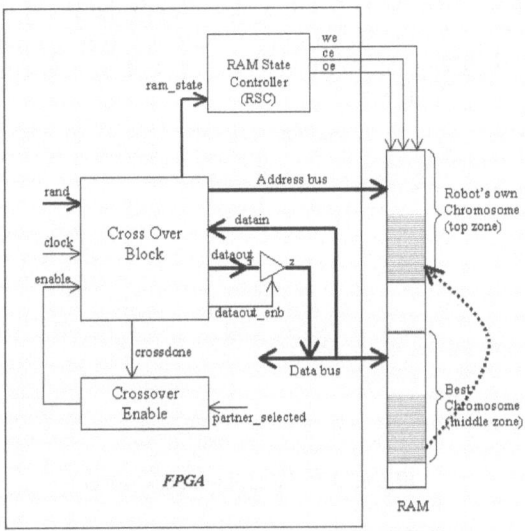

Fig. 3. Crossover Block

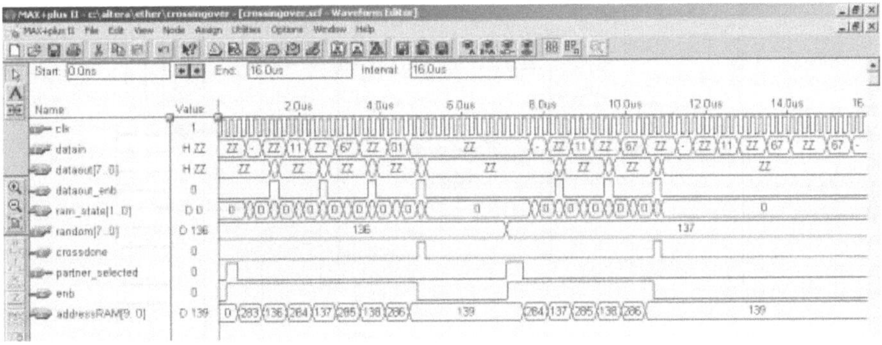

Fig. 4. Simulation of Crossover Block

Similarly more sub functions can be added to access the performance of the robot. In the above example the first function encourages the exploration and second and third functions punish the robot to discourage collision and spinning. A fitness block was modelled in VHDL to calculate the fitness value. This calculated fitness value is stored in a 17-bits register which can be called as fitness register. The robot life starts with a sufficient fitness value such that even if the robot collides all of its life time, the fitness number comes out at the end of the life should not be less than zero. So for a robot with a life time of five minutes the starting point could be in the range of 90,000. It loses 3 points for every collision, loses 2 points if it starts turning for more than five seconds and in every second it gains one point if it moves forward. These timings were measured by a counter. Signals from microswitches (ms1 to ms8) are fed into this block to detect collisions. To know the direction of the motor the direction command (m_dirc in Figure 6) from the motor control block are also fed into this block. Once a generation is finished the calculated fitness value is written in a particular block of the RAM. The state (read, write, idle) of the RAM is determined by the 'ram_state' signal sent to the RAM State Controller (RSC). Fitness block has three states: Initialisation, Evaluate, Writing. Two control signals (fwenb, rset) actually define the state of this block.

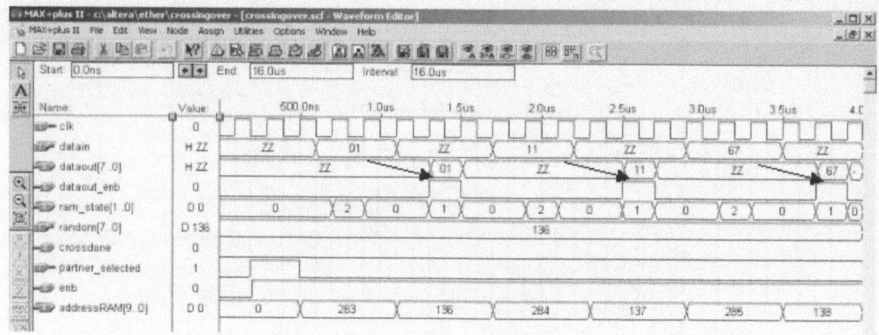

Fig. 5. Data Swapping in Crossover

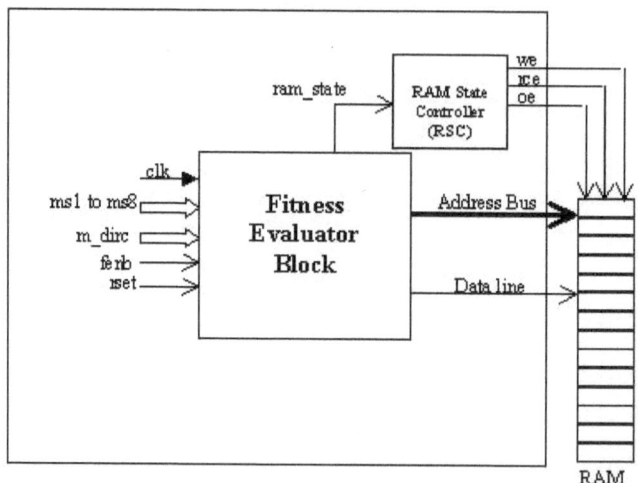

Fig. 6. Fitness Block

Simulation results of this block are shown in Figure 7. The affect of collision and spinning on fitness value is shown in the figure. When the 'm_dirc' is '01' it indicates that the robot is spinning and the 'd' variable starts counting up (Figure 7) and when it reaches a value of 500 (which indicates 5 sec when a clock with a period of 10ms is supplied) then 2 is subtracted from the fitness value. So in the Figure 7 at point 'B' value of the 'regfit' changes from 90001 to 89999. The register named 'regfit' was used to store the fitness value. When the fitness block is initialised the 'regfit' starts with a value 90000. At the point 'A' in the simulation a 1 is added with 90000 after the robot moves forward successfully for 1 second. This 1 second is determined by a variable 'c' when it counts to 100 with a 100Hz (10ms period) clock input. At 'C' in the simulation the regfit is again initialised to 90000 as the control inputs are rset = 1 and fenb= '0'. At 'D' a collision is detected and it continues for a while. Due to this collision fitness value is decreased by 3 in every 10 ms, so if a collision stays for 1 sec then 300 will be subtracted. This simulation is just an example of a fitness block. Other fitness functions can be added in this block.

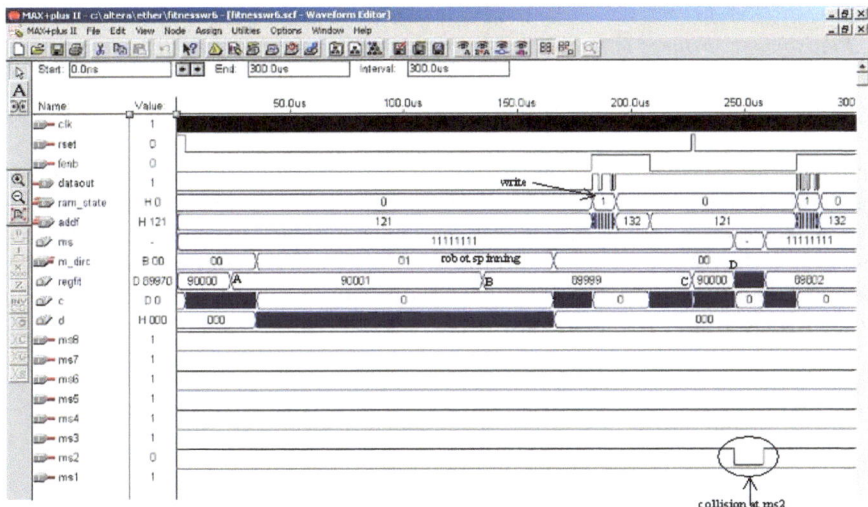

Fig. 7. Simulation of Fitness Block

6   Training on Genetic Evolution

The objective of this experiment was to find out how the training in robot's life time could effect the evolution of the neural network. The evolution was realised among three robots. In the typical system the partner selection and the crossover operation of the genetic process were realized inside the PC, but the fitness value was calculated inside the FPGA. At the end of each generation the fitness value and the neuron bits were transferred from the FPGA to the PC and the PC then took over the control of the rest of the genetic control. Once the new offspring was created, the neuron bits were downloaded to the robot's RAM. To save design and construction time proto-type boards (with Altera's FLEX EPF8452ALC84-4 chip) were used to house the robot's controller modelled in VHDL. Each prototype board has a limited number of I/O pins which led us to employ three boards to house the complete controller. The board containing the ANN was 80% full using 269 logic cells of Altera's EPF8452ALC84-4 chip and the board containing the fitness block was 66% full using

224 logic cells of a similar FPGA chip. The third FPGA was carrying the sensor and motor control blocks and it was only 30% full utilizing 102 logic cells of the FLEX chip. As one set of FPGA boards were designed, offspring were downloaded one by one. Two distinct evolution were examined: trained and untrained. In trained evolution the robot was trained in every generation. The untrained evolution was purely genetic without training. An environment was created with obstacles and walls where each robot was evaluated one by one. The robot's performance was measured by a fitness function which was as follows:

1. *Each generation starts with 90000 points.*
2. *Reward: Fitness is increased by 300 points in every 1 second robot moves forward without any collision.*
3. *Punishment: Fitness is decreased by 3500 points in every 10 seconds robot is turning left or right.*
4. *Punishment: Fitness is decreased by 5 points in every 10.95 milliseconds if robot has a collision.*

Different points in the fitness function were carefully chosen so that it gave more or less the same weight for reward and punishment. Though it seems that only 5 is subtracted for a collision but actually a huge number will be subtracted if a collision stays for a while. This is because 5 is subtracted in every 10.95 ms (operating cycle) so if a collision stays for 10 second then accumulated subtracted number will be 4550. Similarly if the robot moves forward for 10 seconds without collision then the fitness value can increase by up to 3000 points. The settings of the experiment are summarized in Table 1.

Table 1. Summary of Experiment Settings

Parameter	Definition
ANN Controller for Trained evolution	ANN in [3]; 3 classes-5 neurons per class-8 bits per neuron
ANN Controller for Untrained evolution	ANN in [3]; 3 classes-5 neurons per class-8 bits per neuron
Initial Fitness Value	90,000
Generation time	2 mins
Crossover	One point
Mutation	0%
Training Algorithm	As described in [3]
Initial Neural Contents	Random
Motor Speed	Fixed at maximum
Sensors Enabled	3 sensors on the left, right and front were enabled.

6.1 Discussion

This experiment demonstrates that the trained evolution converges to the optimum solution much faster than the untrained evolution. Both evolutions were started with the same random chromosomes and after five generations all three robots in the trained evolution achieved higher fitness values than the untrained robots of the same generation. A chromosome which is biased in one generation can be a very good

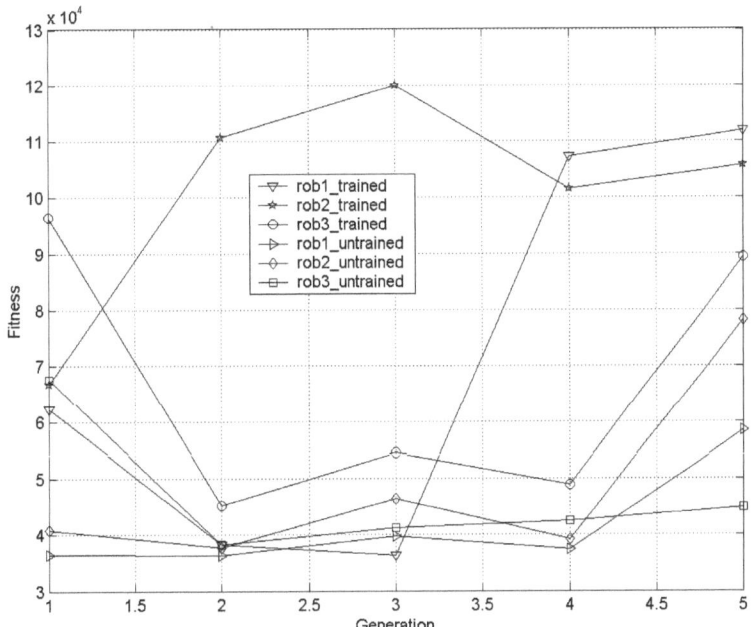

Fig. 8. Robot's Fitness in Trained and Untrained Evolution

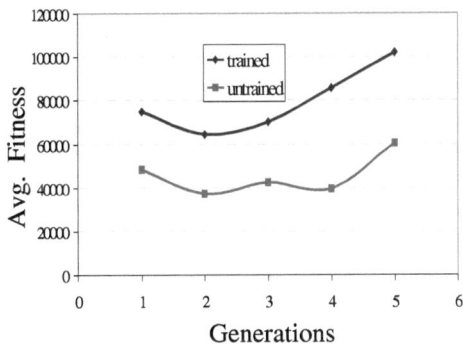

Fig. 9. Higher Fitness in Trained Evolution

chromosome in the next generation. A sharp rise of the fitness value of robot1 from generation-3 to generation-4 explains this phenomenon. In generation-3 robot2 had the best chromosome, highest fitness value, and at the end of this generation the crossover between the chromosomes of robot1 and robot3 improves the biased chromosome of robot1 to a very good chromosome in generation-4. After that robot1 dominated and it also helped robot3 to improve its chromosome abruptly from generation-4 to generation-5 and this explains the sharp rise of fitness value of robot3 from generation-4 to generation-5 in Figure 8. So when training is applied there is always a good possibility of having a robot with a very good chromosome at a much earlier stage than in untrained evolution. Once a good robot comes out it will help to improve the chromosomes of other robots in successive generations. In untrained evolution the probability of getting a good robot at early stage is much less. Floreano and Mondada [5] has described that the Khepera robot genetically learns to navigate and avoid obstacles in less than 100 generations and in around 50th generation the best individuals exhibit a near to optimal behaviour. But the current experiment shows that a simple training algorithm can reasonably drop down the number of generations to get the control structure for well navigated robots. There was one biased action found in this experiment that even could not be corrected by either training or evolution. When there were obstacles both on the left and the right all the robots were always turning to the left instead of going forward. This could be corrected by mutation. Mutation flips the bits from 1 to 0s in several random positions of the chromosome. So any biased actions due to many 1s in a particular block of the chromosome can be gradually changed by mutation in a number of generations. Alternatively, when the population is large the probability of getting robots with good chromosomes increases. So chance of getting well navigated robots in a large population is always high. Well navigated robots can improve the robots with bad chromosomes by evolutionary process even without mutation. Another point to be noted in this experiment is the abrupt fall of fitness value of robot3 from generation-1 to generation-2. This was the effect of training in unexpected situations like facing an obstacle below the height of the sensors. The unexpected situations were created on purpose by hand to find out how the training algorithm works in unexpected situations where the object could be too short to be detected by the photodiodes. So upon a collision, the supervisor circuit assumes that there is an object although the detector fails to detect it. This gives wrong address information at the inputs of the neural network and thus the training data will be written in the wrong locations of the ANN. Thus the robot starts behaving strange after the training. Because of this strange behaviour fitness value of robot 3 fell down sharply at generation-2 in the trained evolution. Still through the evolutionary process with the help of other good robots the robot3 again got back a better fitness value at generation-5. It demonstrates that if we disturb the system by pulling it away from its stable state, it will get back to the stable state by the evolutionary process.

7 Conclusion

The experiment in this article explains that the learning process in robot's life time can help to speed up the evolutionary process. It also demonstrated that the evolutionary process helped the system to get back to its stable state even though the system was disturbed. This experiment does not involve the collective behaviours of the robots. A future work could be to constitute a fully distributed control systems embodied in FPGAs for more than one robot at a time. If the controller is entirely decentralized it will be inherently scalable to large numbers of robots in a shared task environment. The current designed controller will definitely provide an appropriate substrate for this and will be an interesting platform for the future work in collective robotics and artificial life.

References

1. Ackley, D.H., M.L. Littman. (1991). Interactions between learning and evolution. In Artificial Life II, edited by C. G. Langton, J. D. Farmer, S. Rasmussen, C. E. Taylor. Addison-Wesley. Reading, Mass.
2. Agarwal, L., Wazlowski, M., Ghosh, S. (1994). An asynchronous approach to synthesizing custom architectures for efficient execution of programs on FPGAs, in Proceedings of the International Conference on Parallel Processing, vol. 2, pp. 290-294.
3. M. A. H. B. Azhar and K. R. Dimond, *"Design of an FPGA Based Adaptive Neural Controller For Intelligent Robot Navigation"*, to appear in proceedings of EUROMICRO Symposium on Digital System Design, IEEE Computer Press, September 4-6, 2002. Dortmund, Germany.
4. Campbell, N. A., Reece, J. B., and Mitchell, L. A., "Biology", 1st ed. California: Addison Wesley Longman,1999.
5. Floreano D., F. Mondada. 1994. Automatic Creation of an Autonomous Agent: Genetic Evolution of a Neural-Network Driven Robot. In From Animals to Animats 3: Proceedings of Third Conference on Simulation of Adaptive Behavior, edited by D. Cliff, P. Husbands, J. Meyer, S. W. Wilson. Cambridge, Mass, MIT Press/Bradford Books.
6. Holland J. H. 1975. *Adaptation in Natural and Artificial Systems*. Ann Arbor, Mich., University of Michigan Press.
7. IEEE Inc., Standard VHDL Language Reference Manual, New York , 1998.
8. Lerner, I. M., and W. J. Libby. 1976. Heredity, Evolution, and Society. 2nd Ed. Freeman & Co., San Francisco,pp.119-174.
9. McLeod, J. (1994). Reconfigurable computer changes architecture, Electronics, p. 5.
10. Mitchell, M., "Genetic Algorithms and Artificial Life", In: Langton C.G.(ed.), Artificial Life- An Overview, MIT Press, pp 267-289,1995.
11. Nawabi Z., VHDL: Analysis and Modeling of Digital Systems, McGraw-Hill, Inc., 1993.
12. Perry, D.L. (1991). VHDL, McGraw-Hill, New York.
13. Wazlowski, M., Smith, A., Citro, R., Silverman, H. F. (1996). Armstrong III: A loosely-coupled parallel processor with reconfigurable computing capabilities, Tech. Rep., Brown University.

Real World Hardware Evolution: A Mobile Platform for Sensor Evolution

Robert Goldsmith

School of Computer Science,
University of Birmingham, England
R.S.Goldsmith@cs.bham.ac.uk
Phone: +44(0)121 414 3707

Abstract. Although hardware evolution is becoming a more popular topic of research, the main focus of this research tends to be with re-configuring electronic circuits using evolutionary techniques. Taking a step back, my research looks at some of the problems of configuring autonomous, mobile systems for varying goals and environments. Concentrating on optical sensors, I am hoping to show that evolving the placement of sensors on the surface of the entity and the frequencies of light these sensors respond to will improve the entity's performance in the environment.

This paper discusses some of the issues of working outside of simulation and presents a hardware platform I consider solves many of these to enable me to carry out the research described above. Unlike other mobile solutions, this platform was designed to be cheap, work in multiple entity environments and cope with large numbers of sensors (in my research, close to 200 sensors per entity) while still being extensible. This paper also covers issues such as processing power, environment management and entity interactions.

1 Introduction

A view slowly growing in popularity among evolutionary hardware researchers is that the richer the medium in which you can evolve solutions, the more effective the evolutionary approach and the 'better' (always an experimentally dependent measure) the resulting solutions [3]. This has lead to research which tries to exploit the inherently complex properties of physical mediums such as doped silicon (using FPGAs and ASICs), radiation damaged silicon and more exotic materials such as liquid crystals and complex polymers [3,1]. However, very little work has been done on a more macro level, working in real (although not necessarily cluttered) environments with real hardware. My research intends to focus on this larger-scale environment and hope to answer many of the questions surrounding the potential benefits of allowing hardware to configure portions of itself in its native working space.

The research will focus on the placement and sensitivity of light sensors on mobile robots interacting with each other and a real environment. Goals for the

A.M. Tyrrell, P.C. Haddow, and J. Torresen (Eds.): ICES 2003, LNCS 2606, pp. 355–364, 2003.

robots will be based on finding 'Food Cubes' which can emit light of different wavelengths and mating by bumping into other robots. Any hardware platform for this research needs to meet a number of criteria and this paper presents a new platform which meets these needs and is flexible enough to be used by others both beyond the scope of my research and in entirely different research areas. In section three the criteria are described and in section four the new design is described in more detail, showing how the criteria have been met. Finally, in section five, some of the restrictions of the platform are covered with possible solutions mentioned for completeness (although they will not be implemented in the current platform). First, however, section two takes a closer look at why real world experiments can be beneficial as well as some of their disadvantages.

2   Simulation vs Real World Experiments

Carrying out any form of research using simulations has a large number of benefits. The researcher is lord and master of the environment they are working in: capable of stopping and rewinding time, moving objects around and even changing the laws of physics at a whim. It is a hugely attractive way to work and there is much that can be gained from working with them. However, simulations should not be treated as identical to the real world and, if you want real-world experimental results, relying on a simulations to provide them is fraught with dangers.

When researching robotics and evolutionary hardware, many researchers do not decide to use simulations because of the benefits they give, but more because it would be more difficult and expensive to do the same experiments in the real world. A simulated environment allows more experiments to be run at once, is often cheaper and allows certain tricks and 'cheats' that you simply could not apply in the real world (such as stopping time so you can calculate reward and fitness values for your population). The problems start with how good the simulation is. Of course, if the research is not really bothered about the real world, a simulation is the best way to carry out the experiments. But as soon as you try to evolve an entity in simulation and then transfer that entity to the real world, building the simulated environment suddenly becomes much more difficult [2]. Evolutionary methods are adept at finding small advantages in unexpected places and using information from the environment to great advantage. Any simulation would have to model the real environment very closely in order to 1) allow the evolutionary processes to find good solutions that take full advantage of the real environment and 2) find solutions that don't rely on quirks of the simulated environment. In a simulated environment designed for robots with only light sensors (for instance) reflection, refraction, shadows, surface textures and imperfections in the environment may all have huge information content that the evolutionary processes could make use of.

Another problem with simulated environments is noise. Real environments are very noisy while simulated environments are not. To overcome this, many simulations introduce white (random) noise into sensor values and movements within the environment. In the real world, there is no such thing as white noise - that is, noise with zero information content. Sensor values may fluctuate and movement may not be accurate but these are not random. They are caused by something either inside or outside the environment. Slippery or bumpy floors could help with location identification. Sensor value interference due to noise from another robot's motors could help a robot find a mate. Even external noise could help in navigation because it is unlikely to be uniformly distributed. Over time, some types of noise will also change. Again, this is very unlikely to be random and could also be beneficial.

3   Why Design a New Platform?

Before committing to the time and expense of designing a new platform, the readily available platforms were considered. However, as is described in more detail below, the requirements for the platform are quite strict and none of the available platforms could meet them all:

Multiple Entities. The research the platform was to be used for is based on an evolutionary approach with fitness evaluation being implicit in the ability of the entity to find food and mates. For this, multiple robots would need to coexist within the same environment. This immediately restricts the methods of communication between robots, their environment and any housekeeping systems. Any technologies implemented would have to be wire free and a number of potentials were considered. Ease of implementation, reliable connection, low cost and reasonable data transmission were all prerequisites.

Processing Power. Each of the robots is a complete unit and needs to be able to run a control system to survive in its environment. Control systems to be tested in the research include neural networks (static and evolved), hand-programmed systems and others. For this, quite considerable processing capabilities would be needed. On top of this, overall control of the experiments was also required. This overall control system will centrally determine, among other things, which potential offspring to initiate and when each entity dies as well as which inactive robot to use next for a new entity (based on battery level).

Expandable Design. The research demands a design that can cope with very large numbers of analogue light sensors. To maintain the usefulness of the robots, the design should also be flexible enough to accommodate other sensors of both a digital and analogue nature as well as allow some expansion facilities for co-processing units.

Low Cost. As was mentioned above, more than one robot will be active in the environment at once. In fact, the aim was to have 6-8 robots active at any one time. Allowing for charging times of the robots, about 10 robots should be needed in total. On top of this, a small number of other units such as the food cubes will be needed. The need for multiple robots forced the need for low cost. Each robot should cost less than £300. For instance, this is, at the current exchange rate at the time of writing, is under one quarter of the cost of a Khepera robot (which would still need further development).

4   Platform Overview

The biggest decision in the design of the platform was to move control processing from the robots to a desktop computer. This centralised all the control processing, sensor data collection, housekeeping and other environmental control and reduced the cost. It also meant that more processing power could be added easily by either replacing the machine or distributing the processing across a cluster. However, this placed an additional load on the communications network as each robot would need to be serviced quickly enough to allow it to react to a dynamic environment. The overall topology of the platform is shown in Fig 1. The host computer communicates via a packet switched radio network to each robot on a polled basis, requesting new sensor values and returning updated motor control values.

Inter-robot communications and those between robots and the environment are handled locally by using IR communications. The result of these exchanges are relayed back to the host computer to enable it to work out what interactions took place between which robots / environment objects.

4.1   Host System Control and Housekeeping

The host computer takes care of almost all of the processing and housekeeping needs of the experiments and its tasks can be broken up into three distinct blocks: robot control, packet radio and environment control and robot interaction tracking. Fig 2 illustrates the three layers. The actual communication is always over the radio network. However, this is hidden from the higher layers and each robot controller is completely unaware of the communications network or any of the other traffic over that network.

Robot Control. The biggest job of the host computer is to run the control system for each of the robots. For each robot, the host computer needs to configure the control system according to specifications (such as the robots genetic makeup), receive sensor information from the robot, process that information and send back motor control commands to adjust the robot's behaviour. Each robot needs to be serviced approximately every twenty milliseconds.

Further to this, when two robots mate, the host computer needs to appropriately create an offspring candidate. Because there is a strict limit on the number of robots in the environment at any one time, the problem of over-populating needs to be dealt with. The solution in this case was to pool all the 'potential offspring' together with a time tag. When a robot becomes free, one of the offspring is selected from the pool based on a distribution biased by the time stamp. This means that, in good times, many offspring will never actually be

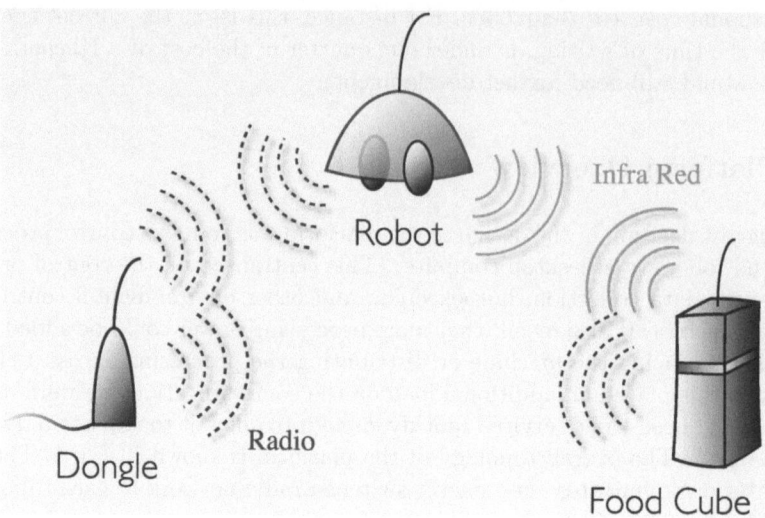

Fig. 1. Overview of the platform communication structure showing overall Packet Switched Radio and local IR communications.

initialised. The size of the offspring pool is therefore quite a good indication of how well the robots have adapted to the current environment.

Environment Control. As well as controlling each of the robots, the host computer needs to know when a robot has interacted with another element in the environment. Passive interactions such as hitting a wall are not recorded but interactions with other robots and food cubes are. When a robot bumps into another active component (robot or food cube) the two components exchange unique identifiers via IR and these are both relayed back to the host computer. The host computer then needs to match the interactions with robots and food cubes and act accordingly. If two robots bump into each other, there is a possibility of offspring being produced (based on definable requirements such as energy level of one or both of the robots). If a robot bumps into a food cube, the robots energy level needs to be adjusted up or down (the level will go down if the food cube is actually a 'poison' the robots should be avoiding). At the same time, the food cube's energy level may need adjusting which may effect the light intensity or colour the food cube emits.

The host computer also monitors the robot's battery level and can therefore judge when a robot will not be able to support an offspring without the batteries going dead. At this point, the robot is marked for recharging and is not used again until its batteries are back to full strength.

Radio Network. At the core of all communications with the host computer is the Packet Switched Radio network. This network is based on a host/client relationship with the host polling each client in turn. This is very similar to USB or BlueTooth. On a physical level, the host computer communicates with a radio

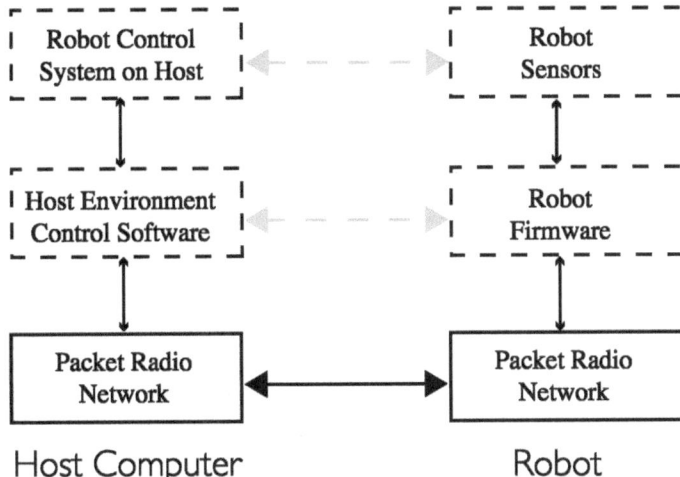

Fig. 2. The layers of communication in the Packet Switched Radio network. The bottom layer is the actual communication layer. On top of this environmental and low level robot control is built and on top of that is the link between each robot and its controller.

'dongle' which acts as the host transmitter for the radio network. This dongle (based around the Motorola HC908[1] 8 bit, 8MHz microcontroller) is connected to the host computer via USB. The host computer controls scheduling of the polling (which robots and food cubes are polled and how often) as well as taking care of re-transmission requests due to collisions and errors on the network.

The radio network itself is based on Radio Packet Controllers (RPCs) by Radiometrix[2]. These modules are capable of 64kbps transmission over the license free 433MHz band. The modules benefit from having a low (3ms) collision detection time before transmission and the ability to schedule packet transmissions. Each packet may be up to 27 bytes long. The dongle, food cubes and robots all have RPC modules that interface directly to their microcontrollers.

4.2   Robot Systems

The robots are based around a Motorola HC908 8 bit 8MHz microcontroller (like the dongle). This microcontroller supplies many of the requirements for the robots including PWM (Pulse Width Modulation) motor control, A/D converting and numerous I/O pins. The microcontroller is linked directly to the RPC and the motor channels (for PWM motor control and feedback) but all other systems are connected via a multiplexed address and data bus. Furthermore, the

---

[1] http://e-www.motorola.com/webapp/sps/site/
   prod_summary.jsp?code=68HC908GP32&nodeId=01M98634

[2] http://www.radiometrix.co.uk

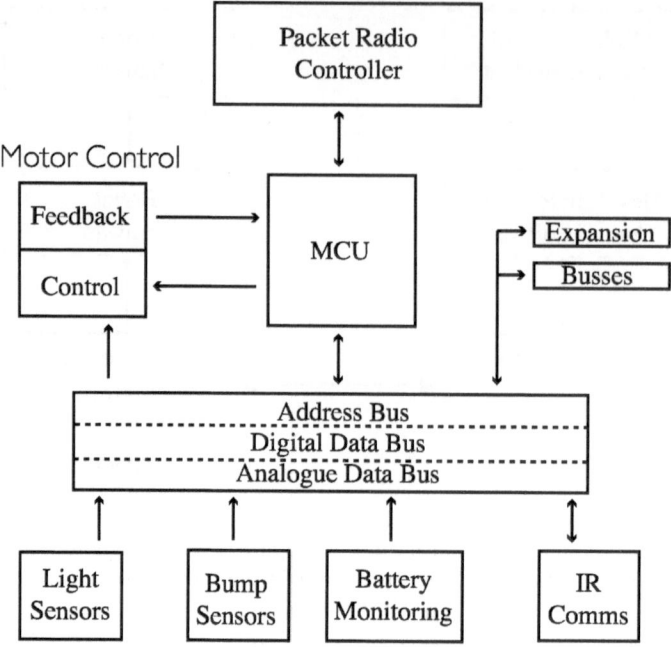

Fig. 3. Robot architecture block diagram.

data part of the multiplexed bus may also be switched between analogue and digital modes. Fig 3 outlines the systems.

Core Systems. The robots core systems consist of the power systems and monitoring, radio communications, motor control and motor speed feedback. They are available whatever sensors are used and whatever expansion modules are added.

The power supply for the robot comes from high capacity NiMH (Nickel Metal Hydride) batteries. These batteries can keep the robot running for about 3 hours on average and can be fast-charged in about 1 hour, 15 mins. They are rated at 3.6 volts (for the motors) with conversion circuitry to supply the 5v regulated line for the electronics. Battery level monitoring and fast-charging circuits are on-board and, during charging, the microcontroller keeps an eye on the batteries, informing the host controller when charging is complete.

The motor control is based on a simple motor driver. The PWM signal is generated internally by the microcontroller and direction control comes from a latch at the low end of the digital address space on the multiplexed address lines. Both motors are independently controllable and feedback comes from two reflective IR opto-encoders (one per wheel). These give an indication of the speed of the wheels which the microcontroller uses to adjust the motor speed using PID (Proportional Integral Differential) control. This feedback also prevents the motors being damaged if they stall when the robot hits a wall or other obstacle.

Radio communications via the RPC are also directly linked to the microcon-troller. The microcontroller looks after configuration of the RPC module and filters out unwanted packets (every RPC receives every packet on the network). Checks for errors in the data are made and valid packets (uncorrupt and destined for that robot) are then processed. Packets may request sensor or status values, set motor speeds or reconfigure the RPC itself.

Sensors. In the experiments the robots were originally designed for, each robot has 192 light sensors (64 each of red, green and blue). These are accessed through the multiplexed address and data lines by first requesting the address of the sensor and then retrieving the data over the analogue data lines. In fact, to cope with so many sensors, a group of three sensors (one red, one green, one blue) are addressed and the relevant analogue data line is then selected for encoding by the microcontroller.

Each light sensor is a phototransistor with a colour filter. These light sensors could be replaced by any component that outputs a voltage change with changing conditions of between 0 and 5 volts. Further, although there are only 64 addresses sensors in the current design (in order to keep the design small and the cost down), 256 could be addressed in theory and each address may have up to 8 analogue channels giving a total of 2048 sensors.

Environmental Interaction. Environmental interaction is defined as bumping into something. As such, the robots have four bump sensors. When a bump sensor is depressed, the robot must have either bumped into a wall, another robot or a food cube (or another robot bumped into it, the order is not relevant). When this event occurs, each robot / food cube transmits their unique ID over IR but only listens for an incoming signal on the side(s) where a bump sensor was activated. The received id, if any, is transmitted back to the host computer so an action may be taken. The robot control system (running on the host computer) often has NO direct knowledge of bumping into anything, only an indirect knowledge of energy levels in the robot going up or down or (in some experiments) the registering of having mated. This is configurable for different experiments.

Expansion. As covered briefly above, the microcontroller accesses most of the systems on the robot via a multiplexed data and address bus which is 8 bits wide. Furthermore, there is an analogue/digital select line and a read/write line. In the current design, only the lowest 8 bytes of the digital address space are used (motor direction, IR transmit and receive [times 4], battery monitoring, bump sensors and misc. indicators) and only the low 64 addresses of the analogue space are used. The rest of the digital space is available for expansion for memory, memory mapped devices or sensors and outputs. The rest of the analogue space may be used for analogue sensors (there is no analogue output capability) as described above.

4.3   Food Cubes

The Food Cubes are based on the same design as the robots although the ana-logue data bus, motor control systems and sensor capabilities are not present.

Instead, the Food Cubes can emit combinations of three frequencies of light (red, green and blue).

Core Systems. As with the robots, the food cubes have packet radio capabilities and battery monitoring facilities. However, they lack motor control, analogue sensor and expansion bus capabilities. The digital data bus is still present to allow the addition of further modules.

Environmental Interaction. The food cubes have the same system of IR exchange as the robots do (and their own unique IDs) and they respond in exactly the same way as the robots when something bumps into them. Like the robots, the food cubes have controlling processes running on the host computer (although these are hand-coded) which allow them to react to environmental factors (such as being bumped into by robots).

Unlike the robots, the food cubes are capable of emitting red, green and blue frequencies of light in differing quantities. Light level is based on PWM style dimming and so may not be usable with some light sensors because they rely on 'persistence of vision' effects (the maximum reaction time of the light sensor being a lot slower than the frequency the emitter is working at). Due to the A/D conversion time on the HC908 processor, this will actually be the determining reaction time in most cases, not the sensor itself. All three frequencies are completely independently controllable allowing for mixtures of colours. Green may represent the level of food available, red the level of poison - a balance the robots would have to learn about. Because the mating criteria for the robots is totally controllable, separate food and mating energy levels could be maintained with green being food and blue being 'mating energy supply'. The brightness of the food cubes is deliberately not very high so that, under normal light levels, robots should have some difficulty spotting food sources from a distance (forcing a search requirement in their repertoire in order to survive).

5   Limitations

Unlike simulations, working in the real world does place a lot of restrictions on what you may do in an experiment and how you set the experiment up. Hardware choices have real consequences and two of the biggest restrictions on experimental setup are covered here.

5.1   Interaction Tracking vs Continuous Tracking

In a simulation, every movement of every entity is known precisely and may be tracked and recorded for later inspection. This continuous tracking of each entity has many advantages when analysing behaviour. In the real world, the precise location, direction and speed of each entity is not so easily measured. The choice was made with this platform to forgo the continuous tracking capabilities due to cost. Instead, only the sensor values and motor commands are loggable, as are interactions with other robots, obstacles and food cubes. This logging still provides a high degree of information but, unlike with continuous tracking, the information is more loosely time-coupled (with only interaction time-stamps being present).

A solution to this may be filming the robots using a digital camera and adding the sensor, command and interaction log information to the video track timeline. This has been considered and it should be possible using an object-based video format (such as Quicktime or Mpeg4) and a high speed video transfer method. However, the equipment would be costly and no huge benefit is perceived currently.

5.2   Radio Network Limits

In choosing off-board control for the robots and the Radiometrix RPC modules for communication, limitations were placed on the expandability of the platform to accommodate more entities. With the relatively low bandwidth available, it is estimated that a maximum of 10 robots and 5 other objects such as food cubes would be able to communicate with the host computer before the bandwidth is saturated.

Alternatives to the Radiometrix RPC modules are available in the form of BlueTooth or IEEE 802.11b wireless Ethernet. However, BlueTooth is only capable of hosting 7 active clients per network so multiple networks would be needed. 802.11b and BlueTooth would also both have required much longer development times and higher cost. It is hoped that the robot design is expandable enough to cope with such modules if they are considered useful in future research.

6   In Summary

This paper presents an overview of a new hardware platform that will allow research in hardware evolution not possible with current platforms. It is cost effective and extensible to allow for future research needs and is not limited to evolutionary hardware experiments but rather any research requiring one or more entities working within the same environment. Although there are some limitations to the current platform and its use in certain experimental environments, these could potentially be overcome with additional technology.

References

1. P. Layzell A. Thompson. *Evolution on Robustness in an Electronics Design*, volume 1801, LNCS, pages 218–228. Springer-Verlag, Berlin, 2000.
2. D. Floreano and F. Mondada. Hardware solutions for evolutionary robotics. In *Proceedings of the first European Workshop on Evolutionary Robotics.*, pages 137–, Berlin, 1998. Springer Verlag.
3. Downing K. Miller J. F. Evolution in materio: Looking beyond the silicon box. In *NASA/DOD Conference on Evolvable Hardware*, pages 167–176. IEEE Computer Society, 2002.

Real-Time Reconfigurable Linear Threshold Elements and Some Applications to Neural Hardware

Snorre Aunet and Morten Hartmann

Department of Computer and Information Science
The Norwegian University of Science and Technology
NO-7491 Trondheim, Norway
{snorre.aunet,mortehar}@idi.ntnu.no
http://caos.idi.ntnu.no/

Abstract. This paper discusses some aspects regarding the use of universal linear threshold elements implemented in a standard double-poly CMOS technology, which might be used for neural networks as well as plain, or mixed-signal, analog and digital circuits. The 2-transistor elements can have their threshold adjusted in real time, and thus the basic Boolean function, by changing the voltage on one or more of the inputs. The proposed elements allow for significant reduction in transistor count and number of interconnections. This in combination with a power supply voltage in the range of less than 100 mV up to typically 1.0 V allow for Power-Delay-product improvements typically in the range of hundreds to thousands of times compared to standard implementations in a 0.6 micron CMOS technology. This makes the circuits more similar to biological neurons than most existing CMOS implementations. Circuit examples are explored by theory, SPICE simulations and chip measurements. A way of exploiting inherit fault tolerance is briefly mentioned. Potential improvements on operational speed and chip area of linear threshold elements used for perceptual tasks are shown.

1 Introduction

Floating-gate CMOS circuits have found their use outside the digital memory use during the last decade, and an introduction to the field can be found in [1]. Floating-gate devices are used as analog memory elements, as part of capacitive-based circuits and as adaptive circuit elements [1]. Non-traditional floating-gate circuitry is also starting to find it's way into commercial use [2]. The circuits presented here are offspring from [3], which might be perceived as having their startingpoints from two basic ideas, namely the multiple-input floating-gate transistor concept [4] and UV-programmable floating-gate circuits [5], using the UV-programming method from [6].

By exploiting the inherent analog amplifying characteristics of the transistors for something more than the traditional switching function similar to the method in [7] one can reduce the number of active elements and interconnect dramatically

A.M. Tyrrell, P.C. Haddow, and J. Torresen (Eds.): ICES 2003, LNCS 2606, pp. 365–376, 2003.

[8]. Combined with operation of the transistors in weak inversion, where the currents are typically down in the sub pA to uA range [8], from experience with a standard 0.6 um CMOS process [9], there is a significant ultra low-power potential whether the building blocks are potentially used for analog, digital or mixed-signal circuits or neural network implementations.

Section 2 introduces one of several [8] real-time reconfigurable linear threshold elements and suggests ways it can be used for implementing a variety of basic Boolean functions. The ultra low power consumption potential and improved manufacturability are then briefly treated. In section 3 strategies possibly increasing fault tolerance is given.

Section 4 gives some pointers towards the potential chip area reduction, and thus production costs reduction, that might be obtained by using linear threshold elements. For some perceptual data processing the speedup of three orders of magnitude compared to a traditional computer implementation is given as an example.

2   CMOS Floating-Gate Circuits in Weak Inversion

2.1   Balancing the Circuit; Setting Switching Point and Current Levels

Our circuit elements need to have their switching point adjusted by an initial UV-programming [6], which is a post-fabrication technique that can be applied to any standard double-poly CMOS technology. The setting of the switching point ensures that the output voltage for the circuit element is at $V_{dd}/2$ for an input at exactly the same voltage level. This is done to achieve symmetric noise margins for digital zeros and ones and the best possible digital operation.

During UV-programming the chip is radiated by UV-C from a lamp. Concurrently the power supply rails and substrates get certain voltages applied that regulate charge transport to the floating gates of the device. Once the UV-exposure is ended, the charge levels on the floating-gates are defined practically indefinitely. Charge transport is done through the UV-activated conductances, which are indicated by extra circles between gate and sources in the transistor symbols in Figure 1.

One and the same UV-programmable elements might be reprogrammed for different current levels, due to for example different needs for operational speed of the basic circuit building blocks [8]. Switching current levels, I_{beq}, might be programmed to 2-3 orders of magnitude [8] due to experience with the process from [9].

2.2   Simple Analysis of a Basic Linear Threshold Element

The following equations can be used to describe the weak inversion currents of the PMOS and NMOS transistors [10]:

$$I_{ds,p} = I_{beq} \prod_{i=1}^{m} exp\left\{ \frac{1}{nU_t}(V_{dd}/2 - V_i)k_i \right\} \tag{1}$$

$$I_{ds,n} = I_{beq} \prod_{i=1}^{m} exp\left\{ \frac{1}{nU_t}(V_i - V_{dd}/2)k_i \right\} \tag{2}$$

Here, $k_i = C_i/C_{tot}$ is the capacitive division factor of the ith input capacitor, C_i, and C_{tot} is the total capacitance seen from the floating-gate. I_{beq} is the balanced equilibrium current, which is the drain current of the transistors when all ordinary input signals and driven nodes are equal to $V_{dd}/2$. I_{beq} is strongly dependent on the V_{dd} level. Here an equal number of capacitively weighted inputs to both the PMOS and NMOS transistor is assumed, and that the intrinsic slope factors in weak inversion, n, are equal for both devices. U_t is the thermal voltage, which is 25.8 mV at room temperature.

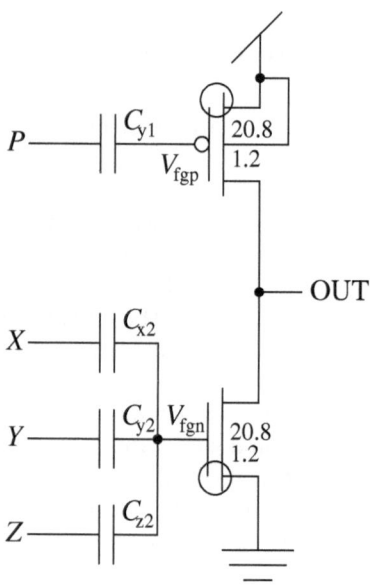

Fig. 1. A CARRY'-,NAND-,NOR and INVERT-circuit [8]. The numbers 20.8 and 1.2 refer to the width and length of the transistors in micrometers, respectively.

To illustrate some traits of the reconfigurable floating-gate circuits, the "P1N3" circuit in Figure 1 is used here. The capacitances between X,Y and Z are all designed for equal size. Also the sum of capacitances coupled to the NMOS (Figure 1) equals the capacitances connected to the PMOS. ($C_{x2} + C_{y2} + C_{z2} = C_{y1}$). The numbers of capacitively weighted inputs have been used to name the circuits, counting ordinary inputs and other inputs used for control of behavior to each two-MOSFET element. The circuit in Figure 1 gets the name P1N3 due to this naming system. The most used inverter may be called P1N1. The circuit has previously been presented as a stand-alone circuit or building block [8]. Using the above equations yields:

$$I_{ds,p} = I_{beq} exp\left\{\frac{1}{nU_t}\left(\frac{V_{dd}}{2} - V_p\right)\right\} \tag{3}$$

$$I_{ds,n} = I_{beq} exp\left\{\frac{1}{nU_t}\left(\frac{1}{3}V_x + \frac{1}{3}V_y + \frac{1}{3}V_z - \frac{V_{dd}}{2}\right)\right\} \tag{4}$$

V_p, for example, means the voltage on the capacitively weighted input to the PMOS transistor in Figure 1. If the voltages V_p, V_x, V_y or V_z are all equal to $V_{dd}/2$ the exponentials in the above equations equal 0, and we have the equilibrium condition with $I_{ds,p}=I_{ds,n}=I_{beq}$. To illustrate how the circuit functions logically, a truth table is used here. As parts of the truth table, e_p and e_n are used; these are the parts of the exponentials directly dependent on input signals and the supply voltage, V_{dd}. For this particular circuit that means

$$e_p = \left(\frac{V_{dd}}{2} - V_p\right) \tag{5}$$

$$e_n = \left(\frac{1}{3}V_x + \frac{1}{3}V_y + \frac{1}{3}V_z - \frac{V_{dd}}{2}\right). \tag{6}$$

When parasitic capacitances are not accounted for, each of the inputs at nodes X,Y and Z are weighted by $1/3$, a more optimistic estimate than would have resulted from including parasitics [8]. When $V_p = V_{dd}/2$ and X, Y and Z are allowed to have the binary values according to the table in Figure 2, Figure 3 results. When $e_p > e_n$ the output approaches the V_{dd} level, since the PMOS is then "strongest". In the opposite case it goes low.

V_{dd}	1	HIGH
V_{ss}	0	LOW

Fig. 2. Meanings of the 3 columns in each row are directly related.

By inspecting the truth table (Figure 3) it is clear that the value of the output depends on the number of 1's and 0's on the inputs only. Inspecting the truth table (Figure 3), just counting 1's and 0's in the input vector, makes it possible to get enough information out from a table with a simpler form, as in Figure 4.

Used this way (Figure 3, Figure 4) the circuit computes the inverted carry for a FULL-ADDER:

$$OUT = CARRY' = (XY + XZ + YZ)' \tag{7}$$

From the truth table, in Figure 3, one can see that by letting any one input be "0", the output is "0" if, and only if, both other inputs are "1". Then the circuit

X	Y	Z	e_p	e_n	OUT
0	0	0	0	$-3V_{dd}/6$	1
0	0	1	0	$-V_{dd}/6$	1
0	1	0	0	$-V_{dd}/6$	1
0	1	1	0	$V_{dd}/6$	0
1	0	0	0	$-V_{dd}/6$	1
1	0	1	0	$V_{dd}/6$	0
1	1	0	0	$V_{dd}/6$	0
1	1	1	0	$3V_{dd}/6$	0

Fig. 3. The table shows parts of the exponentials, e_p, e_n, and output values, for all possible binary values of inputs X,Y,Z when $V_p=V_{dd}/2$. "OUT" provides the CARRY' function for a FULL-ADDER.

P	number of 1's	e_p	e_n	OUT
$V_{dd}/2$	0	0	$-3V_{dd}/6$	1
$V_{dd}/2$	1	0	$-V_{dd}/6$	1
$V_{dd}/2$	2	0	$V_{dd}/6$	0
$V_{dd}/2$	3	0	$3V_{dd}/6$	0

Fig. 4. The table shows parts of the exponentials, e_p, e_n, and output values, for $V_P = V_{dd}/2$ and different numbers of "1's" on ordinary inputs X,Y,Z. Inputs can be either "0" or "1".

implements the 2-input NAND function. If any one input is 1, the output is "1" if and only if both other inputs are "0". Then the circuit works as a 2-input NOR gate. Connecting one input to Vdd or Vss and short-circuiting the other two gives an INVERTER. A 2-input inverting-structure like NAND or NOR is essentially the only function needed to implement any digital function.

If the e_p value is changed, the "threshold" for when, and if, $e_n > e_p$ changes. From the above equations it is easy to see that setting $e_p = -2V_{dd}/6$ makes it necessary to have only 1 binary input at "1" (V_{dd}) to make $e_n = -V_{dd}/6$, which is greater than e_p, and should give a low output. Changing the e_p value to $2V_{dd}/6$ makes it necessary to have all inputs X,Y,Z high, when restricted to Boolean inputs, in order to produce a low output. When perceiving the digital functionality of the linear threshold element only, it is possible to make a table like in Figure 5.

e_p	digital functionality
$-2V_{dd}/6$	NOR3, NOR2, INVERT
0	CARRY', NOR2, NAND2, INVERT
$2V_{dd}/6$	NAND3, NAND2, INVERT

Fig. 5. The table shows part of the exponential, e_p, and the Boolean functions resulting.

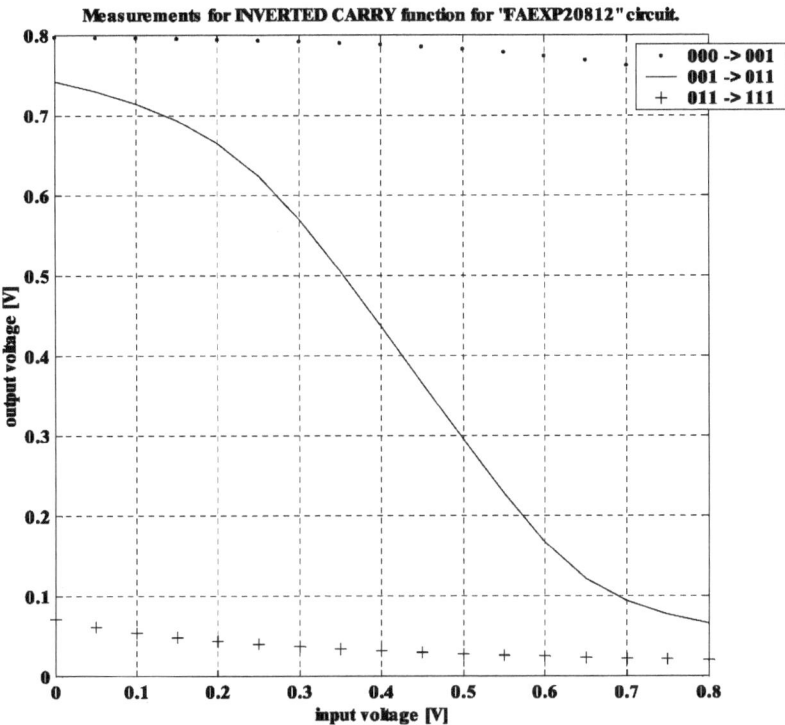

Fig. 6. Actual measurements for the P1N3 circuit, from a prototype chip, when it is producing the CARRY' of a FULL-ADDER [8].

Changing the threshold by adjusting the value of the input capacitively coupled to the PMOS can provide a real-time reconfigurable digital circuit.

2.3   Measured Functionality of P1N3 Linear Threshold Element

The inherent functionality for $e_p = 0$, demonstrated by measurements, can be seen in Figure 6. The voltage on the output goes low if, and only if, 2 or 3 of the inputs X,Y,Z goes high at the same time. Otherwise the output stays high, which in this case is $V_{dd} = 0.8$ V. The three curves (Figure 6) correpond to transitions between the four rows in Figure 4. The continous line shows the output voltage when the number of high inputs moves from 1 to 2, which is the transition between 1 and 0 on the output, for the CARRY' function of binary addition.

The functionality and performance of such circuits will depend on the implementation technology of choice. The number of inputs, their capacitive weights, if an input is coupled both to the PMOS and NMOS are among other factors that can be varied [8]. To increase the voltage gain, the size of the drawn capacitances between inputs and floating-gates should be increased [8].

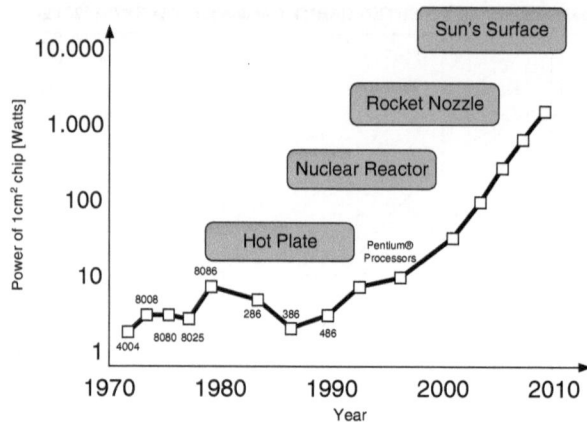

Fig. 7. If nothing is done about power consumption, power will get out of hand and Moore's law cannot continue [13].

2.4   Ultra Low Power Consumption

Dynamic power consumption depends linearly on the physical capacitance being switched [11]. Therefore using less interconnect and fewer active elements for a given function may be attractive for minimizing power consumption. Voltage reduction offers the most direct and dramatic means om minimizing energy consumption [11]. Chip measurements have demonstrated floating-gate circuitry working for supply voltages down to 93 mV [12], while typical supply voltage levels have been in the 300 - 800 mV range [8]. SPICE simulations using the "P5N5" [8] linear threshold elements as building blocks in an 8-transistor FULL-ADDER operating at a V_{dd} of 200 mV compared to a standard cell implementation in the AMS 0.6 CMOS technology [9] showed that the floating-gate implementation used 2500 times less energy, a Power-delay-Product (PDP) of 2.3 fJ while running at a clock frequency of 1 MHz [8]. Power consumption is an increasingly important issue, which is illustrated in Figure 7 based on a slide from the presentation of [13]. The PDP numbers could maybe not be undercut by any known CMOS technology of the same generation.

2.5   Matching and Manufacturability

This technology has some very attractive features, but an important problem that remains is that the UV-programmable circuits have not been demonstrated for circuits containing more than between 10 and 30 transistors transistors yet, at least from the open literature. Hopefully this number can be increased significantly, due to a proposed design method in [8], using as few different building blocks as possible. Since "The Achilles Heel of analog is that every transistor is different" [14] matching [15] of both transistors and passive elements should become the best possible if a larger system could be built from identical building blocks instead of dedicated circuitry for each basic function. Good matching

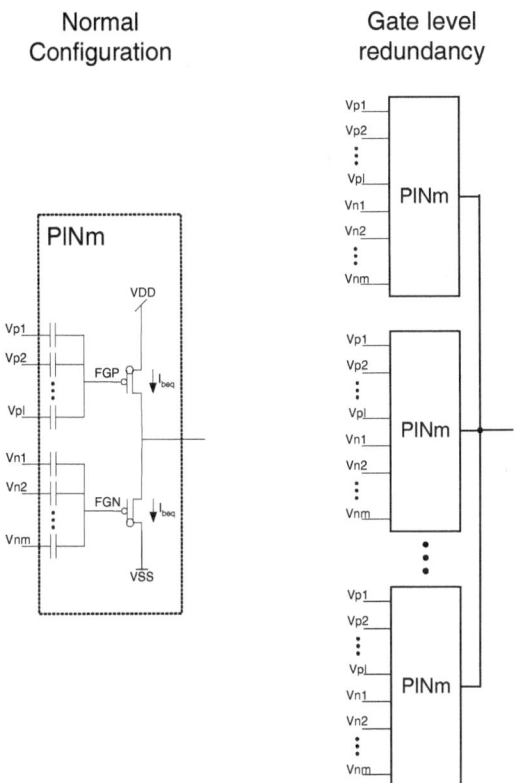

Fig. 8. Fault tolerance improvement scheme.

means a good relative accuracy between identically constructed circuit elements. Certain chip layout techniques [15] improve matching, as well as increasing the size of the drawn circuitry.

3   Increasing Faul Tolerance for Floating-Gate Linear Threshold Elements in Weak Inversion

One way to increase fault tolerance is to introduce redundancy on the gate level by letting, for an example, three identical gates (or more) with similar inputs drive the same output, as is illustrated in Figure 8. Then if one out of the arbitrarely number of redundant gates fails to take part in producing the correct output, the others could still maintain proper functionality. The steep dependency on output current and output resistance of the transistors from the gate voltage, in weak inversion compared to the classical area of operation, might make the principle useful. Letting more than one FGUVMOS circuit element share a common output node has been implemented earlier [7], [16], but for different purposes.

4 Using Linear Threshold Elements for Speeding up Perceptual Processing

The circuit elements herein are linear threshold elements, which are basic processing units in certain neural networks [17]. A classic model of a neuron is a linear threshold device, which computes a linear combination of the inputs, compares the value with a threshold, and outputs +1 or (-1) if the value is larger than the threshold [17].

Real neurons are found in the biological nervous systems. Human brains are by far superior to computers in solving hard problems such as combinatorial optimization and image and speech processing, although their basic building blocks are several orders of magnitude slower, which have boosted interest in the field of artificial neural networks [18], [19].

While neural networks have found wide application in many areas, the limitations and behavior of such networks are far from being understood [20].

Despite the power of digital computers they are not clever enough in the sense of perceptual and "intelligent" processing like seeing an object in the visual field, recognizing what it is, and taking proper action in real time. For biological systems, including humans, in general those are generally effortless tasks [21]. Such tasks are extremely difficult even for state-of-the-art computers. According to [21] the performance gap could never be narrowed by just increasing the clock frequencies of MPU's, integration densities of memories and further sophistication of software programs.

What is sometimes denoted "Intelligent data processing" tasks of today are presented as software programs which is reduced to a series of simple binary operations executed on MPU chips. However, in our brains the algorithms and the hardware are inseparably merged in the system [21]. To carry out the intelligent data processing directly in hardware, "neuron MOS transistors" are exploited in [21]. These transistors are multiple-input floating-gate devices where the floating-gate potential is determined as a weighted sum of multiple input signals via capacitance coupling, and thereby controls the current in the channel. The neuron MOS transistors are so named due to their similarity to the McCulloch-Pitts model of a neuron [21]. An association processor architecture utilizing these principles has been verified to about three orders of magnitude higher performance as compared to typical SISC processors of the time [21].

To which extent the linear threshold elements briefly described in this paper could switch places with the neuron MOS transistors in [21] could be researched further.

5 Using Linear Threshold Elements for Decreasing Chip Area

Assuming that each threshold gate can be built at a cost comparable to that of traditional AND, OR, NOT, termed AON logic, neural networks can be much more powerful than traditional logic circuits [17].

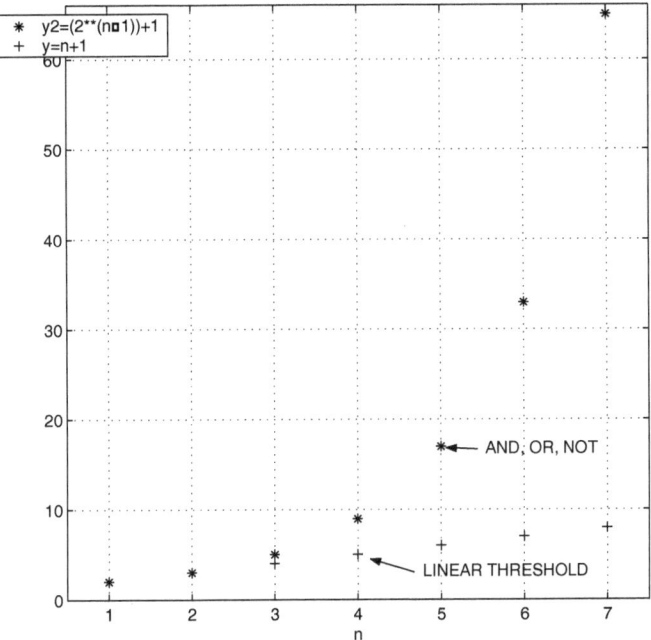

Fig. 9. Number of gates necessary to implement certain functions, as a function of number of bits, n. Number of gates using traditional AND-OR-NOT logic increases exponentially, while the increase can be linear when linear threshold elements are used.

In CMOS the production costs of a chip has a strong dependency on the chip area. Since the area of basic linear threshold elements, like P5N5 [8] and P1N3, are in the same order as normal gates, the costs might be comparable on the gate level.

For some functions, like XOR, the number of elements in a traditional AON circuit will grow exponentially with the number of bits in the input, while when implemented using linear threshold elements the number of gates are linear in the number of input bits [18]. Generally, a depth-2, AON circuit computing XOR of n bits requires at least $2^{n-1} + 1$ gates. A Linear Threshold circuit needs only n+1 gates. Figure 9 illustrates these relationships for n=1, $\cdots$, 7.

Another example of potential use of threshold logic [17] is: Whereas any logic circuit of polynomial size (in n) that computes the product of two n-bit numbers requires unbounded delay, such computations can be done in a neural network with "constant" delay. The product of two n-bit numbers and sorting of n n-bit numbers can be computed by a polynomial-size neural network using only 4 and 5 unit delays, respectively. Unit delay is equal to a "depth" of one for an artificial neural network [20].

Symmetric Boolean functions depend only on the sum of input values, and since the parity function is symmetric it can be computed in two layers of a neural network whereas it takes unbounded delay to compute parity in a logic circuit [17].

For many years the topic of linear threshold logic has been approached in two different ways: theory on computational circuit complexity on one hand, and hardware implementation on the other. There has been very little interaction between the two approaches, as was stated in [18].

6 Conclusion

One type of real-time reconfigurable linear threshold element implemented in floating-gate UV-programmable CMOS technology has been presented, including measurements from a prototype CMOS chip demonstrating functionality. Power-Delay-Product numbers are among the lowest reported, and there are an interesting potential for implementing certain functions using less area then by traditional logic.

To determine the true computational power, due to number of inputs and capacitive weights for an example, further research should be done.

A simple proposal for how to make fault-tolerant circuitry using our linear threshold elements has been proposed.

An example on how similar CMOS circuitry could be used in speeding up a certain perceptual signal processing task about 3 orders of magnitude was briefly mentioned.

Due to the above mentioned aspects the true potential of this immature technology should be explored further.

References

1. P. Hasler, T. S. Lande *Overview of Floating-Gate Devices, Circuits and Systems*, IEEE Transactions on Circuits and Systems II: Analog and Digital Signal Processing, January 2001.
2. *http://www.impinj.com* September, 2002.
3. S. Aunet, Y. Berg, T. Sæther *A New 2-MOSFET Universal Floating-Gate Element for Reconfigurable Digital Logic* Proceedings of the 19th IEEE Norchip Conference, Kista, Sweden, 12-13 November 2001, pp. 240-245.
4. T. Shibata, T. Ohmi *An Intelligent MOS Transistor Featuring Gate-Level Weighted Sum and Threshold Operations*, Technical Digest, International Electron Devices Meeting, 1991.
5. T. S. Lande, D. T. Wisland, T. Sæther, Y. Berg *FLOGIC - Floating-Gate Logic for Low-Power Operation* Proceedings of the 3rd IEEE International Conference on Electronics, Circuits and Systems, 1996, pp 1041-1044.
6. Y. Berg, T. S. Lande, S. Næss *Low-Voltage Floating-Gate Current Mirrors* Proceedings of the the Tenth Annual IEEE International ASIC Conference and Exhibit (ASIC), Portland, OR, USA, 7-10 Sept. 1997, pp 21-24.
7. Y. Berg, D. T. Wisland, T. S. Lande *Ultra Low-Voltage/Low-Power Digital Floating-Gate Circuits* IEEE Transactions on Circuits and Systems II, analog and digital signal processing, Vol. 46, Issue 7, pp. 930-936, July 1999.

8. S. Aunet *Real-time reconfigurable devices implemented in UV-light programmable floating-gate CMOS* Dissertation for the degree of doktor ingeniør, Norwegian University of Science and Technology, ISBN 82-471-5447-1, 2002.

9. Austria Mikro Systeme International AG *0.6 um CMOS CUP Process Parameters* Document no. 9933011, Rev. B, Oct. 1998.

10. Y. Berg, D. T. Wisland, T. S. Lande *Ultra Low-Voltage/Low-Power Digital Floating-Gate Circuits* IEEE Transactions on Circuits and Systems II, analog and digital signal processing, Vol. 46, Issue 7, pp. 930-936, July 1999.

11. J. Rabaey, M. Pedram, P. Landman *Low Power Design Methodologies* in J. Rabaey, M. Pedram (editors), Low Power Design Methodologies, Kluwer Academic Publishers, 1997.

12. S. Aunet, Y. Berg, O. Tjore, Ø. Næss, T. Sæther *Four-MOSFET Floating-Gate UV-Programmable Elements for Multifunction Binary Logic* Proceedings of the 5th World Multiconference on Systemics, Cybernetics and Informatics, Orlando, FL, USA, Volume 3,2001, pp 141-144.

13. P. P. Gelsinger *Microprocessors for the New Millennium: Challenges, Opportunities, and New Frontiers* Digest of technical papers, IEEE International Solid-State Circuits Conference, 2001, pp. 22-25.

14. G. Gilder, B. Swanson *Seattle Sunburst* Gilder Technology Report, 2002.

15. K. R. Laker, W. M. C. Sansen *Design of analog integrated circuits and systems* McGraw-Hill International Editions, ISBN 0-07-113458-1, 1994.

16. R. Bahr *A Design of Linear Four Quadrant Analog Multipliers Using Floating-Gate Transistors* thesis for the cand. scient. degree, University of Oslo, Faculty of Mathematics and Natural Sciences, Department of informatics, May 2001.

17. K. Y. Siu, J. Bruck *Neural Computation of Arithmetic Functions* Proceedings of the IEEE, No. 10, pp. 1669-1675, October 1990.

18. V. Bohossian *Neural Logic: Theory and Implementation* Dissertation for the Ph.D. degree, California Institute of Technology, 1998.

19. D. Hammerstrom *Computational Neurobiology Meets Semiconductor Engineering* Proceedings of the 30th IEEE International Symposium on Multiple-Valued Logic, 2000, pp. 3-12.

20. K. Y. Siu, J. Bruck, T. Kailath, T. Hofmeister *Depth Efficient Neural Networks for Division and Related Problems* IEEE Transactions on information theory, Vol. 39, No. 3, pp. 946-956, May 1993.

21. T. Shibata *Intelligent VLSI Systems Based on a Psychological Brain Model* Proceedings of the 2000 IEEE International Symposium on Intelligent Signal Processing and Systems, pp 323-332, Hawaii, U.S.A., November 5-8, 2000.

Simulation of a Neural Node
Using SET Technology

Rudie van de Haar and Jaap Hoekstra

TU Delft, Subfaculty of Electrical Engineering,
Mekelweg 4, 2628CD Delft, The Netherlands
{R.vandeHaar,J.Hoekstra}@ITS.TUDelft.nl
http://nanocom.et.tudelft.nl
tel: +31 15 2783826

Abstract. A McCulloch and Pitts neuron implemented in nano-tech-nology is simulated. The neuron is created with SET circuits operating in the single-electron current regime. Therefore the extremely low power properties of the SET devices are fully exploited. This neuron is simu-lated with a SPICE description model for a single SET junction.

Keywords: SPICE, Single-electron tunneling (SET), Random Back-ground Charge (RBC) fluctuations, McCulloch and Pitts Neuron, Low Power.

1 Introduction

A very promising device in nano-technology is the single-electron tunneling (SET) junction. This device consists of two conductors, with a very thin insu-lator in between. For example aluminum, aluminum-oxide and aluminum. The distance between the two conductors is in the order of a few nm. It is possible for an electron to tunnel through the insulator, but only when the absolute voltage difference across the junction will be decreased due to the event [1].

The two major problems this device is facing are the random background charge (RBC) fluctuations and the low operating temperature. The last problem is nearly solved. Some SET circuits can already operate at room-temperature [2]. The RBC fluctuation problem is not solved at the device-level yet, but it is possible to design useful circuits with these junctions using redundancy to cope for this type of error [3]. A neural network is a possible way to generate redundancy; this is described in this paper. The combination of neural networks and SET technology has advantages in two ways. First, a powerful neural network requires a large amount of neural nodes. This implies each neuron has to be small in dimensions due to restricted available space. Also, the power dissipation of each neural node has to be extremely low in order to have an acceptable overall power-dissipation. In these aspects, SET devices are very suitable to use in neural networks, because they are extremely small and can they can process information using only a few electrons. Secondly, SET devices introduce errors like RBC fluctuations and show stochastic behavior. A neural network can compensate for

A.M. Tyrrell, P.C. Haddow, and J. Torresen (Eds.): ICES 2003, LNCS 2606, pp. 377–386, 2003.

these small errors and is therefore a suitable architecture for SET devices. In [4] it is stated that when SET circuits operate in a good shielded environment, the RBC fluctuations only appear a few times a day, this error is small enough to compensate for in the neural network architecture. The neural network described in this paper could also be made in CMOS technology, but the power-dissipation would be orders of magnitude higher, even at the lowest possible current. Also the required physical space would have been increased with respect to the SET variant. An example of an McCulloch and Pitts neuron [5] implemented in SET technology is discussed and simulated with a SPICE description model for a single SET junction [6]. With the SPICE model, it is possible to perform a transient analysis for any arbitrary SET circuit operating in the single-electron current regime (SER). From these transient simulations a static/dynamic plot can be extracted. This plot can be a convenient tool for designing SET circuits. In section 2, the SPICE model is discussed in brief. In section 3, the three island SET structure (3IS) is discussed and simulated using SPICE. In section 4, the 3IS is used as a building block to create a McCulloch and Pitts neuron. A small neural network to perform optical character recognition is discussed as an example.

2   The SPICE Model in Brief

The SPICE model is based on the impulse model [1] for the single SET junction, shown in Fig. 1. The model consists of a true capacitor with a current source placed parallel. When the junction voltage U_{C_j} exceeds the critical voltage U^{cr}, the current source is triggered and effectively a charge of exactly the elementary charge e is transferred between nodes (1) and (2), by means of the delta function. Mathematically: $i_t = e\delta(t - t_0)$. Note that the critical voltage is defined as: the maximum allowed voltage across the junction just to remain in the Coulomb blockade [7] region.

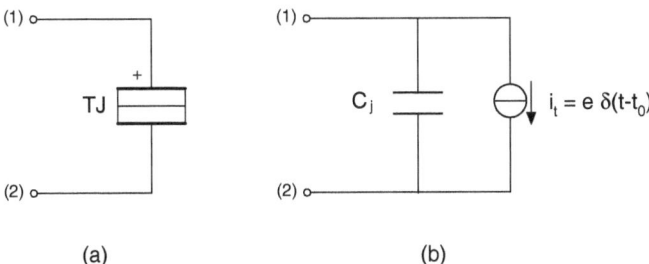

Fig. 1. (a) SET junction network symbol, (b) impulse model of the SET junction.

The SPICE model is a direct implementation of the impulse model, shown in Fig. 2. However, in the SPICE user environment the delta pulse function is not available. The function to transfer the charge e across the junction is chosen to be a rectangular shaped pulse, for which $e = i_t \cdot d_t$ holds. This rectangular shaped pulse is implemented with a one-shot function. This one-shot function is created with programmable SPICE building blocks. All other components of the SPICE model can be found in the standard SPICE libraries.

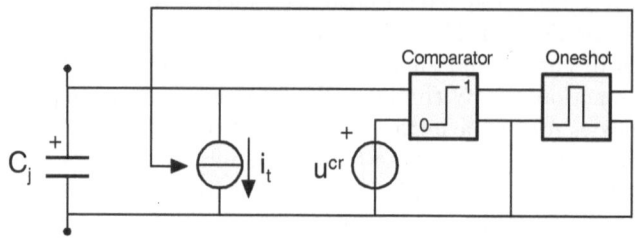

Fig. 2. SPICE implementation of the impulse model.

The values of the capacitors and the tunnel-time are scaled in order to stay within the numerical range of SPICE. The voltage levels are not scaled. The pulse width d_t is chosen to be $1\ ms$ (in the scaled model). This means with this model it is possible to simulate currents up to 1000 electrons per second, which is equivalent to $160 fA$. This current region, from Coulomb blockade region up to $160 fA$ is defined as the single-electron current regime (SER). When SET circuits are simulated outside this region, the current will be clipped to $160 fA$.

With the SPICE model it is also possible to generate a static/dynamic plot for any given SET circuit. To generate a static/dynamic plot, two external sources are swept (or stepped) and one or more outputs of interest are plotted versus the swept external sources. The outputs can be either voltage levels or tunnel event signals taken directly from the one-shot function in the model, or combinations of those signals. The plotted output signals can be either stochastic or deterministic. This static/dynamic plot can be a convenient tool for developing SET circuits. For example, the designer can get direct insight in values to chose for biasing sources for the particular purpose.

The RBC fluctuations are modelled using SPICE by applying a certain charge (fluctuation) to specific circuit nodes.

In section 3, the three island structure, also known as the SET-inverter, is simulated using the SPICE model described. Both a transient simulation and a static/dynamic plot are generated.

3   The Three Island SET Structure

The 3IS was first introduced by Tucker [8]. The main idea behind this structure was to copy the working principle of a CMOS inverter onto SET technology, using 2 SET transistors. The 3IS consists of four tunnel junctions and three islands (a), (b) and (c). The 3IS operates completely in the voltage domain. An advantage of working in the voltage domain only, is the lack of needing high ohmic resistors e.g. for biasing. Another advantage is there is no need for signal transformation to couple a 3IS to one another. The 3IS is shown in Fig. 3.

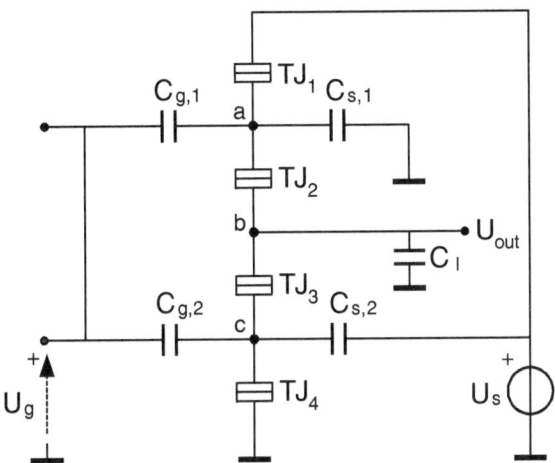

Fig. 3. The three island SET structure, with $C_{g,1} = C_{g,2} = 8$ aF, $C_{s,1} = C_{s,2} = 7$ aF, capacitances of $TJ_1 .. TJ_4$ are $C_{j,1} .. C_{j,4} = 1$ aF and $C_l = 24$ aF for digital operation and 2.4 fF for analog operation, $U_s = 6$ mV.

The 3IS is able to operate in four different modes, namely:

- A analog stochastic mode
- A digital stochastic mode
- A analog deterministic mode
- A digital deterministic mode

The modes can be directly read from the static/dynamic plot, shown in Fig. 4. This plot indicates in which of the four modes the structure operates with respect to the applied external voltage sources U_s and U_g. To obtain this plot, U_g was swept from 0 to 40 mV, and U_s was stepped from 0 to 10 mV in 100 steps. The output quantity plotted, is a combination of the normalized output voltage level U_{out} and the tunnel event signals taken from the one-shots. In this way, the output voltage and the stochastic regions are clearly visible. With this plot, the designer is able to get direct insight in the values to choose for the sources and capacitance values, in order to set the 3IS in the proper operation region.

Both the analog versus the digital mode and the stochastic versus the deterministic mode depend on the values of the external supply voltages and the values of the capacitances. The stochastic mode is the mode where free tunneling will occur. To exploit the extreme low power properties of the 3IS, the stochastic mode should be avoided.

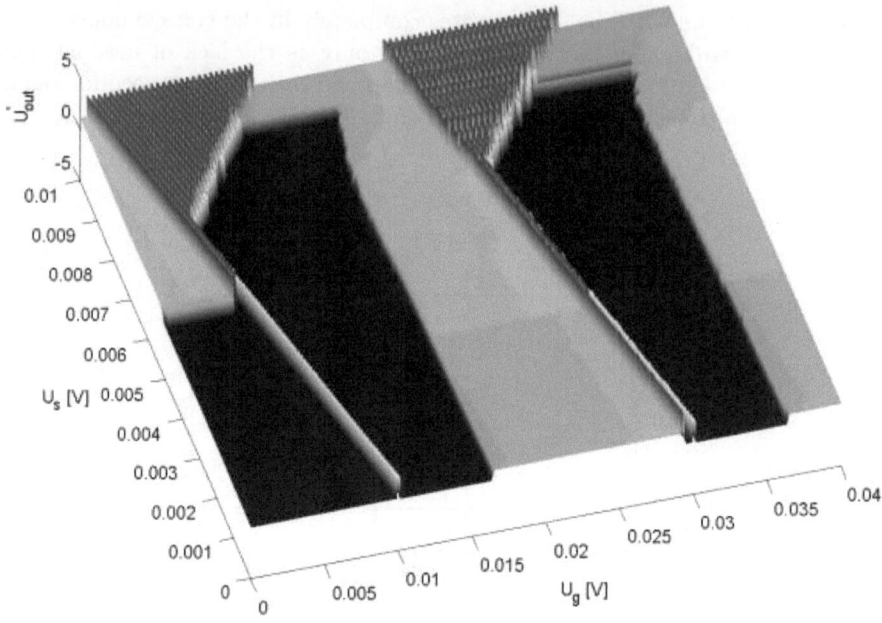

Fig. 4. Static/Dynamic plot of the 3IS. Plotted is a combination of U_{out} (normalized) and the tunnel event signals versus U_s and U_g. The dark area equals output logic level '0', the light area equals output logic level '1' and the speckled area is the stochastic region.

The set of capacitance values are chosen as follows: $C_{j,1}$ .. $C_{j,4}$ are chosen 1 aF, as a starting point. $C_{g,1}$ and $C_{g,2}$ should be an order of 10 times bigger than $C_{j,1}$ to let U_g have more effect on island voltages a and c. $C_{s,1}$ and $C_{s,2}$ are control nodes for the offset voltage of islands (a) and (c), and should be as effective as the gate capacitances, so they should be chosen of the same order of magnitude. For the sake of convenience, we took the values of [8], see Fig. 3.

For the actual design of the neural circuitry, the value of U_s is chosen 6 mV, to stay out of the stochastic region and to have inverter operation at a low gate voltage U_g (around 4 mV), see Fig. 4. For digital operation, C_l is chosen to be 24 aF, because only one electron can effectively be held with 6 mV supply voltage. For analog operation C_l should be chosen as large as possible in order to have many values between $U_{out_{low}}$ and $U_{out_{high}}$. In Fig. 5, a transient SPICE simulation is shown for the 3IS, with $C_{load} = 240$ aF.

It can be seen from Fig. 5 that amplification in the voltage domain is possible; when for example the input voltage is varied between 4 and 5 mV, the output voltage changes approximately 6.5 mV. The reason for choosing the 3IS for the signal-processing instead of simpler SET circuits is because of the buffer-function, which is needed for a high coupling factor, which on its turn is highly preferable in a neural network. The 3IS was used to create a McCulloch and Pitts neuron, this neuron is discussed in section 4.

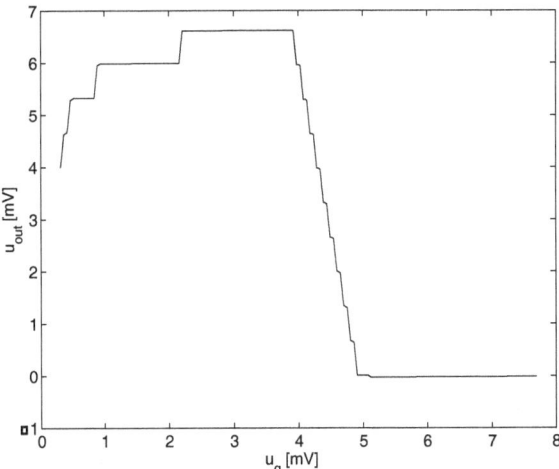

Fig. 5. Plotted is U_{out} versus U_g for the 3IS in the deterministic mode ($U_s = 6$ mV), with $C_l = 240$ aF, which means that at most 10 electrons can be held on C_l.

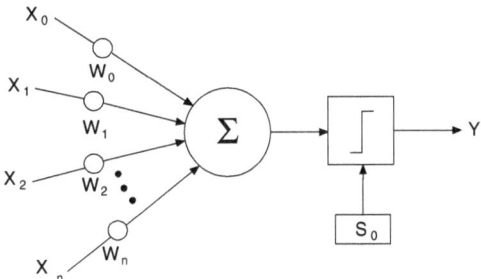

Fig. 6. The basic McCulloch and Pitts neuron.

4   The McCulloch and Pitts Neuron

The basic McCulloch and Pitts neuron is shown in Fig. 6. It requires binary input signals for the input vector $\underline{X}$, and analog input signals for the weight vector $\underline{W}$. The input vector $\underline{X}$ is multiplied by the weight vector $\underline{W}$. These signals are summed and fed through a hard-limiter. (In brief: the inner product $\underline{X} \cdot \underline{W}$ is thresholded). The output signal Y is on its turn also binary. The threshold-level for the hard-limiter can be set by means of signal S_0. The signal-type of S_0 is analog. It is also possible to treat the signal S_0 as an input signal like X_i, which is convenient to use in the SET circuits.

For each neuron the following requirements must be met:

- The neuron must be capable of driving many other neurons, this means: a buffer-function is required
- The overall signal amplification must be greater than, or equal to 1

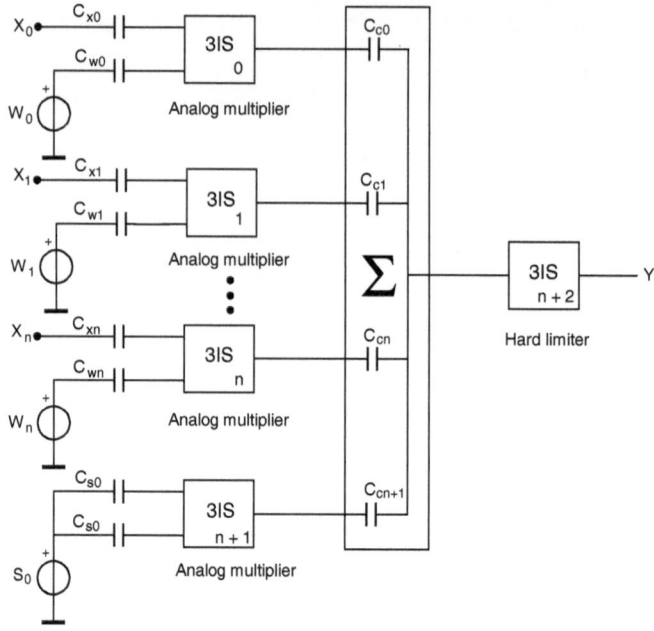

Fig. 7. McCulloch and Pitts neuron in implemented SET technology.

As discussed in the last section, the 3IS is capable of amplifying the signal in the voltage domain. Therefore, the voltage levels in vector $\underline{W}$ have an offset value of 8 mV, which will be divided across C_{xi} and C_{wi}, therefore the working area of the 3IS is shifted to $U_g \geq 4mV$, see Fig. 5. This structure has also a buffering function and is therefore capable of driving many other 3IS, which is needed for a powerful neural network. In Fig. 7 a McCulloch and Pitts neuron implemented in SET technology is shown. In this figure, the coupling capacitors are given by $C_{g,x}$. The value of these capacitors were chosen as follows: In order to have maximum coupling from $3IS_0 .. 3IS_{n+1}$ to $3IS_{n+2}$, $C_{c0} .. C_{cn+1}$ must be chosen as large as possible (∞). On the other hand, to have minimal mutual influence of $3IS_0 .. 3IS_{n+1}$ onto each other, $C_{c0} .. C_{cn+1}$ must be a chosen as low as possible (0). The best tradeoff values were found to let $C_{c0} .. C_{cn+1}$ be smaller than C_l of $3IS_0 .. 3IS_{n+1}$, and larger than $C_{g,i}$ of $3IS_{n+2}$, so 100 aF is taken. With the same arguments $C_{x0} .. C_{xn+1}$ and $C_{w0} .. C_{wn+1}$ were also chosen to be 100 aF. With these values, a 4 input neuron is successfully simulated. Other examples of SET neurons can be found in literature, e.g. [9,10], but they operate in the stochastic mode, and therefore consume more power than the neuron presented in this paper.

As mentioned in the introduction, SET devices are capable of operating with faulty elements. The following example is used to calculate the effect at the output of a disturbance due to a random background charge fluctuation in one of the SET devices.

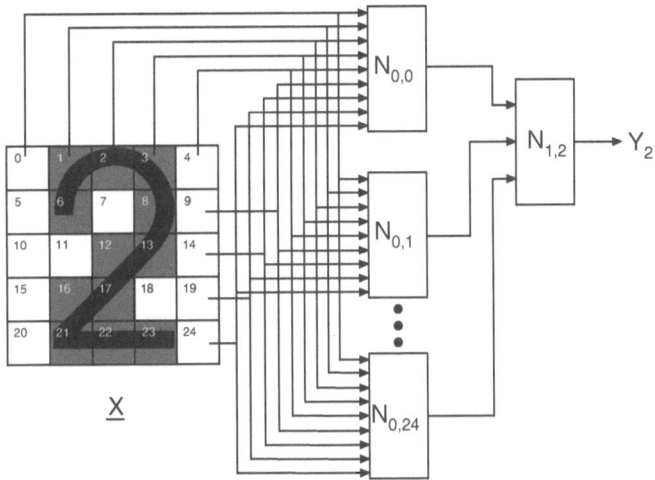

Fig. 8. Neural network for OCR to detect digit '2'.

In Fig. 8, a neural network for optical character recognition (OCR) is shown. With this neural network it is possible to check if the input equals digit '2' or not. The input consists of a binary signal $\underline{X}$, comprised of 25 bits. This is a feedforward network, which consists of two layers. The first layer $(N_{0,i})$ contains 25 neurons and the second layer $(N_{1,i})$ contains 1 neuron. Each neuron has 25 inputs and 1 output. All nodes in the first layer are placed in parallel, this means they all have the same weight vector $\underline{W}$. The node in the second layer performs the function of majority voting element.

Assuming that the neural network is trained, e.g. using the back-propagation learning technique [11], the maximal allowable error, in case of majority voting will be:

$$\epsilon(Y_2) = \frac{\epsilon(\sum_{i=0}^{n} N_{0,i})}{n} + \epsilon(N_{1,2}), \quad n = 24 \tag{1}$$

Where (n+1) equals the number of 3IS in the first layer. As can be seen from equation 1, the errors in the second layer are unacceptable, because when such a node is in error, the output is directly in error. Therefore the nodes in the last layer should not be made in SET technology. Note that this is true for a general neural network made with the nodes simulated in this paper. For the nodes in the other layers, the worst case scenario means 12 out of 25 nodes may be faulty in order to let the network still work properly.

The error per neural node, with respect to the 3IS building blocks, (Fig. 7), is given by equation 2.

$$\epsilon(Y) = \epsilon(3IS)_{n+1} + \epsilon(3IS)_{n+2} + \frac{\epsilon(\sum_{i=0}^{n}(3IS)_i)}{n} \tag{2}$$

This means when an error arises at the threshold 3IS(n+2) or the offset 3IS(n+1), the whole node will be faulty. The other 3IS are operating in the majority voting domain, which means, for an 25 input node, the maximum number of faulty elements for good operation is at maximum 12.

From equations 1 and 2 we can conclude when the last layer neurons are not made in SET technology, that in the worst case scenario, when the errors appear at the 3IS in positions (n+2) and (n+1) in the first layer $(N_{0,i})$, than the maximum number of allowable faulty elements (3IS) equals 12. From the above discussion we can conclude that when this neuron is placed in a good shielded environment, this means the RBC fluctuations will only appear in the order of a few times per day [4] in the worst case, than the circuit can at least operate for 6 days without any error.

5   Conclusions

A McCulloch and Pitts neuron in SET technology is simulated using SPICE. This neuron is based on the 3IS operating in the SER, therefore the extremely low power properties of SET circuits are fully exploited. The 3IS was simulated using a SPICE model for a single SET junction. An example of an OCR network, using the neurons proposed is simulated and discussed. Also a small error analysis for this neuron is given.

References

1. J.Hoekstra, R.H. Klunder, R. van de Haar, E. Rouw, and P.Chand. Circuit design with metallic single-electron tunneling junctions. In *ESSCIRC 2002*, pages 671–674, Firenze, Italy, September 2002. Proceedings of the European Solid-State Circuits Conference.
2. K. Uchida, Junji Koga, Ryuji Ohba, and Arika Toriumi. Programmable single-electron transistor logic for low-power intelligent si lsi. In *IEEE international Solid State Circuits Conference*, pages 206–207. ISSCC, 2002.
3. A. van Roermund and J. Hoekstra. Design philosophy for nanoelectronic systems, from sets to neural nets. *International journal of circuit theory and applications*, 28(6):563–584, 2000.
4. H. Wolf, F.J. Ahlers, J. Niemeyer, H. Scherer, T. Weimann, A.B. Zorin, V.A. Krupenin, S.V. Lotkhov, and D.E. Presnov. Investigation of the offset charge noise in single electron tunneling devices. *IEEE Transaction on Instrumentation and Measurement*, 46(2):303–306, April 1997.
5. W.S. McCulloch and W. Pitts. A logical calculus of the ideas immanent in nervous activity. *Bulletin of Mathematical Biophysics*, (5):115–133, 1943.
6. R. van de Haar, R.H. Klunder, and J. Hoekstra. Spice model for the single electron tunnel junction. In *ICECS*, volume 3, pages 1445–1448, Malta, September 2001. IEEE international conference on electronics, circuits and systems, ISBN: 0-7803-7058-9.
7. H.R. Zeller and I. Giaever. Tunneling, zero-bias anomalies, and small superconductors. *Physical Review*, 181(2):789–799, May 1969.
8. J.R. Tucker. Complementary digital logic based on the 'coulomb blockade'. *Journal of Applied Physics*, 72(9):4399–4413, November 1992.

9. M.J. Goossens, C.J.M. Verhoeven, and A.H.M. van Roermund. Concepts for ultra-low power and very-high-density single-electron neural networks. In *NOLTA '95*, volume 7B-7, pages 679–682, Las Vegas, U.S.A., December 1995. International Symposium on Nonlinear Theory and its Applications.
10. J.F. Jiang, Q.Y. Cai, Z.C. Cheng, B. Shen, and J. Gao. Superconductor based neuron single electron transistor. *Physica C*, (341):1607–1608, 2000.
11. Simon Haykin. *Neural Networks - A Comprehensive Foundation*. Prentice Hall, New Jersey, 1994.

General Purpose Processor Architecture for Modeling Stochastic Biological Neuronal Assemblies

N. Venkateswaran[1] and C. Chandramouli[1]

Waran Research Foundation,No 46B, Mahadevan Street, Mambalam, Chennai,
Tamilnadu, India, 600028.
`warf@vsnl.net, phone:91444899766`

Abstract. Accurate models are needed for understanding the biological neuron in its totality. The need for these accurate models is in a evolvable system like a Brainy Processor, understanding unexplored functionality of the brain and fault simulation of the different cortex regions. The accuracy of the model increases computational complexity. Existing software simulators would be extremely slow in terms of performance for large number of neurons. General-purpose processors would have architecture unsuitable to tackle the heavy floating point computation involved in modeling more accurate neuronal assembly models. To overcome this problem, quite a few ASIC chips to simulate simple neural models like (Leaky integrate and fire, Spiking neuron, Hodgkin and Huxley) have been developed. The ASIC architectures proposed are only for simple neuron models where several characteristics are disregarded to lessen hardware complexity. Suggesting ASIC chips for every type of neuronal assembly would be cost prohibitive. To deal with hardware complexity and different classes of neuronal assembly, a programmable hardware unit or in a wider sense an instruction driven general-purpose processor architecture is proposed. The functional units of this architecture are tailored to evaluate complex stochastic neuronal assembly models.

1 Introduction

Neuron models find extensive application in evolvable systems. Further these models can find application in developing a brainy processor [14]. Understanding the unexplored functionality of the brain and fault simulation of the different cortexes of the brain could be other major applications. These demand accurate and almost complete neuronal assembly models involving wide characteristics.

Several software-based simulators are available for modeling the neuronal activities in an assembly,notable among these is [6]. To overcome the problems of time complexity, these software simulators adopt simple models for a neuron or for a neuronal assembly. Neurons are modeled as point sources thereby greatly reducing the computational complexity and the mode of interaction across the neurons in an assembly [1,2]. To overcome the time constraints existing in software simulation,several hardware implementations have been proposed.

A.M. Tyrrell, P.C. Haddow, and J. Torresen (Eds.): ICES 2003, LNCS 2606, pp. 387–397, 2003.

The present day implementations in VLSI use the integrate and fire model [1,2]. The model is deterministic and does not take care of the inherent stochastic nature of the neuronal activities. Suggesting ASIC chips for every type of neuronal assembly would be cost prohibitive.

To deal with hardware complexity and different classes of neuronal assembly including stochastic assembly models, a programmable hardware unit or in a wider sense an instruction driven general-purpose processor architecture is proposed in this paper. We adopt the model proposed by Gopinath Kallianpur [4] as a thin linear cable upon which the Wiener process is imposed. This model is further extended in [4] to a neuronal assembly, which is represented by coupled SPDEs.

Based on the analysis of the computational structures of different neuronal assembly models including [4], a general-purpose NAM (Neuronal Assembly Modeling) processor architecture is proposed in this paper. The different floating point functional units of this processor are designed for evaluating complex transcendental functions, numerical techniques and for evaluating Wiener processes in space and time. The NAM processor needs very high memory processor bandwidth to match the performance of the functional units and also when dealing with massive neuronal assemblies. The architecture of the General purpose processors will not match with complex floating point functional units needed for a NAM processor. When we try to include these as special purpose modules to a general purpose architecture, the memory processor bandwidth would not be sufficient to match the computational power of these special purpose modules. The architectures of DSP processors are not tailored for solving complex partial differential equations and stochastic processes.

Section two deals with the stochastic model of a single neuron and the neuronal assembly, Section three presents the architecture. The fourth section deals with the applications of the NAM processor

2   A Stochastic Model for a Single Neuron

There is a considerable amount of literature devoted to the implementation of neurons in VLSI. However most of these models utilized the integrate and fire principle. There is a need to consider the more complete model of spatial geometry of the neuron. The stochastic model as in [4] is presented below both for single neuron and neuronal assembly.

$$\frac{\partial u}{\partial t} = -\alpha u + \beta \frac{\partial^2 u}{\partial^2 x}, \ 0 < x < b; t > 0 \tag{1}$$

Where α and β are positive constants such that

$$\frac{\partial u}{\partial x}(0,t) = \frac{\partial u}{\partial x}(b,t) = 0 \ \forall t > 0$$

$$u(x,0) = u_0(x) \tag{2}$$

To equation 1 the random neuronal activities is obtained by including the effect of random impulses, which follow a generalized Poisson process. Accordingly the equation 1 is replaced by

$$\frac{\partial u}{\partial t} = -\alpha u + \beta \frac{\partial^2 u}{\partial x^2} + \dot{W}(t, x), \tag{3}$$

Where $\dot{W}(t, x)$ is the space time gaussian white noise which is a fictitious derivative of the space time wiener process $W(t, x)t \geq 0, 0 \leq x \leq b$. Equation 3 can be rewritten in terms of an Itô stochastic partial differential equation (SPDE).

$$dX(x, t, ; \omega) = -LX(x, t; \omega)dt + dW_t x(\omega), X(0, x, \omega) = u_0(x) \tag{4}$$

By applying concepts of SPDE we can show the solution to equation 4 is given as

$$u(t, x, \omega) = \int_0^t G(t; x, y)u_0(y)dy + \int_0^t \int_0^b G(t - s; x, y)dW_{sy}(\omega), \tag{5}$$

The above single neuron model is extended to study the asymptotic behavior of the voltage potentials of interacting spatially distributed neurons. A rigorous treatment of the above equation is found in [4]. To consider the parallel fiber interactions in a neuronal assembly, a simple and special case of Markin's model is taken into account. The corresponding coupled pair of linear equations converted into SPDEs; is given as.

$$\frac{\partial u_1}{\partial t} = -\alpha u_1 + \frac{r + R}{\gamma c} \frac{\partial^2 u_1}{\partial x^2} - \frac{R}{\gamma c} \frac{\partial^2 u_2}{\partial x^2} + \dot{W}_{tx}^1, \tag{6}$$

$$\frac{\partial u_2}{\partial t} = -\alpha u_2 - \frac{R}{\gamma c} \frac{\partial^2 u_1}{\partial x^2} + \frac{r + R}{\gamma c} \frac{\partial^2 u_2}{\partial x^2} + \dot{W}_{tx}^2,$$

where $\alpha = k/c, g = r(r + 2R)$, and k, c, r, R are all positive constants. For $R = 0$ the equations reduces to 3. Finally the solution is given as

$$U_i(t, x) = \sum_0^\infty A_n^i(t)\varphi_n(x), i = 1, 2. \tag{7}$$

Where for each i, $A_n^i(t), n = 1, 2, \ldots$ is a sequence of independent Ornstein Uhlenbeck Processes given by

$$A_n^1(t) = (\sqrt{2})^{-1}[a_n^1(t) + a_n^2(t)],$$
$$A_n^2(t) = (\sqrt{2})^{-1}[a_n^1(t) - a_n^2(t)],$$

with

$$a_n^i(t) = e^{-\lambda_n^i t} \int_0^b g_i(x)\varphi_n(x)dx + \int_0^b e^{-\lambda_n^i s}dB_n^i(s). \tag{8}$$

the various B_n^i being independent wiener processes.

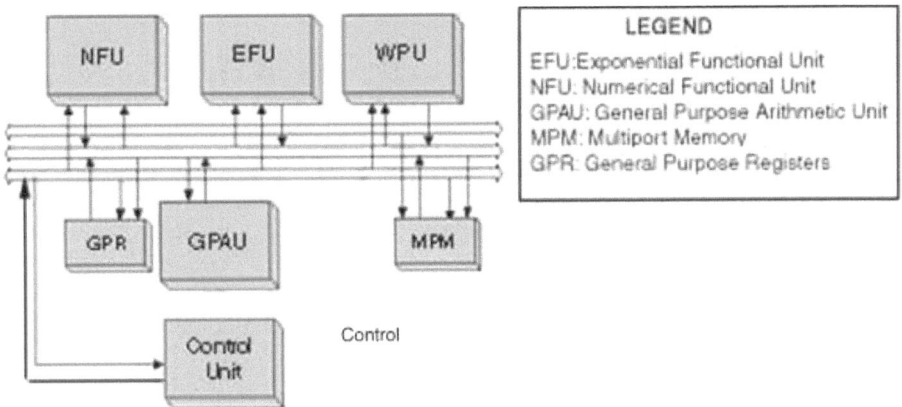

Fig. 1. Architecture of the NAM Processor

3   Processor Architecture

A detailed analysis of the different neuronal models governed by ODE, PDE and
SPDE shows their solutions involve evaluation of complex exponential functions
and need the application of different numerical techniques to arrive at their solutions. In case of SPDEs additionally the Wiener process has to be implemented.
The derivation presented in section 2 towards finding the solution of coupled
SPDEs reveal the above-mentioned facts. The RHS of equation (8) can be split
into two parts one for evaluating the complex exponential functions, application
of appropriate numerical techniques for evaluating the integral and differential
part and the stochastic component by the Wiener process. Hence the concept
behind proposing a processor architecture for modeling complex neural assembly
can be made more of a general purpose in nature applicable to a specific domain,
biological neuronal modeling.

The different functional units of this processor should be, one dedicated towards evaluating complex exponential functions, the second for application of
numerical techniques and the third for evaluating Wiener processes in space
and time. These techniques may include Simpson's Method, Runge Kutta, Gill's
Method [5]. Besides these we need a general-purpose arithmetic unit for performing intermediate processing needed for almost all numeric techniques. Also
the general-purpose arithmetic unit helps in linking computationally the outputs from exponential functional unit, Wiener process unit and the numeric
function unit. Lastly, the ROM as a part of the Multiport memory space is used
to store reciprocals of factorials and other kernel coefficients. The coefficients
from ROM can be loaded from the DRAM and switched of to save power. The
general-purpose processor architecture is shown in Fig. 1.

The connectivity between neurons is extremely high(several 1000s in fact)
[7,8]. This means that the concerned parameter variations are appreciable. Spatial variations along the soma surface are in the wider sub micron range. In

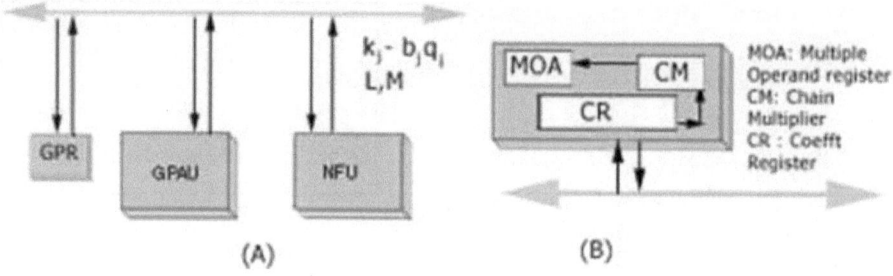

$k_j - b_j q_j$
L, M

MOA ← **CM**

CR

MOA: Multiple
Operand register
CM: Chain
Multiplier
CR : Coefft
Register

GPR

GPAU

NFU

(A)

(B)

Fig. 2. A. Data flow across NFU,GPAU and GPRs for Gill's Method **B**. Architecture
of the NFU

addition, potential variations across the neuron (Including dendrites) are also
appreciable. The computations in the functional units are dominated with ir-
rational and transcendental numbers. All these factors demand floating-point
functional units.

To achieve high performance the functional units should be capable of performing pipelined processing. The arithmetic units used in these functional units
are multiple operand adder (MOA), Parallel Array Multiplier and Chain Multi-
plier(CM). Carry Save Adder(CSA) tree Structures are employed to realize these
pipelined functional units achieving a pipelining rate equal to a full adder delay
[13].

Having finalized the need for floating point pipelined functional units, the
System Level Architecture should possess Instruction Level Pipelining to achieve
overall ultra high performance.

3.1   Architecture of the Numeric Function Unit (NFU)

The analysis of different numeric techniques showed that most frequently used
operations are of the type Multiple Operand Addition and Chain Multiplica-
tion(More than two operands) [5]. Fig 2A shows the data flow for Gill's method
across the GPAU, NFU and the GPRs. Fig 2B shows the functional architecture
of the NFU. The MOA and the CM employed are pipelineable. Based on the
type of numeric techniques applied either partial or complete pipelining may
take place. As an example we show how the GILL's method is implemented.

Gill Algorithm Execution in NFU

```
Input: y1, y2, y3... set of differential equations
Procedure GILL
Begin Set x=x0, y1= y10, y2= y20, y3= y30;
Set j=1
 While (x<x2)
   For I=1 to N  // Done in GPAU
      ki = f (x, y1, y2, y3...)
      Compute C = (ki - bjqi) // done in the GPAU
       N = cj*ki
```

```
        L = h*aj* (C) //Done In NFU
        M = 3* aj* (C)
        yi = yi + L
        qi = qi + M-N
        Store yi,qi in GPR
    Next I
  j = j+1
  If j = 5 Then
      If x >= x2 Then Stop;
      End if
  Else
      j =1
  End if
  End if
  If j = 3 Then
        x = x + h/2
  End if
End While
End Procedure
```

3.2   Architecture of Exponential Unit (EFU)

The architecture shown in the Fig.3. performs pipelined evaluation of an arbitrary power series corresponding to the value of the given input x. It is used to generate the values for the common exponential operations like massive exponential addition or massive exponential multiplication and other equations of the form $c_0 + c_1 x^2 + c_2 x^3 + c_3 x^4 + \ldots$ where $c_0, c_1, c_2, c_3 \ldots$ are coefficients. The Fig. 3 shows the pipelined architecture of the EFU. There are two multiplier stages, $m1, m2, m3$ pipelining along the X-axis and the other $M1, M2, M3$ along the Y axis. Pipelining in the output CSA cluster also takes place along the X-axis. As an example when evaluating an exponential function upto nine terms, a pipelining rate equal to five full adder stages can be obtained. This can be verified from the space time diagram shown in Fig.4 and analyzing the CSA tree based array architecture of the multiplier [13].

The series is evaluated taking three terms at a time. The evaluation of e^x as an example up to 6 terms is is given in the algorithm below. The constant coefficients are factorials and reciprocals or in general a scalar fetched from the SCR (Series Coefficient Register). These coefficients may be a part of the input or a set of intermediate results generated by the GPAU as a part of general processing.

Algorithm used to calculate e^x in EFU

```
Input: x.
Output: S, the value of e^x upto 6 terms M1,M2,M3 are
outputs from multipliers M1,M2,M3 respectively Procedure EXP (x)
For k=0 to 3
        M1 = x^(k+1)
```

```
M2 = x^(k+2)
M3 = x^(k+3)
a = M1 / k1
b = M2 / k2
c = M3 / k3
S = S + a + b + c        // Achieved by the 3*2 Pipelined CSA tree
k = k + 3                // Achieved by the Feedback of x^(k+3)
Next
```

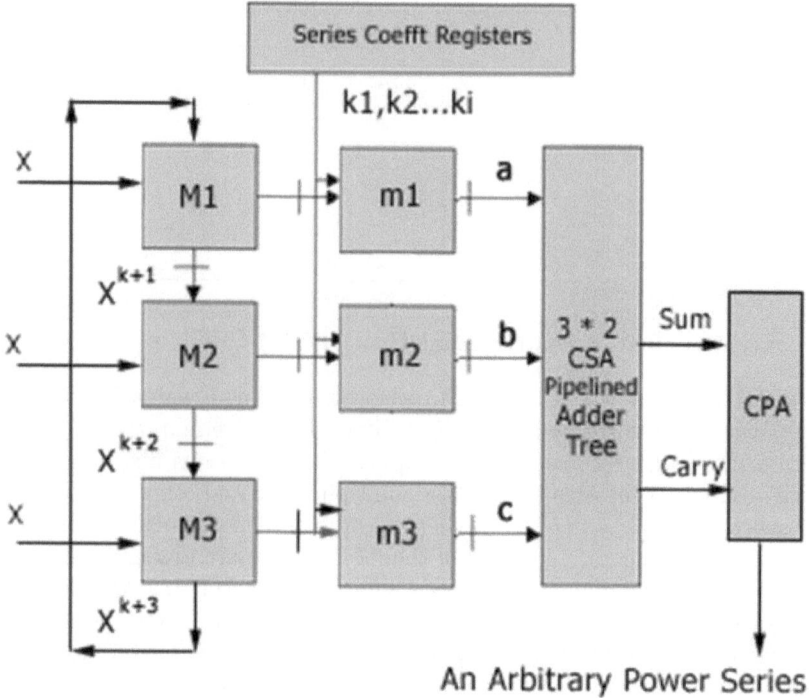

Fig. 3. Pipelined Architecture of the EFU

3.3   Architecture of Wiener Process Unit

The Wiener process is described by a normal distribution with mean zero and variance one and its derivative gives the White noise spectrum. The normal distribution is of the form $e - x^2/2$. To evolve an architecture the Wiener process is discretized as

$$dw = \sqrt{dt}N(0,1)$$

The architecture has a random number generator and a multiplier. The random number generator is designed using the linear congruential method [11], which is given as

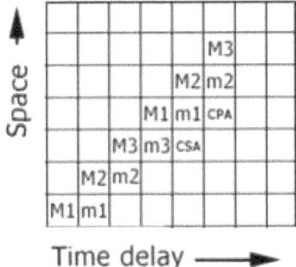

Fig. 4. Space time diagram of the EFU

$$X_{n+1} = (aX_n + c)mod m$$

Where a is the constant multiplier,c is the increment,X_0 is initial value,X_{n+1} the $n+1^{th}$ random number and m the modulus. A functional level architecture is shown in the Fig. 5. There are other algorithms too for generating random numbers. One can design architecture for these algorithms and employ them as functional units [15,16].

```
Procedure Wiener
Begin
     dW(1) = sqrt(dt)*randn;
     W(1) = dW(1);
        // Done in WPU
        // Read from the coefficient registers the values of a,m,c
     WP = a*xn
     SP = WP + c
     Xn+1 = SP mod m
     Compute (Xn+1)^2 / 2
     Store in GPR x1
        // Done in EFU
     R = e-x1^2/2
     Generate the value for the given x
        // Done in WPU
     dw = 0.04472 * R
     Return result dw back to GPAU
End
End Procedure
```

3.4   Memory and Register Architecture

Modeling the neuronal assembly of the brain as a whole is just impossible. A partitioned approach corresponding to the different regions of the brain is favored. The neuronal count even in regions like auditory cortex, cerebral cortex is quite enormous in hundred's of thousands of neurons. This leads us to conclude that even with powerful processor architecture such as the NAM processor, trying to

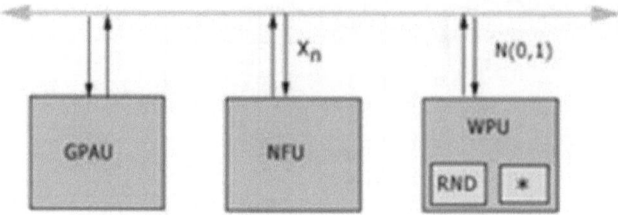

Fig. 5. Architecture of the WPU

simulate neuronal assembly of a particular region is an extremely difficult and time consuming process (Maybe days). Currently no integrated mathematical approach exists to deal with neuronal assembly involving 100s of neurons.

Simulating the models of a realistic neuronal assembly demands an array processor. The node of this array could be the NAM processor. In this context, the processing capability needed of the NAM processor may get reduced to an assembly involving tens of neurons. With this background the memory needed may be in tens of megabytes. The exponential unit needs series coefficient registers for storing the appropriate coefficients relating to a particular series evaluation. At any given instant series are summed taking three terms at a time and this is evaluated for three cycles for generating nine term arbitrary power series. Hence the EFU can have upto a maximum of twelve 80-bit floating point registers according to IEEE standard.

Similarly the NFU needs registers for storing the coefficients corresponding to a given numerical technique like Gill's Method, Runge Kutta, Simpson's method. Based on this it is decided to have sixteen 80 bit floating-point registers. There is a set of registers meant for general processing mostly used by the GPAU. Also, taking the constraint of the compiler into account and the addressing modes, it is fair enough to decide on 16 floating point and 16 fixed-point registers.

3.5   Instruction Set Architecture

The instruction set is a combination of general purpose and special purpose instructions. The general-purpose instruction set includes standard arithmetic; control and data transfer instructions under the standard format.

The special purpose instruction set is grouped into exponential types, numeric types and Wiener types. Fig. 6. Shows the instruction format. The formats for the special purpose instructions are given below.

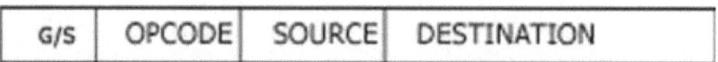

Fig. 6. Instruction Format for the NAM Processor

where G / S :- General or Special Purpose Instruction, Source,Destination : memory, registers;

Addressing Modes: The individual functional units including GPAU are provided with several registers to have more of register oriented data transfer instead of repeatedly accessing the memory. Such a register configuration can well support direct and indirect addressing modes.

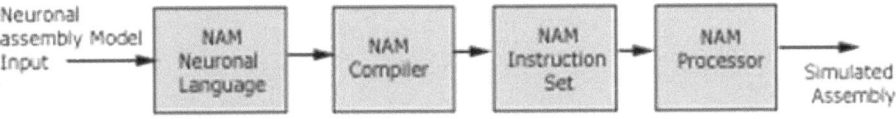

Fig. 7. General Purpose Application of the NAM Processor

4   General Purpose Application of the NAM Processor

In this paper we have presented the architecture of the NAM processor and the NAM instruction set. The Fig. 7 shown includes the NAM compiler and the NAM NL (Neuronal Language). Because of the highly special purpose characteristics of the NAM Processor architecture it is essential to develop special language constructs to specify the complex mathematical model of any kind of neuronal assembly. If existing high-level languages were employed, the programming complexity would be very high. The neuron models could be point source, or linear or spatial, "integrate and fire" or sigmoid. The neuronal assemblies can belong to olfactory, auditory or vision cortexes.

5   Conclusion

A General-purpose processor architecture is proposed for simulating different types of neuronal assembly. This has a definite advantage over an ASIC design targeting a particular model and particular type of neuronal assembly. The instruction set proposed has both general purpose and special purpose instructions.

For modeling very large neural assembly, one has to evolve a mathematical approach for solving large number of coupled stochastic partial differential equations. As of now there are no such mathematical models available. If such models are proposed, it may involve a matrix based solving. To tackle this matrix processing of very high order, The NAM processor may need additional functional units like a 4*4 matrix multiplier or a 4*4 matrix inverter and other matrix oriented operations.

A major impact of this proposed NAM processor is towards ultimate modeling of the complex functionalities of the brain with fault simulations and designing evolvable systems like brainy processors [14]. This NAM processor could be employed by biological neuronal researchers in understanding the mystery of the brain.

Acknowledgments

We are greatly indebted to Prof. Udayabhaskaran who helped us out with the mathematical aspects. The discussions we had with him helped us greatly improve the paper.

References

1. Wulfram Gerstner,"Integrate-and-Fire Neurons and Networks", The Handbook of Brain Theory and Neural Networks, Second edition,(M.A. Arbib, Ed.), Cambridge, MA: The MIT Press, 2002.
2. Pawel kudela,Piotr J Franasczuk, Gregory.K.Bergey, "A simple computer model of excitable synaptically connected neurons". Biological cybernetics, April 1997
3. Ananth Durbha, Murari Sridharan, Arun Krishnamachari, Venkateswaran Alladi, "On the Design of a special purpose VLSI Architecture for Stochastic PDE models for biological neural networks and its simulation Part I and Part II", Project Thesis submitted to university of Madras(1997)
4. Gopinath Kallianpur, "Stochastic Differential Equation Models for Spatially Distributed Neurons and Propagation Of Chaos for Interacting Systems", Mathematical Biosciences, 112, Elsevier 1992, pp 207-224.
5. E V Krishnamurthy, S K Sen, "Numerical Algorithms, Computations in Science and Engineering", East west press private limited, 2000
6. University of Southern California, "Neuron Simulation Language",2002
7. Henry C Tuckwell, "Introduction to Theoretical Neurobiology Vol I: Linear Cable Theory and Dendritic Structure", Cambridge University Press 1988.
8. Henry C Tuckwell, "Introduction to Theoretical Neurobiology Vol II: Non Linear Stochastic Theories", Cambridge University Press 1988
9. Athanasios Papoulis, "Probability Random Variables and Stochastic Processes", 1991, third edition, McGrawHill International publishers
10. Hans Peter Mesmer, "The Indispensable PC Hardware book", Third edition, Addison Wesley,1997
11. Donald.E.Knuth, "The Art of Computer Programming", third edition, Addison wesley, 2000
12. Dean Brettle and Ernst Neibur,"Detailed Parallel Simulation Of a Biological Network", IEEE computational science and Engineering, Winter 1992
13. John L Hennesy, David A Patterson, "Computer Architecture a Quantitative Approach", Morgan Kauffman, 1996
14. N.Venkateswaran,C.Chandramouli, "Stochastic Instruction Sets", WARF Internal report, June 2002,
15. M. Matsumoto and T. Nishimura, "Mersenne Twister: A 623-dimensionally Equidistributed Uniform Pseudorandom Number Generator", ACM Trans. on Modeling and Computer Simulation Vol. 8, No. 1, January pp.3-30 1998
16. Benjamin Jun and Paul Kocher, Intel Random Number Generator Cryptography Research Inc,April 22, 1999

Use of Particle Swarm Optimization to Design Combinational Logic Circuits

Carlos A. Coello Coello[1],
Erika Hernández Luna[1], and Arturo Hernández Aguirre[2]

[1] CINVESTAV-IPN, Evolutionary Computation Group
Depto. Ing. Eléctrica, Sección de Computación
Av. Instituto Politécnico Nacional No. 2508
Col. San Pedro Zacatenco, México, D.F. 07300, Mexico
ccoello@cs.cinvestav.mx
[2] CIMAT, Area de Computación, Callejón Jalisco s/n
Mineral de Valenciana, Guanajuato, Guanajuato 36240, Mexico
artha@cimat.mx

Abstract. This paper presents a proposal based on binary particle swarm optimization to design combinational logic circuits at the gate-level. The proposed algorithm is validated using several examples from the literature, and is compared against a genetic algorithm (with integer representation), and against human designers who used traditional circuit design aids (e.g., Karnaugh Maps). Results indicate that particle swarm optimization may be a viable alternative to design combinational circuits at the gate-level.

1 Introduction

Kennedy & Eberhart [6] proposed an approach called "particle swarm optimization" (PSO) which was inspired on the choreography of a bird flock. The idea of this approach is to simulate the movements of a group (or population) of birds which aim to find food. The approach can be seen as a distributed behavioral algorithm that performs (in its more general version) multidimensional search. In the simulation, the behavior of each individual is affected by either the best local (i.e., within a certain neighborhood) or the best global individual. The approach uses then the concept of population and a measure of performance similar to the fitness value used with evolutionary algorithms. Also, the adjustments of individuals are analogous to the use of a crossover operator. However, this approach introduces the use of flying potential solutions through hyperspace (used to accelerate convergence) which does not seem to have an analogous mechanism in traditional evolutionary algorithms. Another important difference is the fact that PSO allows individuals to benefit from their past experiences whereas in an evolutionary algorithm, normally the current population is the only "memory" used by the individuals. PSO has been successfully used for both continuous nonlinear and discrete binary optimization [4, 6–8].

As far as we know, this paper presents the first attempt to use PSO to design combinational circuits.

A.M. Tyrrell, P.C. Haddow, and J. Torresen (Eds.): ICES 2003, LNCS 2606, pp. 398–409, 2003.

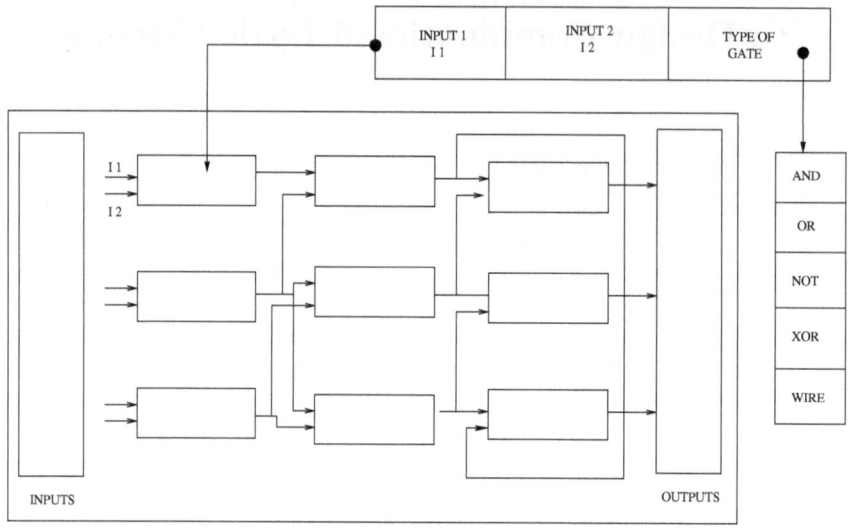

Fig. 1. Matrix used to represent a circuit. Each gate gets its inputs from either of the gates in the previous column. Note the encoding adopted for each element of the matrix as well as the set of available gates used.

2   Problem Statement

We used the same matrix representation to encode a circuit as in some of our previous work [1, 2]. Such representation is shown in Figure 1. This matrix is encoded as a fixed-length string of bits or integers from 0 to $N-1$, where N refers to the number of rows allowed in the matrix (we call it n-cardinality alphabet). In this paper, we will be referring to our GA that uses an n-cardinality alphabet, since we have found in the past that this version of the algorithm consistently produces better results than its binary counterpart [2].

More formally, we can say that any circuit can be represented as a bidimensional array of gates $S_{i,j}$, where j indicates the *level* of a gate, so that those gates closer to the inputs have lower values of j. (Level values are incremented from left to right in Figure 1). For a fixed j, the index i varies with respect to the gates that are "next" to each other in the circuit, but without being necessarily connected. Each matrix element is a gate (there are 5 types of gates: AND, NOT, OR, XOR and WIRE[1]) that receives its 2 inputs from any gate at the previous column as shown in Figure 1. Although our implementation allows gates with more inputs and these inputs might come from any previous level of the circuit, we limited ourselves to 2-input gates and restricted the inputs to come only from the previous level. This restriction could, of course, be relaxed, but we adopted it to allow a fair comparison with our previous GA-based approach.

---

[1] WIRE basically indicates a null operation, or in other words, the absence of gate, and it is used just to keep regularity in the representation used, since otherwise would have to use variable-length strings.

Input 1	Input 2	Gate Type

Fig. 2. Encoding used for each of the matrix elements that represent a circuit.

A chromosomic string encodes the matrix shown in Figure 1 by using triplets in which the 2 first elements refer to each of the inputs used, and the third is the corresponding gate from the available set.

The matrix representation adopted in this work was originally proposed by Louis [9, 10]. He applied his approach to a 2-bit adder and to the n-parity check problem (for $n = 4, 5, 6$). This representation has also been adopted by Miller et al. [11, 12] with some differences. For example, the restrictions regarding the source of a certain input to be fed in a matrix element varies in each of the three approaches: Louis [9] has strong restrictions, Miller et al. [11] have no restrictions and we have relatively light restrictions. The encoding is also different in all cases. Louis [9] only encoded information regarding one input and the type of gate to be used at each matrix position. He also used binary representation. In our case, we have used both an n-cardinality alphabet and a binary alphabet and we encode the gate to be placed at each matrix location plus its two inputs. Miller et al. [11] encode a full Boolean operation using a single integer. This representation is more compact, but it has the problem of requiring that mutation takes the place of crossover to introduce enough diversity in the population, so that the evolutionary algorithm can approach the feasible region.

Finally, the last difference among the three approaches previously mentioned is regarding the fitness function. Louis [9] simply maximizes the number of matches between the outputs produced by the circuit and those indicated in the truth table. We have used a fitness function that works in two stages: first, it maximizes the number of matches (as in Louis' case). However, once feasible solutions are found, we maximize the number of WIREs in the circuit. By doing this, we actually optimize the circuit in terms of the number of gates that it uses. Miller et al. [11] did something similar to Louis until recently (they have recently introduced a two-stage fitness function like the one adopted by us [5]).

Thus, we can say that our goal is to produce a fully functional design (i.e., one that produces all the expected outputs for any combination of inputs according to the truth table given for the problem) which maximizes the number of WIREs.

3   Description of Our Approach

The main motivation for using particle swarm optimization (PSO) to design combinational circuits is that this algorithm has been found to be very efficient in a variety of tasks [8]. Note however, that most of the successful uses of PSO reported in the literature deal with real numbers representations and in this case, we will be using a binary encoding. Although a real numbers representation is possible (and in fact, we are currently working on such a version of the algorithm), the preliminary results reported in this paper were found with a binary representation that worked reasonably well. The use of real numbers to represent a circuit requires a more sophisticated genotype-phenotype mapping. Next, we will describe the details of our implementation.

1. For i $= 1$ to M (M = population size)
   Initialize $P[i]$ randomly
   (P is the population of particles)
   Initialize $V[i] = 0$ (V = speed of each particle)
   Evaluate $P[i]$
   $GBEST$ = Best particle found in $P[i]$
2. End For
3. For i $= 1$ to M
   $PBESTS[i] = P[i]$
   (Initialize the "memory" of each particle)
4. End For
5. Repeat
   For i $= 1$ to M
   $V[i] = V[i-1] + \Phi_1 \times (PBESTS[i] - P[i])$
   $\quad + \Phi_2 \times (PBESTS[GBEST] - P[i])$
   (Calculate speed of each particle)
   (Φ_1 and Φ_2 are upper limits used to
   draw positive random numbers from a uniform distribution
   $POP[i] = P[i] + V[i]$
   If a particle gets outside the pre-defined hypercube
      then it is reintegrated to its boundaries
   Evaluate $P[i]$
   If new position is better then $PBESTS[i] = P[i]$
   $GBEST$ = Best particle found in $P[i]$
   End For
6. Until stopping condition is reached

Fig. 3. Particle swarm optimization pseudocode

The general algorithm of binary PSO is shown in Figure 3. In PSO, there are two types of information available for the particles so that they can make the best decision regarding where to move next. One of these is its own search experience (i.e., a particle has passed through several states and it "knows" which of them has been the best so far). Additionally, it also knows about the performance of the particles in its neighborhood (i.e., it knows which are the best states that its neighbors have reached so far). These two pieces of information correspond to the individual learning and the cultural transmission, respectively [8].

Mathematically speaking, the binary version of PSO is defined such that the probability of an individual's deciding zero or one (i.e., false or true) is [7]:

$$P(x_{id}(t) = 1) = f(x_{id}(t-1), v_{id}(t-1), p_{id}, p_{gd}) \qquad (1)$$

where:

- $P(x_{id}(t) = 1)$ is the probability that individual i will choose 1 for the bit at the d-th position of the binary string.
- $x_{id}(t)$ is the current state of the string position d of individual i.
- t refers to the current iteration.
- $v_{id}(t-1)$ is a measure of the individual's predisposition or current probability of deciding 1.
- p_{id} is the best state found so far.
- p_{gd} is the best state found in the neighborhood so far.

Although the main adjustment expression used by PSO can be seen as a form of mutation, we found out that its explorative power was not enough in circuit design. Therefore, we added a uniform mutation operator such as the one used with traditional genetic algorithms. This operator, however, was only applied to a certain percentage of the population (this is a parameter defined by the user). From our experiments, we determined that a value between 1% and 3% was appropriate to setup the percentage of the population subject to mutation.

Table 1. Truth table for the circuit of the first example.

X	Y	Z	F
0	0	0	0
0	0	1	0
0	1	0	0
0	1	1	1
1	0	0	0
1	0	1	1
1	1	0	1
1	1	1	0

4   Results

We used several examples taken from the literature to test our AS implementation. Our results were compared to those obtained by two human designers and a genetic algorithm with an n-cardinality representation (see [2] for details).

4.1   Example 1

Our first example has 3 inputs and 1 output ant its truth table is shown in Table 1. In this case, the matrix used was of size 5×5, and the length of

Table 2. Comparison of results between our PSO algorithm, the n-cardinality GA (NGA), and two human designers for the circuit of the first example.

NGA	Human Designer 1
$F = Z(X + Y)$	$F = Z(X \oplus Y) + Y(X \oplus Z)$
4 gates	5 gates
2 ANDs, 1 OR, 1 XOR	2 ANDs, 1 OR, 2 XORs

PSO	Human Designer 2
$F = ((Y + Z)X) \oplus YZ$	$F = X'YZ + X(Y \oplus Z)$
4 gates	6 gates
2 ANDs, 1 OR, 1 XOR	3 ANDs, 1 OR, 1 XOR, 1 NOT

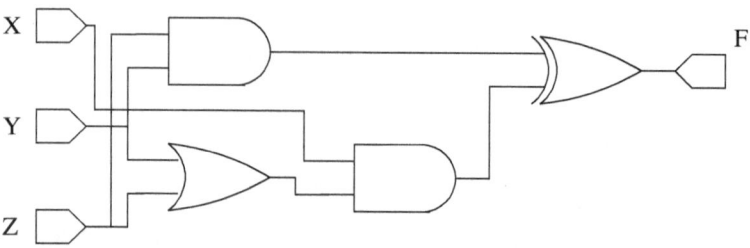

Fig. 4. Graphical representation of the best circuit found by our PSO algorithm for the first example.

each string representing a circuit was 75. Since 5 gates were allowed in each matrix position, then the size of the intrinsic search space (i.e., the maximum size allowed as a consequence of the representation used) for this problem is 5^l, where l refers to the length required to represent a circuit ($l = 75$ in our case). Thefore, the size of the intrinsic search space is $5^{75} \approx 2.6 \times 10^{52}$. Fitness is computed in the following way: 8 (number of outputs that we must match to have a feasible circuit) + 5 × 5 (size of the matrix) - number of gates used (i.e., different of WIRE)). Therefore, a fitness of 29 (the best value produced for this circuit) means that the circuit is feasible (otherwise, its fitness could not possibly be above 8), and that it has 4 gates (i.e., 21 WIREs), because 8 + (25-4) = 8 + 21 = 29.

The graphical representation of the best circuit produced by PSO is shown in Fig. 4. Our PSO algorithm found this solution using the following parameters[2]: 90 particles, 300 iterations (i.e., 27,000 evaluations of the objective function were required), $\Phi_1 = \Phi_2 = 0.8$, $V_{max} = 3.0$, $P_m = 1\%$. The parameters used by the NGA were the following: crossover rate = 0.5, mutation rate = 0.5/75 = 0.0022, population size = 90, maximum number of generations = 300.

The comparison of the results produced by PSO, the NGA and two human designers are shown in Table 2. In this case, human designer 1 used Karnaugh

---

[2] These parameters were empirically derived.

Maps plus Boolean algebra identities to simplify the circuit, whereas human designer 2 used the Quine-McCluskey Procedure.

PSO produced feasible circuits 100% of the time and it found the optimum in 7 out of 20 runs performed (i.e., 35% of the time), reaching a fitness of 28 in all the other runs. The average fitness of the 20 runs performed was 28.35, with a standard deviation of 0.49. The graphical representation of the best solution found by PSO is depicted in Figure 4.

On the other hand, the best solution that the NGA could find using the same population size had also a fitness of 29 (i.e., a circuit with 4 gates), but it appeared only 10% of the time. Also, 20% of the time, the best solution found by the NGA was infeasible. The average fitness of these 20 runs was 21.4, with a standard deviation of 8.438009244. In this example, our PSO algorithm showed a better performance (on average) than the NGA.

Table 3. Truth table for the circuit of the second example.

Z	W	X	Y	F
0	0	0	0	1
0	0	0	1	1
0	0	1	0	0
0	0	1	1	1
0	1	0	0	0
0	1	0	1	0
0	1	1	0	1
0	1	1	1	1
1	0	0	0	1
1	0	0	1	0
1	0	1	0	1
1	0	1	1	0
1	1	0	0	0
1	1	0	1	1
1	1	1	0	0
1	1	1	1	0

4.2   Example 2

Our second example has 4 inputs and one output, as shown in Table 3. In this case, the matrix used was also of size 5×5. Our PSO algorithm used the following parameters: 200 particles, 1000 iterations (i.e., 200,000 evaluations of the objective function were required), $\Phi_1 = \Phi_2 = 0.8$, $V_{max} = 3.0$, $P_m = 3\%$. The parameters used by the NGA were the following: crossover rate $= 0.5$, mutation rate $= 0.0022$, population size $= 200$, maximum number of generations $= 1000$.

The comparison of the results produced by PSO, an n-cardinality GA (NGA), a human designer (using Karnaugh maps), and Sasao's approach [13] are shown in Table 4. Sasao has used this circuit to illustrate his circuit simplification technique based on the use of ANDs & XORs. His solution uses, however, more gates than the circuit produced by our approach.

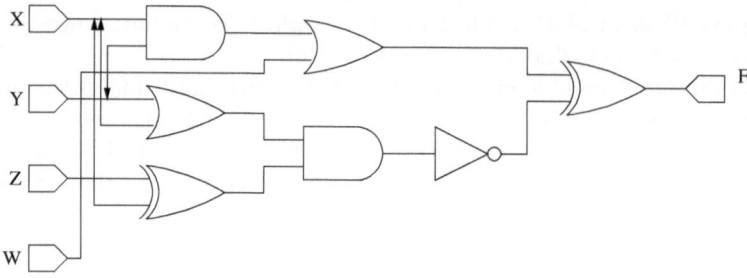

Fig. 5. Graphical representation of the best circuit found by PSO for the second example.

Table 4. Comparison of results between PSO, an n-cardinality GA (NGA), a human designer and Sasao's approach for the circuit of the second example

NGA
$F = (WYX' \oplus ((W+Y) \oplus Z \oplus (X+Y+Z)))'$
10 gates
2 ANDs, 3 ORs, 3 XORs, 2 NOTs

Human Designer 1
$F = ((Z'X) \oplus (Y'W')) + ((X'Y)(Z \oplus W'))$
11 gates
4 ANDs, 1 OR, 2 XORs, 4 NOTs

PSO
$F = (XY + W) \oplus ((Z \oplus X)(X + Y))'$
7 gates
2 ANDs, 2 ORs, 2 XORs, 1 NOT

Sasao
$F = X' \oplus Y'W' \oplus XY'Z' \oplus X'Y'W$
12 gates
3 XORs, 5 ANDs, 4 NOTs

Our PSO algorithm found a solution with a fitness value of 34 (i.e., a circuit with 7 gates) 20% of the time, and feasible circuits were found 67% of the time. The average fitness of the 20 runs performed was 29.35, with a standard deviation of 7.4. The graphical representation of the best solution found is depicted in Figure 5.

The best solution that the NGA could find using the same population size had a fitness of 31 (i.e., a circuit with 10 gates), and it appeared only 20% of the time. Also, 65% of the time, the best solution found by the NGA was infeasible. The average fitness of these 20 runs was 20.25, with a standard deviation of 7.68. In this example, our PSO algorithm showed a better performance than the NGA.

Table 5. Truth table for the circuit of the third example.

A B C D	F₁ F₂ F₃
0 0 0 0	0 0 0
0 0 0 1	0 0 1
0 0 1 0	0 1 0
0 0 1 1	0 1 1
0 1 0 0	0 0 1
0 1 0 1	0 1 0
0 1 1 0	0 1 1
0 1 1 1	1 0 0
1 0 0 0	0 1 0
1 0 0 1	0 1 1
1 0 1 0	1 0 0
1 0 1 1	1 0 1
1 1 0 0	0 1 1
1 1 0 1	1 0 0
1 1 1 0	1 0 1
1 1 1 1	1 1 0

4.3   Example 3

Our third example is a two-bit adder (4 inputs and 3 outputs), and its truth table is shown in Table 5. The matrix used in this case was again of size 5 × 5. Our PSO algorithm used the following parameters: 300 particles, 2000 iterations (i.e., 600,000 evaluations of the objective function were required), $\Phi_1 = \Phi_2 = 0.8$, $V_{max} = 3.0$, $P_m = 1\%$. The parameters used by the NGA were the following: crossover rate = 0.5, mutation rate = 0.0022, population size = 300, maximum number of generations = 2000.

The comparison of the results produced by PSO, an n-cardinality GA (NGA), and one human designer (using Karnaugh maps) are shown in Table 6.

Our PSO algorithm found a solution with a fitness value of 66 (i.e., a circuit with 7 gates) but only once in the 20 runs performed. Feasible circuits were found only 20% of the time. The average fitness of the 20 runs performed was 48.85, with a standard deviation of 6.82. The graphical representation of the best solution found by our PSO algorithm is depicted in Figure 6.

The best solution that the NGA could find using the same number of fitness function evaluations had a fitness of 66 (i.e., a feasible circuit with 7 gates). This solution also appeared only once in the 20 runs performed. However, the NGA could find feasible solutions 75% of the time. The average fitness of these 20 runs was 58.2, with a standard deviation of 7.17. It can be clearly seen that in this example the NGA performed better (on average) than our PSO algorithm.

5   Conclusions and Future Work

We have presented the first formal proposal to use binary particle swarm optimization for designing combinational logic circuits. The approach presented

Table 6. Comparison of results between our PSO algorithm, an n-cardinality GA (NGA), and one human designer for the circuit of the third example.

NGA
$F_1 = B \oplus D$
$F_2 = (A \oplus C) \oplus BD$
$F_3 = AC + BD(A \oplus C)$
7 gates
2 ANDs, 1 OR, 4 XORs

Human Designer 1
$F_1 = A \oplus D$
$F_2 = (A \oplus C)D' + ((A \oplus C) \oplus B)D$
$F_3 = AC + BD(A + C)$
12 gates
5 ANDs, 3 ORs, 3 XORs, 1 NOT

PSO
$F_1 = B \oplus D$
$F_2 = (BD) \oplus (A \oplus C)$
$F_3 = (AC) + ((BD)(A \oplus C))$
7 gates
3 XORs, 3 ANDs, 1 OR

seems promising, since it produced competitive (and in some cases better) results with respect to an n-cardinality genetic algorithm (except for the last example) and it consistently outperformed the solutions produced by human designers. In fact, our experiments[3] indicate that our PSO is very competitive with circuits that have only one output, but the approach is less robust when dealing with multiple-outputs circuits.

One of the current limitations of our approach is the exploratory power of the algorithm which is still not as good as we expected. This is more evident in circuits with several outputs such as the two-bit adder in which our PSO algorithm had a poorer performance than the NGA (on average).

As part of our future work, we are planning to introduce a population-based approach such as the one proposed in [3] to improve the search capabilities of our algorithm. We are also working on an indirect representation that allows us to use real numbers to represent circuits. We hypothesize that PSO will have a better performance if we can manage to produce a real numbers representation, since there is plenty of evidence of such positive behavior in the literature [8].

Acknowledgements

The first author acknowledges support from CONACyT through the NSF-CONACyT project number 32999-A. The second author acknowledges support from

---

[3] There are several more examples available which were not included due to space limitations.

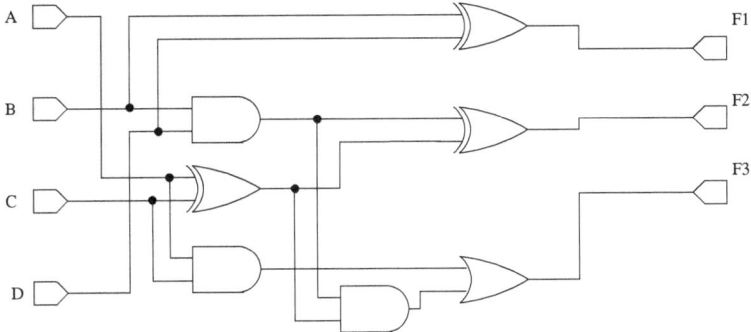

Fig. 6. Graphical representation of the best circuit found by our PSO algorithm for the third example.

CONACyT through a scholarship to pursue graduate studies in Computer Science at the Sección de Computación of the Electrical Engineering Department at CINVESTAV-IPN. The third author acknowledges support from CONACyT through project number I-39324-A.

References

1. Carlos A. Coello Coello, Alan D. Christiansen, and Arturo Hernández Aguirre. Automated Design of Combinational Logic Circuits using Genetic Algorithms. In D. G. Smith, N. C. Steele, and R. F. Albrecht, editors, *Proceedings of the International Conference on Artificial Neural Nets and Genetic Algorithms*, pages 335–338. Springer-Verlag, University of East Anglia, England, April 1997.
2. Carlos A. Coello Coello, Alan D. Christiansen, and Arturo Hernández Aguirre. Use of Evolutionary Techniques to Automate the Design of Combinational Circuits. *International Journal of Smart Engineering System Design*, 2(4):299–314, June 2000.
3. Carlos A. Coello Coello, Arturo Hernández Aguirre, and Bill P. Buckles. Evolutionary Multiobjective Design of Combinational Logic Circuits. In Jason Lohn, Adrian Stoica, Didier Keymeulen, and Silvano Colombano, editors, *Proceedings of the Second NASA/DoD Workshop on Evolvable Hardware*, pages 161–170, Los Alamitos, California, July 2000. IEEE Computer Society.
4. Russell C. Eberhart and Yuhui Shi. Comparison between Genetic Algorithms and Particle Swarm Optimization. In V. W. Porto, N. Saravanan, D. Waagen, and A.E. Eibe, editors, *Proceedings of the Seventh Annual Conference on Evolutionary Programming*, pages 611–619. Springer-Verlag, March 1998.
5. Tatiana Kalganova and Julian Miller. Evolving more efficient digital circuits by allowing circuit layout and multi-objective fitness. In Adrian Stoica, Didier Keymeulen, and Jason Lohn, editors, *Proceedings of the First NASA/DoD Workshop on Evolvable Hardware*, pages 54–63, Los Alamitos, California, 1999. IEEE Computer Society Press.
6. James Kennedy and Russell C. Eberhart. Particle Swarm Optimization. In *Proceedings of the 1995 IEEE International Conference on Neural Networks*, pages 1942–1948, Piscataway, New Jersey, 1995. IEEE Service Center.

7. James Kennedy and Russell C. Eberhart. A Discrete Binary Version of the Particle Swarm Algorithm. In *Proceedings of the 1997 IEEE Conference on Systems, Man, and Cybernetics*, pages 4104–4109, Piscataway, New Jersey, 1997. IEEE Service Center.

8. James Kennedy and Russell C. Eberhart. *Swarm Intelligence*. Morgan Kaufmann Publishers, San Francisco, California, 2001.

9. Sushil J. Louis. *Genetic Algorithms as a Computational Tool for Design*. PhD thesis, Department of Computer Science, Indiana University, August 1993.

10. Sushil J. Louis and Gregory J. Rawlins. Using Genetic Algorithms to Design Structures. Technical Report 326, Computer Science Department, Indiana University, Bloomington, Indiana, February 1991.

11. J. F. Miller, P. Thomson, and T. Fogarty. Designing Electronic Circuits Using Evolutionary Algorithms. Arithmetic Circuits: A Case Study. In D. Quagliarella, J. Périaux, C. Poloni, and G. Winter, editors, *Genetic Algorithms and Evolution Strategy in Engineering and Computer Science*, pages 105–131. Morgan Kaufmann, Chichester, England, 1998.

12. Julian F. Miller, Dominic Job, and Vesselin K. Vassilev. Principles in the Evolutionary Design of Digital Circuits—Part I. *Genetic Programming and Evolvable Machines*, 1(1/2):7–35, April 2000.

13. Tsutomu Sasao, editor. *Logic Synthesis and Optimization*. Kluwer Academic Press, 1993.

A Note on Designing Logical Circuits Using SAT

Giovani Gomez Estrada

Max-Planck-Institute, Heisenbergstraße 3
5Q7, Stuttgart, D-70569, Germany
+49 711 689-3570, giovani@mf.mpg.de

Abstract. We present a systematic procedure to synthesise and minimise digital circuits using propositional satisfiability. After encoding the
truth table into a canonical sum of at most k different products, we seek
its minimal satisfiable representation. We show how to use an interesting
local search landscape for this minimisation. This approach can be very
useful since we can generate exact minimal solutions within reasonably
computational resources.

1   Introduction

This note reviews the synthesis and minimisation of digital circuits using propositional satisfiability (SAT). Here we shall use a SAT encoding for circuit synthesis,
and an incomplete randomised SAT solver for minimisation. That is, we shall
translate the synthesis as a boolean decision problem, and minimisation as finding satisfiability on different encodings. Although the encoding of a truth table
into a logical formula is not a new issue [7, 8], it has been largely unknown in the
circuit design community. Regrettably, they show an over estimation of clauses
and variables, also, perhaps it had little impact because they lack a detailed
analysis of suitable SAT solvers for this encoding.

Here we introduce a corrected estimation for variables and clauses, as well
as to point out a useful behaviour in the SAT solver that allow us to estimate
minimal circuits. By providing both pieces we can thus overcome some of their
initial inconveniences and reintroduce it as an interesting comparison point. By
using this encoding and solving procedure one can generate minimal circuits in
terms of products in canonical form. Notice that, the minimisation to k terms
does not correspond to the minimal number of gates, but to the minimal k-levels
of logic on Programmable Logic Arrays.

The rest of this paper is organised as follows. Firstly, we introduce some
background on SAT, and the transformation used here. Secondly, we explain
how it can be used to extract the minimal number of products. Thirdly, we show
an interesting behaviour of a randomised SAT solver on these encodings. Finally
we give some results, comparisons, and conclusions for this note.

2   Satisfiability and Transformations

The propositional satisfiability problem, or SAT for short, is a key issue in theory of computing. Conceptually, a SAT formula is an ensemble of m clauses in

A.M. Tyrrell, P.C. Haddow, and J. Torresen (Eds.): ICES 2003, LNCS 2606, pp. 410–421, 2003.
© Springer-Verlag Berlin Heidelberg 2003

conjuntive normal form, $\wedge_{i=1}^{m} C_i$. Where every C_i is a disjunction of at least one binary variable, or its negation, from the set $x_1, \ldots, x_n$. Indeed, the question is whether or not exists a truth assignment that satisfies all clauses. In fact every clause can be seen as one restriction to handle, and overall satisfaction is reached whenever *all* clauses are satisfied. This search space is therefore at most $O(2^n)$ in a naïve approach, or $O(1.3302^n)$ by using wise randomisation [5].

Due to this exponential worst case, endeavours in SAT research has centred around two foci, complete and incomplete solvers. Roughly speaking, complete procedures speed-up the inference by applying a set of rules, early conflict detection, etc. These procedures, regrettably, still show painful exponential growth in the number of variables.

Meanwhile, incomplete search procedures heuristically set and refine truth assignments. In this row, most of the incomplete solvers are based in randomised local search. These ideas of heuristic search usually produce very fast results, but nevertheless they are incomplete on whether the formulae is unsatisfied at all. Despite that fact, we should remember one useful property of Monte Carlo simulations, if the algorithm is run repeatedly with independent random choices each time, the failure probability can be made arbitrarily small, regardless of the input. That is why is common to have restarts in many search procedures.

Here we use a well-known and simple but incomplete algorithm, WalkSAT [12]. As we shall see later, WalkSAT does a greedy variable selection, and, with some small probability p, flips a variable from an unsatisfied clause, even if this variable does not gain too much in comparison to others.

2.1   Encodings

The NP-completeness concept, which is based on the Cook-Levin work, deals with the idea of polynomial transformation from a problem P_i to P_j. Notice that such transformations must preserve the essential result, i.e. should P_i returns "yes", also P_j must return "yes" under the same problem input.

Among all possible transformations, we shall use one due to Kamath et al. [7, 8]. They show how to transform a truth table (n inputs, m outputs) in its equivalent boolean decision problem: is it possible to build a circuit with at most k products? The encoding maps any truth table to logic gates AND, OR, and NOT. Optimal circuits can be found by minimising free parameter k, which is the number of different products in canonical form used to represent the circuit. By moving k one may therefore ask for a valid assignment to the formulae E_k up till finding satisfiability. Once a satisfaction has been found, we may decode it to produce a valid and fully functional circuit. This mapping will help us to construct minimal combinatorial circuits in terms of minterms.

Whilst this circuit synthesis technique has been used in a couple of studies on SAT, it has been largely unknown outside that world, even for the circuit design community. We think such fact is perhaps due to an over estimation of variables and clauses, or maybe because their example simply does not follow from their equations.

Here we present the formulation, a reader interested in reading the original derivation should refer to the corresponding article [8]. We shall assume a given truth table T, and a parameter k for the wished number of different minterms. From T we identify A entries or rows as inputs and outputs, x_i^a and y_q^a respectively, where upper index represents the row in T and lower index the column. We also need a set of pairs $\{a, q\}$ such that the output y_q^a has a value "true", as well as the $\{a, q\}$ for "false", denoted as Y^1 and Y^0 respectively. To follow the same notation in [7, 8], we use auxiliary variables σ as:

$$\sigma_{ji}^a = \begin{cases} s'_{ji} & \text{if } x_i^a = 1 \\ s_{ji} & \text{if } x_i^a = 0 \end{cases}$$

Next five sections corresponds to the final formulae E_k:

$n \times k$ clauses type 1:

$$For\ i = 1\ \text{to n}$$
$$For\ j = 1 \quad to\ k$$
$$s_{ji} \vee s'_{ji}$$

$|Y^0| \times k$ clauses type 2:

$$For\ j = 1\ \text{to k}$$
$$For\ \{a, q\}\ \text{s.t.}\ y_q^a = 0$$
$$\bar{w}_{jq} \vee (\bigvee_{k=1}^{n} \bar{\sigma}_{jk}^a)$$

$|Y^1|$ clauses type 3:

$$For\ \{a, q\} \quad \text{s.t.}\ y_q^a = 1$$
$$\bigvee_{j=1}^{k} z_{jq}^a$$

$|Y^1| \times k$ clauses type 4a:

$$For\ j = 1\ \text{to k}$$
$$For\ \{a, q\} \quad \text{s.t.}\ y_q^a = 1$$
$$\bar{z}_{jq}^a \vee w_{jq}$$

$n \times k \times |Y^1|$ clauses 4b:

$$For\ i = 1\ \text{to n}$$
$$For\ j = 1\ \text{to k}$$
$$For\ \{a, q\} \quad \text{s.t.}\ y_q^a = 1$$
$$\bar{z}_{jq}^a \vee \sigma_{ji}^a$$

As can be seen, variables (v) and clauses (c) grow[1] as:

$$v = k \times (2n + m + |Y^1|) \tag{1}$$

$$c = k \times (n + n|Y^1| + |Y^1| + |Y^0|) + |Y^1| \tag{2}$$

---

[1] Previous expressions were $k \times (2n + m(1 + A))$ variables, and $k \times (n + |Y^0|) + |Y^1|(1 + k(2 + n))$ clauses.

2.2   An example

We shall describe the complete formulae for the following small truth table:

x_1	x_2	x_3	y_1	y_2
0	0	0	0	0
0	0	1	0	0
0	1	0	1	1
0	1	1	0	0
1	0	0	0	1
1	0	1	1	1
1	1	0	1	1
1	1	1	1	0

For this specific example we have 3 inputs (n), two outputs (m), and we set k to three, then:

$$Y^1 = \{\{3,1\}, \{3,2\}, \{5,2\}, \{6,1\}, \{6,2\}, \{7,1\}, \{7,2\}, \{8,1\}\}, \quad |Y^1| = 8$$

$$Y^0 = \{\{1,1\}, \{1,2\}, \{2,1\}, \{2,2\}, \{4,1\}, \{4,2\}, \{5,1\}, \{8,2\}\}, \quad |Y^0| = 8$$

The following formulae is derived from the equations:

Type 1	Type 2	Type 3	Type 4a
$s_{1,1} \vee s'_{1,1}$	$\bar{w}_{1,1} \vee \bar{s}_{1,1} \vee \bar{s}_{1,2} \vee \bar{s}_{1,3}$	$z^3_{1,1} \vee z^3_{2,1} \vee z^3_{3,1}$	$\bar{z}^3_{1,1} \vee w_{1,1}$
$s_{2,1} \vee s'_{2,1}$	$\bar{w}_{1,2} \vee \bar{s}_{1,1} \vee \bar{s}_{1,2} \vee \bar{s}_{1,3}$	$z^3_{1,2} \vee z^3_{2,2} \vee z^3_{3,2}$	$\bar{z}^3_{1,2} \vee w_{1,2}$
$s_{3,1} \vee s'_{3,1}$	$\bar{w}_{1,1} \vee \bar{s}_{1,1} \vee \bar{s}_{1,2} \vee \bar{s}'_{1,3}$	$z^5_{1,2} \vee z^5_{2,2} \vee z^5_{3,2}$	$\bar{z}^5_{1,2} \vee w_{1,2}$
$s_{1,2} \vee s'_{1,2}$	$\bar{w}_{1,2} \vee \bar{s}_{1,1} \vee \bar{s}_{1,2} \vee \bar{s}'_{1,3}$	$z^6_{1,1} \vee z^6_{2,1} \vee z^6_{3,1}$	$\bar{z}^6_{1,1} \vee w_{1,1}$
$s_{2,2} \vee s'_{2,2}$	$\bar{w}_{1,1} \vee \bar{s}_{1,1} \vee \bar{s}'_{1,2} \vee \bar{s}'_{1,3}$	$z^6_{1,2} \vee z^6_{2,2} \vee z^6_{3,2}$	$\bar{z}^6_{1,2} \vee w_{1,2}$
$s_{3,2} \vee s'_{3,2}$	$\bar{w}_{1,2} \vee \bar{s}_{1,1} \vee \bar{s}'_{1,2} \vee \bar{s}'_{1,3}$	$z^7_{1,1} \vee z^7_{2,1} \vee z^7_{3,1}$	$\bar{z}^7_{1,1} \vee w_{1,1}$
$s_{1,3} \vee s'_{1,3}$	$\bar{w}_{1,1} \vee \bar{s}'_{1,1} \vee \bar{s}_{1,2} \vee \bar{s}_{1,3}$	$z^7_{1,2} \vee z^7_{2,2} \vee z^7_{3,2}$	$\bar{z}^7_{1,2} \vee w_{1,2}$
$s_{2,3} \vee s'_{2,3}$	$\bar{w}_{1,2} \vee \bar{s}'_{1,1} \vee \bar{s}'_{1,2} \vee \bar{s}'_{1,3}$	$z^8_{1,1} \vee z^8_{2,1} \vee z^8_{3,1}$	$\bar{z}^8_{1,1} \vee w_{1,1}$
$s_{3,3} \vee s'_{3,3}$	$\bar{w}_{2,1} \vee \bar{s}_{2,1} \vee \bar{s}_{2,2} \vee \bar{s}_{2,3}$		$\bar{z}^3_{2,1} \vee w_{2,1}$
	$\bar{w}_{2,2} \vee \bar{s}_{2,1} \vee \bar{s}_{2,2} \vee \bar{s}_{2,3}$		$\bar{z}^3_{2,2} \vee w_{2,2}$
	$\bar{w}_{2,1} \vee \bar{s}_{2,1} \vee \bar{s}_{2,2} \vee \bar{s}'_{2,3}$		$\bar{z}^5_{2,2} \vee w_{2,2}$
	$\bar{w}_{2,2} \vee \bar{s}_{2,1} \vee \bar{s}_{2,2} \vee \bar{s}'_{2,3}$		$\bar{z}^6_{2,1} \vee w_{2,1}$
	$\bar{w}_{2,1} \vee \bar{s}_{2,1} \vee \bar{s}'_{2,2} \vee \bar{s}'_{2,3}$		$\bar{z}^6_{2,2} \vee w_{2,2}$
	$\bar{w}_{2,2} \vee \bar{s}_{2,1} \vee \bar{s}'_{2,2} \vee \bar{s}'_{2,3}$		$\bar{z}^7_{2,1} \vee w_{2,1}$
	$\bar{w}_{2,1} \vee \bar{s}'_{2,1} \vee \bar{s}_{2,2} \vee \bar{s}_{2,3}$		$\bar{z}^7_{2,2} \vee w_{2,2}$
	$\bar{w}_{2,2} \vee \bar{s}'_{2,1} \vee \bar{s}'_{2,2} \vee \bar{s}'_{2,3}$		$\bar{z}^8_{2,1} \vee w_{2,1}$
	$\bar{w}_{3,1} \vee \bar{s}_{3,1} \vee \bar{s}_{3,2} \vee \bar{s}_{3,3}$		$\bar{z}^3_{3,1} \vee w_{3,1}$
	$\bar{w}_{3,2} \vee \bar{s}_{3,1} \vee \bar{s}_{3,2} \vee \bar{s}_{3,3}$		$\bar{z}^3_{3,2} \vee w_{3,2}$
	$\bar{w}_{3,1} \vee \bar{s}_{3,1} \vee \bar{s}_{3,2} \vee \bar{s}'_{3,3}$		$\bar{z}^5_{3,2} \vee w_{3,2}$
	$\bar{w}_{3,2} \vee \bar{s}_{3,1} \vee \bar{s}_{3,2} \vee \bar{s}'_{3,3}$		$\bar{z}^6_{3,1} \vee w_{3,1}$
	$\bar{w}_{3,1} \vee \bar{s}_{3,1} \vee \bar{s}'_{3,2} \vee \bar{s}'_{3,3}$		$\bar{z}^6_{3,2} \vee w_{3,2}$
	$\bar{w}_{3,2} \vee \bar{s}_{3,1} \vee \bar{s}'_{3,2} \vee \bar{s}'_{3,3}$		$\bar{z}^7_{3,1} \vee w_{3,1}$
	$\bar{w}_{3,1} \vee \bar{s}'_{3,1} \vee \bar{s}_{3,2} \vee \bar{s}_{3,3}$		$\bar{z}^7_{3,2} \vee w_{3,2}$
	$\bar{w}_{3,2} \vee \bar{s}'_{3,1} \vee \bar{s}'_{3,2} \vee \bar{s}'_{3,3}$		$\bar{z}^8_{3,1} \vee w_{3,1}$

Type 4b

$$\bar{z}_{1,1}^3 \lor s_{1,1} \quad \bar{z}_{3,2}^5 \lor s'_{3,1} \quad \bar{z}_{2,2}^6 \lor s_{2,2} \quad \bar{z}_{1,2}^7 \lor s_{1,3}$$
$$\bar{z}_{1,2}^3 \lor s_{1,1} \quad \bar{z}_{3,1}^6 \lor s'_{3,1} \quad \bar{z}_{2,1}^7 \lor s'_{2,2} \quad \bar{z}_{1,1}^8 \lor s'_{1,3}$$
$$\bar{z}_{1,2}^5 \lor s'_{1,1} \quad \bar{z}_{3,2}^6 \lor s'_{3,1} \quad \bar{z}_{2,2}^7 \lor s'_{2,2} \quad \bar{z}_{2,1}^3 \lor s_{2,3}$$
$$\bar{z}_{1,1}^6 \lor s'_{1,1} \quad \bar{z}_{3,1}^7 \lor s'_{3,1} \quad \bar{z}_{2,1}^8 \lor s'_{2,2} \quad \bar{z}_{2,2}^3 \lor s_{2,3}$$
$$\bar{z}_{1,2}^6 \lor s'_{1,1} \quad \bar{z}_{3,2}^7 \lor s'_{3,1} \quad \bar{z}_{3,1}^3 \lor s_{3,2} \quad \bar{z}_{2,2}^5 \lor s_{2,3}$$
$$\bar{z}_{1,1}^7 \lor s'_{1,1} \quad \bar{z}_{3,1}^8 \lor s'_{3,1} \quad \bar{z}_{3,2}^3 \lor s'_{3,2} \quad \bar{z}_{2,1}^6 \lor s'_{2,3}$$
$$\bar{z}_{1,2}^7 \lor s'_{1,1} \quad \bar{z}_{1,1}^3 \lor s'_{1,2} \quad \bar{z}_{3,2}^5 \lor s_{3,3} \quad \bar{z}_{2,2}^6 \lor s'_{2,3}$$
$$\bar{z}_{1,1}^8 \lor s'_{1,1} \quad \bar{z}_{1,2}^3 \lor s'_{1,2} \quad \bar{z}_{3,1}^6 \lor s_{3,3} \quad \bar{z}_{2,1}^7 \lor s_{2,3}$$
$$\bar{z}_{2,1}^3 \lor s_{2,1} \quad \bar{z}_{1,2}^5 \lor s_{1,2} \quad \bar{z}_{3,2}^6 \lor s_{3,3} \quad \bar{z}_{2,2}^7 \lor s_{2,3}$$
$$\bar{z}_{2,2}^3 \lor s_{2,1} \quad \bar{z}_{1,1}^6 \lor s_{1,2} \quad \bar{z}_{3,1}^7 \lor s'_{3,2} \quad \bar{z}_{2,1}^8 \lor s'_{2,3}$$
$$\bar{z}_{2,2}^5 \lor s'_{2,1} \quad \bar{z}_{1,2}^6 \lor s_{1,2} \quad \bar{z}_{3,2}^7 \lor s'_{3,2} \quad \bar{z}_{3,1}^3 \lor s_{3,3}$$
$$\bar{z}_{2,1}^6 \lor s'_{2,1} \quad \bar{z}_{1,1}^7 \lor s'_{1,2} \quad \bar{z}_{3,1}^8 \lor s'_{3,2} \quad \bar{z}_{3,2}^3 \lor s_{3,3}$$
$$\bar{z}_{2,2}^6 \lor s'_{2,1} \quad \bar{z}_{1,2}^7 \lor s'_{1,2} \quad \bar{z}_{1,1}^3 \lor s_{1,3} \quad \bar{z}_{3,2}^5 \lor s_{3,3}$$
$$\bar{z}_{2,1}^7 \lor s'_{2,1} \quad \bar{z}_{1,1}^8 \lor s'_{1,2} \quad \bar{z}_{1,2}^3 \lor s_{1,3} \quad \bar{z}_{3,1}^6 \lor s'_{3,3}$$
$$\bar{z}_{2,2}^7 \lor s'_{2,1} \quad \bar{z}_{2,1}^3 \lor s_{2,2} \quad \bar{z}_{1,2}^5 \lor s_{1,3} \quad \bar{z}_{3,2}^6 \lor s'_{3,3}$$
$$\bar{z}_{2,1}^8 \lor s'_{2,1} \quad \bar{z}_{2,2}^3 \lor s_{2,2} \quad \bar{z}_{1,1}^6 \lor s'_{1,3} \quad \bar{z}_{3,1}^7 \lor s_{3,3}$$
$$\bar{z}_{3,1}^3 \lor s_{3,1} \quad \bar{z}_{2,2}^5 \lor s_{2,2} \quad \bar{z}_{1,2}^6 \lor s'_{1,3} \quad \bar{z}_{3,2}^7 \lor s_{3,3}$$
$$\bar{z}_{3,2}^3 \lor s_{3,1} \quad \bar{z}_{2,1}^6 \lor s_{2,2} \quad \bar{z}_{1,1}^7 \lor s_{1,3} \quad \bar{z}_{3,1}^8 \lor s'_{3,3}$$

After examing the formulae or applying the equations (1) and (2), we see it has 48 variables, and 137 clauses. Although the encoding is not exponential in the number of variables, we may not need to find a minimal k, say k^*, by testing all E_k, from $k = 1$ to A. Instead, we can apply a bisection algorithm between boundaries $1 \leq k \leq |Y^1|$. Convergence is assured since there is one and only one root, that is, the encoding is unsatisfiable in $1 \leq k < k^*$, and satisfiable from $k \geq k^*$. Say, the k^* is exactly the turning point or root between unsat to sat. We arrive therefore to the minimal solution in at most $log_2|Y^1| - 1$ steps.

3 Decoding the Circuit

After introducing the formulae to a SAT solver, we found that it is satisfiable, thus values of some variables should be examinated. In order to show the final circuit, we must inspect the values of s, s', and w. First two sets are useful to determine if a literal x_i is part of the j-th product:

$$s_{ji} = \begin{cases} 0 & \text{if } x_i \text{ is in the } j\text{-th term} \\ 1 & \text{otherwise} \end{cases}$$

$$s'_{ji} = \begin{cases} 0 & \text{if } \bar{x}_i \text{ is in the } j\text{-th term} \\ 1 & \text{otherwise} \end{cases}$$

The next table shows a valid assignment for the s and s' set:

Assignment	Meaning
$s_{1,2} = 0$	add x_2 to p_1
$s'_{1,3} = 0$	add $\bar{x}_3$ to p_1
$s_{2,1} = 0$	add x_1 to p_2
$s_{2,3} = 0$	add x_3 to p_2
$s_{3,1} = 0$	add x_1 to p_3
$s'_{3,2} = 0$	add $\bar{x}_2$ to p_3
$s_{1,1} = 1$	do nothing
$s'_{1,1} = 1$	do nothing
$s_{2,2} = 1$	do nothing
$s'_{2,2} = 1$	do nothing
$s_{3,3} = 1$	do nothing
$s'_{3,3} = 1$	do nothing

Therefore, we construct three minterms: $p_1 = x_2\bar{x}_3$, $p_2 = x_1x_3$, and $p_3 = x_1\bar{x}_2$. Now the problem reduces to find out which product corresponds to the l-output. This is done by looking for positive values of w:

$$w_{jl} = \begin{cases} 1 & \text{if } j\text{-th minterm appears in the } l\text{-th output} \\ 0 & \text{otherwise} \end{cases}$$

Next table shows a valid assignment for the w set:

Assignment	Meaning
$w_{1,1} = 1$	add p_1 to y_1
$w_{2,1} = 1$	add p_2 to y_1
$w_{1,2} = 1$	add p_1 to y_2
$w_{3,2} = 1$	add p_3 to y_2
$w_{2,2} = 0$	do nothing
$w_{3,1} = 0$	do nothing

Finally, from the last table we finally rewrite y_1 and y_2 as:

$$y_1 = x_2\bar{x}_3 + x_1x_3$$

$$y_2 = x_2\bar{x}_3 + x_1\bar{x}_2$$

This is the minimal solution, in canonical sum, for the truth table given above. Figure 1 depicts this circuit in logic array format.

4   Local Search Landscape

We have seen how to transform a truth table and decode its solution to produce valid circuits. Now the critical point is the SAT solver algorithm. Although we shall try to avoid an endless list of papers on this area, we should point out that simple randomised strategies plus random walk have been very successful for propositional satisfiability, see e.g. [5, 11].

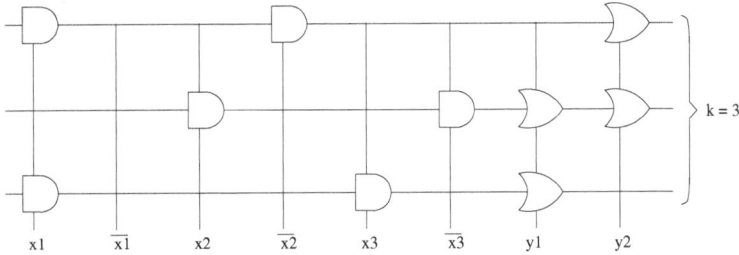

Fig. 1. Logic array representation $(k^* = 3)$ for the example.

GSAT [11] is a randomised greedy search, where we flip and set boolean variables accordingly to its gain. It starts at some initial state (randomised initial assignment) and performs a random walk, by flipping some variables with some probability. We select thus the variable which produces the largest increase in the number of satisfied clauses. The probability to flip one variable is normally assumed to be $\frac{1}{2}$, however this is not mandatory since we can set the probability distribution depending on the input formula [5]. Moreover, with some probability we flip one variable at random, even if such a variable does not satisfy many clauses. GSAT is one technique from a countless mass of successful application of randomised search heuristics.

It should be clear that such a randomised procedure does not confirm that a formulae is unsatisfiable at all. However, here we show how the local search behaviour may indicate, or at least give a clue, whether or not it is satisfiable.

4.1   Computing effort

Figure (2) depicts the local search landscape for a three bits multiplier. We set a cutoff, in the random walk, of 1000 times the number of variables, and two restarts. From the figure we can see the sharp slump down in the local search effort after just crossing the satisfiability barrer. That is, such encoding is unsat from 1 to 29 products, meanwhile is certainly sat from 30. We may thus see that, although it is an incomplete solver, its behaviour is very convenient to identify the satisfiability threshold.

It is also interesting to note the asymptotic tail. Once we have found a satisfiable E_k, every successive $k+1$ has more and more feasible solutions, which in turn helps to the search algorithm to finish faster, i.e. less computing effort. This dramatic slump down in the search for a valid assignment has been observed in every circuit we designed. Unfortunately, such a nice behaviour does not appear in complete SAT solvers, therefore we are still having some uncertainty of the k^*, e.g. this may be $k^* - 1$. Last but no least, we have seen that a cutoff of 1000 times the number of variables is very successful and scales with the problem size, so it does not make sense to try different cutoffs even in larger circuits.

We do produce shorter or similar expressions than Espresso [1] or MisII [2], however our search procedure is slower. It is worth to mention that those

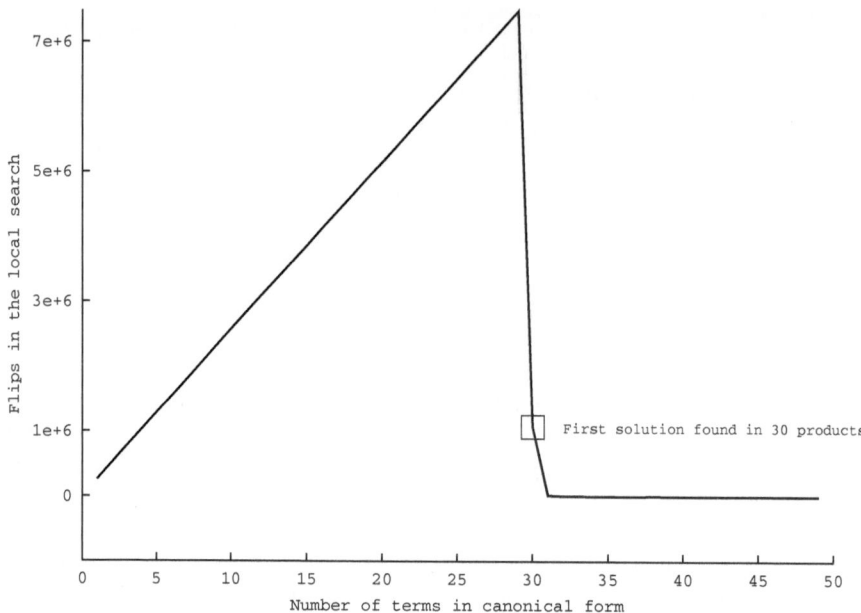

Fig. 2. Local search landscape for a 3bits multiplier. E_k is still unsat until $k \leq 29$ products, thus the local search arrives to the cutoff. Then, in the satisfiability turning point, $k = 30$, the search effort slump sharply.

softwares also suffer scaling factor, whereas the procedure shown before scales polynomially with the number of variables. For circuits with hundreds entries in the truth table we may spend some hours in finding the minimal k. The resulting encoding (E_{30}) for the three bits multiplier has 576 Kb, thus it is not a problem for any current PC nor workstation.

5   Examples and Discussion

Combinational circuit design has been an active area for long time. However, the main drawback has been the problem scale. Whilst it is relatively easy to design and minimise small circuits, search grow too fast to manage large and complex functions. With apologies to many scientists who have been active in this field, we proceed to show a comparison, rather than a comprehensive review.

We selected two representative circuits, adder and multiplier, to compare the underlying boolean circuitry. Multipliers have been argued to grow exponentially with the number of input variables [10]. Please notice that we shall put all examples in canonical form, translating XORs, and avoiding MUX gates. This representation is suitable for standard Programmable Logic Arrays, where k–layers of circuitry are implemented with AND and OR gates, and not restricted to gates having only two inputs. We may say our circuits are minimised in propagation delay, with a worst case delay of k-levels of logic.

5.1   Multipliers

To begin with, we show the minimal function for a two bits multiplier. Table 1 shows its truth table $A * B = C$, where A is $x_1 x_2$, B is $x_3 x_4$, and C is $y_1 y_2 y_3 y_4$:

Table 1. Truth table for a two bits multiplier.

x_1	x_2	x_3	x_4	y_1	y_2	y_3	y_4
0	0	0	0	0	0	0	0
0	0	0	1	0	0	0	0
0	0	1	0	0	0	0	0
0	0	1	1	0	0	0	0
$\vdots$	$\vdots$	$\vdots$	$\vdots$	$\vdots$	$\vdots$	$\vdots$	$\vdots$
1	1	0	0	0	0	0	0
1	1	0	1	0	0	1	1
1	1	1	0	0	1	1	0
1	1	1	1	1	0	0	1

For this small circuit we found that the minimal expression in canonical form has seven different products (i.e. $k^* = 7$), which are shown below. Figure 3 depicts this circuit in logic array format.

$$y_1 = x_1 x_2 x_3 x_4$$

$$y_2 = x_1 \bar{x}_2 x_3 x_4 + x_1 x_3 \bar{x}_4$$

$$y_3 = \bar{x}_1 x_2 x_3 + x_1 \bar{x}_2 x_3 x_4 + x_1 \bar{x}_3 x_4 + x_2 x_3 \bar{x}_4$$

$$y_4 = x_2 x_4$$

This 2bits multiplier is shorter than the expression found by Miller et al. [10]. The comparison was done after replacing one XOR gate to its corresponding AND, OR, and NOT representation. Figure 4 depicts this circuit in logic array format. The following eight different products corresponds to their findings:

$$y_1 = x_1 x_2 x_3 x_4$$

$$y_2 = x_1 \bar{x}_2 x_3 + x_1 x_3 \bar{x}_4$$

$$y_3 = \bar{x}_1 x_2 x_3 + x_1 \bar{x}_2 x_4 + x_1 \bar{x}_3 x_4 + x_2 x_3 \bar{x}_4$$

$$y_4 = x_2 x_4$$

Surprisingly, after changing two XORs from the 2-bits multiplier shown by Coello et al. [3, 4], we arrived to *exactly* the same eight products previously mentioned.

Analogously, a 3-bits multiplier shown by Miller et al. [10] has 36 different terms in sum of products, while the minimal is $k^* = 30$. Notice that this comparison is done after replacing their nine XORs by elementary gates, i.e. AND, OR, NOT.

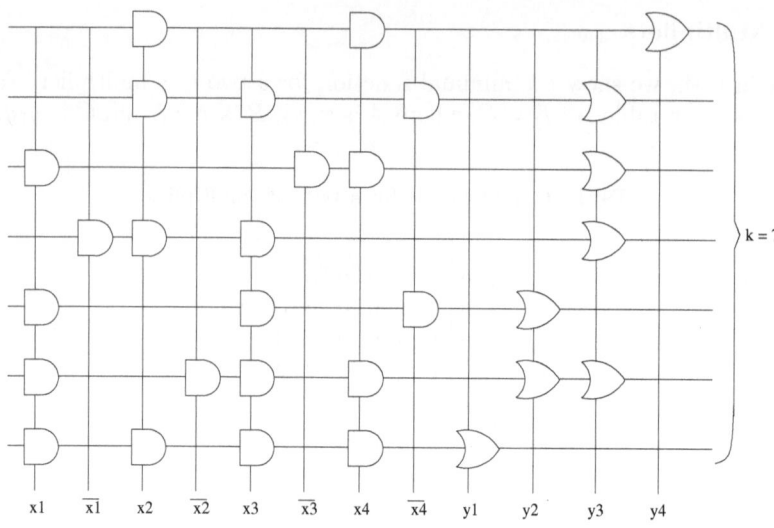

Fig. 3. Minimal 2-bits multiplier in logic array, $k^* = 7$.

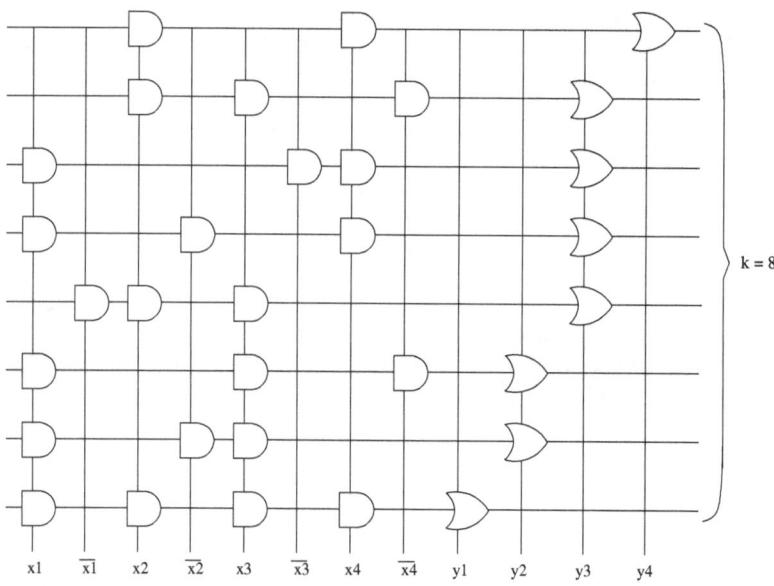

Fig. 4. Two bits multiplier found by Miller et al. and Coello et al. in logic array, $k = 8$.

Both Coello et al. and Miller et al. approaches are based on self-adaptive extrinsic evolution [6], which has produced short and interesting patterns in digital design. Lamentably, after replacing their XORs those designs have larger circuitry in terms of basic components.

5.2   Circuit complexity

Additional information on different multipliers and adders are summarised in table 2.

Table 2. Search space needed to produce some minimal circuits.

Description	k^*	Search space
Two bits multiplier	7	2^{182}
Two bits adder	11	2^{363}
Three bits multiplier	30	2^{3870}
Three bits adder	31	2^{4340}
Four bits adder	75	2^{48975}
Four bits multiplier	121	2^{84942}

For every entry we present their description, k^*, as well as the search space where a first solution was found. The latter number may give us an idea about the circuit hardness. For all circuits we can calculate the number of variables (from eq. 1) needed to encode a minimal circuit in canonical form. This is an unbiased search space where the first, i.e. most difficult, circuit can be found. Moreover, it is not related to any data structure, objectives to minimise, gates, matrix codification nor length allowed to represent a circuit. Hence we may say that synthesising a minimal three bits adder is slightly harder than a three bits multiplier, as can be seen from table (2) under "Search space". Interestingly, same analogy does not hold in four bits.

6   Conclusions and Future Work

We have shown a fresh view on how to synthesise and minimise combinatorial circuits with SAT. This minimisation can be efficiently done by means of a bisection algorithm and a randomised local search. We have found a cutoff in the local search algorithm to one thousand times the number of variables, which works reasonably well, however a complete algorithm, which can assure the exact k^* is still exponential. Our experiments with complete solvers reveal that the exact k^* is at most one product away from the solution found by the randomised algorithm.

We are still missing an exhaustive comparison of complete solvers for this particular encoding. It had been noticed that some solvers perform quite well on certain encodings, whilst very poor in others. Other thing that remains to be investigated is the possibility to improve our current SAT solvers by exploiting any internal structure or pattern of this fascinating encoding.

Acknowledgement

The author wish to thank joint Berlin/Zürich CGC programme, financed by ETH-Zürich and DFG Germany, for supporting this research during 2000/2001.

References

1. R. K. Brayton, G. D. Hachtel, C. T. McMullen, A. L. Sangiovanni-Vincentelli: *Logic minimization algorithms for VLSI synthesis*, Kluwer, 1984.
2. R. K. Brayton, R. Rudell, A. L. Sangiovanni-Vincentelli, A. R. Wang: MIS: A multiple-level logic optimization system. *IEEE Trans. on Computer Aided Design*, vol. 6(6), pp. 1062-1081, 1987.
3. C. A. Coello, R. L. Zavala, B. Mendoza, A. Hernandez: Automated Design of Combinational Logic Circuits using the Ant System. *Engineering Optimization*, vol. 34(2), pp. 109-127, 2002.
4. C. A. Coello, A. Hernandez: Design of Combinational Logic Circuits through an Evolutionary Multiobjective Optimization Approach. *Artificial Intelligence for Engineering, Design, Analysis and Manufacture*, vol. 16(1), 2002.
5. T. Hofmeister, U. Schöning, R. Schuler, O. Watanabe: A probabilistic 3-SAT algorithm further improved. *Proc. of the STACS*, Springer-Verlag LNCS 2285, pp. 192-202, March, 2002.
6. T. Kalganova: *Evolvable Hardware Design for Combinational Logic Circuits*, PhD thesis, School of Computing, Napier University, UK, 2000.
7. A. Kamath, N. Karmarkar, K. Ramakrishnan, M. Resende: A continuous approach to inductive inference. *Mathematical programming*, vol. 57, pp. 215-238, 1992.
8. A. Kamath, N. Karmarkar, K. Ramakrishnan, M. Resende: An interior point approach to boolean vector function synthesis. *Proc. of the MSCAS*, pp. 185-189, 1993.
9. C. M. Li: Integrating Equivalency reasoning into Davis-Putnam procedure. *Proc. of the AAAI*, pp. 291-296, 2000.
10. J. F. Miller, D. Job, V. K. Vassilev: Principles in the evolutionary design of digital circuits - Part I. *Genetic programming and evolvable machines*, vol 1(1/2), pp. 7-35, 2000.
11. B. Selman, H. Levesque, D. Mitchell: A New Method for Solving Hard Satisfiability Problems. *Proc. of the AAAI*, pp. 440-446, 1992.
12. B. Selman, H. Kautz, B. Cohen: Local Search Strategies for Satisfiability Testing. Cliques, Coloring, and Satisfiability: Second DIMACS Implementation Challenge. David S. Johnson and Michael A. Trick, ed. *DIMACS Series in Discrete Mathematics and Theoretical Computer Science*, vol. 26, AMS, 1996.

Using Genetic Programming
to Generate Protocol Adaptors
for Interprocess Communication

Werner Van Belle*, Tom Mens**, and Theo D'Hondt

Programming Technology Lab, Vrije Universiteit Brussel,
Pleinlaan 2, 1050 Brussel, Belgium
{werner.van.belle,tom.mens,tjdhondt}@vub.ac.be
http://prog.vub.ac.be

Abstract. As mobile devices become more powerful, interprocess communication becomes increasingly more important. Unfortunately, this larger freedom of mobility gives rise to unknown environments. In these environments, processes that want to communicate with each other will be unable to do so because of protocol conflicts. Although conflicting protocols can be remedied by using adaptors, the number of possible combinations of different protocols increases dramatically. Therefore we propose a technique to generate protocol adaptors automatically. This is realised by means of genetically engineered classifier systems that use Petri nets as a specification for the underlying protocols. This paper reports on an experiment that validates this approach.

1 Introduction

In the field of evolvable computing, software (and hardware) is developed that adapts itself to new runtime environments as necessary. The runtime environments targetted in this paper are open distributed systems in which interprocess communication forms an essential problem. In these environments an application consists of processes that communicate with other processes to reach specific goals.

With the advent of mobile devices these processes do not necessarily know in which kind of runtime environments they will execute. Therefore they rely on standardised solutions, such as JINI, to find other processes offering a certain behaviour.

Once the other process is known, the real problems start. How can the requesting process communicate with the unknown offered process? Given the fact that those processes are developed by different organisations, the protocols provided and required can vary greatly. As a result protocol conflicts arise.

---

* Corresponding author. He is developing peer-to-peer embedded systems for a project funded by the Flemish Institute for Science and Technology (IWT).
** Tom Mens is a Postdoctoral Fellow of the Fund for Scientific Research - Flanders (Belgium)

A.M. Tyrrell, P.C. Haddow, and J. Torresen (Eds.): ICES 2003, LNCS 2606, pp. 422–433, 2003.

On first sight, a solution to this problem would be to offer protocol adaptors between every possible pair of processes. The problem with this approach is that the number of adaptors grows quadratic to the number of process protocols and as such it simply doesn't scale. The solution is to automate the generation of protocol adaptors between communicating processes.

As a potentially useful technique for this adaptor generation, we explored the research domain of adaptive systems. We found that the combination of genetic programming, classifier systems, and a formal specification in terms of Petri nets allowed us to automate the detection of protocol conflicts, as well as the creation of program code for adaptors that solve these conflicts. This paper reports on an experiment we performed to validate this claim.

2   Prerequisites of Interprocess Communication

Processes communicate with each other only by sending messages over a communication channel (similarily to CSP [1] and the π-calculus [2]). Communication channels are accessed by the process' ports. Processes communicate asynchronously and always copy their messages completely upon sending. The connections between processes are full duplex: every process can *send* and *receive* messages over a port. This brings us in a situation where a process *provides* a certain protocol and *requires* a protocol from another process. A process can have multiple communication channels: for every communication partner and for every provided/required protocol.

We imposed other requirements on the interprocess communication to allow us to generate adaptors:

1. *Implicit addressing.* No process can use an explicit address of another process. Processes work in a connection-oriented way. The connections are set up solely by one process: the connection broker. This connection broker will also evolve adaptors and place them upon the connections when necessary.
2. *Disciplined communication.* No process can communicate with other processes by other means than its ports. Otherwise, 'hidden' communication (e.g., over a shared memory) cannot be modified by the adaptor. This also means that all messages passed over a connection should be copied. Messages cannot be shared by processes (even if they are on the same host), because this would result in a massive amount of concurrency problems.
3. *Explicit protocol descriptions.* While humans prefer a protocol description written in natural language, computers need an explicit formal description of the protocol semantics. A simple syntactic description is no longer suitable.

3   Specifying Protocols

As a running example we choose a typical problem of communicating processes: how processes synchronise with each other. Typically, a server provides a concurrency protocol (often a transaction protocol) [3] that can be used by clients.

The clients have to adhere to this specification or they won't function. Since the clients also expects a certain concurrency behaviour from the server, it is possible that the required interface and provided interface differ.

For example, a client/server can require/provide a full-fledged optimal transaction protocol or it can require/provide a simple locking protocol. When two such protocols of a different kind interact, we can run into an incompatibility problem.

In our example we use a simple locking protocol of the server with which a client can typically lock a resource and then use it. The API for the server is described as follows. (A similar protocol description can be given for the clients.)

```
incoming lock(resource)
  outgoing lock_true(resource)
  outgoing lock_false(resource)
    // lock_true or lock_false are sent back whenever a lock
    // request comes in: lock_true when the resource is locked,
    // lock_false when the resource couldn't be locked.
incoming unlock(resource)
  outgoing unlock_done(resource)
    // will unlock the resource. Send unlock_done back when done.
incoming act(resource)
  outgoing act_done(resource)
    // will do some action on the process.
```

The semantics of this protocol can be implemented in different ways. We will use two kinds of locking semantics [3]:

Counting semaphores allow a client to lock a resource multiple times. Every time the resource is locked the lock counter is increased. If the resource is unlocked the lock counter is decreased. The resource is finally unlocked when the counter reaches zero.

Binary semaphores provide a locking semantics that doesn't offer a counter. It simply remembers who has locked a resource and doesn't allow a second lock. When unlocked, the resource becomes available again.

Differences in how the API considers *lock* and *unlock* can give rise to protocol conflicts. In figure 1 the client process expects a counting semaphore from the server process, but the server process offers a binary semaphore. The client can lock a resource twice and expects that the resource can be unlocked twice. In practice the server just marked the resource as *locked*. If the client unlocks the resource, the resource will be unlocked. Acting upon the server now is impossible, while the client expects it to be possible.

This protocol conflict arises because the API does not specify enough semantic information. Hence, we propose to use a more detailed and generally applicable formalism, namely Petri nets, to offer an explicit description of the protocol semantics. Petri nets [4] offer a model in which multiple processes traverse states by means of state transitions.

In the context of our locking example, this allows us to write a suitable adaptor by relying on: (1) which state the client process *expects* the server to

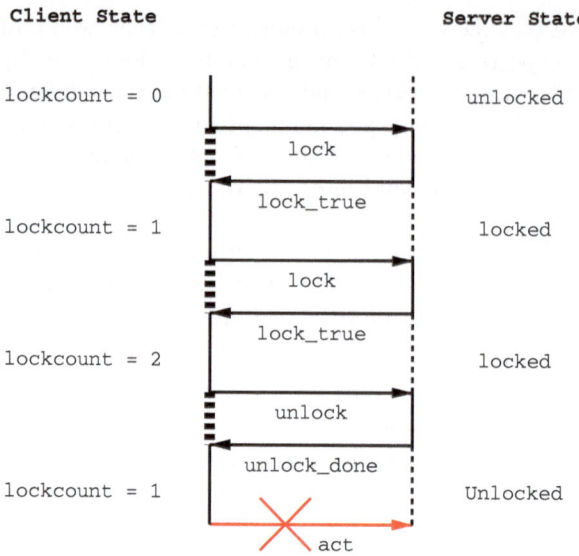

Fig. 1. Conflict between a counting semaphore protocol and a binary semaphore protocol.

be in; (2) in which state the server process is. Both kinds of information are essential: If we don't know that the client thinks that it still has a lock, and we don't know that the state of the server is unlocked (see figure 1), no learned algorithm can make a correct decision.

As an example Petri net that offers the needed semantics, the left part of figure 2 specifies a binary semaphore locking strategy. The current state is *unlocked*. From this state the process requiring this protocol, can choose only one action: *lock*. It then goes to the *locking* state until *lock_true* or *lock_false* comes back. We can also use this Petri net to model the behaviour of a process that *provides* this protocol. It is perfectly possible to offer an protocol that adheres to this specification, in which case, the incoming *lock* is initiated from the client, and *lock_true* or *lock_false* is sent back to the client when making the transition.

4   Evolving Protocols

Protocol adaptors overcome the semantic differences between two processes. We propose to use a genetic algorithms [5] with classifier systems to generate adaptors. Classifier systems are known to work very well in cooperation with genetic algorithms, they are Turing complete and they are *symbolic*. This is important because our Petri net description is in essence a symbolic representation of the different states in which a process can reside.

If this representation would be *numerical,* techniques such as neural networks [6], reinforcement learning and Q-learning [7] could probably be used.

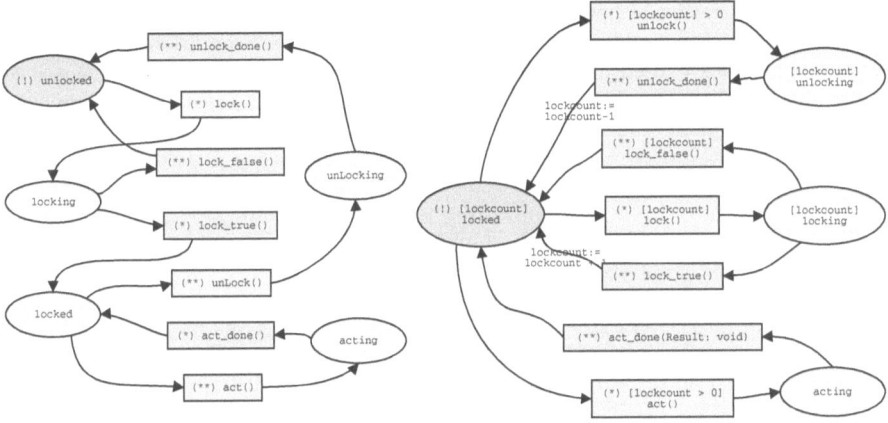

Fig. 2. Two Petri-net descriptions of process protocols. Ellipses correspond to states. Rectangles correspond to transitions. The current state (marked with '!') is coloured yellow. The red transitions (marked with '*') represent incoming messages. The blue ones (marked with '**') represent outgoing messages.

The standard questions before implementing any genetic programming technique are: What are the individuals and their genes? How do we represent the individuals? How do we define and measure the fitness of an individual? How do we initially create individuals? How do we mutate them and how do we create a cross-over of two individuals? How do we compute a new generation from an existing one? For a quick overview of the parameters of our genetic program we refer to table 4.

In our implementation, the individuals will be protocol adaptors between communicating processes. The question of how to represent these individuals is more difficult. We could use well-known programming languages to represent the behaviour of the adaptor. Unfortunately, the inevitable syntactic structure imposed by these languages complicates the random generation of programs. Moreover, these programming languages do not offer a uniform way to access memory.

An alternative that is more suitable for our purposes is the Turing complete formalism of *classifier systems* [5]. A classifier system is a kind of control system that has an input interface, a finite message list, a classifier list and an output interface. The input and output interface put and get messages to and from the classifier list. The classifier list takes a message list as input and produces a new message list as output. Every message in the message list is a fixed-length binary string that is matched against a set of classifier rules. A classifier rule contains a number of (possibly negated) conditions and an action. These conditions and actions form the genes of each individual in our genetic algorithm. Conditions and actions are both ternary strings (of 0, 1 and #). '#' is a pass-through character that, in a condition, means 'either 0 or 1 matches'. If found in an action, we simply replace it with the character from the original message.

Table 1. Illustration of how actions produce a result when the conditions match all messages in the input message list. ˜ is negation of the next condition. A disjunction of two conditions is used for each classifier rule. The second rule does not match for input message 001. The third rule does not match because the negated condition is not satisfied for input message 001.

Input message list = { 001, 101, 110, 100 }

Condition		Action	Matches	Result
00#	101	111	yes	111
01#	1##	000	no	/
1##	˜00#	###	no	/
1##	###	1#0	yes	100, 110

Output message list = { 111, 100, 110 }

Table 1 shows a very simple example. When evaluating a classifier system, all rules are checked (possibly in parallel) with *all* available input messages. The result of every classifier evaluation is added to the end result. This result is the output message list. For more details, we refer to [5].

A classifier system needs to reason about the actions to be performed based on the available input. In our implementation, the rules of a classifier system consist of a ternary string representation of the client state and server state (as specified by the Petri net), as well as a ternary string representing the requested Petri net transition from either the client process or the server process[1]. With these semantics for the classifier rules, translating a request from the client to the server requires only one rule. Another rule is needed to translate requests from the server to the client (see table 2)[2].

The number of bits needed to represent the transitions depends on the number of possible state transitions in the two Petri nets of figure 2. The number of bits needed to represent the states of each Petri net depends on the number of states in the Petri net as well as the variables that are stored in the Petri net (e.g., the *lockcount* variable requires 2 bits if we want to store 4 levels of locking).

Although this is a simple example, more difficult actions can be represented. Consider the situation where the client uses a counting-semaphores locking strategy and the server uses a binary-semaphores locking strategy. In such a situation we don't want to send out the lock-request to the server if the lock count is larger than zero. Table 3 shows how we can represent such a behaviour.

---

[1] Currently we are investigating whether we can drop this translation and generate an adaptor from the Petri net representation directly.

[2] This method of using full classifier systems as individuals is known as the Pittsburgh approach [8]. The Michigan approach, whereby a set of classifier rules evolve together to reach a solution[9], is not suitable for our purposes because one rule doesn't cover the behaviour of an adaptor. As such, cross-over between single rules would not help that much.

Table 2. Blind translation between client and server processes. The last 5 characters in column 1 represent the corresponding transition in the Petri net. The characters in the second and third column represent the states of the client and server Petri net, respectively. The fourth column specifies the action to be performed based on the information in the first four columns.

classifier condition			action	rule description
requested transition	client state	server state	performed action	
00####	####	###	11#...#	Every incoming action from the client (00) is translated into an outgoing action on the server (11)
01####	####	###	10#...#	Every incoming action from the server (01) is translated into an outgoing action to the client (10)

Table 3. Translating a client process lock request to a server process lock action when necessary.

classifier condition			action	rule description
requested transition	client state	server state	performed action	
00 001	~##00	###	10 010 ...	If the client wants to lock (001) and already has a lock (~##00) we send back a lock_true (010)
00 001	##00	###	11 001 ...	If the client wants to lock (001) and has no lock (##00) we immediately send the message through (001).

The genetic programming we implemented uses a full classifier list with variable length. The classifier list is an encoding of the Petri nets, as representation for the *individuals*. Every individual is initially empty. Every time an individual encounters a situation where there is no matching gene a new gene (i.e., a new classifier rule) will be added with a condition that covers this situation and a random action that is performed on the server and/or the client. This way of working, together with the use of Petri nets guarantees that the genetic algorithm will only search within the subspace of possible solutions. Without these boundaries the genetic algorithm would take much longer to find a solution.

Fitness of an individual is measured by means of a number of test scenarios. Every test scenario illustrates a typical behaviour the client requests from the server. The fitness of an individual is determined by how many actions the scenario can execute without yielding unexpected behaviour. Of course this is not enough; we should not have solutions that completely shortcut the server. For example, the algorithm could return *lock_true* every time a request comes in from the client, without even contacting the server. To avoid this kind of behaviour our algorithm provides a *covert channel* that is used by the test scenario to contact the server to verify its actions.

The genetic programming uses a steady-state GA, with a ranking selection criterion: to compute a new generation of individuals, we keep (*reproduce*) 10% of the individuals with the best fitness. We throw away 10% of the worst individuals (not fit enough) and add *cross-overs* from the 10% best group[3]. To create a *cross-over* of individuals we iterate over both classifier lists and each time randomly

---

[3] These values were taken from [10] and gave good results during our experiments.

Table 4. Parameters and characteristics of the genetic program

parameter	value
individuals (genotype)	variable-length classifier system represented as bitstring
population size	100
maximum generations (100 runs)	11
parent selection	ranking selection (10 % best)
mutation	bitflip on non ranked individuals
mutation rate	0.8
crossover	uniform
crossover rate	0.1
input/output interfacing	Petri net state/transition representation
actions	message sending
fitness	number of successfully executed actions

select a rule that will be stored in the resulting individual. It should be noted that the individuals that take part in cross-over are never mutated. The remaining 80% of individuals are *mutated*, which means that the genes of each individual are changed at random: for every rule, a new arbitrary action to be performed on server or client is chosen. On top of this, in 50% of the classifier rules, one bit of the client and server state representations is generalised by replacing it with a #. This allows the genetic program to find solutions for problems that are not presented yet.

5   The Experiment

We will now present the experiment that shows the feasibility of the above techniques to automatically learn an adaptor between incompatible locking strategies.

The experiment is set up as a connection broker between two processes. The first process contacts the second by means of the broker. Before the broker sets up the connection it will generate an adaptor between the two parties to mediate semantic differences. It does so by requesting a running test process from both parties. The client will produce a test client and test scenarios. The server will produce a test server. In comparison with the original process, these testing processes have an extra *testing* port, over which we can reset them. Furthermore, this *testing* port is also used as the covert channel for validating actions at the server.

The genetic program, using a set of 100 individuals (i.e., adaptor processes), will deploy the test processes to measure the fitness of a particular classifier system. Only when a perfect solution is reached, i.e., a correct adaptor has been found, the connection is set up. For reliability reasons we have repeated the experiment (i.e., the execution of the genetic program) 100 times.

The scenarios offered by the client are the ones that determine what kind of classifier system is generated. We have tried this with three scenarios, as illustrated on the left of figure 3. Scenario 1 is a sequence: [lock(), act(), unlock()]. Scenario 2 is the case we explained in figure 1. Scenario 3 is similar

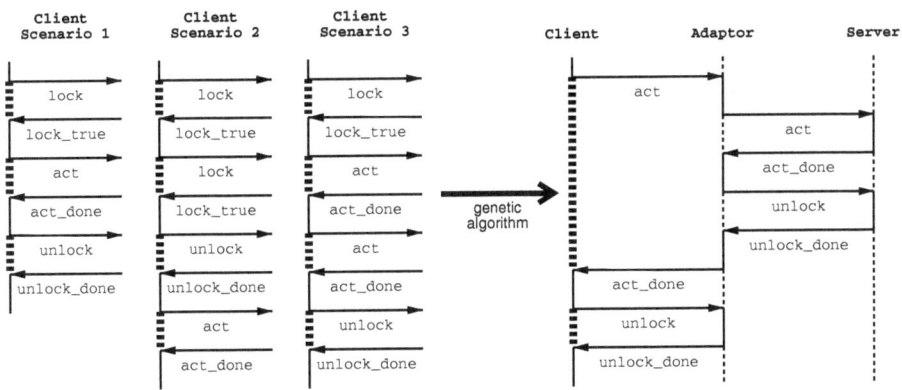

Fig. 3. The three test scenarios we used as initial input to the genetic programming. The dashed vertical lines are waits for a specific message (e.g., lock_true). When using only scenarios 1 and 2 as input, one of the generated adaptors behaved as specified on the right hand side.

to scenario 1: [lock(), act(), act(), unlock()]. The reason why we added such a look-alike scenario will become clear in our observations. In all three scenarios, we issue the same list of messages three times to ensure that the resource is unlocked after the last unlock operation.

5.1   Observations

An examination of the results of several runs of our genetic programming algorithm lead to the following observations:

When we used the covert channel to measure the fitness, we found that (for all 100 runs of the GA) a perfect solution was found within at most 11 generations. When we didn't use the covert channel to check the actions at the server side, the genetic algorithm often (30%) created a classifier that doesn't even communicate with the server. In such a situation the classifier immediately responds *lock_true* whenever the client requests a lock.

Occasionally a classifier system was generated with a strange behaviour. Its fitness was 100%, but it worked asynchronously. In other words, the adaptor would contact the server process before the client process even requested an action from the server process. It could do so because the adaptor knew that the client would request that particular action in the given context. This is illustrated on the right of Figure 3. It implies that a learned algorithm can anticipate certain kinds of future behaviour.

Initially, we only used scenario 1 and 2 to measure the fitness of each individual. We encountered the problem that sometimes, the adaptor anticipates too much and after the first *act*, keeps on acting. This problem was solved by assigning a zero fitness to such solutions.

Table 5. The generated classifier system for a single run.

requested transition	client state	server state	performed action	description
00 00 01	10 00 #1	11 ####	00 10 01	client?lock() & client=locked → client.lock_true()
00 00 01	10 00 1#	11 ####	00 10 01	client?lock() & client=locked → client.lock_true()
00 00 01	10 00 00	~ 11 010	00 11 01	client?lock() & client ≠ locked → server.lock()
00 00 10	10 00 1#	11 ####	00 10 11	client?unlock() & clientlock>2 → client.unlock_done()
00 00 10	10 00 01	11 ####	00 11 10	client?unlock() & clientlock=1 → server.unlock()
00 00 11	10 #####	11 ####	00 11 11	client?act() → server.act()
00 01 10	10 #####	11 ####	00 10 10	server?act_done() → client.act_done()
00 01 00	10 #####	11 ####	00 10 00	server?lock_false() → client.lock_false()
00 01 01	10 #####	11 ####	00 10 01	server?lock_true() → client.lock_true()
00 01 11	10 #####	11 ####	00 10 11	server?unlock_done() → client.unlock_done()
00 00 01	10 00 00	11 010	00 10 00	client?lock() & server=locked & client ≠ locked → client.lock_false()

As a last experiment we measured the fitness by combining information from all three scenarios. This allowed the genetic algorithm to find a perfect solution with less generations because separate individuals that developed behaviour for a specific test scenario were combined in a later generation using cross-over. This illustrates the necessity for a cross-over operator. A random search would take considerably more time to find a combination of both behaviours.

One of the classifier system that is generated by the genetic algorithm, when providing as input all three test scenarios, is given in table 5. The produced classifier system simply translates calls from client to server and vice versa, unless it is about a lock call that should not be made since the server is already locked. The bit patterns in the example differ slightly from the bit patterns explained earlier. This is because we need the ability to make a distinction between a 'transition-message' and a 'state-message'. All transition messages start with 00 and all state-messages start with 10 for client-states and 11 for server-states.

5.2   Discussion

In our approach, the problem of 'writing correct adaptors' is shifted to the problem of 'specifying correct test sets': whenever the developer of a process encounters an incompatibility, he needs to specify a new test scenario that avoids this behaviour. This test scenario is given as additional input to the genetic algorithm, so that the algorithm can find a solution that avoids this incompatibility. The result of this approach is that the programmer does not have to implement the adaptors directly, but instead has the responsibility of writing good test sets (i.e., a consistent set of scenarios that is as complete as possible). This is a non-trivial problem, since the test set needs to cover all *explicit* as well as *implicit* requirements. The main advantage of test sets over explicit adaptors is that we would need a new adaptor for every pair of communicating processes, while we only need one test set for each individual process. As such, test sets are more robust to changes in the environment: when a process needs to communicate with a new process, there is a good chance that the test set will already take into account potential protocol conflicts. Another important advantage of test sets is that they can help in automatic program verification. Bugs in the formal specification (the Petri net) can be detected and verified at runtime. As such, this approach helps the developer to stay conform to the program specification. This clearly helps him in his goal to write better software.

Below we discuss some strengths and weaknesses of our approach:
Automatically generated adaptors can be better than hand-crafted adapters since they can reorder incoming and outgoing messages as necessary. This can result in anticipated behaviour that boosts performance.
The genetic algorithm we proposed has the problem that it needs to learn a certain behaviour based on a very small set of examples. Therefore, the learning algorithm will automatically generate more general or more specific adaptors when offered test sets. This tendency to generalize matches does not always correspond to how a programmer tries to generalise. If there is a close correspondence, the programmer simply needs to write test sets that will naturally be generalised to the desired adaptor. On the other hand, if the generalisation (or specialisation) does not fit the developer's way of thinking, the algorithm will generate seemingly illogical adaptors.
An ideal test scenario should cover all the actions that will be invoked upon the server in all possible combinations. How can we write good tests that do not leave any open holes for the programmer? And if we can write such tests, are the Petri nets still necessary? In other words, is it possible to learn the adaptor automatically just by looking at the interaction between the processes?
Some protocol conflicts that seem simple at first sight cannot be solved (not even by humans). For example, the locking protocol example could be used to lock simple (x, y) - specified cells on a checker board. It is impossible to protocol this with a locking strategy that locks and unlocks the whole board at once. A simple solution such as 'lock all fields' will not work because other communication partners can enter the field and lock a single position. This indicates that the approach presented here is good in solving 'control flow' problems but is bad at converting 'data representations'. However, this is a general AI problem [11] for which no solution yet has been found.

In this paper we presented only a simple example with small protocols (binary versus counting semaphore locking strategies) for the sake of the presentation. We are currently investigating how we can write adaptors by generating a suitable petri-net instead of a classifier system. More elaborate experiments and technical details can be found on http://borg.rave.org/adaptors/.

6   Conclusion

We proposed an automated approach to create intelligent protocol adaptors to resolve incompatibilities between communicating processes. Such an approach is indispensable to cope with the combinatorial explosion of protocol adaptors that are needed in an open distributed setting where processes interact with other processes in unpredictable ways.

Our approach uses a genetic programming tecnique that evolves classifier systems. These classifier systems contain classifiers that react upon the context they receive from both client process and server process. The context is defined as a combination of the possible client-side and server-side states as given by a user-specified Petri net. To measure the fitness of an adaptor, the user needs to provide test scenarios as input. This enables the user to avoid undesired behaviour in interprocess communication.

Acknowledgements

Thanks to Johan Fabry, Tom Lenaerts, Anne Defaweux, Tom Tourwé and the anonymous referees for reviewing this paper.

References

1. Hoare, C.: Communicating Sequential Processes. International Series in Computer Science. Prentice Hall (1985)
2. Milner, R.: Communicating and Mobile Systems: the π-calculus. Cambridge University Press (1999)
3. Lea, D.: Concurrent Programming in Java (2nd edition) Design Principles and Patterns. The Java Series. Addison Wesley (2000)
4. Reisig, W.: An Informal Introduction To Petri Nets. Proc. Int'l Conf. Application and Theory of Petri Nets, Aarhus, Denmark (2000)
5. Glodberg, D.E.: Genetic Algorithms in Search, Optimization, and Machine Learning. Addison-Wesley (1989)
6. Kröse, B., van der Smagt, P.: An introduction to neural networks. University of Amsterdam (1996)
7. Sutton, R.S., Barto, A.G.: Reinforcement Learning – An Introduction. MIT Press (1998)
8. Smith, S.: A Learning System Based on Genetic Adaptive Algorithms. PhD thesis, Department of Computer Science, University of Pittsburgh (1980)
9. Bull, L., Fogarty, T.: Co-evolving communicating classifier systems for tracking. Proc. Int'l Conf. Neural Networks and Genetic Algoriths (1993)
10. Koza, J.R.: Genetic Programming; on the programming of computers by means of natural selection. MIT Press (1992)
11. Morgenstern, L.: The problem with solutions to the frame problem. In Ford, K.M., Pylyshyn, Z., eds.: The Robot's Dilemma Revisited: The Frame Problem in Artificial Intelligence. Ablex Publishing Co., Norwood, New Jersey (1996) 99–133

Using Genetic Programming and High Level Synthesis to Design Optimized Datapath

Sérgio G. Araújo, A. Mesquita, and Aloysio C.P. Pedroza

Electrical Engineering Dept.
Federal University of Rio de Janeiro
C.P. 68504 - CEP 21945-970 - Rio de Janeiro - RJ - Brazil
Tel: +55 21 2260-5010 - Fax: +55 21 2290-6626
{granato,aloysio}@gta.ufrj.br,   mesquita@coe.ufrj.br

Abstract. This paper presents a methodology to design optimized electronic digital systems from high abstraction level descriptions. The methodology uses Genetic Programming in addition to high-level synthesis tools to automatically improve design structural quality (area measure). A two-stage, multiobjective optimization algorithm is used to search for circuits with the desired functionality subjected additionally to chip area constraints. Experiment with a square-root approximation datapath design targeted to FPGA exemplifies the proposed methodology.

1 Introduction

Recently, integrated circuit (IC) design has focused on meeting time-to-market and costumers specific requirements for more and more complex systems [1]. To address this new trend, automation of the entire design process from conceptualization to silicon or a describe-and-synthesize methodology [2] has become necessary. Using a top-down design approach, the system functionality is specified through a behavioral model, simulated to verify correct operation and synthesized using high-level synthesis (HLS) tools.

A chip design process using HLS involves several steps, the first being the specification of the system functionality. To this end a variety of conceptual models can be used as discussed in detail in [3]. Finite-state machines (FSM), finite-state machines with datapath (FSMD) and Petri nets are some examples. Programming languages are conceptual models of particular interest since they can support heterogeneous modeling, i.e., they can describe data, activity and control simultaneously. Hardware description languages (HDLs) such as VHDL [4], [5], are able to capture the system functionality in a machine-readable and simulatable form. The system functional specification step is followed by the HLS one, where the behavioral model is translated into a register-transfer level (RTL) structural model. Logic and layout synthesis steps follow in the sequence.

Although more and more EDA tools are available providing high abstraction level design facilities, the market still resists to the automatic behavioral synthesis approach. Gajski et al. [6] point out two reasons for this: poor-quality results and lack of interactivity during the synthesis process. High-level behavioral specifications are

A.M. Tyrrell, P.C. Haddow, and J. Torresen (Eds.): ICES 2003, LNCS 2606, pp. 434–445, 2003.

usually considered not quite "synthesizable" since at this abstraction level many low-level implementation details are masked (by definition) from the designer [7]. To cope with this problem most behavioral synthesis tools constrain the design space by assuming a "target architecture" to which all high level specifications are mapped. The result is a reasonable average solution locally optimized for each design problem.

On the other hand, layout synthesis and hardware implementations are considerably simplified by FPGA (Field Programmable Gate Arrays) prototyping. By eliminating the need to cycle through an IC production facility, both time-to-market and financial risk can be substantially reduced. Recent statistics [8] indicate that, today, approximately one half of all chip designs are started using FPGAs. This trend is followed in modern synthesis tools that offer a smooth migration path between FPGA and ASIC technologies, providing direct mapping of an FPGA design into ASIC libraries [9].

This work presents a digital IC design methodology based on Evolvable Hardware (EHW) techniques [10], [11]. Genetic Programming is used to automatically explore the design space and improve the design quality. To accomplish this, a multi-criteria fitness function was defined to rank candidate solutions according to their area sizes. An experiment with a square-root approximation datapath design targeted to FPGA is detailed. Simulation results confirm the automatic generation of optimized structures starting from a high level specification.

The paper is organized as follows: Section 2 presents related works in order to establish the context in which the proposed methodology is applied. In Section 3 the Evolutionary Computation and Genetic Programming concepts of interest to the proposed methodology, are discussed. The design methodology is detailed in Section 4 and the example of a datapath design is presented in Section 5. Conclusions are drawn in Section 6.

2   Related Works

Evolutionary Algorithms have already been used in lower levels of electronic design, including routing, partitioning and placement [1], [12], and more recently in the automation of the structural design of digital circuits, as well as in the search for quality implementations, as related below.

Recent works [13] - [16] present evolved structures entirely in software using computer simulations (*extrinsic* evolution) to evaluate intermediate solutions. In [13] an algorithm capable of evolving 100% functional arithmetic circuits is presented. Based on a rectangular array of uncommitted logic cells of FPGAs, the algorithm is able to re-discover conventional optimum designs for one and two-bit adders, eventually improving the gate count of human produced designs of the two-bit multiplier.

Kalganova & Miller [14] addressed a two-stage *multiobjective* fitness function. In the first stage a 100% functional solution, which occurs when the truth table is matched, is sought. In the second stage the number of gates actually used in each candidate solution is taken into account in the fitness function. This allows circuits to evolve with the desired functionality minimizing their number of active gates. The authors limited their focus to combinational logic circuits, containing no memory elements and no feedback paths. A similar multiobjective optimization technique was proposed by Coello et al. [15], the main difference being the use of a population-based technique to split the search task among several sub-populations. In this approach *each* objective is assigned to a small sub-population, which is merged with the rest of the individuals when one of the objectives is satisfied. The merged populations contribute to minimize the total amount of mismatches produced between the encoded circuits and the truth table.

Hounsell & Arslan [16] addressed an EHW environment called *Virtual Chip* in which timing constraints are taken into account. In this environment not only gates but also functional macros, such as half-adders and compound gates, are available to the evolutionary process. In order to speed up the overall process a *single* VHDL program describing all candidate circuits is used in the simulations. In this work a *phased* approach was introduced to ease the evolution of complex systems. The phased evolution consists of evolving a circuit in stages: the system initially evolves a "sub-circuit" for each output of the desired circuit; in the sequence, redundant logic between the evolved sub-circuits is removed as they are combined to generate the required circuit. Using this approach a 3-bit multiplier was evolved with results as good as CAD based circuits.

An alternative to the above approaches is the use of formal grammars to represent the evolution of hardware designs, as in the work of Hemmi et al. [17]. In this approach, hardware specifications, which produce not only hardware structures but also behaviors, are automatically generated as HDL programs. To create HDL-descriptions the authors used Genetic Programming (GP) to evolve trees from productions (rephrased rules in the HDL grammar). The fitness is evaluated using simulation tools and hardware implementation is created *after* the learning task is completed. This system was used to automatically generate a sequential binary adder.

The approach to be presented in this paper uses the main techniques reported above, together with new improvements. Similar to [17], formal grammar of an HDL language was used to build candidate solutions. However, the place-and-route task was placed *inside* the evolutionary process in order to supply the fitness function with chip area information. The multiobjective fitness function adopted closely parallels the two-stage approach suggested in [14], although applied in a different framework. Also, an experience was made using a single VHDL program, as suggested in [16], in order to reduce computing effort.

3   Evolutionary Computation and Genetic Programming

Evolutionary Computation (EC) consists of a class of probabilistic algorithms inspired by the principles of Darwinian natural selection and variations that can greatly benefit from the increased simulation power for complex design, control and knowledge discovery applications [18]. Genetic Algorithm (GA) and Genetic Programming (GP) are the instances of EC most widely used in Evolvable Hardware [19].

GA [20] is based on the evolutionary process occurring in nature where a population is shaped by the survival of best-fit individuals. GA usually works with strings of fixed length in which one or more parameters are coded. GP is a branch of GA, the main difference being the solution representation. Unlike GA, GP can easily code genotypes of different sizes, which increases the capacity in structure creation.

Koza introduced canonical GP [21] with tree-based genome using LISP language. Since that, other genome types have been proposed like linear and graph genomes [22]. Other GP systems were developed to automatically create programs in arbitrary languages (other than LISP) [23], [24]. An alternative way to encode domain-knowledge is by formal grammars [25]. Programs may be produced by combining a context-free grammar (CFG) with GP [26] - [29], known as G³P (Grammar-Guided Genetic Programming). G³P provides a framework for automatic creation of error free programs in arbitrary language. Genetic Programming Kernel (GPK) [28] was used as the core engine for the GP system used in this paper. GPK is a complex system that evolves programs in any language once a Backus-Naur form (BNF) is provided as input. However, the initialization procedure was changed to overcome GPK difficulty associated with creating the first generation, as pointed out in [29].

4 Design Methodology

The proposed design methodology evolves independent VHDL programs trying to breed a solution (hardware description) that implement the desired functionality *and* satisfies design constraints for area. This methodology uses a mixture of C code and VHDL, with HLS tools. The execution flow of the proposed methodology is drawn in Fig. 1. This figure shows the two main components of the methodology: the GP core and the Valuation function. Also, it can be seen that, differently to other methodologies [16], [17], synthesis and place-and-route are performed for each candidate solution, which gives an exact value for the area.

The whole methodology is based on the use of the GP algorithm. Fig. 2 shows the flowchart of the GP system used, adapted from [21]. In this figure, the index "i" refers to an individual in the population of size M. The variable "Gen" is the number of the current generation. A population of M individuals (programs) is randomly created with the restriction of a pre-defined BNF containing a subset of VHDL grammar obeying the *sufficiency* property [21]. The system runs until the termination criterion is satisfied. Since the proposed methodology is concerned with unknown solutions, the termination is achieved when the number of generations exceeds a maximum pre-defined number G (*generational predicate*).

To create VHDL source codes, the entity declaration and the architecture declarative part of the VHDL program are maintained fixed, as they are common to all individuals. The architecture statement part of the VHDL code will evolve.

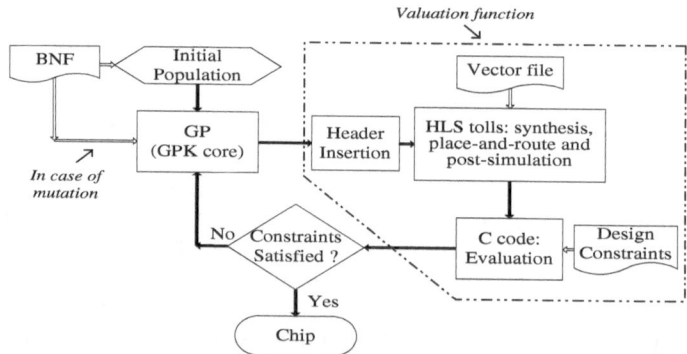

Fig. 1. Execution flow of the methodology

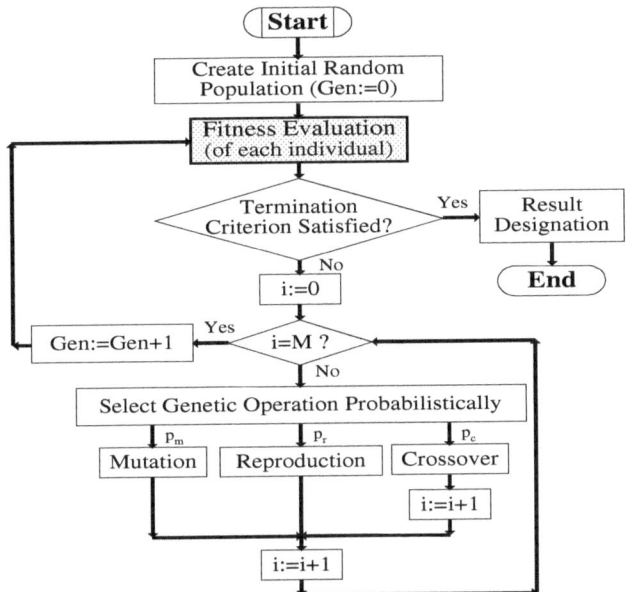

Fig. 2. Flowchart of the GP system

The evaluation of the objective function is a key procedure of the GP system. The steps involved in this operation are depicted in Fig. 3. After a new population is created it is necessary to complete the VHDL source code by inserting the entity declaration and the architecture declarative part of each individual. Then, the individuals are synthesized using Altera MAX+PLUS II compilation tool [30]. To effectively evaluate the fitness of an individual it is necessary to simulate each of the fitness cases and check the results against target values.

The objective function to be minimized is a multi-parameter error function combining functionality with the area parameter in a two-stage approach similar to that proposed in [14]. In this process the functional specification (F_1 *fitness*) must be first matched to trigger the search for an optimal area implementation (F_2 *fitness*). Let the average functionality error, fe_{av}, be defined as:

$$fe_{av} = \frac{1}{N} \sum_{i=1}^{N} w_i \left| D_i - S_i \right| \tag{1}$$

Where, D_i is the desired value for fitness case i, S_i is the simulated value obtained for fitness case i, w_i is a weighting factor and N is the number of fitness cases.

The fitness function F is defined as follows:

$$F = \begin{cases} F_1 = fe_{av} & \text{if } \exists i \in \{1, \cdots, N\} \mid \left| \dfrac{D_i - S_i}{D_i} \right| > te \\ F_2 = fe_{av} * k(lu) & \text{otherwise} \end{cases} \tag{2}$$

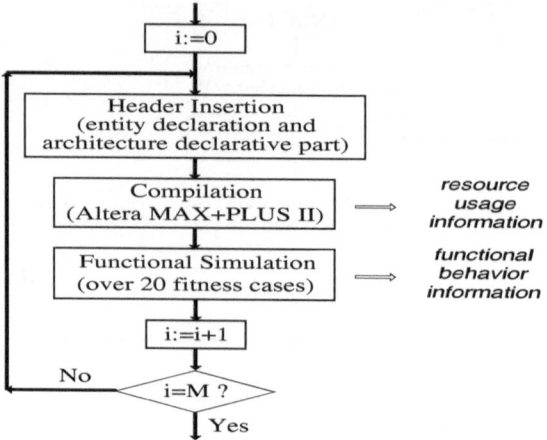

Fig. 3. Valuation block of the Fig. 2

Where, *te* is the target functional specification error for the fitness cases, *lu* is the number of used logic elements (Cf. Subsection 5.3) and *k* is a function of *lu*. The *k(lu)* function is a weighting factor that considers the effect of the area used to implement the desired functionality. It is an exponential function of *lu* that privileges the individuals containing the lowest quantities of logic elements. The *k(lu)* function will take the form:

$$k(lu) = ae^{b*lu} \tag{3}$$

Where, *a* and *b* are constants that must be defined by solving the following system:

$$k_{min} = ae^{b*lu_{min}}$$
$$k_{max} = ae^{b*lu_{max}} \tag{4}$$

Where, lu_{min} *and* lu_{max} are minimum and maximum quantities of logic elements and k_{min} and k_{max} are the minimum and maximum values of *k(lu)*, respectively.

5 Experiment

To exemplify the proposed methodology, it was selected a datapath design taken from [6]. The aim is to compute the square-root approximation (SRA) of two integers, *a* and *b*. In [6] the following approximation formula is given:

$$\sqrt{a^2 + b^2} \cong max\,((0.875x+0.5y),x) \tag{5}$$

Where $x = max$ (|a|, |b|), and $y = min$ (|a|, |b|).

   In this problem of *symbolic regression* the goal is to find a function in symbolic form that is a good, or an exact, fit to a group of 20 sets of numerical data points. The *black box* of the SRA is given in Fig. 4. In this figure, the "start" input and "done" output signals are present for interface control issues.

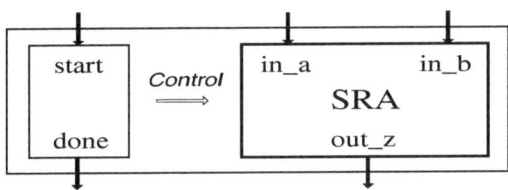

Fig. 4. *Black box* of the SRA

5.1 BNF Definition

A BNF grammar describes admissible structures of a language through a 4-tuple $\{S,N,T,P\}$ where S denotes the *start* symbol, N the set of *non-terminal* symbols, T the set of *terminal* symbols and P the *productions*, i.e., rewritten rules that map the elements of N to T. The BNF that defines the syntax of the VHDL statement part of the SRA description is defined as:

```
S       ::= <fe>;
<fe>    ::= "PROCESS BEGIN WAIT UNTIL (clk'event and clk='1');
            IF in_a>=in_b THEN x<=in_a; y<=in_b; ELSE y<=in_a;
            x<=in_b; END IF; t6<=" <expr> "; IF t6>=x THEN
            t7<=t6; ELSE t7<=x; END IF; out_z<=t7; done<='1';
            END PROCESS;";
<expr>::= "(" <expr> ")" <op> "(" <expr> ")" | <var> |
            <var> "(7 downto " <shf_size> ")";
<op>    ::= " + " | " - ";
<var>   ::= " x " | " y " | " w ";
<shf_size> ::= "1" | "2" | "3" | "4" | "5" | "6" | "7";
```

In the above BNF description the signals "x", "y", "w", "t6" and "t7" are internal signals of the VHDL architecture body. The signal "w", assigned zero, is present to make ease the insertion or removal of a terminal. *max* and *min* operations are already fixed in BNF through *IF* clause. The original problem was restricted to use input and output operands as unsigned type instead of integer type, reducing computing effort.

5.2 GP Parameters

The main control parameters of the GP for the *first stage* (F_1 *fitness*) of the proposed experiment are shown in Table 1. In the *second stage* of the evolution (F_2 *fitness*) the GP parameters are the same, except for the objective and raw fitness rows of Table 1 that change to incorporate the search for a design with optimal area.

5.3 Results

The EPF8282A FPGA from Altera FLEX 8000 family, with up to 282 logic elements (LEs), was used as target device. Each LE consists of a look-up table (LUT), a function generator that computes any function of four variables, and a programmable register. In this experiment, area will be measured in number of LEs. The target functional specification error (*te*) was set to 2.5%. It establishes for *each* fitness case the maximum relative error between the simulated value of the dependent variable, produced by the synthesized VHDL programs, and the desired value.

Table 1. GP tableau for the first stage of the SRA design

Objective:	Find a VHDL program that implements a two argument function that fits a sample of 20 I/O sets {a_i, b_i, $F(a_i,b_i)$}, where the target function is $F(a,b) = sqrt(a^2 + b^2)$.
Terminal symbols:	x, y, w, +, −, 7 downto (*logical shift-right operator*), 1, 2, 3, 4, 5, 6, 7 (*index of shift-right operator*) and the sentences between quotation marks on the right-hand side of <fe>.
Non-terminal symbols:	<fe>, <expr>, <op>, <var>, <shf_size>.
Fitness cases:	Sample of 20 sets of data points {a_i, b_i, $F(a_i, b_i)$}, a_i and b_i randomly chosen from the interval [0, 255] and satisfying the restriction $sqrt(a_i^2 + b_i^2) \le 255$.
Raw fitness:	The average, taken over 20 fitness cases, of the absolute value of the difference between the value of the dependent variable produced by simulating synthesized VHDL program and the target value of the dependent variable (Cf. eq.(1)).
Selection Method:	Fitness-proportionate.
Main GP Parameters:	M=100, G=51, p_c=0.73, p_m= 0.12, without elitism.

A population size of 100 individuals was used, evolving up to a maximum of 51 generations. Experiments have been carried out with N=20, w_i=1 and $k(lu) = 0.25e^{(0.0139*lu)}$, which guarantees $k(lu) \in$ [0.5,1.0] for $lu \in$ [50,100] (Cf. eq.(3) and eq.(4)). The initial population was created using the ramped half-and-half generative method [21], replacing GPK initialization procedure.

The first functional individual, i.e., one that met all fitness cases evaluations with the functional specification target error of 2.5% was created at generation 18. The system started, then, the search for an implementation with optimal area. During the second stage of the evolution a variety of individuals agreeing with the specified *te* value emerged, although it happened with a non-optimal area use. Functional individuals with 95, 92 and 90 LEs have been reported. After 51 generations the best individual was on hand. It was selected at generation 29. Its main fitness (*F*) was 0.827 and the average functionality error (*fe$_{av}$*), 1.103. This individual was synthesized with 78 LEs. The statement part of this VHDL program was:

```
BEGIN
 w <="00000000";
 PROCESS BEGIN WAIT UNTIL (clk'event and clk='1');
  IF in_a >= in_b THEN x <= in_a;   y <= in_b;
  ELSE y <= in_a;   x <= in_b; END IF;
  t6 <=((y(7 downto 1))+((x)-(x(7 downto 3))))+(y(7 downto 6));
  IF t6 >= x THEN t7 <= t6;
  ELSE t7 <= x; END IF;
  out_z <= t7;   done <= '1';
 END PROCESS;
```

Fig. 5 shows (a) the evolution of the fitness of the best individual and (b) its area versus generation curve. In Fig. 5 (b) it can be noted that at generation 13 the best individual increases its area in order to achieve better accuracy. At generation 18 it reaches the desired accuracy and begins to look for optimal area, which is found at generation 29.

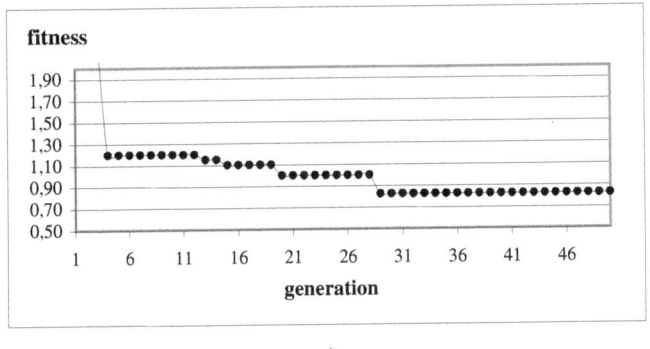

a)

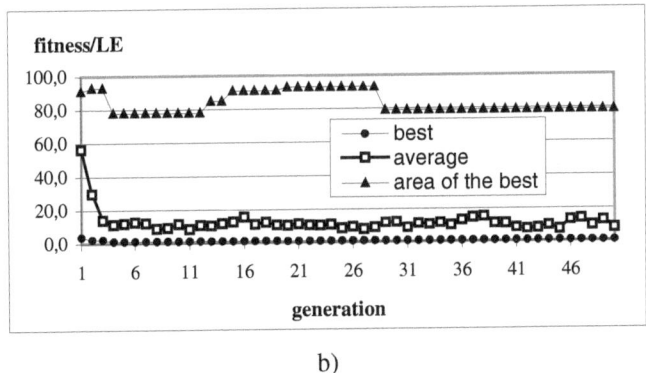

b)

Fig. 5. (a) Histogram of the fitness and (b) of the area of the best individual

Table 2 reports optimal area implementation (LE column) obtained for other values of the functional specification target error (*te*) with the respective hardware description. An interesting result shown in Table 2 refers to the individual with optimal area for *te*=25% that do not use the internal signal *y* (minimum between the two inputs). This indicates that the system actually looks for optimal area, taking into account route and resource issues of the synthesis tool.

In some independent runs it could be noted that the first functional individual was also the one with optimal area. Nevertheless, this is not always true, as was shown in the SRA design for *te*=2.5%. Also, an experience was made adding the multiplier and shift-left operators on the right-hand side of *<op>* and *<expr>* non-terminal symbols, respectively (Cf. Subsection 5.1). The system showed to be robust by eliminating individuals with these two operators from the early generations.

Table 2. LE quantities for other *te* values with respective hardware descriptions

te	LE	Hardware description
25%	51	x + x(7 downto 3)
15%	58	x + y(7 downto 2)
12%	59	x + y(7 downto 1)
5%	70	x - x(7 downto 3) + y(7 downto 1)

The drawback of the proposed methodology is its excessive time consumption. Currently available tools for compiling VHDL programs are slow, particularly the place-and-route ones [31]. To overcome this problem a pre-stage, to be run before the start of the main optimization process shown in Fig. 1, was introduced. In the pre-stage the system evolves a single VHDL program as in [16], each individual is described in a single VHDL PROCESS to bring the whole population close to the target functionality. No place-and-route is accomplished at this time. The pre-stage finishes when the first functional individual comes out.

With this revised approach the system runs three times faster while remaining at the pre-stage. It was observed that the performance of the overall design process improves considerably when compared with the original approach (Fig. 1).

6   Conclusion

A methodology based on Genetic Programming paradigm to evolve optimized implementations was presented. It implies the use of a multiobjective evolutionary technique. The methodology is fully automatic in the sense that it is not necessary to write code but define a BNF grammar and specify the design constraints. Cycling through a place-and-route process makes it possible to consider the exact area used by each implementation in the optimization.

The proposed methodology is capable of synthesizing combinational circuits, using basic digital or mathematical operators declared in a BNF, which approximates a given function with a specific accuracy while looking for optimal area. From this work it can be concluded:

i)   It is not necessary to exactly specify the basic operators in the BNF.
ii) Performance optimization may be achieved instead of (or in addition to) area if one considers timing parameters in the objective function.
iii) The methodology is synthesis-tool independent.

Finally, the methodology can be used to implement a new function in a semi-filled FPGA, i.e., in addition to an already described function (in a VHDL PROCESS of the VHDL program). In this case the new function is described in a new VHDL PROCESS. The code of the already described function is maintained fixed while the new code will evolve. The system will look for synthesizing the new function using any common construct between both functions in order to reduce area usage.

References

1. Y-L Lin, *Recent Development in High Level Synthesis*, ACM Transactions on Design Automation of Electronic Systems, Vol. 2, No. 1, 1997.
2. D. Gajski, N. Dutt, A. Wu and S. Lin, *High Level Synthesis: Introduction to Chip and System Design*, Kluver Academic Publishers, 1992.
3. D. Gajski, J. Zhu and R. Dörner, *Special Issues in Codesign*, Technical Report ICS-97-26, 1997.
4. IEEE Std. 1076-1993, *VHDL Language Reference Manual*, 1993.
5. Z. Navabi, *VHDL: Analysis and Modeling of Digital Systems*, McGraw Hill, 2^d Edition, 1998.
6. D. Gajski, I. Tadatoshi, V. Chaiyakul, H. Juan and T. Hadley, *A Design Methodology and Environment for Interactive Behavioral Synthesis*, Technical Report 96-29, 1996.
7. K. Kuusilinna, T. Hämäläinen and J. Saarinen, *Practical VHDL Optimization for Timing Critical FPGA Applications*, Microprocessors and Microsystems 23, pp. 459-469, 1999.
8. D. Landis, *Programmable Logic and Application Specific Integrated Circuits*, Handbook of Components for Electronics, Chapter II, Vol. 1, 1995.
9. Synopsys Inc., *Synopsys Online Documentation*, 2001.
10. H. DeGaris, *Evolvable Hardware: Genetic Programming of a Darwin Machine*, McGraw-Hill, Artificial Neural Nets and Genetic Algorithms, Springer-Verlag, NY, 1993.
11. R. Zebulum, M. Pacheco and M. Vellasco, *Evolvable Systems in Hardware Design Taxonomy, Survey and Applications,* Evolvable Systems: From Biology to Hardware, (ICES96), pp. 344-358, 1996.
12. F. Bennett III, J. Koza, J. Yu and W. Mydlowec, *Automatic Synthesis, Placement and Routing of an Amplifier Circuit by Means of Genetic Programming*, Proc. of the Third Int. Conf. on Evolvable Systems, ICES 2000, pp. 1-10, Edinburgh, Scotland, UK, 2000.
13. J. Miller, P. Thomson and T. Fogarty, *Designing Electronic Circuits Using Evolutionary Algorithms. Arithmetic Circuits: A Case Study*, Genetic Algorithms and Evolution Strategies in Engineering and Computer Science, edited by D. Quagliarella et al., Publisher: Wiley, 1997.
14. T. Kalganova and J.Miller, *Evolving More Efficient Digital Circuits by Allowing Circuit Layout Evolution and Multi-Objective Fitness*, Proc. of the First NASA/DoD Workshop on Evolvable Hardware, pp. 54-63, Los Alamitos, CA, IEEE Computer Society Press, 1999.
15. C. Coello, A. Aguirre and B. Buckles, *Evolutionary Multiobjective Design of Combinational Logic Circuits,* Proc. of the Second NASA/DoD Workshop on Evolvable Hardware, pp. 161-170, IEEE Computer Society, Los Alamitos, CA, 2000.
16. B. Hounsell and T. Arslan, *A Novel Evolvable Hardware Framework for the Evolution of High Performance Digital Circuits*, GECCO-2000, pp. 525-532, 2000.
17. H. Hemmi, J. Mizoguchi and K. Shimohara, M. Tomassini, *Development and Evolution of Hardware Behaviors*, Towards Evolvable Hardware: The Evolutionary Engineering Approach, Int. Workshop, Lausanne, edited by E. Sanchez and M. Tomassini, Springer-Verlag, LNCS 1062, pp. 250-265, 1996.
18. J. Rosca, *Hierarchical Learning with Procedural Abstractions Mechanisms*, PhD Thesis, University of Rochester, New York, 1997.
19. M. Tomassini, *Evolutionary Algorithms*, Towards Evolvable Hardware: The Evolutionary Engineering Approach, Int. Workshop, Lausanne, edited by E. Sanchez and M. Tomassini, Springer-Verlag, LNCS 1062, pp. 19-47, 1996.
20. J. Holland, *Adaptation in Natural and Artificial Systems*, The University of Michigan, 1st Edition, 1975.
21. J. Koza, *Genetic Programming: On the Programming of Computers by Means of Natural Selection*, MIT Press, 1992.
22. W. Banzhaf, P. Nordin, R. Keller and F. Francone, *Genetic Programming: An Introduction,* San Francisco, CA, Morgan Kaufmann and Heidelberg, 1997.

23. R. Keller and W. Banzhaf, *Genetic Programming Using Genotype-Phenotype Mapping from Linear Genomes into Linear Phenotypes*, Proc. of Genetic Programming 1996, pp. 116-122, MIT Press, 1996.
24. N. Paterson and M. Livesey, *Evolving Cache Algorithms in C by GP*, Genetic Programming 1997, pp. 262-267, MIT Press, 1997.
25. A. Ratle and M. Sebag, *Genetic Programming and Domain Knowledge: Beyond the Limitations of Grammar-Guided Machine Discovery,* Parallel Problem Solving from Nature, 2000.
26. M. Wong and K. Leung, *Applying Logic Grammars to Induce Sub-functions in Genetic Programming*, Proc. of 1995 IEEE Conf. on Evolutionary Computation, pp. 737-740, USA:IEEE Press, 1995.
27. P. Whigham, *Grammatically-Based Genetic Programming*, Proc. of the Workshop on Genetic Programming: From Theory to Real-World Applications, pp. 33-41, Morgan Kauffmann Publishers, 1995.
28. H. Hörner, *A C++ Class Library for Genetic Programming*, Release 1.0 Operating Instructions, Viena University of Economy, 1996.
29. C. Ryan and M. O'Neill, *Grammatical Evolution: Evolving Programs for an Arbitrary Language,* LNCS 1391, Proc. of the First European Workshop on Genetic Programming, pp. 83-95, Springer-Verlag, 1998.
30. Altera Inc., *Altera Digital Library Databook 2001*, 2001.
31. D. Montana, R. Popp, S. Iyer and G. Vidaver, *EvolvaWare: Genetic Programming for Optimal Design of Hardware-Based Algorithms*, Genetic Programming 1998, pp. 869, 1998.

The Effect of the Bulge Loop upon the Hybridization Process in DNA Computing

Fumiaki Tanaka[1], Atsushi Kameda[2],
Masahito Yamamoto[3], and Azuma Ohuchi[4]

[1] Graduate School of Engineering, Hokkaido University
North 13, West 8, Kita-ku, Sapporo 060-8628, Japan
fumiaki@dna-comp.org
http://ses3.complex.eng.hokudai.ac.jp/
[2] Japan Science and Technology Cooperation (JST)
Honmachi 4-1-8, Kawaguchi 332-0012, Japan
kameda@dna-comp.org
[3] PRESTO, Japan Science and Technology Cooperation (JST) and
Graduate School of Engineering, Hokkaido University
North 13, West 8, Kita-ku, Sapporo 060-8628, Japan
masahito@dna-comp.org
[4] CREST, Japan Science and Technology Cooperation (JST) and
Graduate School of Engineering, Hokkaido University
North 13, West 8, Kita-ku, Sapporo 060-8628, Japan
ohuchi@dna-comp.org

Abstract. To improve the efficiency of DNA computing, we need to prevent hybridization errors. In this paper, we focus on the bulge loop structures, which cannot be prevented by current sequence design methods. To estimate the formation of bulge loop structure, we measured the melting temperature (Tm) and the binding intensity of sequences with a single bulge loop in some simple experiments. Based on the experimental results, we mainly discuss the effect of the base type and the position of the loops. We also discuss the possibilities for modeling the formation of bulge loops based on these experimental results.

Keywords: DNA computing.

1 Introduction

Hybridization processes are the most important processes in DNA computing and the efficiency needs to be improved for use in all applications. To control the chemical reactions of DNA, we must design the sequence set adequately.

Recently, two design strategies have been proposed for preventing a hybridization error called mis-hybridization. One expresses the binding intensity between two sequences as the number of Watson-Crick complementary base pairs [7,8,9], and the other is based on a thermodynamics model [10]. The first strategy is represented by the "H-measure" function to take the minimum Hamming distance obtained by successively shifting and lining up the WC-complement of one sequence against another [9]. However, these strategies do not

A.M. Tyrrell, P.C. Haddow, and J. Torresen (Eds.): ICES 2003, LNCS 2606, pp. 446–456, 2003.
© Springer-Verlag Berlin Heidelberg 2003

prevent the formation of a secondary structure such as a bulge loop or a pseudo-knot. Understanding how a secondary structure forms is more important in the case where long sequences are created from several short sequences. The latter strategy calculates the intensity of a secondary-structure formation based on free energy. Andronescu et al. proposed an algorithm to prevent sequences from forming a secondary structure [10]. To estimate this formation they formulated three functions, which evaluate the free energy of a stacked pair, a hairpin loop, and an internal or a bulge loop. Related to the thermodynamics approach, predicting the melting temperature (Tm) is useful for constructing the criteria (namely, the evaluation function) for a sequence design that excludes sequences that may potentially form a secondary structure. As a practical method, we can use the nearest-neighbor model (NN-model) to predict the Tm value and the thermodynamics parameters, namely, the enthalpy and the entropy of all dimers in a sequence [1,2,3,4,5,6].

For example, to predict the Tm of duplex 5'-AGTG-3'/3'-TCAC-5' at a concentration of 4×10^{-6} M, according to the NN model, the enthalpy and entropy of 5'-AGTG-3'/3'-TCAC-5' are calculated with the following equation:

5'-AGTG-3'/3'-TCAC-5' = initiation + symmetry + 5'-AG-3'/3'-TC-5' + 5'-GT-3'/3'-CA-5' + 5'-TG-3'/3'-AC-5'

Here, the *initiation* term describes the contribution whether terminal base pair is closed by an A-T pair or a G-C pair. The *symmetry* term describes the contribution whether sequence is palindrome or not.

By using of Hatim's parameters [5], the enthalpy of 5'-AGTG-3'/3'-TCAC-5' is calculated as (2.3 + 0.1) + 0 + (-7.8) + (-8.4) + (-8.5) = -22.3 kcal/mol. Identically, the entropy is (4.1 - 2.8) + 0 + (-21.0) + (-22.4) + (-22.7) = -64.8 cal/molK.

Finally, the Tm of a non-palindrome sequence is calculated with the following equation [1,2,3,4,6]:

$$Tm = \frac{\Delta H}{RlnCt/4 + \Delta S} \tag{1}$$

Therefore, the Tm of 5'-AGTG-3'/3'-TCAC-5' is approximately 299 K(=26°C).

The NN-model is currently only applicable to a completely complementary duplex or to a duplex containing single mismatches or dangling ends.

In this paper, we focus on the bulge loop structure. Sequence design based on the H-measure cannot exclude sequences that form a bulge loop. As mentioned above, the NN-model cannot calculate a duplex with a bulge loop because the thermodynamics parameters of the bulge loop are still unknown. Although Andronescu et al. proposed a design strategy, which also considers the bulge loop, they modeled the binding intensity of the bulge loop according to the length of the loop only [10]. We investigated the effect of the base type of the loop and the loop position (i.e., a loop near the end or in the middle). To accomplish this, we measured the Tm of the sequences that had a single bulge by varying the base type and the loop position.

In addition, we investigated the relation between the Tm and the binding intensity in simple experiments. Based on the results, we discuss the importance of the Tm in DNA computing.

Furthermore, the reaction of DNA proceeds autonomously, which is called "self-assembly". For example, by designing DNA sequences adequately, they can form a two-dimensional structure called DNA tile by self-assembling. This process is part of the computational capacity of DNA as an evolvable system. Although DNA computes by self-assembling, an unintended bulge loop may cause hybridization errors. Understanding the tendency of bulge-loop formation is useful for all DNA computing models including evolvable computation.

2   Measuring the Tm

2.1   Measuring the Tm with "SYBR Green I"

We measured the Tm by using the smart cycler system (Cepheid), which uses a thermal cycler to control the temperature and to measure the fluorescence intensity of a solution containing DNA. The SYBR Green I is a non-sequence specific intercalator, which binds with the double strand region and fluoresces. As the ratio of double strands in the solution decreases, the fluorescence intensity decreases. Therefore, by raising the temperature, the fluorescence intensity can be reduced. The fluorescence intensity of SYBR Green I versus temperature profiles (melting curves) were acquired by ramping the temperature from 35 to 95 °C. The Tm is determined as the point where the gradient of the melting curve is the steepest. The buffer used for measuring the Tm was 20 mM Tris-HCl (pH 7.4), 8 mM $MgCl_2$, 7.5 mM DTT, and 0.05 $\mu g/\mu l$ BSA.

2.2   Reproducibility of the Smart Cycler System

The reproducibility of the smart cycler system was examined by measuring the Tm between the *ligand* and the non-mutated *analyte* (see Table 1). Figure 1 shows the two curves, one is the frequency distribution curve plotted with 57 Tm values we measured, and the other is the ideal normal distribution curve derived from the average and the standard deviation of the measured Tm. The distribution of the measured Tm approximately follows the ideal normal distribution curve, though more experimental data is needed. Because the average is 71.0 °C and the standard deviation 0.44 °C, the Tm of a sequence that is more than 72.32(71.0+3·0.44) °C or less than 69.68(71.0-3·0.44) °C is statistically valuable with a 0.997 probability.

3   Sequence Design

3.1   Design Strategy

As a subject for analysis, we focus on a sequence pair, whose one sequence complements the other. We call one sequence the *"ligand"* and the other the

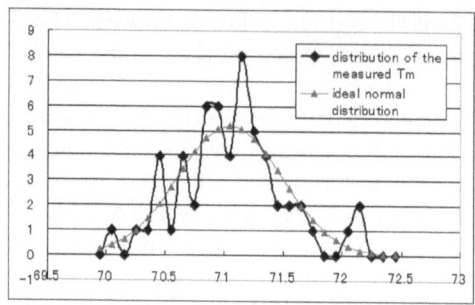

Fig. 1. Reproducibility of the smart cycler system

"*analyte*". Then, we designed sequences with a single bulge loop by changing the *analyte*. We tested the effect of sequences with a single bulge loop by comparing the binding of the *ligand* to the mutated *analytes* that have a single bulge loop with that of the *ligand* to a non-mutated *analyte*.

To measure the Tm of the sequences accurately, we need to design sequences that demonstrate an "all-or-none" two-state transition (i.e., random coil ⟷ double helix) [2,3,4,5]. For example, self-complementary sequences should be modeled on a three-state transition model instead of on a two-state one because self-complementary sequences have the potential to form both a double helix and a hairpin structure. To obtain this property, sequences must be designed to minimize the possibility of forming shift hybridization duplexes or stable alternative secondary structures.

We evaluated the degree of likelihood of a shift hybridization forming between the *ligand* and the *analyte* by using the following formula based on H-measure [9].

$$f_{Shift} := \max_{-n<k<0,0<k<n} \{n - H(x, \sigma^k(\overline{y}))\} \tag{2}$$

where n is the sequence length ($=20$), x denotes the *ligand* and y the *analyte*, $H(*, *)$ denotes the Hamming distance, $\overline{y}$ denotes the complementary sequence of y, σ^k denotes the right (left) shift in case of $k > 0$ ($k < 0$), respectively, and k denotes the number of shifts. Note that in the case of $k = 0$, $f_{Shift} = 20$ because an *analyte* is a completely complementary sequence of a *ligand*. As the value of this formula is large, the possibility of a shift hybridization forming is large based on the hypothesis that the binding intensity of a hybridization can be approximated by the number of complementary base pairs.

Similarly, the possibility of an alternative secondary structure forming is evaluated by the following formula obtained by modifying eq (2).

$$f_{Self} := \max_{-n<k<n} \{n - H(x, \sigma^k(x^R))\} \tag{3}$$

where x^R denotes the reverse sequence of x and all remaining variables are the same as that of eq (2).

A consecutive base, for example 'GGG', tends to facilitate secondary structure formations such as hairpin loops and pseudo knots. We thus prefer to avoid sequences containing a consecutive base.

In addition, in DNA computing research, most sequences are designed to have a 50% GC content to prevent secondary structures forming. We also imposed this constraint to the *ligand*.

Finally, the *ligand* has the following features:

1. $f_{Shift} = 6$
2. $f_{Self} = 8$
3. no consecutive base
4. 50% GC content

The *analytes* with a single bulge were obtained by deleting a 3'-end base and inserting a new base into various positions in the *analytes*. We decided not to the base, which was located in the loop position, in the consecutive base in sequence, because it may have changed the loop position (Table 1). For example, 3'-GATGATATGCGTCTGCGATG-5'/5'-CTACTATACGCA G GACGCTA-3' (G is located in the loop position) could form an alternative bulge loop, 3'-GATGATATGCGTCTGCGATG-5'/5'-CTACTATACGCAG G ACGCTA-3'. Therefore, to investigate the effect of the position of a single bulge exactly, we decided to insert the base as described above.

3.2   Evidence of Forming a Bulge Loop

Do the *ligand* and the mutated *analyte* with a single bulge really form a bulge loop structure? In other words, can they form the base pairs on one side sequence of a loop position and dangle on the other side sequence? To verify this, we compared the Tm between the *ligand* (3'-GATGATATGCGTCTGCGATG-5') and mutated *analytes* that had an internal single bulge (e.g. 5'-CTACTATACG T CAGA CGCTA-3') with that between the *ligand* and "half-complementary" *analytes* (5'-CTACTATACG **TCTCGCGAGT**-3' or 5'-**TGTAATGCGA**CAGACGCTAC-3'), which are complementary to the *ligand* located halfway from the 5' or 3'-end of the sequence, respectively (see Table 1).

The Tm between 3'-GATGATATGCGTCTGCGATG-5' and 5'-CTACTATAC G T CAGACGCTA-3' is 62.0 °C, while that between 3'-GATGATATGCGTCTGCG ATG-5' and 5'-CTACTATACG**TCTCGCGAGT**-3' is 38.2 °C and that between 3'-GATGATATGCGTCTGCGATG-5' and 5'-**TGTAATGCGA**CAGACG CTAC-3' is 52.5 °C. These results show an interaction between a *ligand* and a mutated *analyte* with an internal single bulge on both sides of the loop position in the sequence. Therefore, we confirmed that at least one bulge loop is formed in the middle.

3.3   Duplex Stability of DNA with a Single Bulge

To investigate the duplex stability of DNA with a single bulge, we compared the Tm between the *ligand* and the mutated *analyte* with a single bulge to

Table 1. Designed Sequences: The underlined base is the mismatched base pair, while the boxed base is the bulge loop. †These sequences were used to investigate the binding intensity of sequences with a single bulge by varying the loop position. ‡These sequences may form hairpin structures.

Name	Sequence	Tm with the *ligand* (°C)
Ligand	3'-GATGATATGCGTCTGCGATG-5'	-
Analyte	5'-CTACTATACGCAGACGCTAC-3'	71.0
	5'-**A**TACTATACGCAGACGCTAC-3'	68.6
	5'-**T**TACTATACGCAGACGCTAC-3'	69.7
	5'-**G**TACTATACGCAGACGCTAC-3'	68.8
	5'-CTAC**A**ATACGCAGACGCTAC-3'	65.5
	5'-CTAC**C**ATACGCAGACGCTAC-3'	65.8
mutated *analyte*	5'-CTAC**G**ATACGCAGACGCTAC-3'	65.2
with a single mismatch	5'-CTACTATAC**A**CAGACGCTAC-3'	60.9
	5'-CTACTATAC**T**CAGACGCTAC-3'	58.2
	5'-CTACTATAC**C**CAGACGCTAC-3'	59.4
	5'-CTACTATACGCAGAC**A**CTAC-3'	62.7
	5'-CTACTATACGCAGAC**T**CTAC-3'	61.3
	5'-CTACTATACGCAGAC**C**CTAC-3'	63.9
	5'-CTACTATACGCAGACGCTA**A**-3'	68.3
	5'-CTACTATACGCAGACGCTA**T**-3'	69.1
	5'-CTACTATACGCAGACGCTA**G**-3'	69.5
	5'-CT **C** ACTATACGCAGACGCTA-3' †	68.7
	5'-CT **G** ACTATACGCAGACGCTA-3'	67.0
	5'-CTAC **A** TATACGCAGACGCTA-3'	64.4
	5'-CTAC **G** TATACGCAGACGCTA-3' †	66.2
	5'-CTACTA **C** TACGCAGACGCTA-3' †	65.7
	5'-CTACTA **G** TACGCAGACGCTA-3'	64.3
	5'-CTACTATA **T** CGCAGACGCTA-3'	64.0
	5'-CTACTATA **G** CGCAGACGCTA-3' † ‡	66.3
mutated *analyte*	5'-CTACTATACG **A** CAGACGCTA-3'	63.5
with a single bulge	5'-CTACTATACG **T** CAGACGCTA-3' †	62.0
	5'-CTACTATACGCA **T** GACGCTA-3'	65.3
	5'-CTACTATACGCA **C** GACGCTA-3' †	63.9
	5'-CTACTATACGCAGA **T** CGCTA-3'	61.7
	5'-CTACTATACGCAGA **G** CGCTA-3' † ‡	59.5
	5'-CTACTATACGCAGACG **A** CTA-3'	64.7
	5'-CTACTATACGCAGACG **T** CTA-3' †	64.3
	5'-CTACTATACGCAGACGCT **C** A-3' †	69.5
	5'-CTACTATACGCAGACGCT **G** A-3'	69.7
Half-	5'-CTACTATACG**TCTCGCGAGT**-3'	38.2
complementary	5'-**TGTAATGCGA**CAGACGCTAC-3'	52.5

that between the *ligand* and the mutated *analyte* with a single mismatch (Table 1). For example, the Tm of a middle bulge 5'-CTACTATACG T CAGACGCTA-3' is 62.0 °C, while that of a middle mismatch 5'-CTACTATAC**T**CAGACGCTAC-3' is 58.2 °C. In terms of the stability defect in the middle of the sequence, a single bulge loop is more stable than a single mismatch. This means that sequence design strategies, which overlook the formation of an undesired bulge loop structure, lead to serious errors in the hybridization process.

In terms of the stability defect near the end of a sequence, more experimental data are needed.

4   Experiments

4.1   The Tm of Sequences with a Single Bulge

We investigated the effect of the base type and the position of the loop by measuring the Tm. The results are shown in Figure 2. The Tm differed 2.3 °C maximum with varying base type and 9.0 °C with varying loop position. Therefore, the effect of the loop position is greater than that of the base type. In addition, the distribution of the effect of the base type is the same level in relation to the loop position.

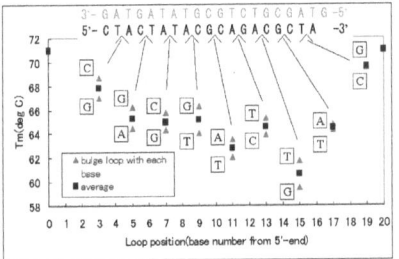
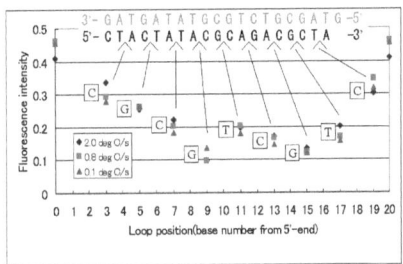

Fig. 2. Tm of sequences with a single bulge with varying base types and positions of the loop. Note that loop positions 0 and 20 are completely complementary duplexes (not bulge loop structures).

Fig. 3. Fluorescence intensity of sequences with a single bulge with varying loop positions. Note that loop position 0 and 20 are completely complementary duplexes (not bulge loop structures).

4.2   The Binding Intensity of the Sequence with a Single Bulge

We investigated the binding intensity of a sequence with a single bulge by varying the loop position and applying polyacrylamide gel electrophoresis.

First, we mixed the *ligand*, the non-mutated *analyte*, and each mutated *analyte* with a single bulge by varying loop positions labeled with a fluorescence label (FITC) at 5'-end into the solution at a concentration ratio of 1:1:1. In other words, each mutated *analyte* with a single bulge bound to the *ligand* competed with the non-mutated *analyte*. The mutated *analytes* used for this experiment were selected by arbitrarily picking out one sequence per loop position, so we had nine sequences in total (see Table 1). The buffer was 20 mM Tris-HCl (pH 7.4), 8 mM MgCl$_2$, 7.5 mM DTT, and 0.05 μg/μl BSA.

Second, the mixture was heated to denature the DNA and make it single-stranded. Then, the temperature was lowered to either 2.0, 0.8, or 0.1 °C/s.

Third, each mixture was analyzed on a 10% polyacrylamide gel by measuring the fluorescence intensity of the FITC (Fig. 4). Note that only the mutated *analytes* were labeled with the FITC. Therefore, the fluorescence intensity showed the amount of mutated *analytes* with a single bulge.

Each mixture was separated into a double-strand (upper band) and a single-strand (lower band) by the gel electrophoresis. The thickness of the upper band corresponded to the amount of DNA forming a bulge loop structure (i.e., the duplex between the *ligand* and the mutated *analytes*). On the other hand, the thickness of the lower band corresponded to the amount of mutated *analytes* that could not form a duplex. Therefore, the ratio of the fluorescence intensity of the upper band in each lane represented the binding intensity of each bulge loop structure.

A plot of the ratio of the fluorescence intensity of the upper band vs. the loop position is shown in Fig. 3. Roughly speaking, the effect of decreasing the Tm of the bulge loop in a middle position is greater than that of a bulge loop near the end. We did not observe any particular effect of the temperature gradient during the hybridization process. This indicates that the hybridization time may be reduced.

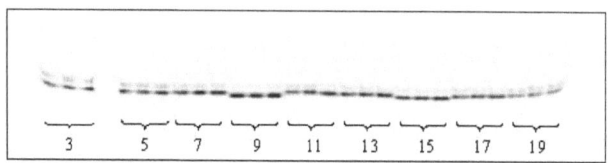

Fig. 4. Gel electrophoresis on a 10% polyacrylamide gel. The upper band is a double-strand DNA, and the lower band is a single-strand DNA with an FITC. The lanes can be divided every three lanes by varying the temperature gradient in hybridization. The number in this figure corresponds to the position of the bulge loop. The temperature gradient during the hybridization is 2.0 °C/s, 0.8 °C/s and 0.1 °C/s from left to right in each group. The lower bands of lanes 9 and 15 are lower than others. This suggests the possibility that 5'-CTACTATA G CGCAGACGCTA-3' and 5'-CTACTATA G CGCAGACGCTA-3' form a hairpin structure.

5   Discussion

The experimental results confirmed that the Tm correlates with and seems to be a good indicator for estimating the binding intensity (correlation coefficient = 0.82). Both the base type of the loop and the loop position influence the destabilization effect of the bulge loop. Therefore, at least two factors should be considered when modeling the formation of the bulge loop structure. Because the distribution of the effect of the base type was the same level in relation to the loop position in this experiment, we could consider the effect of the base type and the position of the loop independently assuming that the two factors do not influence each other. In the future, we will model the binding intensity of the bulge loop structure by applying the NN-model and incorporating the effect of the loop position.

In Figure 4, the lower bands (containing a single-strand) of lanes 9 and 15 were lower than the others. This implies the formation of an intramolecular structure, such as a hairpin. However, the Tm of the bulge loop, which was positioned at the 9-th base from the 5'-end was not extremely low, while the Tm of the bulge loop positioned at the 15-th base from the 5'-end was extremely low (see Table 1). Considering these results, we hypothesized that the sequence in lanes 9 formed an intramolecular structure that was different from the structure formed by the sequence in lanes 15 and that only that sequence formed the destabilizing structure.

The constitutive question in this study is whether the experimental results are sequence-independent or not. To answer this, we performed the following experiments. We again tested the relationship between the loop position and the Tm with (1)the same *analyte* and mutated *ligand* (not *analyte*) with a single bulge or (2)a different *ligand* and *analyte* set (called "*ligand2*" and "*analyte2*", expediently). If the above experimental results are sequence-independent, the results of these experiments should show a similar trend. Figure 5 shows the results of testing the same *analyte* and mutated *ligand* with a single bulge and figure 6 shows those of the *ligand2* and *analyte2* set. Both results indicate that the bulge loop near the end is stabilized, while the one located in the middle is relatively destabilized. The correlation coefficient is 0.93 between the results of the *ligand* and the mutated *analyte* with a single bulge and those of the *analyte* and mutated *ligand* with a single bulge. The correlation coefficient is 0.84 between the results of the *ligand* and the mutated *analyte* with a single bulge and those of *ligand2* and mutated *analyte2* with a single bulge, while it is 0.89 between the results of the *analyte* and the mutated *ligand* with a single bulge and those of *ligand2* and mutated *analyte2* with a single bulge. These results prove that the relation between the loop position and the Tm is sequence-independent.

6   Conclusion

In this paper, we examined the Tm and the binding intensity of sequences with a single bulge loop. Based on these data, we discussed some factors of the formation

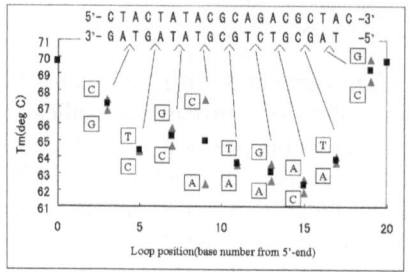

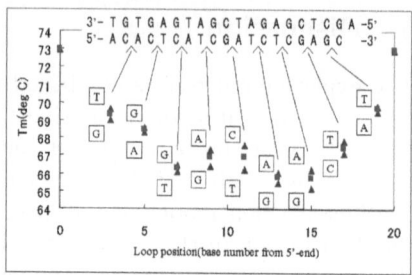

Fig. 5. Tm of sequences with a single bulge with varying base types and positions of the loop in *ligand*. Note that loop positions 0 and 20 are completely complementary duplexes (not bulge loop structures).

Fig. 6. Tm of *ligand2* and mutated *analyte2* with a single bulge with varying base types and positions of the loop. Note that loop positions 0 and 20 are completely complementary duplexes (not bulge loop structures).

of bulge loops in hybridization processes. The Tm correlated with the binding intensity with a correlation coefficient of 0.82. The base type of the loop and the loop position influenced the Tm and the binding intensity. Roughly speaking, the bulge loop near the end was stabilized, while the bulge loop located in the middle was relatively destabilized. The Tm measurements suggested that the effect of the base type of the loop was not influenced by the loop position and vice versa. This result implied that the binding intensity of the bulge loop could be properly modeled by considering the effects of the base type and the position of the loop independently. The modeling enables estimating the binding intensity of the bulge loop. Though this study revealed the qualitative relationship between the loop position and the binding intensity, more detailed experiments will enable estimating the relationship quantitatively. The quantitative data of the destabilized effect of a loop position can be used for the evaluation function of the sequence design. DNA sequences generated with this function will enable reducing hybridization errors in DNA computing.

References

1. Nicolas Peyret, P. Ananda Seneviratne, Hatim T. Allawi, and John SantaLucia, Jr.: "Nearest-Neighbor Thermodynamics and NMR of DNA Sequences with Internal A·A, C·C, G·G, and T·T Mismatches", Biochemistry 1999, 38, 3468-3477
2. Hatim T. Allawi and John SantaLucia, Jr.: "Nearest-Neighbor Thermodynamics of Internal A·C Mismathces in DNA: Sequence Dependence and pH Effects", Biochemistry 1998, 37, 9435-9444
3. Hatim T. Allawi and John SantaLucia, Jr.: "Nearest Neighbor Thermodynamic Parameters for Internal G·A Mismathces in DNA", Biochemistry 1998, 37, 2170-2179
4. Hatim T. Allawi and John SantaLucia, Jr.: "Thermodynamics of internal C·T mismatches in DNA", Nucleic Acids Research, 1998, Vol. 26, No. 11, 2694-2701
5. Hatim T. Allawi and John SantaLucia, Jr.: "Thermodynamics and NMR of Internal G·T Mismatches in DNA", Biochemistry 1997, 36, 10581-10594

6. Salvatore Bommarito, Nicolas Peyret, and John SantaLucia, Jr: "Thermodynamic parameters for DNA sequences with dangling ends", Nucleic Acids Research, 2000, Vol. 28, No. 9, 1929-1934

7. Fumiaki Tanaka, Masashi Nakatsugawa, Masahito Yamamoto, Toshikazu Shiba and Azuma Ohuchi: "Towards a General-Purpose Sequence Design System in DNA Computing", Proc. of the 2002 IEEE Congress on Evolutionary Computation, vol. 1, pp. 73-78, 2002

8. Satoshi Kobayashi, Tomohiro Kondo, and Masanori Arita: "On Template Method for DNA Sequence Design", Preliminary Proc. of Eighth International Meeting on DNA Based Computers, pp. 115-124, 2002.

9. M. Garzon, R. Deaton, L.F. Nino, E. Stevens: Encoding Genomes for DNA Computing, Proc. of the Third Annual Genetic Programming Conf., pp. 684-690, 1998.

10. Mirela Andronescu, Danielle Dees, Laura Slaybaugh, Yinglei Zhao, Anne Condon, Barry Cohen, and Steven Skiena, "Algorithms for testing that DNA word designs avoid unwanted secondary structure", Preliminary Proc. of Eighth International Meeting on DNA Based Computers, pp. 92-104, 2002.

Quantum versus Evolutionary Systems.
Total versus Sampled Search

Hugo de Garis, Amit Gaur, and Ravichandra Sriram

Brain Builder Group, Computer Science Dept,
Utah State University, Logan, UT 84322-4205, USA
degaris@cs.usu.edu, +1 435 512 1826
http://www.cs.usu.edu/~degaris

Abstract. This paper introduces a quantum computing algorithm called "QNN" (Quantum Neural Networks) which measures quantum mechanically and *simultaneously* the fitness values of all 2^N possible chromosomes of N bits used to specify the structure of the networks. Previous attempts to apply quantum computing algorithms to evolutionary systems (e.g [1-4]) applied classical computing evolutionary algorithms to the choice and sequence of quantum operators, which is a hybrid approach (i.e. the EAs were classical, and the applications were quantum mechanical). Our QNN algorithm, on the other hand, is fully quantum mechanical, in the sense that the fitnesses are calculated quantum mechanically as well, thus allowing the fitness values of all possible 2^N chromosomes to be measured simultaneously. Evolutionary algorithms (EAs) are a form of *sampled search* in a huge search space (of 2^N points). If N is large, then 2^N is astronomically large and computationally intractable. The QNN algorithm thus undermines the implicit basic assumption applicable to the field of evolutionary systems (ES), namely that one must employ a sampled search approach (i.e. an evolutionary algorithm (EA)) to explore the huge search space. The QNN algorithm is a form of what is called in this paper a "total search" algorithm. The whole space is searched and is done simultaneously, which makes the adjective "evolutionary" in the term "evolutionary systems" redundant. One can speculate that as the number of qubits implemented in real systems increases (currently the state of the art is 7 [5]), then it is likely that the current emphasis on evolutionary approaches to optimization problems and complex system building, will fade away and be replaced by the "total search" approach allowed by quantum computational methods.

Keywords: Quantum Algorithm, Neural Networks, Evolutionary Computation, Total versus Sampled Search, Quantum System Building

1   Introduction

As stated in the abstract above, the primary aim of this paper is to show that it is possible to use a quantum algorithm (that we call "QNN") to measure *simultaneously* the fitness values of all 2^N chromosomes in the population of an evolutionary algorithm (EA), hence potentially revolutionizing the field of evolutionary computation (EC) and hence those fields that rely upon EC, such as evolutionary system (ES) building. This quantum algorithm shows that the "evolutionary" element in EC is no longer needed. EAs are a form of "sampled search", whereas QNN is a form of "total

A.M. Tyrrell, P.C. Haddow, and J. Torresen (Eds.): ICES 2003, LNCS 2606, pp. 457–465, 2003.

search", i.e. testing every possible 2^N points in the search space and doing this simultaneously, using the quantum computational phenomenon of superposition.

This paper consists of the following sections. Section 1 is this introduction. Section 2 gives some brief background to the classical approach to neural network evolution. Section 3, the largest, presents the QNN algorithm is reasonable detail. Section 4 discusses some ideas for future work, and section 5 summarizes briefly.

2 Classical Evolution of Neural Networks

This section gives a brief overview of classical neural networks and how our group has evolved them in the past. The basic ideas of this classical approach to neural network evolution are carried over to the quantum approach in the QNN algorithm.

Our group's classical approach employs fully connected neural networks of N neurons (hence N^2 connections), each of which is "weighted" by an M-bit binary fraction. These weights are evolved using a Genetic Algorithm (GA), to generate desired signaling behaviors for given external inputs. The signal values are reals that range between +1.0 and -1.0. An incoming signal vector (to each neuron) is scalar-producted with its weight vector, and an external signal value is added to this product. The sum is applied to a sigmoid function whose output is the output signal of the neuron that is transmitted to all other neurons. The "fitness" value is defined to be the measure of how closely the neural net output follows the given target. It is these fitness values that drive the evolutionary process.

3 The QNN Algorithm

It is assumed in this paper that readers are already familiar with the basic principles of quantum computing. If not, then interested readers may consult the references [8-11]. This section describes a quantum-computing-based algorithm called QNN that generates neural networks, which will serve as a model (or proof of concept) for future work. Our group aims to develop such quantum algorithms further, and to combine quantum generated neural circuit modules into quantum artificial brains. (Note: The first author's primary research is in building artificial brains). By a "quantum artificial brain" is meant that its components can be generated using future quantum based algorithms such as QNN, and that the resulting brain can be run using quantum mechanical principles. The construction of a quantum artificial brain would be somewhat analogous to the construction of a classical artificial brain except that the former approach would use quantum computing and the latter would use classical computing.

3.1 Basic Ideas

Considering Fig. 1,

   a) W11, W22, W12, W21 are the (binary) weights of the specific links in the directions indicated by the arrows. W11, W22, W12, W21 = {0,1}.

   b) NS1 and NS2 are the output signals of the neurons. NS1, NS2 = {0,1}. At the 0th clock tick, NS1 and NS2 are 0. For all further clock ticks, the outputs of the neurons are NS1 and NS2 (for neuron 1 and 2 respectively).

c) EXS is the external signal given to each neuron. EXS = {1}
d) The desired (or target) behavior of this extremely simple neural net circuit is to output the minimum number of zeros (i.e. the maximum number of 1's) for all clock ticks.

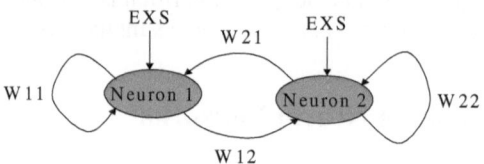

Fig. 1. A Simple 2 Neuron Network

3.2 Description of the Qnet Circuit

Here we describe the Qnet Circuit (see Fig. 2), operator by operator, for one clock tick.

Step 1) Initialization of the States

In Fig. 2 the left most boxes represent the qubits of the Qnet Circuit.

i) The first four boxes W11, W22, W21, W12 represent the weights of the neural net that are prepared in a superposed state using Hadamard gates. These superimposed weights represent the whole search space of all possible chromosomes in genetic algorithm terms.
ii) The fifth and sixth boxes NS1, NS2 represent the signals output by the two neurons from the previous clock tick. Initially NS1 and NS2 are |0>.
iii) The seventh box EXS represents the external signal given to each neuron initially to start the signaling of the neural net.
iv) The remaining boxes represent the qubits in the pure state as indicated by each box, i.e. either |0> or |1>.

Step 2) Signaling through the Qnet Circuit

Operators 1–4: The first step in implementing the neural net is to calculate the scalar (or dot) product of the weights and the signals using the equations

$$(W11*NS1 + W21*NS2) \tag{1}$$
$$(W22*NS2 + W12*NS1) \tag{2}$$

We applied four Controlled Controlled Not Gates (CCNot), and fixed the target bit at |0>. The circuit will then behave like an AND gate (i.e., (a,b,0) => (a,b,a&b)). We used an AND gate to implement the scalar product because we have binary weights and signals (which can be either 0 or 1).

Operators 5–8: After calculating the scalar product, we add the external signal EXS to equation (1) above to get NS1 for the next clock tick. These operators when combined, perform the addition of the scalar product and the external signal. They take the external signal as the carry in and output the sum of the three terms and the carry. This quantum addition is based on the work of Vedral et al [4].

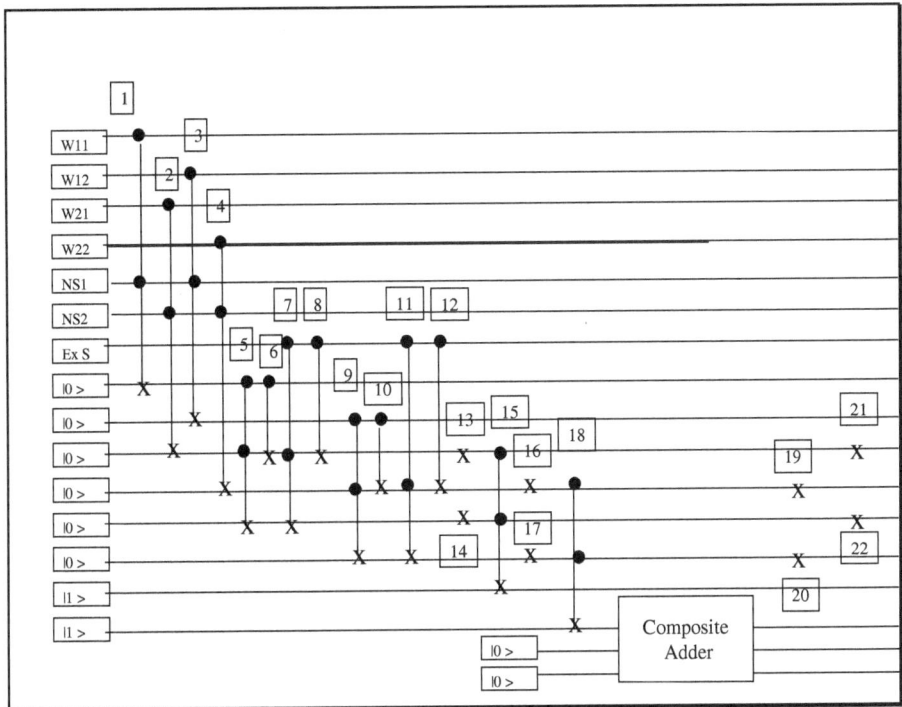

Fig. 2. The 2-Neuron Qnet Circuit

Operators 9–12: These operators do for NS2 and EXS what operators 5-8 do for NS1 and EXS.

Step 3) Threshold Calculation

After computing the output signal for each neuron, we decide next whether the neuron will fire or not by comparing the output signals we calculated from equations (1) & (2) (+ EXS) with some threshold value. By threshold value we mean that if the output signal value (two bits) of the neuron is greater than or equal to the threshold value, the neuron outputs a 1, otherwise a 0, that will be input to the circuit at the next clock tick. We take the threshold value to be 1 because then our input signal has only one bit (after obtaining the two signals we ORed them and accordingly we passed a 1 or a 0).

Operators 13–18: These operators when combined, perform the ORing of the two signal bits. We built an OR gate with the help of NAND gates and when we analyzed the Toffoli gate, we found that if we keep the previous target bit at 1 we get results similar to a NAND gate. To build an OR gate we used de Morgan's law, namely (A'* B')' = A + B, i.e. 2 NOT gates and a NAND gate.

Step 4) Fitness Calculation

Here the fitness is defined to be the minimum number of 0's output, for all the clock ticks. After comparing the signal with the threshold we need to store the output signals (i.e. the NS1 output signals of the neural net) to perform the fitness calculation afterwards, as well as pass them as neuron signals for the next clock tick. As the output signal of the network, only one output neuron signal is chosen (for all clock ticks). This raises the issue of copying signal values, which is not allowed according to the "No-Cloning Theorem". (Since our signals are in superimposed states, there is no way to store them). We therefore place a **composite adder** in Fig. 2 which gives not only the sum of the signal outputs from all previous clock ticks, but also outputs the same signal that was input to it. This signal is the neuron output signal for the next clock tick. The "composite adder" consists of 6 half adders, as shown in Fig. 3.

Fig. 3 shows the internal circuit of the composite adder of Fig. 2. In this circuit the first block of six half adders represents the composite adder at clock T and the second block of six half adders represents the composite adder at clock T+1. Fig.3 shows how incoming information is both passed through the circuit and is added to the fitness sum. Six half adders are taken because we use 64 clock ticks. When we add 64 bits serially we get a 6 bit binary number ($2^6 = 64$). In Fig. 3 the first half adder on the left side will be given the output signal NS1 of a neuron as an input from clock T-1.

Signals S0 to S5 are the sum bits from clock T. The input NS1 is passed through a NOT gate so that if we get a 1 as output we give a 0 as input. We thus add the number of zero's in the output signals, to find the fittest chromosome (with lowest fitness value). All the other inputs are qubits prepared in the pure state |0>. The output of this composite adder at clock T is given as input to the composite adder at clock T+1. We obtain some unused bits that we label GB (garbage bits). We repeat this process for 64 clock ticks, thus obtaining the sum of the output signals for each chromosome. We can then apply "Durr's Algorithm" [12] to find the index position of the minimum value in a given set. This index maps to the index of the weights. We can therefore find the weights of the fittest chromosome and use the corresponding neural network to perform the desired function.

Step 5) Recycling of the Qubits

The operators 19 - 22 in Fig. 2 and 23 – 34 in figure 4 are used to reverse the operations that were performed earlier, by running them in reverse order. This saves circuitry by allowing us to use the same wires.

3.3 Extending the QNN Algorithm to a 3-Neuron Neural Network

The circuits for a 3-neuron neural network differ essentially only in the calculation of the fitness. Here we present the fitness function circuit for 3-neuron neural network. Extending manually such circuits to a greater number of neurons (>3) is tedious but doable. (See the idea in the next section for a quantum compiler to perform such tedious tasks).

The remaining circuits for the 3-neuron Qnet circuit can be developed easily, based on the 2-neuron Qnet, as shown above. The 3-neuron Qnet obviously requires more qubit lines and quantum operators.

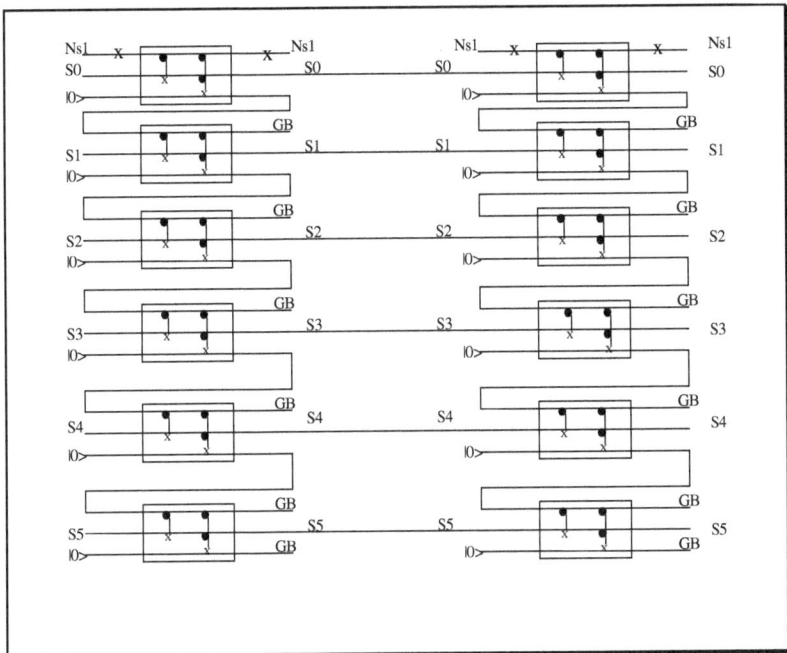

Fig. 3. Internal Circuit of Composite Adder (Fitness Calculation)

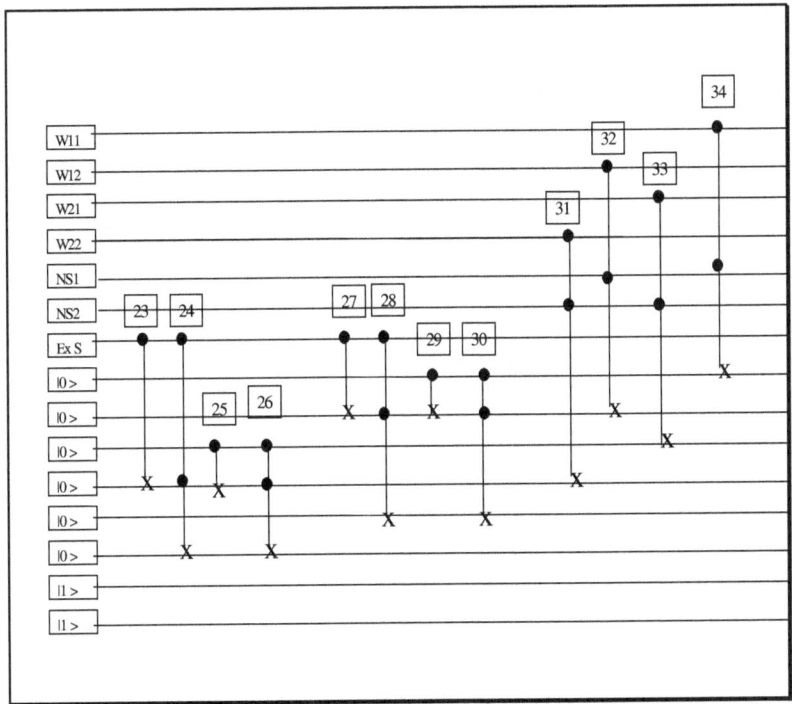

Fig. 4. A Circuit for Recycling the Qubits

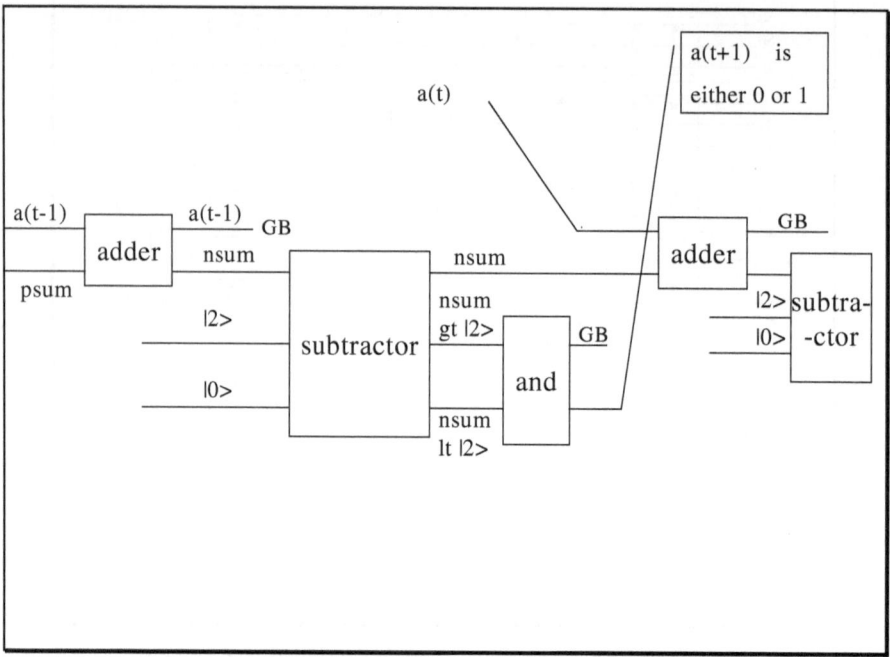

Fig. 5. The Fitness Function Circuit for a 3-Neuron Neural Network

4 Future Work

This section introduces some ideas for future work on quantum algorithms as applied to the generation of neural networks, quantum artificial brains, etc.

a) More Complex Algorithms

Our group would like to devise more elaborate QNN-like algorithms for less toy-like neural net evolution applications. In our classical neural net evolution work, we evolve neural net modules that control motions, detect patterns, make decisions, etc., and use them as components towards building artificial brains [3]. We aim to devise similar algorithms using quantum techniques, as far as this is possible.

b) Quantum Multi-Module Systems & Quantum Artificial Brains

If one can use a QNN-like algorithm successfully to generate a quantum neural circuit module with a given function, then this can probably be done for many different functions. We intend studying "quantum multi-modules" in order to build larger systems such as "quantum artificial brains".

c) A "Quantum Compiler"

Figs. 2–5 show clearly that the Qnets corresponding to quantum neural network circuits that have a large number of neurons are tedious to design by hand. What is needed is a "Qnet Compiler" or "Quantum Compiler" to generate Qnets automatically. A quantum compiler would accept high-level English-like statements written in a special "Quantum Neural Network (QNN) Language" (call it "qC") and translate them into Qnets. Such quantum compilers would also need several additional tools and features, such as the following -

1. **Qubit Initialization Tool:** As in ordinary quantum computing, we will need to prepare input qubits in superimposed states (in order to take advantage of quantum computing's massive parallelism due to state superposition).

2. **Parallelism:** Quantum operations on the superimposed qubits are performed in parallel (in Hilbert Space). This feature would need to be included in the new language "qC".

3. **Loop Restrictions:** Loops are not allowed in quantum computing. A Qnet is therefore required to be run sequentially. A quantum language such as "qC" would not allow any kind of **For** or **While** loops.

4. **No Fan-out and Fan-in:** Fan-out and fan-in are not allowed in quantum computing, hence **If Then Else** statements would similarly not be allowed in the new qC language either.

5. **No Copying:** Since an exact copy of a superimposed state is not allowed (according to the "No Cloning Theorem" of quantum computing), no intermediate calculation at a particular iteration will be allowed. All such calculations should be performed at the end.

6. **Only One Result:** Any algorithm written in qC should produce only one result. We know from quantum computing principles that as soon as we perform a measurement, we destroy the superimposed state. We should therefore save all information until we perform the measurement at the last moment.

5 Brief Summary

We believe that the QNN quantum algorithm above shows that it is possible to perform the fitness measurements of all possible 2^N points in the search space simultaneously, thus making traditional evolutionary algorithms (EAs) rather redundant. Since evolutionary systems (ESs) use EAs as their basic tools, ESs too will be affected. Admittedly the QNN algorithm treats only very primitive neural networks, but with tools such as quantum compilers to aid the creation of Qnets of great complexity and sophistication, elaborate and powerful neural networks could be generated (rather than evolved) that could be used as modular components in more complex multi-module systems, such as quantum artificial brains.

References

1. NASA/DoD Conference on Evolvable Hardware July 15 - 18, 2002, Washington DC, USA. http://cism.jpl.nasa.gov/ehw/events/nasaeh02/
2. Xilinx Inc., XC Virtex2 6000 programmable chip, http://www.xilinx.com
3. Hugo de Garis, Michael Korkin, *THE CAM-BRAIN MACHINE (CBM) An FPGA Based Hardware Tool which Evolves a 1000 Neuron Net Circuit Module in Seconds and Updates a 75 Million Neuron Artificial Brain for Real Time Robot Control*, Neurocomputing journal, Elsevier, Vol. 42, Issue 1-4, February, 2002. Special issue on Evolutionary Neural Systems, guest editor : Prof. Hugo de Garis.

4. Vlatko Vedral, Adriano Barenco, and Artur Ekert. *Quantum networks for elementary arithmetic operations,* PHYSICAL REVIEW A, 54, number 1, july 1996. http://prola.aps.org/pdf/PRA/v54/i1/p147_1
5. Tammy Menneer and Ajit Narayanan. *Quantum-inspired Neural Networks,* Submitted to NIPS 95, Denver, Colorado, December 1995. http://citeseer.nj.nec.com/cache/papers/cs/28/ftp:zSzzSzftp.dcs.ex.ac.ukzSzpubzSzconnect zSz329.pdf/menneer95quantuminspired.pdf
6. A. Barenco, C. H. Bennett, R. Cleve, D. P. DiVincenzo, N. Margolus, P. Shor, T. Sleator, J. Smolin, and H. Weinfurter, *Elementary gates for quantum computation,* Phys. Rev. A, 52, pp. 3457-3467 (1995). http://prola.aps.org/pdf/PRA/v52/i5/p3457_1
7. Lov K. Grover. *Rapid sampling through quantum computing.* Proceedings of *32th Annual ACM Symposium on Theory of Computing (STOC),* 2000, pages 618-626. quant-ph/9912001. http://arxiv.org/PS_cache/quant-ph/pdf/9912/9912001.pdf
8. Eleanor Rieffel. *An Introduction to Quantum Computing for Non-Physicists,* ACM Computing Surveys, Vol. 32, No.3, September 2000, pp. 300-335. http://xxx.lanl.gov/PS_cache/quant-ph/pdf/9809/9809016.pdf
9. Michael Brooks. *Quantum Computing and Communications,* Springer-verlag London Limited 1999, Chapter 12.
10. Jozef Gruska. *Quantum Computing,* McGraw-Hill International (UK) Limited 1999.
11. Michael A. Nielsen, Issac L. Chuang. *Quantum Computation and Quantum Information,* Cambridge University Press 2000.
12. Christoph Durr, Peter Hoyer, *A Quantum Algorithm for Finding the Minimum,* arXiv:quant-ph/9607014 v2 7 Jan 1999

Author Index